INTRODUCTION TO COMPUTING FOR ENGINEERS

INTRODUCTION TO COMPUTING FOR ENGINEERS

SECOND EDITION

Steven C. Chapra, Ph.D.

Professor of Civil Engineering
University of Colorado, Boulder

Raymond P. Canale, Ph.D.

Professor of Civil Engineering
The University of Michigan, Ann Arbor

McGRAW-HILL, INC.

New York   St. Louis   San Francisco   Auckland   Bogotá   Caracas   Lisbon   London
Madrid   Mexico City   Milan   Montreal   New Delhi   San Juan   Singapore   Sydney   Tokyo   Toronto

This book is printed on recycled, acid-free paper containing a minimum of 50% total recycled fiber with 10% postconsumer de-inked fiber.

INTRODUCTION TO COMPUTING FOR ENGINEERS

This book is printed on acid-free paper.

1 2 3 4 5 6 7 8 9 0 DOH DOH 9 0 9 8 7 6 5 4 3

P/N 010922-2

This book was set in Times Roman by Beacon Graphics Corporation.
The editors were B. J. Clark, Kiran Kimbell, and Scott Amerman;
the production supervisor was Friederich W. Schulte.
The cover was designed by Joan Greenfield.
R.R. Donnelley & Sons Company was printer and binder.

Photo Credits for Part Openers: Part 1, Courtesy IBM Corporation; Part 2, Courtesy The Tandy Corporation; Part 3, Jeff Zaruba/Stock Market; Part 4, Dale O. Dell/ Stock Market; Part 5, Courtesy Steven Beekman; Part 6, Courtesy MAGI Synthavision Solid Modeling System; Part 7, Courtesy Michael Cetaruk/The Lotus Co.

Library of Congress Cataloging-in-Publication Data

Chapra, Steven C.
   Introduction to computing for engineers / Steven C. Chapra,
Raymond P. Canale. —2nd ed.
     p.   cm.
   Includes bibliographical references and index.
   P/N 010922-2
   1. Engineering—Data processing.   I. Canale, Raymond P.
II. Title.
TA345.C46   1994
620′.00285—dc20   93-29400

ABOUT THE AUTHORS

STEVEN C. CHAPRA teaches in the Civil, Environmental and Architectural Engineering Department at the University of Colorado. His other books include *Numerical Methods for Engineers* and two graduate texts on mathematical modeling.

Dr. Chapra received engineering degrees from Manhattan College and the University of Michigan. Before joining the faculty of the University of Colorado, he worked as an engineer and computer programmer for the Environmental Protection Agency and the National Oceanic and Atmospheric Administration, and was an Associate Professor at Texas A&M University. In addition to his books, Dr. Chapra has published several computer programs and numerous journal articles. All deal with mathematical modeling and the application of computers for engineering problem solving.

Dr. Chapra is active in a number of professional societies. He has received a number of awards for his teaching and research, and was the 1987 recipient of the Meriam/Wiley Distinguished Author Award by the American Society for Engineering Education for his textbooks on computing and engineering.

RAYMOND P. CANALE received his Ph.D. in 1968 from Syracuse University. As a professor of civil and environmental engineering at the University of Michigan, he has taught courses on computers and numerical methods for over 20 years. He also performs extensive research in the area of mathematical and computer modeling of environmental systems. He has authored books on mathematical modeling of aquatic ecosystems and is a coauthor of *Numerical Methods for Engineers*. He has published over 100 scientific papers and reports, and designed and developed personal computer software for engineers.

With his extensive experience as a practicing professional engineer, Professor Canale has been an active consultant for engineering firms as well as industry and governmental agencies. In addition, he has served as an expert technical witness on numerous occasions.

THIS BOOK IS DEDICATED TO OUR STUDENTS

CONTENTS

Preface   **xxi**

Preface for the Second Edition   **xxvii**

PART ONE   **COMPUTING FUNDAMENTALS**   **1**

Chapter 1   **Computing and Engineering**   **3**

1.1   **Software**   **4**

1.1.1   Homemade Software   **5**

1.1.2   Generic Software   **6**

1.1.3   Engineering-Specific Software   **7**

1.2   **Applications**   **8**

1.2.1   Computations, or "Number Crunching"   **9**

1.2.2   Graphics   **9**

1.2.3   Design, Testing, and Production   **9**

1.2.4   Control and Automation   **10**

1.2.5   Business and Management   **10**

1.2.6   Communication   **11**

1.2.7   Education   **11**

1.3   **An Introductory Course on Computing and Engineering**   **12**

1.3.1   Computer Languages and Types of Machines   **12**

Problems   **13**

Chapter 2   **Computer Hardware**   **14**

2.1   **History of Computing**   **14**

2.1.1   Preelectronic Computing Devices   **14**

2.1.2   How Computers "Think" (Hardware)   **16**

	2.1.3	Early Electronic Computers	**17**
	2.1.4	The Modern Computer Age	**19**
2.2	**Overview of Computer Hardware**		**24**
	2.2.1	The Central Processing Unit (CPU)	**24**
	2.2.2	Input and Output Devices	**26**
	2.2.3	Secondary Memory	**28**
2.3	**Which Type of Hardware Is Best?**		**30**
	Problems		**31**

Chapter 3   Internal Representation and Systems Software   **33**

3.1	**Preliminary Concepts**		**33**
	3.1.1	Number Systems	**33**
	3.1.2	Word Size	**38**
3.2	**Internal Representation of Information**		**39**
	3.2.1	Integers	**39**
	3.2.2	Real (or Floating-Point) Numbers	**43**
	3.2.3	Characters and Symbols	**47**
	3.2.4	Summary	**47**
3.3	**Machine, Assembly, and High-Level Languages**		**50**
3.4	**Systems Programs**		**51**
	3.4.1	The Operating System	**51**
	3.4.2	Utility Programs	**54**
	3.4.3	Language Translators	**54**
	Problems		**55**

Chapter 4   Algorithm Design and Structured Programming   **57**

4.1	**The Program-Development Process**		**57**
4.2	**Algorithm Design**		**59**
4.3	**Flowcharting**		**60**
	4.3.1	Flowchart Symbols	**61**
	4.3.2	Levels of Flowchart Complexity	**63**
	4.3.3	Selection and Repetition in Flowcharts	**65**
	4.3.4	Accumulators and Counters in Flowcharts	**65**
	4.3.5	Structured Flowcharts	**66**
4.4	**Pseudocode**		**67**
4.5	**Structured Design and Programming**		**72**
	4.5.1	Modular Design	**72**
	4.5.2	Top-Down Design	**73**
	4.5.3	Structured Programming	**73**
4.6	**Final Comments Before You Begin to Program**		**75**
	Problems		**76**

PART TWO   STRUCTURED BASIC   **83**

Chapter 5   Fundamentals of Structured Basic   **85**

    5.1   Which Basic?   **86**

    5.2   A Simple Basic Program   **88**

    5.3   Numeric Constants, Variables, and Assignment   **90**

       5.3.1   Numeric Constants   91

       5.3.2   Variables and Assignment   93

       5.3.3   Algebraic Equations versus Computer Assignment Statements   95

       5.3.4   Variable Names   96

       5.3.5   String Constants and Variables   97

       5.3.6   Variable Types in QuickBASIC   98

       5.3.7   CONST Statements   99

    5.4   Introduction to Input-Output   **100**

       5.4.1   The CLS Statement   100

       5.4.2   Input   100

       5.4.3   Output   104

    5.5   Quality Control   **106**

       5.5.1   Errors or "Bugs"   108

       5.5.2   Debugging   109

       5.5.3   Testing   109

    5.6   Documentation   **111**

       5.6.1   Internal Documentation   111

       5.6.2   External Documentation   111

    5.7   Storage and Maintenance   **113**

    Problems   114

Chapter 6   Computations   **116**

    6.1   Arithmetic Expressions   **116**

       6.1.1   Rules for Arithmetic Operators   117

       6.1.2   Exponentiation in BASIC   122

    6.2   Intrinsic, or Built-In, Functions   **123**

       6.2.1   General Mathematical Functions   124

       6.2.2   Trigonometric Functions   126

       6.2.3   Miscellaneous Functions: INT, FIX, SGN, and RND   129

       6.2.4   Other Library Functions   135

    6.3   User-Defined Statement (One-Line) Functions   **136**

    Problems   139

Chapter 7   Selection   **144**

    7.1   The GOTO Statement   **144**

    7.2   The Single-Line IF/THEN Statement   **146**

7.3 Formulating Logical Expressions   **147**

7.4 Compound IF/THEN Statements and Logical Operators   **149**

    7.4.1   Conjunction   **150**

    7.4.2   Disjunction (the Inclusive "OR")   **150**

    7.4.3   Logical Complement   **150**

7.5 Structured Programming of Selection   **154**

    7.5.1   Improper Selection with the IF/THEN Statement and GOTOs   **154**

    7.5.2   The Block-Structured IF/THEN/ELSE Construct   **154**

    7.5.3   Nesting of Block IF/THEN/ELSE Constructs   **157**

    7.5.4   The CASE Construct   **158**

Problems   **163**

Chapter 8   Repetition   **167**

8.1 GOTO, or INFINITE, Loops   **167**

8.2 Decision Loops   **169**

    8.2.1   The DO WHILE (or PreTest) Loop   **169**

    8.2.2   The DO UNTIL (or PostTest) Loop   **172**

    8.2.3   The EXIT DO (Break or MidTest) Loop   **173**

    8.2.4   Logical Loops and "What If?" Computations   **174**

    8.2.5   The INKEY$ Function   **178**

8.3 Counter (FOR/NEXT) Loops   **179**

8.4 Nesting Control Structures   **182**

    8.4.1   Nesting Counter Loops   **182**

    8.4.2   Nesting Decision and Counter Loops   **184**

    8.4.3   Nesting Repetition and Selection Structures   **187**

    8.4.4   The EXIT FOR   **188**

Problems   **189**

Chapter 9   Modular Programming   **194**

9.1 The GOSUB/RETURN   **194**

9.2 Sub Procedures   **195**

9.3 Passing Parameters to Procedures   **199**

9.4 Function Procedures   **202**

    9.4.1   Comparison between SUB and FUNCTION Procedures   **204**

9.5 Local and Global Variables   **205**

    9.5.1   SHARED Statement   **205**

    9.5.2   The STATIC Statement and Attribute   **207**

9.6 Modular Programming   **209**

9.7 Recursive Procedures   **213**

Problems   **214**

Chapter 10 Data Structure   **219**

    10.1 READ AND DATA STATEMENTS   **219**

    10.2 Arrays   **223**
          10.2.1 Dimensioning Arrays   **225**
          10.2.2 Two-Dimensional Arrays   **227**
          10.2.3 How Arrays Are Stored in Memory   **229**
          10.2.4 Arrays in Procedures   **232**

    10.3 Records   **236**

    Problems   **240**

Chapter 11 Advanced Input-Output   **245**

    11.1 Formatted Output   **245**
          11.1.1 Print Using   **245**
          11.1.2 Screen Design   **251**
          11.1.3 The LOCATE Statement   **252**

    11.2 String Manipulations   **253**
          11.2.1 Concatenation   **253**
          11.2.2 String Manipulation Functions   **253**

    11.3 Files   **259**
          11.3.1 File Types   **259**
          11.3.2 BASIC File Statements   **261**
          11.3.3 Standard File Manipulations   **264**

    Problems   **267**

PART THREE STRUCTURED FORTRAN   **271**

Chapter 12 Fundamentals of Structured Fortran   **273**

    12.1 Why Fortran?   **273**

    12.2 Which FORTRAN?   **274**

    12.3 "Talking" to the Computer in Fortran   **275**

    12.4 How to Prepare and Execute Fortran Programs   **280**
          12.4.1 Creating and Editing Programs   **280**
          12.4.2 Compiling Programs   **281**
          12.4.3 Linking Programs   **281**
          12.4.4 Executing or Running Programs   **281**

    12.5 A Simple Fortran Program   **282**
          12.5.1 Adding Descriptive Comments: The Comment Statement   **282**

    12.6 Numeric Constants, Variables, and Assignment   **283**
          12.6.1 Numeric Constants   **284**
          12.6.2 Variables and Assignment   **285**
          12.6.3 Algebraic versus Computer Assignment Statements   **287**

	12.6.4	Variable Names	288
	12.6.5	String Constants and Variables	289
	12.6.6	Variable Types	289
	12.6.7	PROGRAM and PARAMETER Statements	292
12.7	**Introduction to Input-Output**		**293**
	12.7.1	The List-Directed READ Statement	293
	12.7.2	The List-Directed PRINT Statement	295
	12.7.3	Descriptive Input	296
	12.7.4	Descriptive Output	297
12.8	**Quality Control**		**299**
	12.8.1	Errors or "Bugs"	299
	12.8.2	Debugging	300
	12.8.3	Testing	301
12.9	**Documentation**		**302**
	12.9.1	Internal Documentation	302
	12.9.2	External Documentation	303
12.10	**Storage and Maintenance**		**304**
	Problems		**305**

Chapter 13   Computations   **307**

13.1	**Arithmetic Expressions**		**307**
	13.1.1	Rules for Arithmetic Operators	308
	13.1.2	Exponentiation in FORTRAN	313
	13.1.3	Integer and Mixed-Mode Arithmetic	314
	13.1.4	Complex Variables	315
13.2	**Intrinsic, or Built-In, Functions**		**316**
	13.2.1	General Mathematical Functions	316
	13.2.2	Trigonometric Functions	320
	13.2.3	Truncating and Rounding Functions	322
13.3	**User-Defined Statement (One-Line) Functions**		**325**
	Problems		**328**

Chapter 14   Selection   **334**

14.1	**Spaghetti Code**		**334**
	14.1.1	The GO TO Statement	335
	14.1.2	The CONTINUE Statement	336
	14.1.3	"Pasta" Statements	336
14.2	**The Single-Line IF Statement**		**337**
14.3	**Formulating Logical Expressions**		**338**
	14.3.1	Logical Variable	338
	14.3.2	Relation between Expressions	338

14.4	**Compound Logical IF Statements and Logical Operators**		**341**
	14.4.1	Conjunction	341
	14.4.2	Disjunction	341
	14.4.3	Logical Complement	342
14.5	**Structured Programming of Selection**		**345**
	14.5.1	"Spaghetti" Selection	345
	14.5.2	The Block-Structured IF/THEN/ELSE Construct	345
	14.5.3	Nesting of Block IF/THEN/ELSE Constructs	348
	14.5.4	The CASE Construct	349
14.6	**Fortran 90 Enhancements**		**352**
	Problems		354

Chapter 15   Repetition   **360**

15.1	**GO TO, or Infinite, Loops**		**360**
15.2	**Decision Loops**		**361**
	15.2.1	The DO WHILE (or PreTest) Loop	362
	15.2.2	The DO UNTIL (or PostTest) Loop	366
	15.2.3	The Break (or MidTest) Loop	368
	15.2.4	Logical Loops and "What If?" Computations	371
15.3	**Counter (DO) Loops**		**375**
15.4	**Nesting Control Structures**		**380**
	15.4.1	Nesting Counter Loops	380
	15.4.2	Nesting Decision and Counter Loops	381
	15.4.3	Nesting Repetition and Selection Structures	384
15.5	**Repetition in Fortran 90**		**384**
	Problems		389

Chapter 16   Modular Programming   **395**

16.1	**Subroutine Subprograms**		**395**
16.2	**Passing Parameters to Subprograms**		**397**
16.3	**Function Subprograms**		**403**
	16.3.1	Comparison between Functions and Subroutines	405
16.4	**Local and Global Variables**		**406**
	16.4.1	The COMMON Statement	406
16.5	**Modular Programming**		**409**
16.6	**Fortran 90**		**413**
	Problems		415

Chapter 17   Data Structure   **419**

17.1	**The Data Statement**		**419**

17.2   Arrays   **421**

17.2.1   Dimensioning Arrays   **423**

17.2.2   Specification Statements   **425**

17.2.3   Implied DO Loops   **426**

17.2.4   Two-Dimensional Arrays   **428**

17.2.5   How Arrays Are Stored in Memory   **431**

17.2.6   Two-Dimension Arrays and DATA Statements   **434**

17.2.7   Arrays in Subprograms   **434**

17.3   Fortran 90 Records   **441**

Problems   **445**

Chapter 18   Advanced Input-Output   **449**

18.1   List-Directed I/O   **449**

18.2   Formatted Input and Output   **452**

18.2.1   Format Specification   **452**

18.2.2   Codes to Control Spacing   **455**

18.2.3   Codes for Character Data   **457**

18.2.4   Codes for Integers   **460**

18.2.5   Codes for Real Numbers   **461**

18.3   Files   **463**

18.3.1   File Types   **463**

18.3.2   FORTRAN Sequential Files   **465**

18.3.3   Standard File Manipulations   **466**

Problems   **468**

PART FOUR   DATA ANALYSIS   **473**

Chapter 19   Graphics   **475**

19.1   Line Graphs   **475**

19.2   Rectilinear Graphs   **477**

19.2.1   Tabulating Data   **479**

19.2.2   Choosing Graph Paper   **480**

19.2.3   Positioning the Axes   **481**

19.2.4   Graduating and Calibrating the Axes   **482**

19.2.5   Plotting Points and Drawing Lines   **483**

19.2.6   Adding Labels, an Identification Key, and a Title   **484**

19.3   Other Types of Graphs   **486**

19.3.1   Logarithmic Graphs   **486**

19.3.2   Polar Graphs, Nomographs, and Charts   **491**

19.4   Computer Graphics   **494**
    19.4.1   Graphics Equipment and Terminology   **496**
    19.4.2   The Electronic TOOLKIT   **497**
    19.4.3   Computer Graphics and Engineering   **500**
Problems   **501**

Chapter 20   Sorting and Searching   **505**

20.1   Sorting   **505**
    20.1.1   Selection Sort   **505**
    20.1.2   Bubble Sort   **506**
    20.1.3   Shell Sort   **507**
    20.1.4   Tabular Data and Sorting in Place   **507**
    20.1.5   Comparisons and Advanced Algorithms   **510**

20.2   Searching   **512**
    20.2.1   Sequential Search   **512**
    20.2.2   The Binary Search   **512**
    20.2.3   Comparisons   **513**
Problems   **513**

Chapter 21   Significant Digits and Error Analysis   **516**

21.1   Reliability of Numerical Quantities   **516**
    21.1.1   Rounding   **517**
    21.1.2   Significant Figures   **518**
    21.1.3   Significant Digits and Calculations   **521**

21.2   Accuracy and Precision   **523**

21.3   Error   **524**
    21.3.1   Bias   **525**
    21.3.2   Uncertainty   **527**
    21.3.3   Total Error   **529**
Problems   **529**

Chapter 22   Statistics   **531**

22.1   Populations and Samples   **532**

22.2   The Distribution of Data and Histograms   **532**
    22.2.1   Histogram Software   **536**

22.3   Measures of Location   **536**
    22.3.1   The Mean   **537**
    22.3.2   Medians and Percentiles   **540**
    22.3.3   The Mode   **541**
    22.3.4   Comparison of Central Location Statistics   **541**
    22.3.5   The Geometric Mean   **541**

22.4   **Measures of Spread**   **542**

    22.4.1   Range and Average Deviation   **543**

    22.4.2   Variance and Standard Deviation   **543**

    22.4.3   The Coefficient of Variation   **546**

    22.4.4   Algorithm for Descriptive Statistics   **547**

22.5   **The Normal Distribution**   **549**

22.6   **Prepackaged Statistics Software**   **552**

Problems   **553**

PART FIVE   **COMPUTER MATHEMATICS**   **557**

Chapter 23   **Curve Fitting: Regression**   **559**

23.1   **Linear Regression**   **561**

23.2   **Quantifying the "Goodness" of the Least-Squares Fit**   **564**

23.3   **Algorithm for Linear Regression**   **567**

23.4   **Alternative and Advanced Methods**   **568**

Problems   **569**

Chapter 24   **Curve Fitting:Interpolation**   **575**

24.1   **Linear Interpolation**   **577**

24.2   **Parabolic Interpolation**   **579**

24.3   **The General Form of the Lagrange Polynomial**   **581**

24.4   **Algorithm for Lagrange Interpolation**   **583**

24.5   **Pitfalls and Advanced Methods**   **584**

Problems   **586**

Chapter 25   **Equation Solving: Roots of an Equation**   **589**

25.1   **The Graphical Method**   **590**

25.2   **The Bisection Method**   **591**

25.3   **Algorithm for Bisection**   **596**

25.4   **Pitfalls and Advanced Methods**   **599**

    25.4.1   Pitfalls   **599**

    25.4.2   Advanced Methods   **599**

Problems   **600**

Chapter 26   **Equation Solving: Linear Algebraic Equations**   **604**

26.1   **Gauss Elimination**   **607**

    26.1.1   Pivoting   **610**

26.2   Algorithm for Gauss Elimination   611

26.3   Pitfalls and Advanced Methods   611
    26.3.1   Pitfalls   611
    26.3.2   Advanced Methods   613
    Problems   613

Chapter 27   Computer Calculus: Differentiation   617

27.1   The Derivative   617

27.2   Numerical Differentiation   620

27.3   Pitfalls and Advanced Methods   622
    27.3.1   Pitfalls   622
    27.3.2   Advanced Methods   623
    Problems   623

Chapter 28   Computer Calculus: Integration   627

28.1   The Trapezoidal Rule   630

28.2   Algorithm for the Trapezoidal Rule   635

28.3   Advanced Methods   636
    Problems   637

Chapter 29   Computer Calculus: Rate Equations   642

29.1   What Is a Rate Equation?   642

29.2   Where Do Rate Equations Come From?   644

29.3   Euler's Method   647

29.4   Computer Programs for Euler's Method   649

29.5   Advanced Methods   650
    Problems   651

PART SIX   ENGINEERING APPLICATIONS   655

Chapter 30   Organizing Principles in Engineering   657

30.1   Conservation Laws   658

30.2   Balances in Engineering   661

30.3   The Engineering Problem-Solving Process   661
    Problems   662

Chapter 31   Chemical Engineering: Mass Balance   664

**31.1   Using Linear Algebraic Equations in the Steady-State
Analysis of a Series of Reactors**   664

31.2 Using Rate Equations in the Transient Analysis of a Reactor   667

31.3 Using Roots of Equations in the Determination of a Reactor's Response Time   671

31.4 Using Integration in the Determination of Total Mass Input or Output   671

Problems   672

Chapter 32   Civil Engineering: Structural Analysis   676

32.1 Using Linear Algebraic Equations in the Analysis of a Statically Determinant Truss   676

32.2 Using Linear Regression in the Calculation of Elongation of Members   680

32.3 Using Integration in the Determination of Total Force and Line of Action for Distributed Loads   682

Problems   683

Chapter 33   Mechanical Engineering: Vibration Analysis   687

33.1 Using Roots of Equations in the Design of an Automobile Shock Absorber   687

33.2 Using Regression in the Determination of the Spring Constant   690

33.3 Using Integration in the Calculation of Work with a Variable Force   690

Problems   693

Chapter 34   Electrical Engineering: Circuit Analysis   695

34.1 Analyzing a Resistor Network with Linear Algebraic Equations   695

34.2 Designing an *RLC* Circuit Using Bisection   698

34.3 Analyzing a Nonideal *RL* Circuit Using Interpolation and Rate Equations   701

Problems   702

PART SEVEN   SPREADSHEETS   705

Chapter 35   Fundamentals of Spreadsheets   707

35.1 The Origin of Spreadsheets   707

35.2 The Spreadsheet Environment   709

35.3 Menu Operation   711

35.4 Numbers, Labels, and Formulas   714

35.5 Editing Cell Quantities   717

		35.6	**Engineering Applications of Spreadsheets**	**719**
		35.7	**The Electronic Toolkit**	**724**
		Problems		**727**
Chapter 36	**Table Operations**			**732**
		36.1	**Tables as Databases**	**732**
			36.1.1 Sorting	**733**
		36.2	**Sensitivity Analyses with Static Tables**	**736**
			36.2.1 One-Way Sensitivity Tables	**736**
			36.2.2 Two-Way Sensitivity Tables	**739**
		36.3	**Cell Moving and Copying**	**741**
		36.4	**Dynamic Tables**	**743**
		Problems		**746**
Chapter 37	**Data Analysis**			**751**
		37.1	**Graphics**	**751**
		37.2	**Statistics and Regression**	**756**
			37.2.1 Spreadsheet Statistics	**756**
			37.2.2 Spreadsheet Regression	**758**
		37.3	**Equation Solving**	**760**
			37.3.1 Matrix Algebra	**761**
			37.3.2 Spreadsheet Matrix Manipulations	**763**
		37.4	**Electronic Toolkit**	**767**
		Problems		**775**
Chapter 38	**Macros**			**781**
		38.1	**Keyboard Macros**	**781**
			38.1.1 Command Macros	**783**
		38.2	**Macro Programs**	**787**
		38.3	**To Macro or Not to Macro?**	**792**
		Problems		**792**
Appendix A	**Derived Trigonometric Functions That Can Be Developed from the Intrinsic Functions of High-Level Computer Languages**			**796**
Appendix B	**QuickBASIC Functions**			**797**
Appendix C	**FORTRAN 77 Library Functions**			**800**
Appendix D	**Electronic Toolkit User's Manual**			**803**
		Index		**811**

PREFACE

Engineering practice has always been inextricably tied to the evolution of new tools and technologies. Whether developing the levers, ramps, and pulleys of ancient times or the machines of the industrial revolution, engineers have exploited new devices and concepts that have extended their capabilities. Today, computer technology is providing a host of powerful new tools that will have a profound impact on the profession. Because the strength of the computer lies in its ability to manipulate and store data, engineering, with its dependence on mathematics and its need for accurate information, is an ideal candidate for exploiting the potential of these new computing tools.

Although its potential is just beginning to be tapped, modern computing technology can enhance the engineering student's education in a variety of ways. In the most immediate sense, the computer significantly increases the efficiency with which students can perform computations and process information, liberating them to devote more time to the conceptual, creative aspects of problem solving. Other benefits range from the use of computers for simulation and tutorial instruction to enhanced visualization through computer graphics.

We believe that the student's first computing course is critical to the effective exploitation of these opportunities. Admittedly, this is not an uncontroversial position. Some educators feel that a first course may be unnecessary because students attain computer literacy in high school. Although secondary education has ventured into this area, high school efforts often prove inadequate for a number of reasons, particularly where student engineers are concerned.

Primary among the reasons is the critical shortage of competent teachers in all secondary technical areas, particularly in computer science. As of 1990, our student polls continue to confirm that about 40 percent of our incoming students have had no computing in high school, 40 percent have been exposed to BASIC (usually Applesoft or Microsoft), and the remaining 20 percent having taken some other language (usually Pascal). Consequently, we have concluded that, at least for the near term, we cannot rely on secondary education to provide computer-literate students.

Despite these difficulties, there are still students who, because of innate curiosity or exceptional school systems, acquire extensive computer skills during their high school years. However, this does not mean that these students have skills that are compatible with or attuned to engineering. For one thing, many of our incoming "hackers" are extremely game-oriented, and although games are excellent vehicles for developing creativity and programming cleverness, they do not necessarily foster the discipline required in engineering. For instance, we rarely encounter an incoming student of any skill level who can satisfy one of our most important criteria for computer literacy—being able to compose a well-structured program that can be readily used and modified by a colleague.

The state of precollege computer education today is similar to the state of high school sex education during the 1950s and 1960s. It was the rare and fortunate individual who received an accurate, comprehensive introduction to that subject. More often than not, information was gathered piecemeal in the street or through haphazard experimentation. Today's adolescents face a similar situation with respect to computer literacy; that is, many are acquiring their computer skills in the technological gutter.

Because so few freshmen are oriented toward engineering computing and because such a disparity of computer skills exists among incoming students, a freshman or sophomore course provides a degree of standardization. The roots of such a course are already in place in the form of the two courses traditionally available to provide orientation. The first type—introduction to computer programming—is intended to orient the student to the locally available hardware and to develop programming skills. Most courses of this type have, in the past, been configured around mainframe computers and FORTRAN. The second type—introduction to enineering—evolved from the old "slide-rule course," which dealt with the rudiments of the slide rule, engineering drawing, and descriptive geometry. In time, this course incorporated engineering computations and principles, pocket calculators, modeling, and design. Most recently, material on digital computers and programming has been added to some versions. This book blends elements of both courses. Early chapters are aimed at developing sound programming skills, and the latter parts deal with engineering applications and problem-solving software.

Although this book derives from the courses just described, our objective is to forge a new approach. The major thrust of this approach is to bring freshmen and sophomore engineers to a common level of computer literacy. In this spirit, the text is organized around five capabilities that we believe are the essential elements of engineering computer literacy. These capabilities are:

1. *Engineers must have a general understanding of what computers are and how they operate.* From their first day on the job until they retire, engineers will rely on computers as one of their primary tools. No area of engineering is immune. Just as a little mechanical knowledge can hold you in good stead when servicing or buying a car, knowing something about computers lets student engineers use them more effectively. Also, a familiarization with hardware and its operation goes a long way toward rendering computers less mysterious and intimidating.

The first three chapters of this book constitute a short orientation to computing. Chapter 1 provides illustrations of how computers are applied in engineering. Chapter 2 offers a brief history of computing, particularly with regard to the remarkable advances of the past 20 years. It also includes an overview of current computer hardware and operating systems. This material is slanted toward personal computers because students are more likely to have hands-on experience with PCs than with mainframes. However, pertinent aspects of workstations and mainframe systems are also discussed because of their relevance to engineering practice. Chapter 3 focuses on the internal representation of numbers and character information by computers. We have found this material essential for subsequent applications.

2. *Engineers must be able to design, test, and document a structured computer program.* This is perhaps the most controversial aspect of our definition of computer literacy. Some educators contend that programming skills are unnecessary because of proliferation of packaged or "canned" computer software. Although this contention may be true in a broad societal context or in other disciplines, we believe that engineers must have a working knowledge of programming. No matter how many good packaged programs are available, problems will invariably arise that will require tailor-made software. Additionally, a vast "literature" of algorithms has been developed over the past 20 years that has great utility in engineering. However, these algorithms are usually written in a high-level computer language, and engineers who cannot access these codes will be unable to tap this valuable resource.

Moreover, one of the truly justifiable fears related to extensive application of canned software is that users will treat them as "black boxes." That is, they will apply them with little regard for their inner workings and attendant subtleties and limitations. We contend, however, that computer-literate engineers will be more discriminating about packaged programs and will be less naive about their capabilities and limitations. In the same spirit, although some of the innovative engineering software of the future may originate from professional programmers, a great deal will undoubtedly be developed by engineers themselves. Computer literacy among engineers will stimulate and hasten the process.

Most important though, students learn to think critically and logically when they develop computer software. If a program is written incorrectly, it will not work. Students are forced to go back, retrace their steps in a logical fashion, and locate their mistakes. In the process, cognitive abilities, discipline, and other design skills are exercised that might otherwise lie dormant. Consequently, students are not merely trained to program; they can also learn how to think and to design.

In fact, the first computing course can provide the student with a meaningful and relevant experience in engineering design. This is particularly true when the student views program development as a process that ultimately delivers a quality product to a satisfied customer.

In this regard, a program must be written so that it is easy to understand and maintain. Therefore, this text stresses the use of structured programming techniques. Years ago learning to program was more craft than art. Today structured programming has elevated computer programming to a higher plane. Structured programming is a set of rules that prescribes good style habits for the programmer. Although structured programming is flexible enough to allow considerable creativity and personal expression, its rules impose enough constraints to render resulting codes far superior to unstructured versions. In particular, the final product is more elegant and easier to understand. The clarity of structured code is particularly important because a great deal of engineering work is performed by teams, where programs must be shared and modified by several persons.

Languages included in this book were selected in large part for their widespread availability, BASIC being the choice for personal computers and FORTRAN for mainframes. Because today's engineer must function effectively in either of these environments, major sections are devoted to each. The other important factor in our choice of BASIC and FORTRAN was structure. Early BASIC dialects were deficient in this regard, but the modern BASIC dialect used herein (specifically Microsoft's QuickBASIC) includes many of the essential features of a structured language: indention, long variable names, modular procedures, and excellent selection and repetition constructs. Although structure is not required, modern BASIC can be employed to develop structured modular programs that are virtually as satisfactory as those developed with a language such as Pascal, which enforces structure. The same is true for the FORTRAN 77 employed herein.

Fifteen chapters are devoted to developing programming skills and language capabilities. Chapter 4 offers an overview of the general elements of structured programming style. This material places a strong emphasis on algorithm design using flowcharts and pseudocode.

Chapters 5 to 11 cover QuickBASIC, and Chapters 12 to 18 deal with FORTRAN 77. The sections on BASIC and FORTRAN are similar in organization; thus after mastering one language, students should be able to study the other in an efficient manner. However, an effort is made to emphasize the unique strengths of each language. This is done so that students recognize how each might be preferable in different problem contexts.

3. *Engineers must recognize the ways computers can be applied to solve problems.* Engineers are professional problem solvers, so this facet of computer literacy is essential to a freshman or sophomore computing course. Among other things, the applications provide motivation and make the subject come alive. Chapters 19 to 29 consider computer-oriented mathematics and cover the general areas of data analysis (graphics, sorting, and statistics), curve fitting (regression and interpolation), equation solving (root location and elimination), and computer calculus (integration, differentiation, and rate equations). The latter material is presented in simple introductory fashion in order to be compatible with the mathematical sophistication of freshman engineers.

All these chapters include structured flowcharts or pseudocode to implement the methods. Thus, students can begin to build software libraries that will be applicable in their subsequent coursework. In addition, all these chapters include examples and problems to illustrate the application of the techniques to engineering. These applications are reinforced by Chapters 30 to 34, which include problem-solving examples that focus on mathematical modeling in each of the major areas of engineering.

4. *Engineers should be able to use packaged software.* In addition to their own programs, engineers should be able to utilize software packages. We have, therefore, chosen to devote Chapters 35 through 38 to spreadsheets. Chapter 35 deals with the

rudiments of spreadsheets and the idea of "what if" calculations, which are valuable for building insight into computations and their sensitivity. Chapters 36 and 37 deal with table operations and data analysis aspects such as statistics, regression, and sorting. Finally, Chapter 38 introduces the notion of macros.

In addition to explicit coverage of spreadsheets, we have developed software called the Electronic TOOLKIT as a freeware supplement to the text. This package includes generic programs for several numerical methods, descriptive statistics, a graph generator, and a simple spreadsheet. These programs provide a vehicle for introducing the student to this type of tool. The software has been designed to be compatible with this book; however, because the programs are generic, they can be applied in other academic and professional contexts as well.

5. *Engineers should be capable of recognizing which computing tool is preferable in different problem contexts.* For many years, FORTRAN and mainframe computers were the sole computing tools available to engineers. Today, a wide variety of hardware, languages, and software packages are at their disposal. With this freedom comes choice. Unfortunately, we notice that some of our students become enamored with particular computer tools and often try to apply them to any and all problem settings. For example, just as a screwdriver is a poor choice for hammering a nail, a spreadsheet is inappropriate for most large-scale numerical computations. Consequently, an underlying theme of this book is that different problems require different tools. We reinforce this theme continuously by stressing the pros and cons of the hardware, languages, and software packages we review.

This book contains a great amount and range of material. In fact, we believe that two semesters could be comfortably spent on its contents; the first on programming, and the second on applications and canned software. However, this mass of material was also included to give instructors the flexibility to choose among a variety of topics when designing a one-semester version of the course. In addition, we hope that the book will serve not only as a text but also as a reference book for use in later classes.

We have included material on personal, workstation, and mainframe computing because we believe that today's engineer must be comfortable working in each of these environments. All the material in Chapters 19 through 34 can be applied on personal, workstation, or mainframe computers. However, we expect that students will first use this text with one of these environments, but not all. Thus, when students must inevitably operate in one of the others, the book can serve as a reference to facilitate their efforts. The same sort of strategy underlies our inclusion of both BASIC and FORTRAN.

The breadth of material in this text is also related to the fact that computing has experienced dramatic growth in the last decade. Twenty years ago, computing was monolithic; there were only mainframes, and these environments were dominated by large companies such as IBM and Control Data. Today's situation offers a great many more options and capabilities, but the rapid rate of change and the sheer number of choices can lead to confusion and frustration.

We hope our book helps young engineers successfully navigate these turbulent waters. We believe this generation of young engineers to be extremely important to our profession, to the nation, and to the world. Countries such as Japan and Germany have made great strides and investments in engineering and technological development.

Concurrently, the results of U.S. scholastic aptitude scores suggest a deterioration in the quality of American education. We believe quality application of computing technology represents an opportunity to upgrade engineering education and perhaps mitigate the impacts of these alarming trends.

This generation of young students is better suited to accomplish the needed upgrading than any previous generation of engineers. Today's freshmen and sophomores were TV babies who received the rudiments of their education on public television and spent their adolescence in video arcades. They are not afraid of computers and are, in fact, more comfortable sitting in front of a screen pushing buttons than almost anywhere else. If the enthusiasm and talent of this first generation of computer-oriented youth can be channeled productively, the benefits will be reaped into the next century. We hope that this book is one positive contribution in this direction.

ACKNOWLEDGMENTS

We would like to acknowledge reviews by Professors Jasper Avery, University of Idaho; Jerry R. Bayless, University of Missouri; Steve Beyerlein, University of Idaho; Peter J. Bosscher, University of Wisconsin; Rocky Durrans, University of Alabama; Jack D. Foster, Broome Community College; Roy W. Hann, Texas A&M University; P. F. Pasqua, University of Tennessee; Jack Seltzer, University of Delaware; Masoud Tabatabai, Pennsylvania State University; Thomas F. Wolff, Michigan State University; and Lt. Col. James Holt, Air Force Institute of Technology.

A number of persons have been instrumental in creating the type of academic computing environments in which this project flourished. Associate Dean Dave Clough of the University of Colorado has demonstrated a commitment to computing and engineering education that we found inspiring. The contributions of Professors Jim Heaney and Jean Hertzburg to computing at our institutions are appreciated. In addition, we would like to thank Sandra Braitberg, who helped us prepare graphic material for the book.

Thanks are also due to our friends at McGraw-Hill. In particular, B. J. Clark provided the leadership and vision that underlies McGraw-Hill's commitment to computing and engineering. On the editorial side, Kiran Kimbell did much to expedite the project and make it a pleasurable experience. Scott Amerman, editing supervisor, Joan Greenfield, designer, and Friederich W. Schulte, production supervisor, did an admirable job in seeing the manuscript through to completion of the bound book. Last but not least, the McGraw-Hill sales force continues to demonstrate to us why they are considered the best in the business.

Finally, we would like to thank our families, friends, and students for their enduring patience and support. In particular, Cynthia and Christian Chapra were always there to provide the understanding, perspective, and love that helped us up and over this particular mountain.

                                                              Steven C. Chapra
                                                              Raymond P. Canale

PREFACE FOR THE SECOND EDITION

It has been over six years since the original publication of *Introduction to Computing for Engineers*. Although we have maintained the overall plan of the original book, we have made several modifications to improve upon it.

Our decision to keep both BASIC and FORTRAN was not made lightly. We discussed several other possibilities, such as an all-FORTRAN edition and a Pascal/FORTRAN edition. We stuck to BASIC for two reasons. First, we still find that BASIC is the friendliest context within which the freshman engineer can be introduced to computing. As mentioned above, the majority of our incoming students are not computer literate and BASIC provides the most pleasant and nonthreatening first exposure to computer programming. Second, because of new dialects such as QuickBASIC, we are able to teach BASIC in a highly structured fashion. Thus, we can teach the algorithmic aspects of programming in a way that is similar to Pascal.

We have retained FORTRAN because it is still such a force in engineering. At our institutions, it remains the language of choice in many upper-level courses.

The major difference between the two editions is that we have revised the section on generic programs so that it is totally devoted to spreadsheets. We did this for both positive and negative reasons. On the negative side, we just could not do justice to canned programs such as word-processing and database management systems in the limited space available. On the positive side, we find that of all the tools we introduce in this book, the spreadsheet seems to be the most immediately useful to our students. In a way, the spreadsheet is supplanting the pocket calculator as our student's principal tool for routine computing tasks such as homework and laboratory assignments. As a consequence, we wanted to expand our material on spreadsheets to encompass more information concerning graphics, data analysis, and macros.

Aside from these general aspects, we have made a number of specific modifications that we would like to call to your attention. These are listed on page xxviii:

1. The material on computing and computers has been upgraded to reflect changes in technology such as the growing importance of engineering workstations.

2. We have included a more complete discussion of number representation and round-off error. We have found such material useful in our recent teaching of the course and believe that it should be stressed.

3. We have improved the FORTRAN section. In particular, we have included aspects of the new FORTRAN 90 standard that are beginning to show up in popular dialects such as Microsoft FORTRAN.

4. Our software package, the Electronic TOOLKIT, has been included at the back of the book at no additional cost to the student. The material on data analysis and computer mathematics is now more closely integrated with the Electronic TOOLKIT. We use the TOOLKIT primarily for examples and, by describing the underlying algorithms, to reinforce why software packages should not be used as "black boxes."

These are the major features of the second edition. Your comments and suggestions would be welcomed.

Steven C. Chapra
Raymond P. Canale

INTRODUCTION TO COMPUTING FOR ENGINEERS

COMPUTING FUNDAMENTALS

Before launching into an extensive discussion of programming, it is essential to review some background material on computers and how they work. This part of the book is devoted to such introductory material. *Chapter 1* is designed to orient you to the ways in which computers are applied in engineering. *Chapter 2* deals with the actual equipment that constitutes a computer—that is, the hardware. Chapter 2 presents the historical development of these machines, as well as the components and functions of the modern computer. *Chapter 3* begins our discussion of the instructions or programs (that is, the software) that must be developed to allow the computer to perform tasks such as data processing, numerical computations, and graphics. Our emphasis in this chapter is on the way in which information is represented within a computer. Finally, *Chapter 4* reviews the algorithm design process and structured programming.

COMPUTING AND ENGINEERING

An engineer carries a portable personal computer into a courtroom to present testimony regarding the pollution of a water-supply and recreational reservoir in Virginia. By projecting computer graphics onto a large screen, the engineer presents complex technical information to the court in a highly communicative fashion. In addition, the computer is used interactively to make predictions concerning the effect of anticipated land development on the reservoir's future water quality. The judge and both attorneys pose hypothetical scenarios, and the engineer uses the computer to generate immediate predictions of possible outcomes.

An engineer in Michigan uses special electronic pens and graphic display devices to prepare alternative design drawings for a new automobile. Each design can be tested on her workstation to evaluate its structural strength and behavior under simulated road conditions. When the optimal version has been selected, the computer is used to prepare blueprints from the stored design. Finally, tapes are generated to control robots and automatic machine tools during the car's manufacturing phase.

A supertanker runs aground off the Texas coast. On shore, engineers from a spill-response team enter the latest water current and wind data into a microcomputer which makes hourly predictions of the point at which the spill is likely to hit the coastline. This information is transmitted to cleanup crews who must be on the spot in order to minimize damage to the beaches. It is also relayed to aircraft which drop bacteria that are capable of "eating" the oil.

A professional engineer from California who had formerly been employed by a large firm has started a private consulting practice. Because of a lack of capital, he must initially work alone out of a small office. However, because of the acquisition of a personal computer, the engineer can perform many of the normal functions that in the past required a staff of support personnel. Among other things, the computer allows the engineer to perform design calculations, prepare reports and correspon-

TABLE 1.1 Major Categories of Engineering Software
and Applications for Computers

Software
 Homemade software
 Generic software
 Engineering-specific software
Application
 Computations
 Graphics
 Design, testing, and production
 Control and automation
 Business and management
 Communication
 Education

*dence, and handle the accounting and billing needs of the firm. These capabilities
provide the initial cash flow that will eventually allow the engineer to hire additional staff as the business expands.*

Although the foregoing scenarios suggest that computers (and especially personal computers[1]) are already having an effect on engineering, their potential is just beginning to
be tapped. Interestingly, the first major applications of personal computers originated in the business community. This might at first seem surprising because one would
have assumed that the lion's share of applications would have originated in the seemingly more computer-oriented fields of engineering and science. But for a variety of
reasons, business firms have been quicker to sense and to capitalize on the competitive
edge provided by computers.

This situation is changing. Because of their speed, memory, and computational
capabilities, computers have been allied with engineering for years. However, the enhanced accessibility offered by personal computers has recently stimulated major increases in the use of these tools by our profession.

The present chapter explores some of the ways that computers serve the engineering community now and in the foreseeable future. Table 1.1 shows some major categories of computer software and applications. Notice that there is some overlap among
the groupings. For example, one type of generic software—computer graphics—is
used in nearly all the other categories in Table 1.1. Despite such overlaps, each of the
groups is distinct enough to merit separate discussion.

1.1 SOFTWARE

The instructions, or programs, needed to direct the computer to perform tasks are
called *software.* In order to use computers effectively, engineers require a wide variety

---

[1]By "personal computers" we mean microcomputers. We will elaborate further on this classification, along with other
types of computers, when we discuss hardware in Chap. 2.

FIGURE 1.1
Professor John Kemeny of
Dartmouth College.
Twenty-five years ago,
Professor Kemeny
stated that, in the future,
"knowing how to use the
computer will be as important
as reading and writing."
Today's microelectronic
revolution is transforming
Professor Kemeny's
prophecy into a reality.

of software, which can be generally divided into three major categories: homemade, generic, and engineering-specific software.

1.1.1 Homemade Software

About 25 years ago, Professor John Kemeny (Fig. 1.1), one of the developers of the BASIC computer language, stated that "knowing how to use a computer will be as important as reading and writing." At that time, Prof. Kemeny's statement was greeted with skepticism. This reaction was partially justified by the inaccessibility of computers at the time he made the statement. Today, in the light of the microelectronic revolution, Kemeny's words sound prophetic. In particular, it is increasingly evident that practicing engineers must attain computer literacy to maintain professional competence.

In a broad societal context, *computer literacy* can be defined as the ability to recognize applications where computers are appropriate and to be able to employ computers for such applications. We believe that for engineers computer literacy should also include the capability of composing problem-solving programs in at least one computer language. Although many engineers will employ packaged or "canned" programs for major portions of their work, their ability to write their own "homemade" programs has several benefits.

First, no matter how many canned programs are available, there will inevitably be problems that will require tailor-made computer programs. In many of these cases, engineers must be capable of developing problem-specific software to fill their immediate needs.

Second, one of the truly justifiable fears related to extensive application of canned software is that users will treat the package as a "black box." That is, they will apply the software with little regard for its inner workings and attendant subtleties and limitations. Computer-literate engineers will be more likely to look critically at software designed by others and will be less naive regarding canned software and its possible limitations. In addition, they will be much better prepared to understand and communicate with software engineers.

Finally, although some of the innovative engineering software of the future may originate from professional programmers, a great deal more will undoubtedly be developed by engineers themselves. Computer literacy among engineers will stimulate and hasten this process.

In their excellent little book, *The Beginner's Guide to Computers* (1982), Bradbeer, DeBono, and Laurie make the comparison that the present state of computer software is similar to that of literature when Gutenberg and Caxton developed the first printing presses in the fifteenth century. Prior to that time, books were the exclusive domain of a select group of clergymen and nobility. The vast majority of the populace had never seen a book, let alone learned how to read one. After the invention of the printing press, the situation changed and books began to pervade society as a whole. A century later, Shakespeare wrote *Hamlet!*

By analogy, the microelectronic revolution represents a quantum leap in the computer capabilities of the general public. Today, the Shakespeares of software are creating their masterpieces. The premier creations in engineering will likely arise from the present pool of computer-literate student and professional engineers.

1.1.2 Generic Software

There are certain computer applications that are so general that they can be brought to bear on a wide variety of engineering and other problems. Rather than compose a new program every time one of these applications is required, generic software packages are available to implement these tasks.

Prime examples are the *word-processing packages* that are available for composing and editing documents. The utility of these packages does not necessarily depend on the particular problem context where they are applied. Thus we classify them as general or "generic" tools.

Other examples are listed in Table 1.2. *Mathematical and graphics packages* are designed to perform calculations and to produce visual representations of information. The most notable examples of the latter category are the *computer-aided drafting,* or *CAD,* packages that are increasingly employed in engineering.

Spreadsheets are a special type of mathematical software that allow the user to enter and perform calculations on rows and columns of data. As such, they are a computerized version of a large piece of paper or worksheet on which an involved calculation can be displayed or spread out. Because the entire calculation is automatically updated when any number on the sheet is changed, spreadsheets are ideal for "what if?" sorts of analyses. *Database management systems* are employed to store and manipulate large quantities of data efficiently.

Geographical information systems, or *GIS,* are programs that allow the manipulation and display of spatially distributed information. Engineers are playing a major role in the development and application of these tools to a variety of problem contexts. For example, large-scale projects ranging from highway design to dam construction rely increasingly on GIS.

It should be noted that specific versions of all the packages can be tailored to the

TABLE 1.2 Some Examples of Generic Software That Are Used by Engineers

Type	Examples
Mathematical	
General	Matlab, GAMS
Symbolic manipulations	Mathematica, Derive, Macsyma
Statistics	SAS, SPSS, SYSTAT, STATGRAPHICS, S+
Numerical methods	IMSL, NAG
Optimization	MINOS, ZOOM, LINGO, LINDO
Graphics	
General	Harvard Graphics, Freelance, GKS, Sigma Plot, DeltaGraph, PHIGS, GRAPHER, SURFER, XLIB/OPENLOOK, XLIB/MOTIF
Computer-aided drafting	AUTOCAD, CADKEY
Spreadsheets	Lotus 1-2-3, Quattro Pro, Excel
Word processing	WORD, WordPerfect, Frame Maker
Database management	DBASE IV, Paradox, Oracle, Ingres, Postgres
Geographical information systems (GIS)	Arc Info, GRASS, IDRISI, Inter Graph, SPANS, MOSS

special needs of any particular profession. For instance, most early word-processing software was not designed to handle mathematical symbols and formulas easily. Today, because of demands from engineers and scientists, this capability has become a standard part of most word-processing packages. Similarly, a spreadsheet package could be customized for the specific requirements of an engineering problem context. In most cases, generic software is flexible enough to be applied effectively in a variety of settings.

1.1.3 Engineering-Specific Software

Although there are a great many areas that can be addressed with generic software, there are many other engineering applications that require tailor-made programs. Some of these programs will be so specific that they will be limited to a single problem setting. However, others will be similar in spirit to the generic software discussed in the previous section in the sense that they can be applied broadly. For example, engineers working in different locales could successfully apply the same structural design or electrical network analysis programs. Some other examples are listed in Table 1.3.

1.2 APPLICATIONS

The microelectronic revolution has stimulated a variety of novel applications of computers in engineering. Some of the more important areas in which computers are playing an increased role are graphics, design, automation, communication, and education. However, before reviewing these more recent applications, we must first discuss the area where computers have traditionally shown their greatest utility, that is, in large-scale calculations, or "number crunching."

TABLE 1.3 Some Examples of Engineering-Specific Software That Are General Enough for Broad Application

Field	Software
Chemical engineering	Process control
	Distillation tower design
	Integrated process design
Civil engineering	Structural analysis
	Surveying
	Pollutant fate and transport models
Electrical engineering	Electrical power system control
	Microprocessor design
	Circuit design
Mechanical engineering	Pipe network design
	Finite-element analysis for mechanical parts
Industrial engineering	Computer-aided design/manufacturing (CAD/CAM)
	Statistical quality control
General engineering	Computer-aided drafting (CAD)
	Risk analysis
	Critical path method (CPM)
	Program evaluation and review technique (PERT)

1.2.1 Computations, or "Number Crunching"

From the earliest days, computers have been used to implement large-scale calculations. The earliest electronic computers—the ABC and the ENIAC—were expressly developed to perform otherwise arduous and tedious calculations. To this day, "number crunching" remains one of the primary applications areas for computers in engineering.

The advent and the widespread availability of personal computers is doing much to encourage and broaden these applications. In fact, as they increase in memory and speed, personal computers are enhancing not only our capabilities but also the manner in which calculations are performed. Their impact is directly attributed to their accessibility— that is, their low cost, their portability, and their graphic and interactive capabilities.

For example, in the past, engineers had to go to great expense and suffer some inconvenience to employ large mainframe systems. Today, a personal computer can perform many of the same tasks for a relatively small capital investment and minimal operating costs. Additionally, as seen in the oil-spill vignette at the beginning of this chapter, computational power can be transported out into the field.

Beyond these examples, there are more subtle changes linked to the graphic and interactive qualities of the machines. On mainframe systems, there is usually a lag between the time that a program is entered into the computer and the time that the results are obtained. On personal computers, the response is immediate for small- to moderate-sized calculations. The engineering workstation further expands this capability. Thus, the engineer has the capability of performing successive runs to gain immediate insight into the calculation and its implications. Such interactive or exploratory computations are extremely useful for developing intuition and holistic awareness regarding the particular engineering problem under study. This is especially effective when computer graphics are employed to express the results visually.

It is presently impossible to forecast where these changes will lead and how they will ultimately influence the profession. However, present trends suggest an overall upgrading and enhancement of our capabilities to perform and interpret computations.

1.2.2 Graphics

As stated previously, most computer applications can be enhanced by graphical output. For example, as just mentioned, computations come alive when expressed in a visual format.

However, beyond their utility in communicating results effectively, graphics capabilities are being developed in their own right. For example, high-resolution computer monitors allow the presentation of very sharp images. Thus, photographs, video, and satellite imagery can become integrated into computer displays.

Beyond the quality of the images, special visually oriented software is being produced. For example, CAD systems are presently available for use in preparing engineering drawings.

1.2.3 Design, Testing, and Production

Today most of the basic arithmetic, logic, and storage elements of a computer can be contained on a single integrated circuit chip called a *microprocessor*. Because of their size, speed, memory, and portability, microprocessors are used increasingly to perform tasks connected with the effective design and operation of engineered works and products. One of the more prominent applications is in *computer-aided design and manufacturing* (CAD/CAM)[2] of the type described in the second scenario at the beginning of this chapter. This application involves the use of microprocessors to formulate the design, to perform preliminary testing, and to control manufacturing and assembly operations. A significant aspect of CAD/CAM is that it is an integrated process. The computer serves as the focus for this integration.

Important tools of CAM are robots that can more efficiently perform manufacturing tasks formerly accomplished manually. The application of CAM and robotics is one reason for the explosive growth of industrial productivity in the Far East and Europe. In the United States, engineers are applying these tools to a wide variety of manufacturing processes.

1.2.4 Control and Automation

Microprocessors have numerous potential applications throughout engineering and society at large. As is the case for robotics, many of these applications involve the control and automation of processes formerly handled by humans. For example, many engineers use computer-controlled sampling devices to collect and analyze test data. Integrated

---

[2]Note that the acronym "CA-" for "computer-aided" or "computer-assisted" is widely used in engineering. This often results in the same acronym being used to represent two different meanings, as in "CAD" standing for "computer-aided drafting" and "computer-aided design."

computer networks are increasingly employed for other control functions ranging from electric power to traffic to aerospace systems. Some structures, called "smart buildings," are being built that use computers to regulate their heating and cooling. Control and automation will be one of the major areas where microprocessors will have an immense, direct impact on society. Engineers will play a key role in the design and implementation of such applications.

1.2.5   Business and Management

Many engineers own firms or function professionally as private consultants. As such, they have a need for the many business-oriented software packages that are presently available. These programs perform functions—such as accounting, billing, and payroll management—that are essential for the effective operation of a company. Although they require an initial capital investment, their long-term payoff in increased speed and efficiency makes them an essential component of the modern engineering firm.

Beyond this direct use, one of the major tasks of the practicing engineer is to complete projects in a cost-effective manner. Engineers must therefore always be aware of the economic aspects of their work. Certain business-oriented software can serve these efforts. In addition, aids such as CPM (critical path method) and PERT (program evaluation and review technique) are available in computerized form to facilitate the scheduling and implementation of projects.

1.2.6   Communication

One definition of engineering labels it "the art of executing a practical application of scientific knowledge." As such, engineers stand at the midpoint between science and the needs of society. Because of this, our work involves a substantial amount of societal interaction, and the ability to communicate effectively is a hallmark of successful engineers. Whether marketing a project to a prospective client or presenting a report to a government agency, communication of technical information to nontechnical audiences is a critical ingredient in a large amount of engineering work.

One of the most obvious examples is report and proposal generation. Many of the generic programs discussed in Sec. 1.1.2 can contribute greatly to quality publication of engineering documents. Integrated software has been developed that contains several capabilities such as word-processing, spreadsheet, and graphics programs in one package. Such integrated software greatly enhances the report-generation process. Together with electronic mail and FAX, such software provides a quantum leap in our capability to package and transmit technical information.

Another context in which personal computers enhance communication is in the area of presentations. Of course, they can serve the obvious function of preparing effective graphics. However, as seen in the courtroom vignette at the beginning of the chapter, they can also be directly utilized as communication tools. In particular, they can be used to establish a dialogue between the engineer and the client. In the courtroom example, the personal computer allowed the judge and the lawyers to pose questions that the computer could evaluate and answer. This sort of interactive involvement is a potent vehicle for making technical results less intimidating and more accessible to clients and the public. Such involvement can yield positive benefits as the client actively participates in the technical endeavor.

The ultimate extension of the interactive presentation is the computer game. Most of us are familiar with these from our experiences with arcade and home-entertainment video games. Beyond these recreational applications, computer games are powerful communication devices. In a sense, the flight simulators used to train airplane pilots are sophisticated games. In the coming decades, engineers will employ games for training technicians and for communicating technical results to clients. As discussed in the next section, another area of application will be in the classroom.

1.2.7 Education

Communication is of critical importance to engineering. Nowhere is this more important than in the training of engineers. In Sec. 1.1.1 we made the case that computer literacy should be an absolute necessity for all practicing engineers. Additionally, related topics such as computer math (for example, numerical methods and statistics), CAD/CAM, and other computer-oriented subjects will undoubtedly gain more prominence in engineering curricula. As prices drop, more and more student engineers will possess their own personal computers. Courses are being modified to capitalize on this increased computational firepower.

The benefits will be great. For example, personal computers will remove some of the drudgery associated with large-scale calculations. Thus more time will be freed for studying problem formulation and solution interpretation. In this way, insight and creative intuition can be nurtured.

Beyond these impacts, computers may also be employed directly as learning tools. To date, two primary modes are used for this purpose: tutorials and simulations. *Tutorials* set up an interactive dialogue between the student and the machine. In their most basic form, tutorials present material and ask questions. Depending on the answer to the question, the computer either reiterates the material or advances to another topic. In this way, concepts are clarified and students progress at their own pace.

Simulation is another name for computer games. In this case, the computer can be used to simulate a system for the student to study. Thus the student can explore the behavior of the system in response to a variety of conditions and in this way build up competence and intuition.

Both tutorials and simulations can be effective learning tools. This is particularly true because most students today have grown up in a video environment. Their first exposure was to the somewhat passive medium of television. Most students probably acquired the rudiments of reading, writing, and arithmetic from preschool educational television programs. In the late 1970s, video arcades came to provide a version of the technology that demanded active participation on the part of the player. In the late 1980s, home-entertainment video games became the rage. Because of these experiences, most students today are very comfortable with video technology. Although there is a certain justification for those who lament the negative effects of television, computer education is one effort to glean something positive from this technology. By bridging the gap between video and engineering education and stressing the interactive nature of the medium, we can capitalize on the attraction and accessibility of television rather than continuing to lament some of its acknowledged faults.

We have now come full circle from voicing a plea for computer literacy at the beginning of the chapter to stressing the direct use of computers as educational tools. On this note, we move along to outline how this book can serve as the basis for a first course on computers for engineers.

1.3 AN INTRODUCTORY COURSE ON COMPUTING AND ENGINEERING

As stated above, the present book is designed to support a first computing course for engineers. The book has two major thrusts. First, it provides sufficient information for you to attain computer literacy in a high-level programming language. Second, it illustrates a wide variety of computer applications in engineering.

Part One deals with fundamental information related to computer hardware and software. Just as automobile owners find it useful to learn a little about cars, we believe that it will be beneficial for you to know something about how computers work.

This introductory material is followed by *Parts Two* and *Three,* which are devoted to programming in structured BASIC and FORTRAN, respectively. These parts include a sufficient level of detail to allow you to fulfill our definition of computer literacy. That is, they will provide you with enough information to write your own high-quality problem-solving programs.

After you have learned to program, the next step is to investigate how this skill can be applied to engineering. Thus the following parts of the book are devoted to a variety of problem-solving skills and engineering applications related to the computer. *Parts Four* and *Five* are devoted to data analysis and computer mathematics. Here we discuss topics such as sorting and statistics that contribute to the effective communication and analysis of information. We also introduce some rudimentary numerical methods. These are general techniques for solving mathematical problems with computers. *Part Six* illustrates how the numerical methods are used for engineering problem solving. This material concentrates on problems that would be intractable without the computer. Finally, *Part Seven* deals with spreadsheets. Here we again focus on the engineering applications of this important tool.

1.3.1 Computer Languages and Types of Machines

Today's engineer has a wide variety of computing tools at his or her disposal. During the course of your career, you will undoubtedly have occasion to make use of a spectrum of devices ranging from your own personal computer to large institutional machines. For this reason, and also because we anticipate that students will be using different types of machines in conjunction with their first computing course, we have designed this book so that it can be used with many of these systems. Most parts of the book are written so that the material can be implemented on personal computers, engineering workstations, or large mainframe computers.

This approach extends to our handling of computer algorithms. Aside from the specific sections devoted to computer programming languages, the algorithms in the remainder of the book are all expressed in the generic format of flowcharts or pseudocode.

Although we have gone to great pains to ensure that this book is compatible with many types of computers, we have tried wherever possible to acknowledge and explore some of the exciting new applications that are directly associated with personal computers. This is why we have devoted Part Seven to the spreadsheet, which represents a powerful new tool for our profession. Even those who use a mainframe for this course can cover this material under the realistic assumption that most engineers will own or have ready access to a personal computer in the foreseeable future. In addition, at the time of the printing of this edition, spreadsheets are starting to become available on workstations and mainframes.

In the same spirit, we have designed a software package, called the Electronic TOOLKIT, that can be used in conjunction with the text. This package is available for the IBM PC and compatible computers. Although its use is optional, this software provides you with immediate capability to explore problem solving with graphics, statistics, spreadsheets, and several numerical methods. As such, it can serve to illustrate how canned software can be applied in engineering practice.

With this as background, we can now begin to learn something about computing. We will start by investigating the history of the computer itself.

PROBLEMS

1.1 Distinguish between homemade, generic, and engineering-specific software.

1.2 Define computer literacy for engineers.

1.3 Give three reasons why engineers should learn to write computer programs.

1.4 What do we mean when we say that personal computers are "accessible"?

1.5 What is CAD/CAM? What does it involve?

1.6 Distinguish between the two primary modes of educational software: tutorials and simulations.

1.7 Visit your local computer store.
   (a) Compare two or three of the computers. Classify them according to type, cost, and capabilities.
   (b) Find several engineering and business software packages. Classify them according to their application.

1.8 Determine what types of computers are used at your school or at a local business or engineering firm. Prepare a report that contains a list and description of the different types of hardware, software, and application areas.

COMPUTER HARDWARE

Computers are like automobiles in the sense that you need not know everything about them to operate them effectively. However, as with automobiles, a little knowledge about their workings can often help you to utilize them more intelligently and efficiently.

The present chapter is an introduction to the physical equipment, or *hardware,* that constitutes a typical computing system. However, before we discuss modern computers, we will first present an overview of their historical development.

2.1 HISTORY OF COMPUTING

Today's microelectronic revolution is due in large part to the accelerating development of computer hardware over the past 25 years. The present section is intended to place these recent advances in a broader historical perspective. Beyond satisfying your innate curiosity, this material will help you to appreciate the magnitude of today's advances as well as to anticipate things to come. In addition, it should deepen your understanding of existing computers, because much of today's hardware evolved from the concepts and devices of the past.

2.1.1 Preelectronic Computing Devices

Humans have been interested in calculations and keeping track of information from the earliest days of recorded history. As will be clear from the following pages, the ability to calculate has always been closely related to the technology and energy supplies of the day. For example, one of the earliest "computers" was the simple muscle-powered device called the *abacus.* Developed in ancient China and Egypt, this device consists of rows of beads strung on wires in a rectangular frame (Fig. 2.1*a*). The beads are used to keep track of powers of 10 during the course of a computation. In the hands of a skilled operator, the abacus can rival a pocket calculator in speed.

In the early seventeenth century, the Scottish mathematician John Napier invented

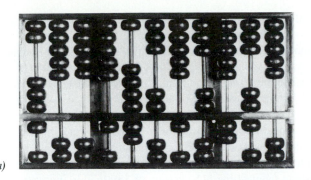

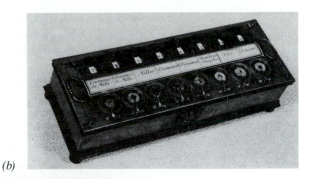

(a)   (b)

(c)   (d)

FIGURE 2.1
Preelectronic computing
devices: (*a*) an abacus, (*b*) the
pascaline, (*c*) Babbage's
difference engine, and
(*d*) the Mark I. *(Courtesy of
IBM Corporation.)*

the *logarithm* and a manual computing device called Napier's bones. The former simplified multiplication and division by reducing them to addition and subtraction. The latter consisted of calculating rods that allowed multiplication and division to be implemented manually. Because they were sometimes constructed from bones, they were dubbed *Napier's bones*. These influenced the design of the *slide rule* and other subsequent calculating machines.

Although manual devices such as the abacus and Napier's bones certainly speed up computations, machines provide an even more powerful means for extending human calculating capability. Stimulated by the industrial revolution, seventeenth century scientists developed the first such mechanical computing devices. The French scientist and philosopher Blaise Pascal invented an adding machine in 1642. Called the *pascaline* (Fig. 2.1*b*), the device used gears with teeth to represent numbers. In the late 1600s, Gottfried Leibniz developed a similar mechanical calculator that could also multiply and divide.

The next major advances in computing technology came in the 1800s. In the early nineteenth century, Joseph Jacquard invented a loom that employed punched cards to control the production of patterned cloth automatically. As recently as the 1970s, punched cards were still the primary way that information was transmitted to and from computers. In addition, Jacquard's idea of storing a sequence of instructions on cards is conceptually similar to modern computer programs.

Although Jacquard's loom introduced some important new ideas, Charles Babbage's steam-driven calculators were more closely related to modern computers. Babbage was motivated by the large numbers of errors he found in mathematical tables. He developed a *difference engine* (Fig. 2.1*c*) in the early 1800s that could compute and print out tables automatically. He also conceived, but never built, an *analytical engine* that incorporated many features of modern computers, including punched card instructions, internal memory, and an arithmetic unit to perform calculations.

A final major contribution of the nineteenth century occurred when Herman Hollerith used punched cards to tabulate the U.S. census of 1890. This innovation reduced the time of tabulation from a decade to 3 years. Hollerith went on to found the Tabulating Machine Company, which later merged with several other firms to form International Business Machines, now called IBM.

During the early part of the twentieth century, IBM and other manufacturers produced a variety of computing devices for business use. These were all *electromechanical;* that is, they were powered by electricity and had moving parts. The ultimate of these devices was the Mark I developed by Howard Aiken in collaboration with IBM. Although an awesome piece of technology (Fig. 2.1*d*), the slowness of the Mark I's electromechanical components made it quickly obsolete in the face of a more advanced technology—electronics.

2.1.2 How Computers "Think" (Hardware)

In contrast to electromechanical devices, electronic machines have no moving parts. Their major components operate on the basis of electricity. In order to appreciate the development of electronic computers, we must digress momentarily to describe the way in which these machines store and process information.

A mechanical device such as the pascaline kept track of quantities by a series of coupled gears and levers. But how can purely electrical instruments perform similar tasks? The answer lies in the two fundamental ways in which quantities can be handled: measuring and counting. Entities that exist on a continuous scale such as height, velocity, or mass are *measured*. In contrast, quantities that are treated as indivisible units such as people, rivets, or beans are *counted*. The distinctions can be appreciated by realizing that if you measure an individual's height as 71 in, he or she could just as well be 70.9999 or 71.0001 in. However, if you count 15 rivets in a package, there are 15 and only 15 rivets.

Electronic computers that measure are called *analog computers*. These devices use the level of an electrical quantity, such as voltage, to correspond to a numerical value that is to be measured. An example would be your car's gas gauge. As the float in your gas tank moves up or down, an electrical quantity is increased or decreased accordingly. The resulting changes are transmitted electrically to the gauge on your dashboard. Analog computers use a similar scheme to keep track of quantities and their interactions. It should be noted that the slide rule is an example of a mechanical analog calculator.

Although analog computers have their place in engineering and science, they do

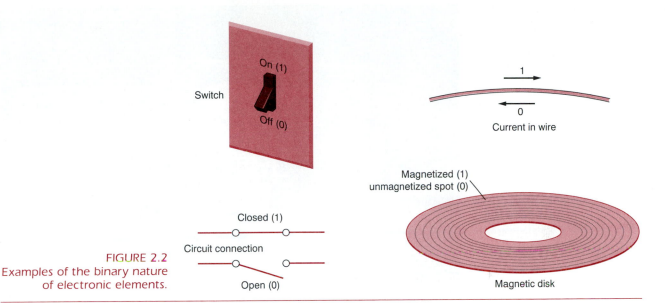

FIGURE 2.2
Examples of the binary nature of electronic elements.

not represent the primary way that electronic computers handle quantities. Most modern computers count. For this reason they are called *digital computers* after the primary way that humans count, with their fingers or "digits." However, because they have neither hands nor feet, computers count in their own unique way—that is, with electrical elements that are either "on" or "off."

Several examples are shown in Fig. 2.2. Note that by convention, the "on" state corresponds to the number 1, whereas the "off" state corresponds to 0. Each element taking on a value of 1 or 0 is referred to as a *bit,* which is a contraction of *binary* dig*it.* Electronic computers represent all their information in binary form. We will provide a description of how this is done when we discuss software in Chap. 3. For the time being, the important notion to understand is that all computer operations can be boiled down to a series of elements that are either "on" or "off." The importance of this notion to the present discussion is that the evolution of computers is closely linked to the type of electronic device or *logic element* used to keep track of these "ons" and "offs."

2.1.3 Early Electronic Computers

The first electronic computers employed vacuum tubes as their primary logic elements (Fig. 2.3). These machines were one-of-a-kind devices built for experimental purposes. For example, the ABC (Atanasoff-Berry computer) was developed at Iowa State University to solve simultaneous equations in the late 1930s.

A few years later, a more general purpose machine, the ENIAC (*e*lectronic *nu*merical *i*ntegrator *a*nd *c*alculator), was developed during World War II. Designed by J. Presper Eckert, Jr., and John W. Mauchly at the University of Pennsylvania, it contained tens of thousands of vacuum tubes, consumed large quantities of electricity, gen-

FIGURE 2.3
Logic elements used in
electronic computers: vacuum
tubes (first generation),
transistors (second
generation), and integrated
circuits (third and fourth
generations). *(Courtesy of
IBM Corporation.)*

FIGURE 2.4
The ENIAC. Built for military
purposes, this early computer
filled a room, generated large
quantities of heat, and
consumed great amounts of
electricity. However, it
represented the first
successful general-purpose
electronic computer.
*(Courtesy of the Sperry
Corporation.)*

erated much heat, and nearly filled an 800-ft^2 room (Fig. 2.4). Although by present standards it was slow and unreliable, it represented the first successful general-purpose electronic calculator.

Although the ENIAC represented a major advance, it had to be rewired every time a new task was performed. A way to circumvent this problem was developed by the mathematician John von Neumann. He suggested that instructions for processing information could be stored in the computer's memory. Thus, when the operator needed to perform a new task, a set of instructions could be accessed by the machine along with the data. Two such *stored-program computers,* the EDSAC (*e*lectronic *d*elay *s*torage *a*utomatic *c*alculator) and the EDVAC (*e*lectronic *d*iscrete *v*ariable *a*utomatic *c*omputer), were developed at the end of the 1940s. With these machines, the stage was set for the modern computer age.

2.1.4 The Modern Computer Age

Early electronic computers were for the exclusive use of the military and researchers. In the early 1950s, computers began to be sold. The development of a commercial computer industry truly represents the beginning of today's computer revolution.

The modern era can be conveniently broken down into four distinct generations distinguished by the primary electronic component or circuit element within the computer (Fig. 2.5). Each new logic unit has led to computers that are faster, smaller, and

FIGURE 2.5
The evolution of the modern computer era can be conceptualized as a succession of generations closely linked to technological advances. In the third generation, a divergence emerged with the advent of machines expressly designed for the needs of individuals. Today, a broad range of computers is available. The progression seems to be moving toward a fifth generation in which artificial intelligence and parallel processing could figure prominently.

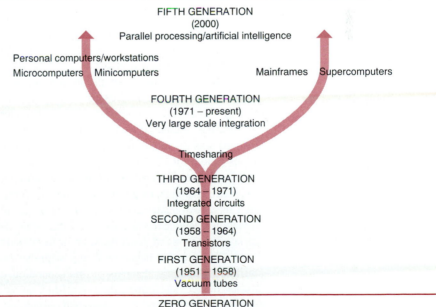

FIFTH GENERATION
(2000)
Parallel processing/artificial intelligence

Personal computers/workstations
Microcomputers   Minicomputers   Mainframes   Supercomputers

FOURTH GENERATION
(1971 – present)
Very large scale integration

Timesharing

THIRD GENERATION
(1964 – 1971)
Integrated circuits

SECOND GENERATION
(1958 – 1964)
Transistors

FIRST GENERATION
(1951 – 1958)
Vacuum tubes

ZERO GENERATION
(pre-1951)
Manual and mechanical

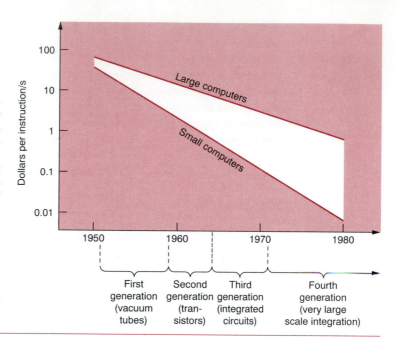

FIGURE 2.6
The progress in computer prices and performance. The four generations of the modern computer age, along with their major circuit elements, are noted on the abscissa. The average rate of improvement for the large general-purpose computers have been less than that for comparable small machines because the newer large computers offer added functions at the cost of more computing power. *(Adapted from Branscomb, 1982.)*

cheaper than previous devices (Fig. 2.6). The magnitude of this improvement has been remarkable. It has been stated (Cougar and McFadden, 1981) that "if the automobile industry had been able to advance its technology as rapidly as the computer industry, a Rolls-Royce today would cost $2.50 and would get 2 million miles per gallon!"

The *first generation* began in 1951 when the U.S. Census Bureau purchased a UNIVAC I computer from the Remington-Rand Corporation. This computer, as did other early electronic devices, used *vacuum tubes* as its primary logic elements. Although the tubes were a major advance over electromechanical parts, they had a number of disadvantages, including excessive heat generation, size, and unreliability. Punched cards or tape was the major medium for information transmission and storage. Most applications were limited to business-oriented data processing such as payrolls and accounting. The machine was usually run by a trained operator. Computer programs were written in obscure languages and were extremely complicated, requiring expert programmers. All the above characteristics should make it obvious that, although the first-generation machines demonstrated utility for large-scale data processing, they were not yet appropriate for the broad general use of professionals such as practicing engineers.

This situation began to change with the *second generation* of the early 1960s. The primary electronic elements of this generation were *transistors*. These solid-state electronic devices performed the same function as vacuum tubes but were smaller, faster, and more reliable and had lower heat output and power requirements. Consequently, the computers of this generation were cheaper to operate than the earlier machines. Along with lower cost, the development of easier-to-use programming languages such as

FORTRAN made computers accessible to large numbers of practicing engineers for the first time in history.

The next technological breakthrough, which ushered in the *third generation* of the late 1960s, was the integrated circuit. An *integrated circuit,* or *IC,* consists of a tiny silicon chip on which many individual transistors are fabricated. As with the previous advances, the IC resulted in large-scale computers that were faster, more efficient, and more reliable. However, other changes occurred that went beyond mere increases in computer power and that truly heralded the beginning of a microelectronic revolution.

Up to the mid-1960s only organizations such as corporations, universities, government agencies, and the military could afford to acquire and maintain a large-scale, or *mainframe,* computer (Fig. 2.7*a*). Because of a growing demand from smaller organizations and some individuals, *timesharing* systems were developed whereby several independent users could simultaneously utilize a large mainframe computer. General Electric offered the first widespread timesharing service. Access was via telephone cables, and users were billed in a way similar to that by which a utility charges a customer.

Stimulated by the availability of these systems, programmers developed software to capitalize on the interactive nature of timesharing computers. This software allowed users to submit programs and data to the computer and receive a prompt reply. Thus an interactive dialogue between the user and the machine was established. Among the developers of this software were John Kemeny and Thomas Kurtz of Dartmouth College, who at the same time developed the first major interactive programming language—BASIC.

Although timesharing boomed for a while and certainly advanced the cause of the small user, the real breakthrough came in the late 1960s with the introduction of the minicomputer by the Digital Equipment Corporation (DEC). Made possible by the integrated circuit, the *minicomputer* was essentially a scaled-down version of the large mainframe machines. Although less powerful, minicomputers were more than adequate for most engineering computations. Because of their relatively lower cost and convenience, they were widely used in small businesses, including many engineering firms. Today, the minicomputer has evolved into what is called the *workstation* (Fig. 2.7*b*).

The great increase in user control and access ushered in by timesharing and the minicomputer has continued and has actually accelerated in the *fourth generation* of the modern computer age. Although the IC is still the primary electronic element, *microminiaturization* has led to chips holding more and more circuits. This process, which started in the early 1970s, is called *very large scale integration (VLSI).* The most important product of these advances is the *microprocessor.* This is a single silicon chip that contains the complete circuitry of the computer's central processing unit (CPU). As such, the microprocessor is the heart of the *microcomputer* or, as it is called today, the *personal computer* (Fig. 2.7*c*).

As might be expected, the evolution of the microprocessor and of the personal computer is intertwined. The first microprocessor, the Intel 4004, was introduced in 1971. It was a 4-bit microprocessor, so called in reference to the general manner in which it handled information—that is, in chunks of 4 bits at a time. For the purposes of the present discussion, you can tentatively accept the notion that the power of a personal computer is generally related to the number of bits handled by its microprocessor.

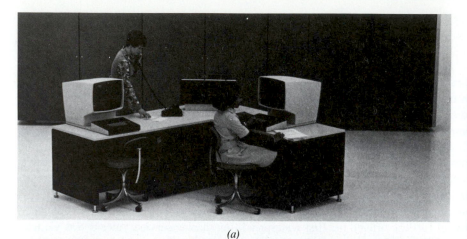

(a)

(c)

(b)

FIGURE 2.7
The three classes of
computers used by today's
engineer: (*a*) the mainframe
(*b*) the workstation, and (*c*) the
personal computer. *(Courtesy of IBM,
Sun Microsystems, Inc., and
Texas Instruments Corporation,
respectively.)*

     Although 4-bit microprocessors such as the Intel 4004 could be used for devices such as calculators, they could not serve as the basis for a computer. It was only with the development of 8-bit processors that microcomputers became feasible. The first of these, the Intel 8008, was introduced in 1972.

     The actual idea of the microcomputer originated in a talk by Alan Kay in 1972 (Kay, 1972). In mid-1974, an article was published in a magazine called *Radio Electronics* that described the technique for constructing a personal computer with a microprocessor at its heart. Soon after, in early 1975, the first microcomputer, the MITS Altair, appeared. This machine used an improved version of the Intel 8008, the Intel 8080, as its microprocessor.

The beginning of a commercial microcomputer industry can probably be dated to mid-1977 when Steve Wozniak and Steve Jobs put their first Apple computer on the market. From that start, the company boomed and has made available updated models, as well as other products such as the Macintosh. The Apple, as did the other early microcomputers, used an 8-bit microprocessor as its major component.

Another major watershed in the evolution of the microcomputer market was the entrance of IBM in 1981. Its major product, the IBM PC, has dominated the market to the extent that it can justifiably be considered the industry standard. The original IBM PC used a 16-bit version of the Intel 8080 microprocessor—the Intel 8088. Although the Intel 8088 was generally more powerful than the earlier 8-bit chips, it was not a true 16-bit chip. This is because, although it operated with data in 16-bit chunks, it transmitted data 8 bits at a time. True 16-bit microprocessors, such as the Intel 8086, were eventually produced; and today 32-bit personal computers are widely available. At the time this edition was printed, even more powerful PCs were being produced.

As we move into the twenty-first century, more and more information and capability are being condensed onto silicon chips (Fig. 2.8). The miracle of the microelec-

FIGURE 2.8
The evolution of computer capabilities since World War II. (Courtesy of USA Today.)

PCs pack more power

Today's desktop PC is nearly 200 times more powerful than the 30-ton ENIAC, the world's first computer. And progress hasn't stopped: Dell's 486P50 is four times as powerful as a high-powered model IBM sold in 1988 and one-sixth the cost.

	ENIAC	IBM System 370 Model 168	IBM PS/2 Model 70	Dell 486P50
Date	1946	1975	1988	1992
Size	30 tons, about the size of a boxcar	Refrigerator size	Desktop size	Desktop size
Instructions processed per second	100,000	2 million	5 million	19.4 million
Cost[1]	$3,223,846	$8,828,625	$13,840	$2,188

Source: Dell, Intel, University of Pennsylvania, USA TODAY research.   1–1992 dollars

tronic revolution is that, for an affordable price, today's practicing engineer can have ready access to a personal computer or workstation that is superior in performance and capability to most of the first- and second-generation mainframe machines.

2.2 OVERVIEW OF COMPUTER HARDWARE

Now that we have provided some historical perspective, we can begin to focus on specific aspects of modern computer hardware. The emphasis of the following discussion will be on the personal computer. We have chosen to emphasize these machines because they are the ones that you will likely own and operate in the coming years.

Although every type of computer will appear different at first glance, all include the major components shown in Fig. 2.9. The present chapter is devoted to a brief description of each of these components.

2.2.1 The Central Processing Unit (CPU)

When most people think of a computer, they usually visualize input and output devices such as the keyboard and the monitor screen. This is understandable because these are the devices that are the primary communication interfaces between the user and the machine. However, the heart of the system, and by some definitions the actual computer, is the *central processing unit,* or the *CPU.*

FIGURE 2.9
A schematic of the major components of a modern computer.

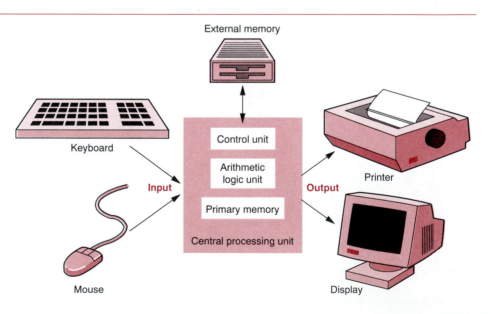

The CPU is the control center or "brain" of the computer. Its functions can be divided into three categories:

1. *The control unit*   This controls the electric signals passing through the computer. It performs the necessary function of coordination; that is, it directs and monitors the operation of the entire system.

2. *The arithmetic/logic unit*   This is where the arithmetic and logical operations are performed.

3. *The primary memory*   This unit serves as a temporary retention device for information either being input to the system or resulting from internal calculations. In addition, it is used to store programs.

Primary memory is commonly referred to as *random-access memory,* or *RAM.* "Random access" means that the computer can go directly to the programs and data it wants, and the access time for any point in the memory is essentially the same. This is in contrast to *serial access,* where the computer must check each memory location in sequence to locate what it wants.

Whereas a few systems have a fixed amount of RAM, most allow expansion to meet the needs of the user. Along with other advances, memory chips have been developed to allow expanded RAM capacities in relatively inexpensive units. Many of the original personal computers had RAMs of 32Kb and 64Kb. The "Kb" stands for kilobytes, where a *byte* corresponds to 8 bits. Although the prefix "kilo" designates 1000, in computer technology it actually represents 1024.[1] Therefore "64Kb" signifies that the computer has enough room to store $64 \times 1024 \times 8$ (or 524,288) bits of information. Interestingly, this is just about the memory capacity of many of the original first-generation computers that took up entire rooms. Personal computers with capacities of many megabytes (1 million bytes or 1Mb) are currently available. Such capacity is critical for a significant number of modern engineering applications.

It should be noted that, in many definitions, primary memory is not included as part of the CPU. In fact, only a portion of the primary memory (registers and cache memory) is part of the CPU. A large segment, called the main RAM, is usually considered an independent entity. For the purposes of the present discussion, however, we will refer to them collectively as the primary memory to distinguish them from secondary disk storage.

Aside from RAM, there is another type of memory that bears mention. *Read-only memory,* or *ROM,* holds computer programs that are so common that manufacturers install them permanently into the computer's hardware. In this sense, ROM is a combination of hardware and software, and as such, the programs that are stored there are sometimes called *firmware.*

The actual innards of a personal computer are displayed in Fig. 2.10. Although each model has its own unique layout, the components in this illustration are found in most machines.

---

[1] As far as the computer is concerned, 1024 is a nice round number. This is because computers use bits to store information as will be described subsequently in Chap. 3. The quantity 1024 is equal to 2 raised to the power 10.

One additional feature of the personal computer that bears mention is that it contains an electronic clock. This has a number of functions, the primary of which is to synchronize the computer's operations. The computational speed of a microprocessor is related to the frequency of this electronic clock. The trend in personal computers is toward clocks with higher frequencies—that is, with more cycles per second, or hertz. The typical unit of measurement is expressed in *megahertz* (MHz), or 10^6 cycles per second. Clock speeds of typical personal computers in the early 1990s range from 20 to 70 MHz. Workstations generally have faster clocks.

We should mention that electronic circuits have limits to how fast they can operate. Thus, the clock speed cannot require them to work faster, for they will fail to produce correct results. Consequently, as clock speeds in computers get faster, the speed of the electronic circuits themselves must improve.

2.2.2 Input and Output Devices

Although the CPU itself is sometimes referred to as "the computer," it is useless without peripheral equipment to allow communication with the outside world. From the standpoint of human use, the vehicles for such communication are the input and output devices. The development of new devices each year precludes a complete description here of all available technology. However, four types are, and will continue to be, the primary equipment whereby engineers interact with personal computers. These include two types of devices for input—keyboards and pointers— and two for output—monitors and printers.

The keyboard.   The *keyboard* is traditionally (and still) the primary input device for the personal computer. Most keyboards are patterned after a conventional typewriter. However, other formats are available for some machines. For example, the keyboard of the conventional typewriter, the *QWERTY* keyboard, was deliberately designed to slow the typist down. This was because the original mechanical typewriters would jam if operated too quickly. Two alternatives, the *Dvorak simplified keyboard (DSK)* and the *American simplified keyboard (ASK),* have been developed which result in error reductions and speed increases of about 50 to 60 percent. Because a strong market demand for these innovations has yet to develop, they are difficult to obtain. However, their benefits could possibly stimulate wider availability in the future.

Finally, there are a number of factors that influence the ease of keyboard use. For example, keyboards vary with respect to touch and spacing. In addition, most keyboards are now separate units from the rest of the computer and can be moved about at the user's convenience.

The pointing device.   Imagine if every time you wanted to turn your automobile, you had to rapidly type in the words "left" or "right" on a keypad by your dashboard. Clearly some activities like steering a car are more effectively executed with analog (that is, a steering wheel) rather than digital instrumentation.

Similarly, as graphics and graphically based interfaces have become prevalent on PCs and workstations, means other than the conventional keyboard are being devised to interact with the computer. Among the most popular are a class of devices called

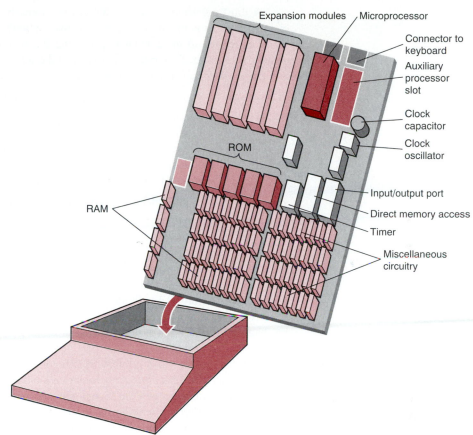

FIGURE 2.10
The innards of a personal computer—the "system board." The board shown here is patterned after the generral layout of the original IBM PC. The inside of your own personal computer may look different, but it should contain many of the same types of components.

pointers. The most successful to date is the hand-held device called the *mouse.* This is a palm-sized pointing device that you move around a flat surface. These movements are communicated to the computer, which induces corresponding movement of the cursor on the monitor screen. The mouse has one or more buttons which are employed for a variety of purposes such as clicking on menu selections, drawing, and so forth. Although keyboards are still superior for certain activities (such as text entry), the mouse is better suited for analog activities such as computer graphics. There are a variety of other activities that can be implemented effectively by both means. For these cases, the chosen mode of interaction usually reduces to personal preference.

Another pointing device which has been in existence for many years is the electronic stylus or *pen.* Pen devices are surging in popularity today because they are being combined with small portable computers with flat liquid-crystal display screens (see the description below). New software is being developed which integrates the pen-pointing device directly into the operation of the computer. Some even allow the computer to interpret handwritten information input by the user.

A third pointinglike device that has been employed in engineering and commercial applications is the *touch screen*. The user actually touches the display screen and a signal is relayed to the computer specifying the coordinates of the location where the contact was made. Software is written to take advantage of this type of information and trigger such operations as menu selection.

The monitor.   The *monitor* is a televisionlike unit that allows you to display your input as you enter it and your output as the computer responds. The most commonly available monitor, one related to the common television set, is the *cathode-ray tube (CRT)*. However, other technology such as the *liquid-crystal display (LCD)* is available. LCDs have the advantage that they require less power than CRTs. As a consequence, they have found application for small portable personal computers.

The printer.   Monitors have several limitations: the output is not permanent or portable, and only a small amount of information may be displayed at any one time. Therefore there are frequent occasions when you will require a *printer* to generate a permanent record, or *hard copy,* of your output.

Printers can be categorized in a number of different ways. One method is according to the mechanisms whereby the characters are formed. *Dot-matrix printers* use a series of dots to form the individual characters. In contrast, *solid-font* mechanisms produce fully formed characters. Because the images they produce are sharper, solid-font mechanisms are usually considered superior to dot matrix and are preferred when "letter-quality" output is required.

Another way in which printers are categorized is by the manner in which the print is transferred to the page. *Impact printers* create characters by striking the type against an inked ribbon that presses the image onto the paper. The conventional typewriter is of this kind. *Nonimpact printers* do not involve physical contact but transfer the image to the paper using ink spray, heat, xerography, or laser.

Although there are some exceptions, nonimpact printers are usually quieter and faster than impact printers because they have fewer moving parts. In addition, they have other advantages including the quality of their graphics capability and their ability to switch fonts automatically. As you might expect, they are also more costly.

2.2.3   Secondary Memory

Because the contents of the CPU's primary memory are lost when you turn off the machine, you will require some means of storing programs or data for later use. This is accomplished by storing the information on secondary memory media such as magnetic disks. Two of the most common media on personal computers are diskettes and hard disks.

Diskettes.   These media, which are called *diskettes* or *floppy disks,* come in two sizes: $5\frac{1}{4}$-in (Fig. 2.11*a*) and $3\frac{1}{2}$-in (Fig. 2.11*b*). Diskettes are made of a flexible plastic that can be magnetized (recall Fig. 2.2). Data is stored on the disk as magnetized spots on concentric tracks. The greater the number of tracks there are per inch, or *disk density,* the greater the amount of information that can be stored. At present, diskettes usually come in two densities: double- and high-density.

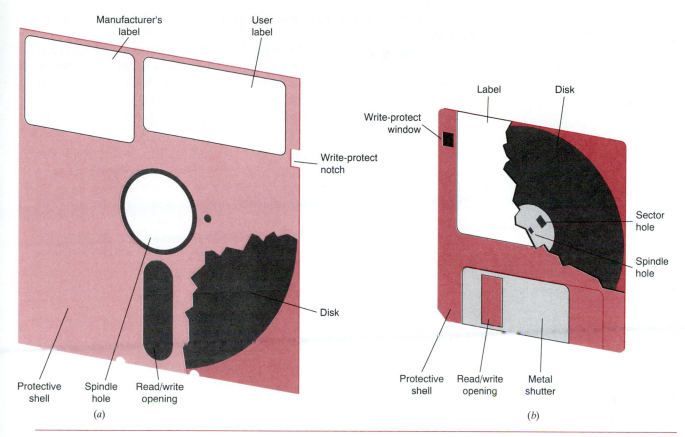

FIGURE 2.11
The two types of diskettes:
(*a*) $5\frac{1}{4}$-inch and (*b*) $3\frac{1}{2}$-inch.

The larger $5\frac{1}{4}$-in version has a flexible (or soft-sided) casing and is protected by a paper envelope when not in use. In contrast, the $3\frac{1}{2}$-in are encased in a rigid plastic shell. They are conventionally capable of storing more information than soft-sided disks. Their smaller size, their larger capacity and durability, as well as the fact that they have faster access times make them generally superior to the $5\frac{1}{4}$-in type. In addition, there are some emerging technologies such as $3\frac{1}{2}$-in quad density and 2-in disks that could see widespread use in the coming years.

Diskettes usually require formatting for a particular system, and a disk formatted for one system may not be compatible with another. Care must be exercised when handling diskettes, particularly the $5\frac{1}{4}$-in versions, and they eventually need to be replaced. Backup copies of all important disks should always be maintained to prevent their loss in the event of damage. We will return to diskettes when we discuss program storage in later parts of the book.

Hard-drive disks. For years, hard disks have been common secondary storage media on mainframe computers. Recently, hard disks and drives are being marketed with most personal computers. These have a number of advantages, including more reliability, larger storage, and faster access times. However, in contrast to diskettes, the hard disk is permanently contained in its disk drive and does not have the portability of a floppy disk.[2]

The significance of hard-disk technology lies in the fact that some personal computers are limited by the speed and capacity of their disk storage devices. The addition of a hard drive elevates the personal computer to a higher plane. With a hard drive, the machine is capable of holding a major database and implementing some of the larger applications software used by engineers. As such, it approaches a comprehensive computing tool for the professional engineer.

Tape and optical disk storage.   Finally, there are other media which have utility as secondary storage. For example, cartridge tapes are increasingly employed as backups of hard drives.

Optical disks, such as compact disks (CDs) and video disks, offer the capability to store immense quantities of data. The CD technology is often used to store large databases of permanent information. Hence, it is referred to as *CD-ROM*, which is short for "compact disk with read-only memory." Optical disks can hold pictures and video sequences.

2.3   WHICH TYPE OF HARDWARE IS BEST?

This chapter has been designed to provide general background on computer hardware. Because of the differences between computer models, we have had to limit our treatment to a general discussion of most common devices. Thus we may not have included some important and emerging technologies that could prove useful to you. Nevertheless, we hope that this introduction gets you started, and if you are employing a personal computer along with this book, we hope that you will take this opportunity to make an in-depth study of its configuration and workings. This will probably involve study of the machine's user's manual. We cannot stress strongly enough that the more you learn about your hardware, the more you will get out of your computer.

Beyond personal computers, today's workstations and mainframe machines have developed and evolved into more powerful and efficient tools. Although personal computers will probably play a major role in your professional and personal computing, workstations and mainframes will continue to have great significance in engineering. In particular, these machines are still the best vehicles for implementing the large-scale data processing and computations that engineers sometimes confront.

In addition, it must be stressed that computer classifications are not always hard and fast. For example, a more advanced class of mainframe called the *supercomputer* is available as a powerful tool in engineering and scientific research environments (Fig. 2.12). Furthermore, the distinction between the personal computer and the workstation is sometimes fuzzy. Although workstations are generally larger and faster, PCs can be upgraded so that they exhibit comparable performance.

In conclusion, we must stress the fact that there is no "best" computer. Just as

---

[2]Although most hard drives are fixed, it should be noted that some are removable.

FIGURE 2.12
A supercomputer. These
machines allow engineers and
scientists to perform
computations that were
impossible in the past.
(Courtesy of Los Alamos
National Laboratory).

good artisans would never limit themselves to a single tool, so also a good engineer usually employs a variety of computers. The beauty of today's situation is that we have potential access to all the types of computers. For example, preliminary work on a large-scale computation might be performed on a personal computer or workstation. Then a telecommunications link can be used to transfer the program to a mainframe or supercomputer for fast implementation. In this way, the strengths of each type of computer are exploited. In the coming decades, engineers with access to a range of machines will have enormous capability at their command.

PROBLEMS

2.1 Explain the fundamental difference between electromechanical and electronic computing devices. Provide an example of each type.

2.2 Define hardware and software.

2.3 (*a*) Explain the difference between analog and digital computers.
(*b*) Characterize the following as either counted or measured:

Cities	Mass	Temperature
Voltage	Stars	Trees
Pressure	Elevation	Cars
Current	Words	Force

2.4 What does the word "bit" stand for? How does it relate to the way that a computer "thinks"?

2.5 What are the primary electronic logic elements corresponding to the four generations of the modern computer age?

2.6 What is a microprocessor? Give examples of 4-, 8-, and 16-bit microprocessors.

2.7 Define the terms "supercomputer," "mainframe computer," "workstation," and "personal computer." Who would tend to own each type?

2.8 What are the components of the CPU, and what are their primary functions?

2.9 What is RAM? How is it distinguished from serial-access memory? From which type do you think that information could be accessed faster? Why?

2.10 Distinguish between a bit and a byte. How many bits of information are contained in a 640Kb personal computer? Can you guess what 4 bits are called?

2.11 How does ROM differ from RAM?

2.12 Distinguish between dot-matrix and solid-font printing mechanisms, and between impact and nonimpact printers.

2.13 The advertised storage capacity of a personal computer is 2Mb. Determine the storage capacity in the following units:
(*a*) bits, (*b*) bytes, and (*c*) kilobytes.

2.14 Visit your local computer store. What kinds of diskettes are sold? How much do they cost, and how is their capacity characterized? What type of disk is the least expensive, measured in terms of kilobytes stored per dollar.

INTERNAL REPRESENTATION AND SYSTEMS SOFTWARE

When most people think of computers, they usually visualize the hardware described in the previous chapter. However, without human direction, hardware is a useless amalgam of metal and plastic. The human direction is provided by what are called computer programs. A *program* is merely a set of instructions that tells the computer what to do. All the programs that are necessary to run a particular computer are collectively called *software*.

To understand how computers are able to process a program's instructions and data, the first part of the present chapter is devoted to the *internal representation* of information within the computer. In addition, we will provide an overview of the *systems programs* which, among other things, control the computer's internal activities. As such, they will allow you to operate your computer in an effective manner. In addition, systems programs are necessary to produce the *applications programs* that are employed for engineering problem solving.

3.1 PRELIMINARY CONCEPTS

Before describing how information is stored, we must introduce some preliminary ideas. First, because computers use a rather unfamiliar scheme to store and manipulate data, we will describe the concept of a number system. Then we will present the concept of word length as a unit of information.

3.1.1 Number Systems

Recall that in Sec. 2.1.2 we introduced the notion that computer intelligence boils down to whether a bit is 0 or 1. Now we will see how such a simple idea can be used to represent more complicated information on a computer.

The most elementary type of information that we might want to represent is a numerical quantity. To understand how a binary digit, or bit, is used for this purpose we have to understand how number systems work.

A *number system* is merely a convention for representing quantities. Because we have 10 fingers and 10 toes, the number system that we are most familiar with is the *decimal,* or *base 10,* number system. A *base* is the number used as the reference for constructing the system. The base 10 system uses the 10 digits—0, 1, 2, 3, 4, 5, 6, 7, 8, and 9—to represent numbers. By themselves, these digits are satisfactory for counting from 0 to 9.

For larger quantities, combinations of these basic digits are used, with the position or *place value* specifying the magnitude. The rightmost digit in a whole number represents a number from 0 to 9. The second digit from the right represents a multiple of 10. The third digit from the right represents a multiple of 100 and so on. For example, if we have the number 86,409, then we have eight groups of 10,000, six groups of 1000, four groups of 100, zero groups of 10, and nine more units, or

$$(8 \times 10^4) + (6 \times 10^3) + (4 \times 10^2) + (0 \times 10^1) + (9 \times 10^0) = 86{,}409$$

Figure 3.1*a* provides a visual representation of how a number is formulated in the base 10 system. This type of representation is called *positional notation.*

Now, because the decimal system is so familiar, it is not commonly realized that there are alternatives. For example, if human beings happened to have eight fingers and

FIGURE 3.1
How the (*a*) decimal (base 10) and (*b*) binary (base 2) systems work. In *b*, the binary number 10101101 is equivalent to the decimal number 173.

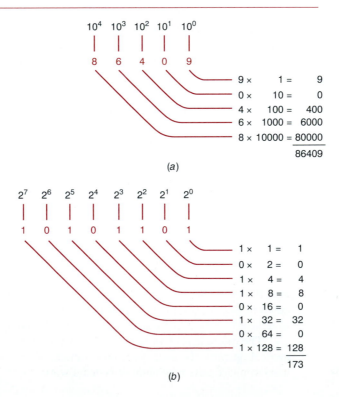

toes, we would undoubtedly have developed an *octal,* or *base 8,* representation. In the same sense, our friend the computer is like a two-fingered animal that is limited to two states—either 0 or 1. This relates to the fact that the primary logic units of digital computers are on/off electronic components. Hence, numbers on the computer are represented with a *binary,* or *base 2,* system. Just as with the decimal system, quantities can be represented using positional notation. For example, the binary number 11 is equivalent to $(1 \times 2^1) + (1 \times 2^0) = 2 + 1 = 3$ in the decimal system. Figure 3.1*b* illustrates a more complicated example.

EXAMPLE 3.1   *Transforming Base 10 Numbers to Base 2*

PROBLEM STATEMENT: Convert the base 10 number, 41, to binary representation.

SOLUTION: Although conversion of a base 2 number to base 10 is simple (Fig. 3.1*b*), the reverse process is not so straightforward. In the present example, we will illustrate two ways to convert from decimal to binary form.

One way, called *successive approximation,* is based on determining the largest power of 2 that "fits into" (that is, is less than or equal to) the number. In the case of 41, the largest power that fits is $2^5 = 32$. Thus, we know that there is a 1 in the 2^5 place in the binary representation. The 32 can then be subtracted from the original number, 41, to yield a residual of $41 - 32 = 9$. Next, we can determine that $2^3 = 8$ fits within 9, and therefore, there is a 1 in the 2^3 place of the binary representation. This number can be subtracted from the residual to give $9 - 8 = 1$. Because $2^0 = 1$ fits exactly into this remainder, we know that there is a 1 in the 2^0 place. Putting all these operations together yields the final result:

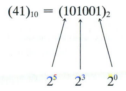

$$(41)_{10} = (101001)_2$$

An alternative approach, which is less intuitive and somewhat more mechanical, is to divide the number by 2 successively and keep track of the remainders. For the present example,

$$41/2 = 20, \quad \text{rem} = 1$$
$$20/2 = 10, \quad \text{rem} = 0$$
$$10/2 = 5, \quad \text{rem} = 0$$
$$5/2 = 2, \quad \text{rem} = 1$$
$$2/2 = 1, \quad \text{rem} = 0$$
$$1/2 = 0, \quad \text{rem} = 1$$

The remainders, when read from the bottom up, give the base 2 form of the number:

101001. If you are curious why this works, a nonsense exercise might help you understand. Convert the number 86409 back into base 10 (that's the nonsense part!) using successive division by 10.

Aside from binary representation, you can devise other number systems. An interesting example that has been used in computing is the *hexadecimal* or *base 16* system. Because there are only 9 decimal digits, the first letters of the alphabet are employed for the remaining digits in this system as in

0 1 2 3 4 5 6 7 8 9 A B C D E F

It takes a little mental adjustment to feel comfortable considering "A" a number. Similarly, accepting valid hexadecimal numbers such as B43F or even BEEF requires some intellectual flexibility on your part.

The relationship of this system to the binary, octal, and decimal systems is summarized in Table 3.1. Look at the column for the hexadecimal numbers. Notice that because F is the last available symbol, the following number is 10. So, just as 10 follows 9 in the decimal system, 10 follows F in hexadecimal. However, it should be noted that $(10)_{16}$ is called "one-zero," not "ten" to make it clear that it is actually equal to the decimal quantity $(1 \times 16^1) + (0 \times 16^0) = 16 + 0 = 16$. People who are turning 40 years old enjoy being told that their age is only 28 in base 16.

It should also be recognized that the octal and hexadecimal systems are used in digital computers because 8 and 16 are integral powers of 2. Consequently, base 8 and

TABLE 3.1   Comparison of the Decimal, Binary, Octal, and Hexadecimal Number Systems

Decimal	Binary	Octal	Hexadecimal
0	0	0	0
1	1	1	1
2	10	2	2
3	11	3	3
4	100	4	4
5	101	5	5
6	110	6	6
7	111	7	7
8	1000	10	8
9	1001	11	9
10	1010	12	A
11	1011	13	B
12	1100	14	C
13	1101	15	D
14	1110	16	E
15	1111	17	F
16	10000	20	10

base 16 representations offer convenient vehicles for expressing binary numbers in an abbreviated or shorthand format (see Prob. 3.4 at the end of this chapter). These are called *commensurate number bases.*

The base 10, or decimal, system is not commensurate with bases 2, 8, and 16, since 10 is not an integral power of 2 ($2^{3.322} \cong 10$). This suggests that decimal quantities are not completely compatible with the internal representation of computers (more on that later).

Aside from transformations between number bases, arithmetic operations can also be performed on numbers written in other bases. As in the following example, the operations are identical with the ones that you are familiar with involving base 10 quantities.

EXAMPLE 3.2   *Performing Arithmetic Operations in Binary*

PROBLEM STATEMENT: Addition is simple in base 10 arithmetic. For example,

$$
\begin{array}{r}
35 \\
+15 \\
\hline
50
\end{array}
$$

Perform the same operation in base 2.

SOLUTION:
The first step is to convert the numbers into binary. Using either of the approaches outlined in Example 3.1 yields

$$(35)_{10} = 100011$$
$$(15)_{10} = 001111$$

When you sum base 10 numbers, you merely sum each column starting in the *least significant digit* (farthest right place[1]), being careful to carry values when the total for a column exceeds 9. Similarly, when summing binary numbers, you perform the same procedure but carry values when the total for the column exceeds 1. Consequently, for the present case, the first step would yield

$$
\begin{array}{r}
1 \longleftarrow \quad \text{carry} \\
100011 \\
+001111 \\
\hline
0
\end{array}
$$

In binary addition, the sum of $1 + 1 + 1 = 11$. Therefore the result for the second least significant digit is

---

[1] The reason that the farthest right digit is called the "least" significant is that it represents the smallest component of the number's total value.

$$1 \longleftarrow \text{carry}$$
$$100011$$
$$+001111$$
$$10$$

The process can be continued to yield the final result:

$$100011$$
$$+001111$$
$$110010$$

The result can be checked by recognizing that $(110010)_2 = (1 \times 2^5) + (1 \times 2^4) + (1 \times 2^1) = 32 + 16 + 2 = (50)_{10}$, which is the correct answer. You will observe that generally there are many more carries in binary addition than its decimal counterpart.

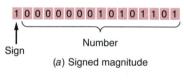

Sign   Number

(a) Signed magnitude

1 1 1 1 1 1 1 1 0 1 0 1 0 0 1 1

(b) 2's complement

FIGURE 3.2
The representation of the base 10 integer, −173, using a 16-bit word and (a) the signed magnitude method and (b) the 2's-complement method. For the signed magnitude approach (and for other systems in general), a "0" is used conventionally to designate a positive value and a "1" specifies a negative.

3.1.2 Word Size

At this point, it should be evident how a computer could store an integer number. It could simply transform the quantity into binary format and then set a series of its electronic elements "on" (corresponding to 1) or "off" (corresponding to 0) according to the sequence of the binary digits. An example is shown in Fig. 3.2a.

Although such a scheme would be straightforward, it could pose a problem when the computer desired to transmit this information. This is because numbers of different magnitude could have different-sized binary representations. For example, the decimal number 14 requires a binary number that is 4 bits long, 1110; whereas 173 requires 8 bits, 10101101. Suppose that the computer transferred such information serially (that is, 1 bit at a time) along a single wire. Such a method of transmission would be analogous to passing the bits through a pipe. If 14 and 173 were transmitted in binary form, the result at the other end would come out as 111010101101. The numbers are of different length, but there is no clue as to where one ends and the other begins. Thus the receiver would be unable to decipher their meaning.

In fact, numbers are actually transmitted through a computer's circuitry with parallel signal connections. If the unit of transmission were a byte, this would be analogous to having a connection consist of eight separate pipes that are bound together into a single entity called a *bus*. Clearly, passing numbers consisting of different numbers of bits would still pose a problem for such a scheme. This is due to the fact that the number of required connections or wires would vary depending on the size of the number.

For this reason (as well as for other considerations), computers typically use a fixed number of bits to represent information. This fundamental unit of information is formally referred to as a *word*. The fixed length is employed to facilitate transmission, manipulation, and storage of data. For example, if an 8-bit (that is, a *byte*) code is employed, then a "bus" of eight parallel connections would work nicely.

3.2   INTERNAL REPRESENTATION OF INFORMATION

Now that we have reviewed the concepts of word size and binary representation, we can actually describe how information is represented within the computer. This information consists of three types: integers, real numbers, and characters. As described next, each has its own representation scheme.

3.2.1   Integers

As already illustrated in Fig. 3.2*a* it should be simple to conceive how integers could be represented in the computer. The most straightforward approach, called the *signed magnitude method,* employs the first bit to indicate the sign, with a 0 for positive and a 1 for negative. The remaining bits are used to store the number. For example, the integer value of -173 could be stored using 16 bits (Fig. 3.2*a*).

Although the signed magnitude method is certainly straightforward, it is not the method used in modern computers. Early computer developers decided that it was not prudent to use a separate bit for a number's sign. Instead, they came up with a scheme, called the *2's-complement method,* to embed the sign in the binary representation itself.

To understand how this is done, suppose that we use an 8-bit word to store our integers. What happens when you add 00000001 to the number 11111111 binary? you get 00000000 back. Of course, there is a final carry of 1, but we cannot represent that because it does not "fit" into our 8-bit system. Another way to look at that operation is to add 1 to some number and obtain 0 as a result:

$$00000001 + x = 00000000$$

We have already deduced that, for an 8-bit system, the number x has to be equal to 11111111; but x also satisfies our intuitive definition of the quantity -1, the additive inverse of 1. That is, it is the number which when added to 1 yields 0.

To carry this one step further, add 00000010 binary to 11111110. Again, you get 00000000; therefore, 11111110 appears to be equivalent to -2 decimal.

A nice way to visualize the scheme is to construct a *number circle* (Fig. 3.3). The positive numbers are on the right semicircle, and the negative values are on the left semicircle. Because we have an even number of values in our system ($2^8 = 256$), and we single one out to be zero, there has to be one left over at the bottom of the circle—in this case, 10000000. Note that this number, which is admittedly strange, is its own negative. (Try adding it to itself and see what happens.) What do we do with this number? It is convention to consider it as an extra negative number—in this case, equivalent to -128 decimal. This convention was adopted because its most significant bit (the leftmost bit) is a 1, just like the rest of the negative numbers. Therefore, our convention of interpreting 8-bit words as signed integers gives us a range of -128 through 0 to $+127$.[2]

This brings up a couple of important points. The first is that the 2's-complement representation is somewhat arbitrary. It is an accepted convention, but there could be

[2]Another convention that is sometimes used is to leave out the -128 completely.

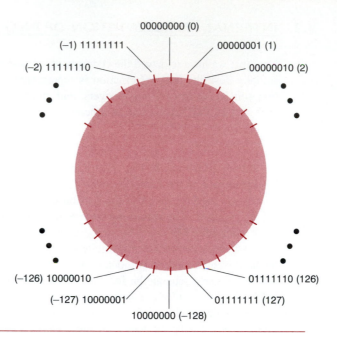

FIGURE 3.3
Number circle for 8-bit
2's-complement
representation of integers.

other ways to arrange it. The second is that an 8-bit byte does not allow us a very large range of integers.

Before we go on with other representations, let's see how we can easily compute the 2's-complement binary representation of the negative of a number. There are two steps:

1. Take the complement of the original number; that is, change all 0's to 1's and all 1's to 0's.
2. Add 1 to the result.

Check this procedure with the numbers on the circle.

EXAMPLE 3.3   *Subtracting by Adding the 2's Complement*

PROBLEM STATEMENT: Perform the following subtraction by adding the 2's complement using 8-bit integers.

$$
\begin{array}{r}
35 \\
-15 \\
\hline
20
\end{array}
$$

SOLUTION: The first step is to convert the numbers into binary. As was done in Example 3.2, $(35)_{10} = (00100011)_2$ and $(15)_{10} = (00001111)_2$. Now the subtrahend can be converted to its 2's complement by forming the complement

$$00001111 \xrightarrow{\textit{complement}} 11110000$$

and adding 1 to yield 11110001. This number can be added to the minuend as in

$$
\begin{array}{r}
00100011 \\
+11110001 \\
\hline
00010100
\end{array}
$$

Remember that the last carry is discarded because it does not fit into our 8-bit integer representation. The answer can be verified as $(00010100)_2 = (1 \times 2^4) + (1 \times 2^2) = 16 + 4 = 20$.

As we stated earlier, one advantage of the 2's-complement representation is that it does not require a separate digit for the sign. The foregoing example illustrates a second advantage. Because subtraction is equivalent to adding the 2's complement of the subtrahend, it means that the computer's arithmetic unit does not have to have an internal program to perform integer subtractions. All it needs is a single routine to add integers.

In addition, integer multiplication merely requires repeated additions and integer division repeated subtractions. To see this clearly, think of doing multiplication and long division by hand (if you can remember how). Consequently, all integer arithmetic can be based on simple addition! This is one of the reasons why the 2's-complement approach was devised originally.

To increase the range of the integer representation, we consider the use of a 16-bit word. This allows a range of

$$-2^{15} \quad \text{to} \quad 2^{15} - 1$$

or

$$-32768 \quad \text{to} \quad +32767$$

which is suitable for many calculations involving integers. It should be noted that most programming systems do not even allow the use of byte-length integers. In most cases, the simplest integer is a word-length representation of 16 bits.

If we require even greater range in integers, we can "paste" two 16-bit words together to make a 32-bit representation which has a range of

$$-2^{31} \quad \text{to} \quad 2^{31} - 1$$

or

$$-2{,}147{,}483{,}648 \quad \text{to} \quad +2{,}147{,}483{,}647$$

or about

$$-2 \times 10^9 \quad \text{to} \quad 2 \times 10^9$$

Note that a 32-bit integer is made up of 4 bytes and will require 4 bytes of storage in memory, since memory storage is organized by bytes.

As can be derived from these examples, a general relationship for the range is

$$-2^{n-1} \quad \text{to} \quad 2^{n-1} - 1$$

where n is the number of bits in a word. The significance of such ranges is explored in the following example.

EXAMPLE 3.4   *"Wraparound" of Integers*

PROBLEM STATEMENT: Investigate what happens when the following addition is evaluated using 8-bit integers:

$$\begin{array}{r} 127 \\ +2 \\ \hline 129 \end{array}$$

SOLUTION: The first step is to convert the numbers into binary. As was done in Example 3.2, $(127)_{10} = (01111111)_2$ and $(2)_{10} = (00000010)_2$. Therefore the addition is

$$\begin{array}{r} 01111111 \\ +00000010 \\ \hline 10000001 \end{array}$$

which is equal to -127! This odd outcome arises from the fact that the result of our addition exceeds the acceptable range for the 8-bit integer (that is, -128 to 127). This phenomenon is referred to as *wraparound*. The reason for this nomenclature can be understood by inspecting the number circle from Fig. 3.3. Notice that adding 2 to 127 carries us past the upper limit of the positive integers. Rather than obtaining the correct answer, we "wraparound" to the erroneous result -127.

This type of error can occur when an integer operation yields a result that exceeds the allowable range. It is particularly dangerous because it yields a result that is not totally outlandish (as would happen, for example, if you tried to divide by zero). Therefore, it is conceivable that the computation could continue in a seemingly reasonable fashion when, in fact, it would be dead wrong.

As in the previous example, it is important to avoid exceeding the range of integers that can be represented in your computer. This and other insights are the payoff from making the effort to understand internal representation. We now turn to how real numbers are represented.

3.2.2 Real (or Floating-Point) Numbers

Real numbers are often called "floating point" numbers. This terminology comes from earlier times when, in some computer systems, real numbers were represented by schemes where the "number-system point" was understood to fall at a fixed position between 2 digits. This latter system was called "fixed point" representation. It is rarely used today.

There are many ways that a real number can be represented by a series of binary digits. The convention reviewed here is that used by Microsoft for its language products, but you should be aware of the fact that there are others. We have chosen the Microsoft approach in part because it follows the *IEEE*[3] *convention,* which is the standard for number representation on a wide variety of computers.

We are accustomed to exponential notation in the decimal system, for instance,

41.26 represented as 4.126×10^1

In this example, 4.126 is referred to as the *mantissa* or *significand* and 1 is called the *exponent* or *characteristic.* It is convenient to use the same scheme in binary representation, but we must remember that we are no longer dealing with a number base of 10. So first, we must convert our decimal number to a binary one:

41.26 decimal = ??? binary

The 41 is straightforward to convert. Using successive approximation (Example 3.1) yields 101001.

Now, how can we handle the fractional part, the 0.26? In other words, as depicted in Fig. 3.4, how do we determine the 1's and 0's for the places to the right of the binary

[3]Institute of Electrical and Electronic Engineers.

FIGURE 3.4
Converting real numbers from base 10 to base 2 representation involves the determination of the 1's and 0's that must be entered to the right of the binary point.

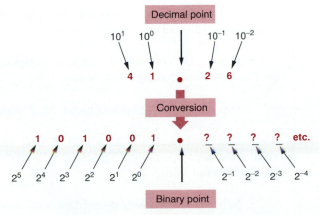

STEP 1: Let i represent the ith binary place to the right of the binary point, and start with $i = 1$. Let the decimal fractional part be `fract`. Go to step 2.

STEP 2: Is 2^{-i} less than or equal to `fract`, the fractional part? If yes, go to step 3. If no, go to step 4.

STEP 3: Subtract 2^{-i} from `fract` to form a new value for `fract`, and set the digit in the ith binary place to the right of the binary point to be 1. Go to step 5.

STEP 4: Set the digit in the ith binary place to the right of the binary point to be 0. Go to step 5.

FIGURE 3.5
A step-by-step procedure or "algorithm" for transforming a fractional base 10 number to binary.

STEP 5: Do you have a sufficient binary precision yet? If yes, stop. If no, increment i by 1 and go back to step 2.

point (that is, the positions for 2^{-1}, 2^{-2}, 2^{-3}, and so on)? An approach for doing this, which is similar to the successive approximation method described before, is outlined in Fig. 3.5 and is illustrated in the following example.

EXAMPLE 3.5   Converting a Decimal Fractional Number to Binary

PROBLEM STATEMENT: Employ the step-by-step procedure delineated in Fig. 3.5 to transform $(0.26)_{10}$ to 10 binary digits.

SOLUTION:

- *Step 1* $i = 1$ and `fract` = 0.26.

- *Step 2* $2^{-1} = 0.5$ is greater than 0.26. Therefore, go to step 4.

- *Step 4*   Set the first position to the right of the binary point to 0. Therefore the first approximation of the answer is $(0.0)_2$.

- *Step 5*   We obviously do not have sufficient precision yet, so we increment i by 1 to yield $i = 2$ and go back to step 2.

- *Step 2* $2^{-2} = 0.25$ is less than 0.26. Therefore we go to step 3.

- *Step 3*   `fract` = 0.26 − 0.25 = 0.01, and set the second position to the right of the binary point to 1. Therefore, the improved approximation of the answer is $(0.01)_2$. Proceed to step 5.

- *Step 5*   We obviously do not have sufficient precision yet, so we increment i by 1 to yield $i = 3$ and go back to step 2.

At this point we could continue the algorithm until we attained 10 binary places of precision. The result would be 0.0100001010 binary, which can be verified by summing the individual terms:

$$1 \times 2^{-2} = 0.25$$
$$1 \times 2^{-7} = 0.0078125$$
$$1 \times 2^{-9} = \underline{0.001953125}$$

$$0.259765625$$

Note that the binary equivalent determined in Example 3.5 is not really an equivalent but an approximation, which is in error by -0.000234375. If we require more precision than this, it will be necessary to use more binary digits. Therefore, 41.26 decimal is approximately equal to 101001.0100001010 binary.

This result can in turn be expressed in binary exponential notation using a process called *normalization*. This consists of moving the binary point of the mantissa to the position immediately to the right of the most significant digit and adjusting the exponent accordingly. For the present case, this yields

$$1.010010100001010 \times 2^5$$

The process of expressing the number in this way is the basis for the internal representation of the real number 43.26 in the personal computer.

If we normalize the real number in the way shown, there will always be a 1 digit to the left of the binary point. Consequently, it has no information value. Therefore, as long as the computer understands that it is "really there," this digit can be omitted from our explicit representation. Consequently, we need only store the remainder of the mantissa 010010100001010 and the characteristic or exponent 5 decimal or 101 binary.

That brings us to the way that real quantities are stored in the computer using the IEEE convention. Simple real numbers are represented by a 4-byte 32-bit arrangement, as shown in Fig. 3.6.

The mantissa (or significand) is 23 bits long and takes up most of the rightmost

FIGURE 3.6
IEEE format binary
representation of the decimal
number: 41.26.

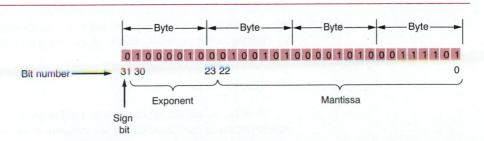

3 bytes. The exponent or characteristic is 8 bits long and uses most of the leftmost byte and 1 bit from the next byte. The sign bit indicates explicitly whether the mantissa is plus or minus, with a 1 value meaning negative and 0 positive. The significand stored is just like the one we computed above, with the exception that it is always taken out to the full 23 bits; ours was only 15 bits long. This 23-bit mantissa is shown in Fig. 3.6.

The exponent field is a bit tricky. To make calculations more efficient in the computer's CPU, the convention uses what is called a *biased exponent*. Since the exponent field is 8 bits wide, there are $2^8 = 256$ possible exponent values. The convention used is shown in Table 3.2.

The bias value is 01111111 binary or 127 decimal. Therefore, our exponent above, which was 5 or 101, would have the representation of $5 + 127 = (132)_{10} = (10000100)_2$; and we can now "paste" the real number together as shown in Fig. 3.6.

The resolution of this number representation is one part in 2^{24} or about one part in 10^7 for the mantissa, that is, 7 significant decimal digits. The limits on the exponent are

$$2^{-127} \quad \text{to} \quad 2^{128}$$

or about

$$10^{-38} \quad \text{to} \quad 10^{39}$$

In summary, we see that the standard "single precision" IEEE floating-point format for more recent Microsoft software products allows for real numbers with 7 significant digits in a range of approximately

$$10^{-38} \quad \text{to} \quad 10^{39}$$

The single-precision IEEE format is sufficient for most engineering computations. However, if we require more precision, or larger exponent values, more bytes are used in the representation. There is one other standard representation which does this: it is called double precision and uses 8 bytes. It provides about 15 significant digits and an exponent range of approximately

$$10^{-308} \quad \text{to} \quad 10^{308}$$

Aside from the common representations, there are special chips that are added to personal computers to increase their precision and speed in handling real numbers. For example, the Intel iAPX-87 (or 8087 or 80287 or 80387) numeric data coprocessor chip, which does floating-point math in hardware, uses an *80-bit* representation for floating-point numbers, which gives 19 significant digits and an exponent range of about

$$10^{-4900} \quad \text{to} \quad 10^{4900}$$

Finally, we should again stress that although it has been widely adopted, the IEEE representation is not standard on all computers or with all software products. It is

TABLE 3.2

The Convention for the Biased Exponent Employed in IEEE Format for Floating-Point Numbers

Binary Value	Exponent Value
11111111	128 decimal
.	.
.	.
.	.
10000000	1
01111111	0
01111110	1
.	.
.	.
.	.
00000000	−127

necessary at times, when transferring numerical data from one product's environment to another, to rearrange (or convert) from one representation to another.

You should now appreciate that, although it is beneficial to know how the personal computer represents real quantities (and particularly its limitations in doing so), it is nice that the computer does all the hard work for you. In other words, isn't it great that you don't have to deal with this on a day-to-day basis!

3.2.3 Characters and Symbols

Besides numbers, other characters such as letters and symbols (periods, asterisks, and so forth) must be managed by the computer. In fact, numbers themselves must often be treated as characters. For example, your phone number or social security number could be stored on the computer, but it is highly unlikely that you would employ them in calculations. In such cases, it is most efficient to have the computer treat them differently than "normal" numbers (that is, those that are routinely used for measurement or calculation). Thus, they are treated as characters or, to use computer jargon, as *strings* of characters.

The character set for the personal computer is defined in terms of 128 standard characters, called the *ASCII* character set, and 128 special characters, called the extended ASCII character set. The acronym "ASCII" is pronounced "askey" and stands for the American Standard Code for Information Interchange. The extended set may vary from computer to computer. Because only 256 unique characters are represented, 1 byte can be used for the representation.

Another popular code is called *EBCDIC* (*e*xtended-*b*inary-*c*oded *d*ecimal *i*nter*c*hange *c*ode). The EBCDIC code was developed by IBM and is employed on most of their computers (other than their personal computers). It also uses 8 bits per character.

Table 3.3 lists both the ASCII and the EBCDIC codes for the digits (0 through 9)

TABLE 3.3 ASCII and EBCDIC Representation of the Decimal Digits and Uppercase Letters

Character	ASCII	EBCDIC	Character	ASCII	EBCDIC
0	00110000	11110000	I	01001001	11001001
1	00110001	11110001	J	01001010	11010001
2	00110010	11110010	K	01001011	11010010
3	00110011	11110011	L	01001100	11010011
4	00110100	11110100	M	01001101	11010100
5	00110101	11110101	N	01001110	11010101
6	00110110	11110110	O	01001111	11010110
7	00110111	11110111	P	01010000	11010111
8	00111000	11111000	Q	01010001	11011000
9	00111001	11111001	R	01010010	11011001
A	01000001	11000001	S	01010011	11100001
B	01000010	11000010	T	01010100	11100010
C	01000011	11000011	U	01010101	11100011
D	01000100	11000100	V	01010110	11100100
E	01000101	11000101	W	01010111	11100101
F	01000110	11000110	X	01011000	11100110
G	01000111	11000111	Y	01011001	11100111
H	01001000	11001000	Z	01011010	11101000

and letters (A through Z). Notice how the last four places of the number representations are in fact the true binary values. For the letters, notice that the bit representations are in ascending order. Thus the bit representation for A is less than that for B, that for B is less than that for C, and so on. We will find this useful in later sections when we want to place information in alphabetical order. In addition, although they are not shown in the table, lowercase letters can also be represented in ASCII. The lowercase letters come a fixed number of positions in the ASCII code list after the uppercase. This attribute can be exploited to easily convert text from uppercase to lowercase and vice versa.

3.2.4 Summary

You may be wondering at this point why we have devoted so much material to internal representation. As we mentioned at the beginning of Chap. 2, a little understanding about your system's hardware will go a long way toward making you a more effective user of computers. In the same sense, knowing something about internal representation will often provide you with insights that would not be obvious to someone who is ignorant of these topics.

First and foremost, awareness that the whole system is built on powers of 2 is helpful in itself. Time and time again, you will run into seemingly strange values, such as 256, 1024, 32767, or −2,147,483,648. Hopefully, this chapter has shown you how all these quantities originate naturally from the binary system and the conventions used by hardware and software designers.

In our discussion regarding integers, we learned that, because we are limited to a fixed number of binary digits, there are a finite number of integers on any computer. Therefore, if we try to employ an integer that is beyond the limits, something odd may occur (for example, wraparound as illustrated in Example 3.4).

When we discussed character strings, we saw that alphabetic characters were represented by codes that were in numerical order (recall Table 3.3). Consequently, we can understand how the computer might be able to sort such information in alphabetic order.

Although the above and other insights will hopefully serve you well, the details related to floating-point numbers are certainly the most subtle and probably the most important. For this reason, they bear enumeration:

1. *There is a limited range of quantities that may be represented in floating point.* Just as for integers, there are large positive and negative numbers that cannot be represented. Attempts to use these numbers will result in what is called an *overflow error.*

2. *In addition to large quantities, very small quantities cannot be represented.* There is actually an underflow gap or "hole" between zero and the positive and negative numbers with the smallest magnitudes (Fig. 3.7). When you try to represent such a number, some computers give an error while most just automatically set the quantity to zero. The latter case is usually adequate (unless you want to divide by the resulting

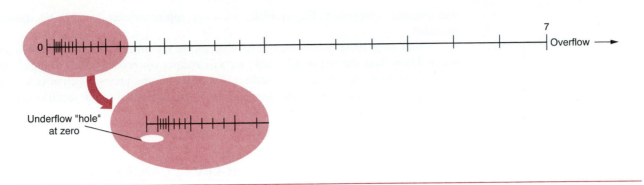

number!). However, there may be cases where you would desire to actually represent a very small number. In those rare cases, the computer would not be able to accommodate your needs.

3. *There is not always a direct correspondence between a base 10 and base 2 number.* In other words, certain quantities that are "round" or exact in our familiar decimal notation cannot be represented precisely in binary. These occur, of course, only with numbers that have fractional parts. A case in point is the quantity 0.1, which is exact in base 10. In conventional floating point, the closest possible representation is 0.09999999, which amounts to an error of 0.00000001. Such discrepancies are called *quantizing errors*.

4. *There are some numbers that by their very nature cannot be represented using a fixed number of digits.* For example, it should be obvious that irrational numbers cannot be represented exactly. Even a seemingly simple quantity such as $\frac{1}{3} = 0.33333\ldots$ cannot be represented exactly because the computer has only a finite number of bits devoted to a single number. Thus, for the 7-digit precision that is commonly available, the computer would represent the value as 0.3333333, which amounts to an error of 0.000000033333.... This is referred to as *round-off error*.

For most cases, such round-off errors are negligible and will not have a significant effect on computations. However, for some engineering calculations, such single-precision numbers can be involved in millions and millions of arithmetic operations. In addition, many of these computations are sequential. That is, each calculation uses the result of previous ones. For such situations, it is possible for even small round-off errors to accumulate. In extreme cases, this can lead to erroneous results. For these situations, using double-precision real numbers can sometimes save the day. By allowing values to be approximated by more significant figures, round-off effects can be greatly mitigated. Thus a computer capable of higher precision can have real benefits in certain situations. In this respect, some mainframes and supercomputers, which employ very high precision representations of numbers, sometimes have an advantage over personal computers. For example, the Cray supercomputer has 128-bit double precision.

Aside from these points, there are other subtle and unexpected results that can occur. Throughout the rest of this book, we will attempt wherever possible to highlight and explain these occurrences. Regardless, we hope that the present section is a first step in providing you with a more than superficial understanding of the significance of computer internal representation.

3.3   MACHINE, ASSEMBLY, AND HIGH-LEVEL LANGUAGES

In the previous section, we showed how a series of 0's and 1's can be strung together to represent numbers, letters, and symbols within a computer. From these basic elements, all sorts of complex structures and commands can be developed. The process is akin to any language. For example, at first glance the English alphabet might seem too simple to be useful. However, from the 26 letters, a progression of words, sentences, and paragraphs can be developed to convey complex and subtle concepts. In the same spirit, a *machine language* composed of a limited number of "words" can be developed from computer bits. These words, which are called the computer's *instruction set,* allow us to instruct and manipulate the computer to perform tasks.

Although machine language allows us to instruct the computer to do our bidding, it is much too inefficient and cumbersome for human communication. A simple human language instruction such as "add A to B and print out the result" requires numerous machine-language words.

We therefore have a fundamental problem: Since computers and human beings speak different languages, how do we get them to communicate? At the beginning of modern computer development, specialists solved the problem by becoming fluent in machine language. This approach was impractical because the complexity of machine language made it indecipherable for all but the most dedicated computer experts. Consequently, intermediate languages were developed that were better suited for human comprehension but which were expressly designed so that they could be efficiently translated into machine language. The first type, called *assembly language,* still required that step-by-step instructions be written for the computer, but it used a more expressive vocabulary than binary digits for that purpose. The second type, called a *high-level language,* condensed groups of machine-language instructions into single statements. These statements were then given descriptive human-oriented labels. Thus the human programmer could write a program using the high-level language, and another program could translate it into machine language for implementation by the computer.

In the remainder of the book, we will deal almost exclusively with *applications programs* that an engineer might write or purchase to solve a particular problem. It is this kind of program that would be written in a high-level language. Before turning to these types of programs, we will review the other major class of programs that you have to deal with when you use computers. These are called *systems programs.*

3.4   SYSTEMS PROGRAMS

Systems software consists of programs which perform background tasks that enable applications programs to run properly. For computers, the systems software can be broken down into three major categories: the operating system, utility programs, and language translators. In the present section, we will briefly introduce the function of each of these components.

3.4.1   The Operating System

In the early years of electronic computing, almost every aspect of operating a computer system required some direct human supervision. Today many of the tasks formerly coordinated by a human operator are handled by the component of the systems software called the operating system.

The *operating system* is a group of programs that coordinates and controls the activities of the machine. In particular, it enables applications software to interface with the hardware of a particular computer. We will start by discussing the operating systems that are typical of mainframes and workstations.

To this point, we have usually divided personal computers, workstations, and mainframes according to their accessibility. For our discussion of operating systems, however, a slightly different perspective will be adopted. This perspective is based on the fact that beyond accessibility, the three types of computers can be classified as to the number of individuals that can use them simultaneously. Employing this classification scheme places the single-user personal computers in a category distinct from the multiple-user workstations and mainframes. Nowhere is this distinction more significant than as it relates to the capabilities of the operating system.

Mainframe and workstation operating systems.   An essential function of a mainframe or workstation operating system is coordination. A large mainframe computer may have upwards of 500 people using it simultaneously. Input, output, and storage of information are mediated by a whole host of diverse hardware. Mainframes can operate with many monitors, card readers, magnetic tapes, and hard disks. They can transmit information via line printers, high-speed page printers, graphics plotters, and different kinds of telecommunications terminals. Once you realize that many of these devices will be needed concurrently by many users, the sheer magnitude of scheduling access to a particular device becomes mind-boggling. It is the job of the operating system to ensure that such coordination proceeds in a smooth and efficient manner.

Now that you have a general idea of the operating system's overall mission, we can take a closer look at its functions. Some of these, which are typical of mainframe computers, are listed here.

1. *Assignment of the system's resources*   The operating system must determine which resources are to be mobilized for any given job. A part of the operating system that contributes to this endeavor is the *job-control program*. This program reads instructions written by the user in job-control language (JCL) that specify the requirements for a particular job. These include the types of input-output devices, the type of

language that is being employed, and other special requirements. In the absence of these instructions, the job-control program makes assumptions regarding some of these needs. Such *default options* will be implemented unless you specify otherwise with your JCL commands.

2. Job and resource scheduling   The operating system must decide not only *what* resources must be used but also *when* they will be used. *Scheduling programs* determine the order in which jobs are implemented. This order is dictated by a number of factors, including the job's priority, the availability of required hardware, and the general level of activity.

There are a variety of sophisticated processing techniques that are employed to ensure that scheduling proceeds as smoothly and efficiently as possible. Timesharing and multiprogamming are two methods that allow the computer to work on several programs concurrently. In *timesharing,* the operating system moves from program to program, giving each a small slice of execution time on every cycle. Thus no single program is allowed to dominate the computer's attention. *Multiprogramming* also involves operating on one program at a time. However, rather than allocating each program a fixed slice of time, it will execute each until a logical stopping point is reached. Because computations and processing are typically much faster than input-output, a strategy might be to operate on one program until a slow input-output step is encountered. The operating system can then switch to another program and return to the first one only when that program is ready for speedy activities again. Today many systems combine aspects of timesharing and multiprogramming in their scheduling schemes.

Because of the slowness of most input-output processes, another technique for expediting the implementation of programs is *spooling.* This involves the temporary holding of input and output in secondary storage. Jobs wait in a queue and are fed in and out of the machine in the order of their priority. In the meantime, the CPU concentrates on quick computations and processing without having to worry about slow input and output.

A final aspect of data processing that contributes to the efficient employment of a computer's resources is virtual storage. In the early days of computing, programs of even moderate length were often too large to fit into primary memory. Consequently, users had to split their programs into smaller pieces (called *overlays*) to be executed separately. Today many mainframe and workstation operating systems perform the same procedure automatically. *Virtual storage* refers to a special area of secondary storage which can be thought of as a large extension of primary memory. The operating system delivers large programs to virtual storage, where they are divided into workable segments that can be executed one at a time in primary memory. In this way, programs can be implemented that are much larger than the actual extent of primary memory.

3. Monitoring activities   The final general function of a large-scale operating system is keeping track of activities. This involves three major tasks: security, bookkeeping, and detecting abnormalities. *Security* is needed to protect a user's data and programs as well as to ensure that individuals who do not belong on the system do not gain access. Among other things, the operating system screens identification information and passwords for this purpose. *Bookkeeping* involves keeping track of the appropriate costs for the services that users obtain. *Detecting abnormalities* entails terminating programs that contain errors or exceed time or storage allocations. For these cases, the operating systems will send a message to the user describing the problem. In addition, the user will also be informed of other abnormalities, such as problems concerning input-output.

Although workstations operate on a smaller scale than mainframes, many of the same types of coordination are required. This is in marked contrast to the single-user operating systems required on most personal computers. Although a variety of different operating systems are employed, *UNIX* is the most prevalent. Developed by Bell Laboratories, it is especially well suited for networks where a number of users simultaneously employ similar workstations. Consequently, it is finding ready application in contexts such as small firms and university computing laboratories.

Personal computer operating systems.

Most personal computers and their operating systems have been designed to meet the needs of one user at a time. Usually the services and peripherals supported directly by the operating system are limited. Typically the operating system controls storage of information on a hard disk, diskettes, tapes, or a combination of these. It can both load the user's program and pass control of the machine to the program.

Usually the functions built into the operating system revolve around the management of files on diskettes or the hard drive. The functions usually included as a part of the operating system are the ability to display the data in a file on a monitor, name files, erase files, list the names of files, move the contents of RAM to a disk file, and run a program. Other tasks can be provided by utility programs that come with the operating system but are in fact distinct programs separate from the operating system. These utility programs reside on the disk and are brought into RAM only when run by the user. In contrast, the core of the operating system resides in RAM at all times when the machine is in operation.

Operating systems for personal computers were initially little more than disk-handling programs. This is why they are traditionally referred to as "disk operating systems," or DOS. Today the functions supplied by mainframe or workstation operating systems are being incorporated into the personal computer operating system. This has now become possible because the size of personal computer RAM has increased as the cost of RAM has decreased. The additional features found on mainframes were always desirable, but the cost of these services prevented the personal computer operating system from offering them.

A number of specific operating systems have been developed for personal computers. Because most operating systems conform to particular microprocessor chips, applications programs that work with one operating system may not work with another. Consequently, applications software for one machine sometimes cannot be run on a different type of computer. Therefore the choice of an operating system is very important. Because so many operating systems are available and because new ones are continually developed, a comprehensive review is inappropriate here. However, some are so ubiquitous that they bear mention.

Of all these machine-specific operating systems, the most important is *Microsoft,* or *MS-DOS.* Because a version of this system is used on the highly popular IBM PC, a number of other computer companies have adopted MS-DOS as their operating system. Consequently, it is presently the most widely employed system on personal computers.

Aside from MS-DOS, another important operating system for today's personal computers is the one employed by the Apple *Macintosh*. Although this machine was originally targeted for business users, it is now widely used in engineering. This is due in part to its innovative design, which included the use of the mouse and $3\frac{1}{2}$-in diskettes. However, from the perspective of operating systems, its greatest innovation was its visual interface based on the concept of a windows environment.[4] A *window* is a rectangular area of a display screen within which an image or file can be displayed. A windows environment is a means of presenting users with views of a number of separate processes, each carrying out separate tasks.

The Macintosh graphical user interface (called *GUI* for short) is so powerful and attractive to users that Microsoft has developed a similar environment called *Microsoft Windows*. This environment is becoming popular because it is compatible with most DOS-based software while having the strengths of the windows environment.

As a final note, we must reiterate our original statement that a comprehensive description of all the most important operating systems is beyond the scope of this book. Consequently, we have omitted such promising developments as the X-windows workstation environment and innovative machines such as the *Next* computer. The key fact to remember is that the existence of so many systems means that software written for one system may not always work on another.

3.4.2 Utility Programs

Some operations are so common that it would be inefficient if every user had to write a program to perform them. Utility programs, which are supplied by the manufacturer or by software companies, are designed to perform these general housekeeping tasks. Some examples are programs for copying, merging data files, and sorting.

3.4.3 Language Translators

As mentioned previously, a number of high-level languages have been developed to facilitate computer programming by users and the development of applications software products. However, before programs written in these languages can be executed, they must be translated into the machine language understood by the computer. The systems software known as a *language translator* performs this task.

There are two common types of language translators—interpreters and compilers. *Interpreters* perform the translation *and* execution of the original high-level program, which is called the *source module,* one line at a time. Thus the translation takes place while the program is being run. In contrast, *compilers* translate the program en masse and essentially generate a new machine-language version called the *object module,* which is then *linked* with support routines from software "libraries" to produce an *executable module.*

Each of the translators has advantages and disadvantages with respect to the other.

---

[4]It should be acknowledged that the windows environment originated at the Xerox Corporation. Apple adopted it and made the approach commercially successful.

Interpreters require much less storage space because they do not generate a stand-alone executable module. Traditionally this has been a positive trait for personal computers, where memory has been at a premium. In addition, interpreters are much easier to use for program development and maintenance because errors can be detected and corrected as they occur.

Although compiled programs are not as efficiently debugged (that is, freed of errors), they have the advantage that execution is much quicker and more efficient. This is because once a program is tested and compiled, the machine-language executable module can be stored on a disk, reloaded into memory, and run very quickly. This is in contrast to an interpreted language, where each line must be reinterpreted on every run.

As might be expected, software developers are trying to devise translators that have some of the advantages of both compilers and interpreters. Today, language developers are accomplishing this by designing compiler-interpreter hybrids that act just like interpreters. These can be used during program development. Once the program is error-free, the finished version can then be formally compiled for efficient execution.

In summary, the intent of this chapter has been to provide you with some background regarding features of computer software that will be "behind the scenes" when you begin to develop your own software. The next chapter will begin to discuss applications programs and how they can be effectively developed to perform engineering tasks.

PROBLEMS

3.1 Express the following base 10 numbers in binary: (*a*) 18191, (*b*) 5389, and (*c*) 66000.

3.2 Express the following binary numbers in base 10: (*a*) 01001101, (*b*) 1110001101, and (*c*) 1000111.

3.3 Express the following base 10 numbers in base 8: (*a*) 87, (*b*) 29, and (*c*) 53.

3.4 The octal and hexadecimal systems provide convenient shorthand notation for binary numbers because 8 and 16 are integral powers of 2. This means that 1 octal digit is the equivalent of exactly 3 binary digits and 1 hexadecimal digit corresponds to exactly 4. This can be seen by inspecting Table 3.1. Consequently, a 16-digit binary number can be conveniently represented by a 4-digit hexadecimal number. For example,

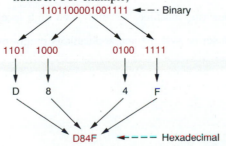

(*a*) Convert the following base 2 numbers to base 8:

011101000010   110101100001

(*b*) Convert the following base 2 numbers to base 16:

1001110100001001   1111010110000100

(*c*) Convert the following base 16 numbers to base 2:

ACDC   F2D1

3.5 Express your age in (*a*) decimal, (*b*) binary, (*c*) octal, and (*d*) hexadecimal.

3.6 Using binary arithmetic, compute the following additions: (*a*) 101 + 011, (*b*) 0011 + 1001 + 0100.

3.7 Using binary arithmetic, compute the following subtractions: (*a*) 1010 − 0101, (*b*) 11000 − 0111.

3.8 Using binary arithmetic, compute the following product: 111011 × 1011.

3.9 Using binary arithmetic, evaluate the following long division:

$$101\overline{)101101}$$

3.10 What are two advantages of using the 2's complement rather than the signed magnitude representation for negative integers?

3.11 Employ 2's-complement 8-bit integers to compute the negative of (*a*) 01011010 and (*b*) 11110000.

3.12 Express the following values in single-precision IEEE format: (*a*) 0.1, (*b*) π, (*c*) *e*, (*d*) −9.81, (*e*) $\frac{1}{3}$, (*f*) 6.626196 × 10^{-27} (Planck's constant), and (*g*) 6.02252 × 10^{23} (Avogadro's number).

3.13 Repeat Prob. 3.7 but perform the subtraction by adding the 2's complement of the subtrahend.

3.14 Distinguish between a mainframe and a personal computer operating system.

3.15 Describe MS-DOS, windows, and UNIX.

3.16 Discuss the advantages and disadvantages of interpreters and compilers.

3.17 Describe machine, assembly, and high-level computer languages.

3.18 Describe systems and applications programs.

ALGORITHM DESIGN AND STRUCTURED PROGRAMMING

The programs that you use to solve problems on a computer are called *applications software*. During your engineering career, there may be periods where your computer work will be limited to applications software developed by someone other than yourself. For instance, these programs might be developed by a software manufacturer or one of your assistants. On these occasions, knowledge of programming and high-level languages may not seem necessary. However, just as some knowledge of automobiles can be invaluable when buying a used car, so will an understanding of programming serve you well when you use someone else's software.

We believe that every engineer should be capable of developing computer programs for small-scale problem solving. Once you have mastered sufficient skill to write such programs, you will have taken a large step toward more intelligent use of software written by others. In addition, you will be able to communicate much more effectively and confidently with software developers.

The present chapter is devoted to topics that should be assimilated before you begin to write computer programs. We will first provide an overview of the program-development process, with special emphasis on ensuring that the program has a strong logical foundation. Then, the remainder of the chapter deals with some general techniques that greatly facilitate the development of high-quality applications software.

4.1 THE PROGRAM-DEVELOPMENT PROCESS

Although the actual writing of a program in a high-level language is of critical importance, it is but one step in the overall development of high-quality software. Figure 4.1 delineates the five steps that constitute the entire process.

We should note that an important element is missing from Fig. 4.1. Before the process can even begin, there is the preliminary step of *problem definition and analy-*

sis. This includes (1) identifying the problem and determining the solution technique and (2) specifying the objectives you want to accomplish with the software.

We have omitted this important step at this point for a number of reasons. First, problem solving is to a large extent an art that is acquired through a long-term process of education and experience. By definition, engineers are problem solvers. Thus, as you mature professionally, your problem-solving skills will be honed as a natural by-product of your development. Second, although problem definition is a critical prerequisite for developing effective software, this skill is not necessary for learning how to write a program in a high-level language. In fact, at this preliminary stage, it would probably complicate matters. Thus we will put off the task of problem solving until after you have mastered the rudiments of programming. Then in the latter parts of the book we can begin to hone your problem-solving skills by using computer programming to approach a variety of engineering problems.

With this as background, we can begin the program-development process. As shown in Fig. 4.1, the first step is designing the underlying logic or algorithm of the program.

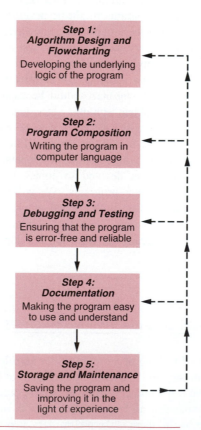

FIGURE 4.1
The five steps required to produce and maintain high-quality software. The feedback arrows indicate that the first four steps can be improved in the light of experience.

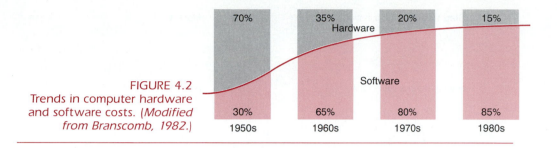

FIGURE 4.2
Trends in computer hardware and software costs. (*Modified from Branscomb, 1982.*)

4.2   ALGORITHM DESIGN

The most common problems encountered by inexperienced programmers can usually be traced to the premature preparation of a program that does not encompass an overall strategy or plan. In the early days of computing, this was a particularly thorny problem because the design and programming components were not clearly separated. Programmers would often launch into a computing project without a clearly formulated design. This led to all sorts of problems and inefficiencies. For example, the programmer could get well into a job only to discover that some key preliminary factor had been overlooked. In addition, the logic underlying improvised programs was invariably obscure. This meant that it was difficult for someone to modify someone else's program. Today, because of the high costs of software development (Fig. 4.2), much greater emphasis is placed on aspects such as preliminary planning that lead to a more coherent and efficient product. The focus of this planning is algorithm design.

An *algorithm* is the sequence of logical steps required to perform a specific task such as solving a problem. Aside from accomplishing its objectives, a good algorithm must have a number of specific properties:

1. Each step must be deterministic; that is, nothing can be left to chance. The final results cannot depend on who is following the algorithm. In this sense, an algorithm is analogous to a recipe. Two chefs working independently and carefully from a good recipe should end up with dishes that are essentially identical.

2. The process must always end after a finite number of steps. An algorithm cannot be open-ended.

3. The algorithm must be general enough to deal with any contingency.

Figure 4.3*a* shows an algorithm for the solution of the simple problem of adding a pair of numbers. Two independent programmers working from this algorithm might develop programs exhibiting somewhat different styles. However, given the same data, their programs should yield identical results.

Step-by-step English descriptions of the sort depicted in Fig. 4.3*a* are one way to express an algorithm. They are particularly useful for small problems or for specifying the broad major tasks of a large programming effort. However, for detailed representations of complicated programs, they rapidly become inadequate. For this reason, more versatile, visual alternatives, called flowcharts, have been developed.

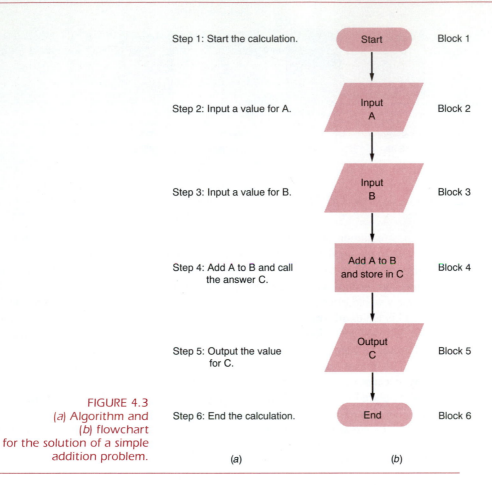

Step 1: Start the calculation.   Start   Block 1

Step 2: Input a value for A.   Input   Block 2
                                                A

Step 3: Input a value for B.   Input   Block 3
                                                B

Step 4: Add A to B and call   Add A to B   Block 4
        the answer C.   and store in C

Step 5: Output the value   Output   Block 5
        for C.   C

FIGURE 4.3
(*a*) Algorithm and
(*b*) flowchart   Step 6: End the calculation.   End   Block 6
for the solution of a simple
addition problem.

                                (a)   (b)

4.3   FLOWCHARTING

A *flowchart* is a visual or graphical representation of an algorithm. The flowchart employs a series of blocks and flowlines (that is, lines with arrow tips indicating direction), each of which represents a particular operation or step in the algorithm. The
flowlines represent the sequence in which the operations are implemented. Figure 4.3*b*
shows a flowchart for the problem of adding two numbers.

   Not everyone involved with computer programming agrees that flowcharting is a
productive endeavor. In fact, some experienced programmers do not even use flowcharts. However, we feel that there are three good reasons for studying them. First, they
are still used for expressing and communicating algorithms. Therefore, you should
learn how to read them. Second, even if they are not employed routinely, there will be
times when they will prove useful in planning, unraveling, or communicating the logic

of your own or someone else's program. Finally, and most important for our purposes, they are excellent pedagogical tools. From a teaching perspective, they are ideal vehicles for visualizing some of the fundamental logic employed in computer programming. Additionally, developing flowcharts provides good basic training in the disciplined logical thought process that lies at the heart of successful computer programs.

4.3.1 Flowchart Symbols

Notice that in Fig. 4.3b the blocks have different shapes to distinguish different types of operations. In the early years of computing, flowchart symbols were not standardized. This led to confusion, because programmers used different symbols to represent the same operations. Today a number of standardized systems have been developed. In this book, we use the set of symbols shown in Fig. 4.4. They are described below.

FIGURE 4.4   Common ANSI symbols used in flowcharts.

SYMBOL	NAME	FUNCTION
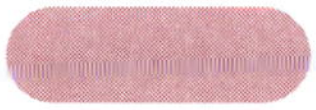	Terminal	Represents the beginning or end of a program.
	Flowlines	Represents the flow of logic. The humps on the horizontal arrow indicate that it passes over and does not connect with the vertical flowlines.
	Process	Represents calculations or data manipulations.
	Input/output	Represents inputs or outputs of data and information.
	Decision	Represents a comparison, question, or decision that determines alternative paths to be followed.
	On-page connector	Represents a break in the path of a flowchart that is used to continue flow from one point on a page to another. The connector symbol marks where the flow ends at one spot on a page and where it continues at another spot.
	Off-page connector	Similar to the on-page connector but represents a break in the path of a flowchart that is used when it is too large to fit on a single page. The connector symbol marks where the algorithm ends on the first page and where it continues on the second.

• *Terminal*   These signify the beginning and end of a program.

• *Flowlines*   These represent the flow of logic, with the arrows designating the direction. A flowchart is read by beginning at the *start* terminal and following the flowlines to trace the algorithm's logic. Thus, as shown in Fig. 4.3*b*, we would begin at block 1 and then follow the arrow sequentially until we arrive at the terminal block 6. If the flowchart is written properly, there should never be any doubt regarding the correct path. Several flowlines can join and lead to the same block, but a flowline cannot split into several lines which lead to several blocks.

• *Process*   Arithmetic and data manipulations are generally placed in these boxes. This symbol can also be used to represent major program segments or modules.

• *Input-output*   These represent either the input or the output of data and information. Thus at one point they might be used to designate the input of information from a keyboard, whereas at other points they might represent the output of data to a printer or to a monitor display.

• *Decision*   This is the only standard symbol that has more than one exit flowline. The proper exit path is determined by the correct answer to a question or conditional statement contained in the block.

• *On-page connector*   Although the flowlines in Fig. 4.4 show humps to depict lines that pass over each other without connecting, it is generally best to avoid crossing flowlines. Circular connectors allow the flow to be discontinued at one section of a flowchart and started again at another location. A number is placed within the exit- and entry-point connectors to specify each pair used on a given page (Fig. 4.5).

• *Off-page connector*   Many flowcharts are large enough that they spread over

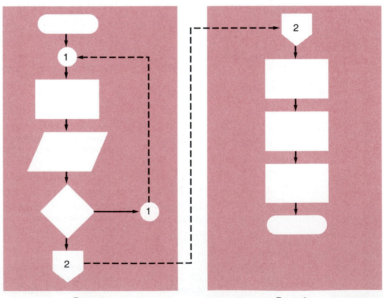

FIGURE 4.5
Illustration of the use of on-page and off-page connectors. The dashed lines would not actually be a part of the charts but are used here to show how control is transferred between connectors.

Page 1   Page 2

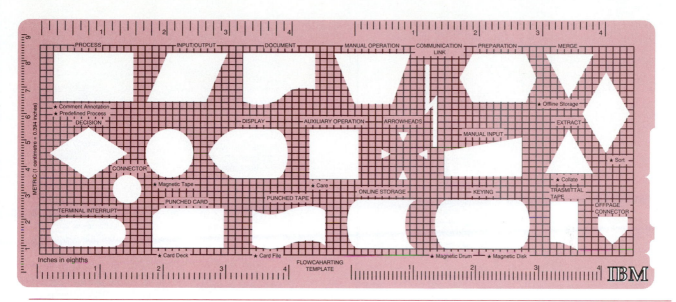

FIGURE 4.6
A flowchart template.

several pages. The spade-shaped symbols in Figs. 4.4 and 4.5 are used to transfer flow between pages in a fashion similar to the circular on-page connectors.

Aside from these basic symbols, we will introduce a few more types in later chapters to represent more advanced programming operations. For the time being, this set is adequate.

The construction of flowcharts is facilitated by the use of a plastic template (Fig. 4.6). This stencil includes the standard symbols for constructing flowcharts. Templates are available at most office-supply shops and college bookstores.

4.3.2 Levels of Flowchart Complexity

There are no set standards for how detailed a flowchart should be. During the development of a major algorithm, you will often draft flowcharts of various levels of complexity. As an example, Fig. 4.7 depicts a hierarchy of three charts that might be used in the development of an algorithm to determine the grade-point average (GPA) or cumulative index of a student engineer. The system flowchart (Fig. 4.7a) delineates the big picture. It identifies the major tasks, or *modules,* and the sequence that is required to solve the entire problem. Such an overview flowchart is invaluable for ensuring that the total scheme is comprehensive enough to be successful.

Next the major modules can be broken down and charted in greater detail (Fig. 4.7b). Finally, it is sometimes advantageous to break the major modules down into even more manageable units, as in Fig. 4.7c. The process of subdividing an algorithm into major segments and then breaking these down into successively more refined modules is referred to as *top-down design.* We will return to this approach later in this chapter.

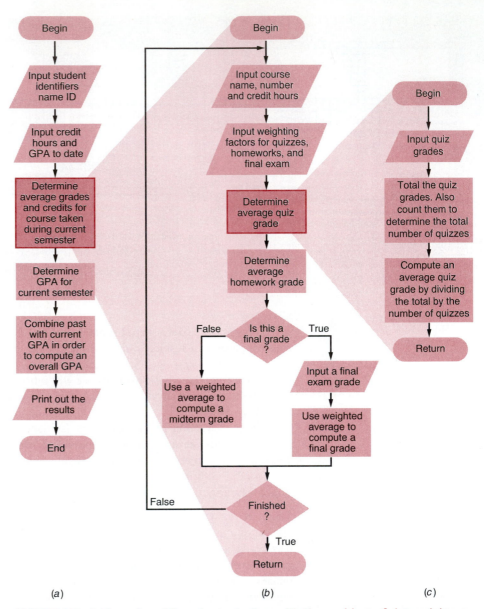

(a)   **(b)**   **(c)**

FIGURE 4.7 A hierarchy of flowcharts dealing with the problem of determining a student's grade-point average. The system flowchart in (*a*) provides a comprehensive plan. A more detailed chart for a major module is shown in (*b*). An even more detailed chart for a segment of the major module is seen in (*c*). This way of approaching a problem is called *top-down design*.

In the following sections we will study and improve the flowchart in Fig. 4.7c. This will be done to illustrate how the symbols in Fig. 4.4 can be orchestrated to develop an efficient and effective algorithm. In the process, we will demonstrate how the power of computers can be reduced to three fundamental types of operations. The first, which should already be apparent, is that computer logic can proceed in a definite *sequence*. As depicted in Fig. 4.7a, this is represented by the flow of logic from box to box. The other two fundamental operations are *selection* and *repetition*.

4.3.3 Selection and Repetition in Flowcharts

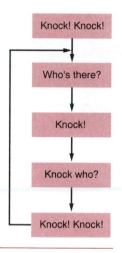

Notice that the operations in Fig. 4.7a and c specify a simple sequential flow. That is, the operations follow one after the other. However, notice how the first diamond-shaped decision symbol in Fig. 4.7b permits the flow to diverge and follow one of two possible paths, or branches, depending on the answer to a question. For this case, if final grades are being determined, the flow follows the right (true) branch, whereas if intermediate or midterm grades are being determined, the flow follows the left (false) branch. This type of flowchart operation is called *selection*.

Aside from depicting a conditional branch, Fig. 4.7b demonstrates another flowchart operation that greatly increases the power of computer programs—*repetition*. Notice that the "false" branch of the second decision symbol allows the flow to return to the head of the module and repeat the calculation process for another course. This ability to perform repetitive tasks is among the most important strengths of computers. The operation of "looping" back to repeat a set of operations is accordingly named a *loop* in computer jargon. The most elementary example is the endless or infinite loop of the riddle in Fig. 4.8. Because it has no exit, this loop will persist interminably. In contrast, the presence of the diamond-shaped decision symbol permits the termination of the loop in Fig. 4.7b once all courses have been evaluated.

FIGURE 4.8
An example of an endless loop. This loop will persist forever because it has no exit. (*Modified from Kelly-Bootle, 1981.*)

The operations of selection and repetition can be employed to great advantage when constructing algorithms. This will be demonstrated in the next section, where we will develop an improved version of Fig. 4.7c.

4.3.4 Accumulators and Counters in Flowcharts

Figure 4.7c employed three sequential blocks to input, total, and count quiz grades and calculate their average. Now that you understand the notions of decisions and loops, we can develop a more detailed flowchart to accomplish these objectives (Fig. 4.9). In order to appreciate the more detailed version, we must introduce the concepts of accumulators and counters.

An *accumulator* is a program-designated storage location that accepts and stores a running total of individual values as they become available during the operation of a program. In Fig. 4.9 the accumulator is given the symbolic name SUM. Notice that the contents of the storage location are initially set to zero. Then with each pass through the loop, a new quiz grade is added to SUM in order to keep a running total.

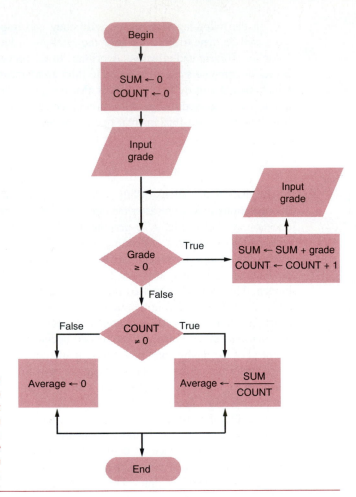

FIGURE 4.9
A detailed flowchart to compute the average quiz grade. Through the use of an accumulator and a counter, this example improves on the simple version shown in Fig. 4.7c.

A *counter* is merely a special type of accumulator that is employed to keep track of the number of passes through the loop. In Fig. 4.9 the symbolic name COUNT is used to represent the counter. As with the accumulator, the counter is initialized to zero. Then with each pass through the loop, 1 is added to COUNT in order to keep a running total of the number of passes. For Fig. 4.9 this corresponds to the number of quizzes that have been summed. Then, upon exit from the loop (which is triggered by entering a negative value for GRADE), SUM is divided by COUNT to determine the average quiz grade. Observe how a decision block is utilized to avoid division by zero.

4.3.5 Structured Flowcharts

Flowcharting symbols can be orchestrated to construct logic patterns of extreme complexity. However, programming experts have recognized that, without rules, flowcharts can become so complicated that they impede rather than facilitate clear thinking.

One of the main culprits that contributes to algorithm complexity is the unconditional transfer, also known as the GOTO. As the name implies, this control structure allows you to "go to" any other location in the algorithm. For flowcharts, it is simply represented by a flowline. Indiscriminate use of unconditional transfers can lead to algorithms that, at a distance, resemble a plate of spaghetti.

In an effort to avoid this dilemma, advocates of structured programming have demonstrated that any program can be constructed with the three fundamental control structures depicted in Fig. 4.10. As can be seen, they correspond to the three operations—sequence, selection, and repetition—that we have just been discussing.

All flowcharts should be composed of the three fundamental control structures shown in Fig. 4.10. If we limit ourselves to these structures and avoid GOTOs, the resulting algorithms will be much easier to understand, implement, and maintain.

4.4   PSEUDOCODE

The object of flowcharting is to develop a quality computer program. As should now be obvious, a program is merely a set of step-by-step instructions to direct the computer to perform tasks. These instructions are also called *code*.

Recall that in top-down design, more detailed flowcharts are developed as we progress from the general overview (Fig. 4.7a) to specific modules (Fig. 4.7c). At the highest level of detail, the module flowcharts are almost in the form of computer code. That is, as seen in Fig. 4.9, each flowchart compartment represents a single well-defined instruction to the computer.

An alternative way to express an algorithm that bridges the gap between flowcharts and computer code is called *pseudocode*. This technique uses English-like statements in place of the graphical symbols of the flowchart. Figure 4.11 displays pseudocode representations for the fundamental control structures.

Keywords such as BEGIN, IF, DO WHILE, and the like, are capitalized; whereas the conditions, processing steps, and tasks are in lowercase. Additionally, notice that the processing steps are indented. Thus the keywords form a "sandwich" around the steps to visually define the extent of each control structure.

Figure 4.12 shows a pseudocode representation of the flowchart from Fig. 4.9. This version looks much more like a computer program than the flowchart. Thus one advantage of pseudocode is that it is more direct to develop a program with it than with a flowchart. The pseudocode is also easier to modify. However, because of their visual form, flowcharts are sometimes better suited for the preliminary design of complex algorithms. Because both are useful, flowcharts and pseudocode will be employed throughout the remainder of this text to communicate algorithms so that you can become familiar with them and capable of using them to develop computer programs.

FIGURE 4.10
The three fundamental
control structures.

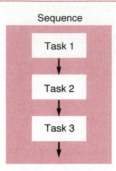

Sequence

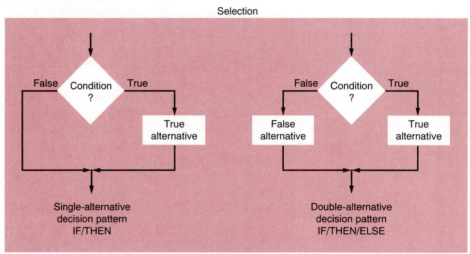

Selection

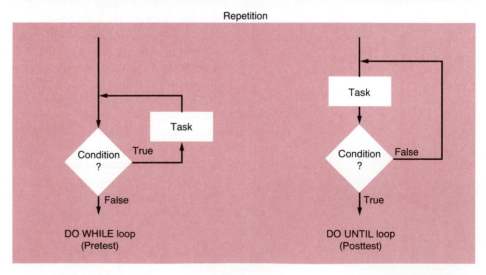

Repetition

FIGURE 4.11
Pseudocode representations
of the three fundamental
control structures.

Sequence

BEGIN task
  Task
END task

Selection

IF condition
  True alternative
END IF

Single-alternative
decision pattern
(IF/THEN)

IF condition
  True alternative
ELSE
  False alternative
ENDIF

Double-alternative
decision pattern
(IF/THEN/ELSE)

Repetition

DO WHILE condition
  Task
LOOP

DO WHILE loop
(Pretest)

DO
  Task
LOOP UNTIL condition

DO UNTIL loop
(Posttest)

FIGURE 4.12
Pseudocode to compute the
average quiz grade; this
algorithm was represented
previously by the flowchart in
Fig. 4.9.

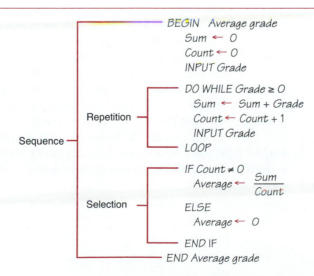

$$Sequence \begin{cases} BEGIN \ Average\ grade \\ Sum \leftarrow 0 \\ Count \leftarrow 0 \\ INPUT\ Grade \\ \\ Repetition \begin{cases} DO\ WHILE\ Grade \geq 0 \\ \quad Sum \leftarrow Sum + Grade \\ \quad Count \leftarrow Count + 1 \\ \quad INPUT\ Grade \\ LOOP \end{cases} \\ \\ Selection \begin{cases} IF\ Count \neq 0 \\ \quad Average \leftarrow \dfrac{Sum}{Count} \\ ELSE \\ \quad Average \leftarrow 0 \\ END\ IF \end{cases} \\ \\ END\ Average\ grade \end{cases}$$

EXAMPLE 4.1   *Developing a Pseudocode*

PROBLEM STATEMENT: Construct pseudocode based on the algorithm outlined in Fig. 3.5.

SOUTION: Recall that the algorithm was written as the following step-by-step series of instructions:

- *Step 1*   Let i represent the ith binary place to the right of the binary point, and start with i = 1. Let the decimal fractional part by fract. Go to step 2.
- *Step 2*   Is 2^{-i} less than or equal to fract, the fractional part? If yes, go to step 3. If no, go to step 4.
- *Step 3*   Subtract 2^{-i} from fract to form a new value for fract, and set the digit in the ith binary place to the right of the binary point to be 1. Go to step 5.
- *Step 4*   Set the digit in the ith binary place to the right of the binary point to be 0. Go to step 5.
- *Step 5*   Do you have a sufficient binary precision yet? If yes, stop. If no, increment i by 1 and go back to step 2.

Close inspection of these instructions indicates that the algorithm already exhibits some structure. For example, step 2 looks like the start of a double-alternative decision pattern. Similarly, if precision is not sufficient, step 5 appears to trigger a DO UNTIL loop back to step 2.

These structures can be seen more clearly by sketching a flowchart. As depicted in Fig. 4.13, we have included some extra instructions to make the algorithm more complete. For example, we add an input block to enter the decimal fraction (fract) along with the number of significant digits (numdig) that are required. We then use this information to make the loop decision more quantitative.

If you study Fig. 4.13 closely, you will soon recognize that it does the job for which it was designed. However, it is not perfectly structured yet. In particular, the DO UNTIL loop is flawed because of the presence of the incrementing assignment block, $i \leftarrow i + 1$, in the loop flowline following the loop decision block. Comparison with Fig. 4.10 clearly shows that this is a violation of the fundamental DO UNTIL structure.

This flaw can be rectified by two modifications. First, the incrementing assignment can be moved to the start of the body of the loop. Second, i should be initialized as 0 rather than as 1 at the start of the program so that the first instance of the incrementing assignment block will then set i equal to 1.

These modifications are included in the finished pseudocode. As depicted in Fig. 4.14, the result is now in a form that can be the basis of well-structured computer code. If you would like to explore further nuances of this example, see Prob. 4.10 at the end of this chapter.

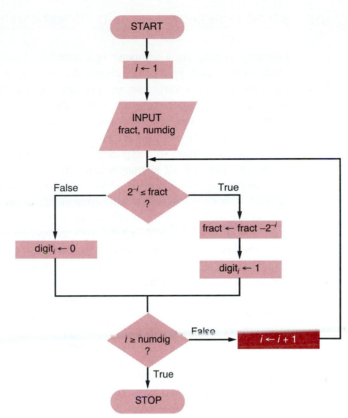

FIGURE 4.13
A "nearly" structured flowchart for the algorithm from Example 4.1. The only major flaw in this design is the presence of the assignment block (highlighted) in the return flowline of the DO UNTIL loop.

```
BEGIN  Fractional binary
    i ← 0
    INPUT fract, numdigits
    DO
        i ← i + 1
        IF 2^{-i} ≤ fract
            fract ← fract − 2^{-i}
            digit_i ← 1
        ELSE
            digit_i ← 0
        ENDIF
    LOOP UNTIL i ≤ numdig
END Fractional binary
```

FIGURE 4.14
Structured pseudocode for the algorithm from Example 4.1

4.5　STRUCTURED DESIGN AND PROGRAMMING

Learning to program a computer is a little like learning to play the guitar. It is relatively easy to pick up the few chords that are required to plunk out a passable tune. However, disciplined training and practice are required to master the effortless technique and expressiveness of the accomplished musician. Before learning to program, it is important that you be aware of some general concepts and principles that will contribute to the quality of your efforts and prepare you for more ambitious endeavors.

In the early years of computing, programming was somewhat more of an art than a science. Although all languages had very precise vocabularies and grammars, there were no standard definitions of good programming style. Consequently, individuals developed their own criteria for what constituted an excellent piece of software.

In the early years, these criteria were strongly influenced by the available hardware. Recall from Chap. 2 that the early computers were expensive and slow and had small memories. As a consequence, one mark of a good program was that it utilized as little memory as possible. Another might be that it executed rapidly. Still another was how quickly it could be written the first time.

The ultimate result was that programs were highly nonuniform. In particular, because they were strongly influenced by hardware limitations, programmers put little premium on how easy the programs might be to use and to maintain. Thus, if you wanted to effectively utilize and maintain a large program over long periods, you usually had to keep the original programmer close by. In other words, the program and the programmer formed a costly package.

Today many of the hardware limitations do not exist or are rapidly disappearing. In addition, as software costs constitute a proportionally larger fraction of the total computing budget (recall Fig. 4.2), there is a high premium on developing software that is easy to use and modify. In particular, there is a strong emphasis on clarity and readability rather than on short, terse, noncommunicative code.

Computer scientists have systematically studied the factors and procedures needed to develop high-quality software of this kind. Although the resulting methods are somewhat oriented toward large-scale programming efforts, most of the general techniques are also extremely useful for the types of programs that engineers develop routinely in the course of their work. Collectively, we will call these techniques *structured design and programming*. In the present chapter, we discuss three of these approaches—modular design, top-down design, and structured programming.

4.5.1　Modular Design

Imagine how difficult it would be to study a textbook that had no chapters, sections, or paragraphs. Breaking down complicated tasks or subjects into more manageable parts is one way to make them easier to handle. In the same spirit, computer programs can be divided into smaller subprograms, or *modules,* that can be developed and tested separately. This approach is called *modular design*.

The most important attribute of modules is that they be as independent and self-contained as possible. In addition, they are typically designed to perform a specific, well-defined function. As such, they are usually short (generally 50 to 100 instructions in length) and highly focused.

One of the primary programming elements used to represent each module is the procedure (also called a subprogram or subroutine). A *procedure* is a series of computer instructions that together perform a given task. A *calling,* or *main, program* invokes these modules as they are needed. Thus a main program orchestrates each of the parts in a logical fashion.

Modules have one entry and one exit point. Decisions can be made within a module that impact the overall flow, but the branching itself should occur in the main program.

Modular design has a number of advantages. The use of small self-contained units makes the underlying logic easier to devise and to understand for both the developer and the user. Development is facilitated because each module can be tested and perfected in isolation. In fact, for large projects different programmers can work on individual parts. In the same spirit, you can maintain your own library of useful modules for later use in other programs. Modular design also increases the ease with which a program can be debugged and tested because errors can be more easily isolated. Finally, program maintenance and modification are facilitated. This is primarily due to the fact that new modules can be developed to perform additional tasks and then easily incorporated into the already coherent and organized scheme.

4.5.2 Top-Down Design

Although the previous section introduced the notion of modular design, it did not specify how the actual modules are identified. The top-down design process is a particularly effective approach for accomplishing this objective.

Top-down design is a systematic development process that begins with the most general statement of a program's objectives and then successively divides it into more detailed segments. Thus the design proceeds from the general to the specific. We have already introduced this approach in our discussion of flowcharting (recall Fig. 4.7).

Most top-down designs start with an English description of the program in its most general terms. This description is then broken down into a few elements that define the program's major functions. Then each major element is divided into subelements. This procedure is continued until well-defined modules have been identified.

Besides providing a procedure for dividing the algorithm into well-defined units, top-down design has other benefits. In particular, it ensures that the finished product is comprehensive. By starting with a broad definition and progressively adding detail, the programmer is less likely to overlook important operations.

4.5.3 Structured Programming

The previous two sections have dealt with effective ways to organize a program. In both cases, emphasis was placed on *what* the program is supposed to do. Although this usually guarantees that the program will be coherent and neatly organized, it does not ensure that the actual program instructions or code for each module will manifest an associated clarity and visual appeal. Structured programming, on the other hand, deals with *how* the actual program code is developed so that it is easy to understand, correct, and modify.

Not everyone agrees as to what constitutes structured programming. In the present context, we define it as a set of rules that prescribes good style habits for the programmer. The following are the most generally agreed upon rules:

1. Programs should consist solely of the three fundamental control structures of sequence, selection, and repetition (recall Fig. 4.10). Computer scientists have proved that any program can be constructed from these basic structures.[1] Each of the structures must have only one entrance and one exit.

2. Unconditional transfers (GOTOs) should be avoided.

3. The structures should be clearly identified with comments and visual devices such as indentation, blank lines, and blank spaces.

Although these rules may appear deceptively simple, they are extremely powerful tools for increasing program clarity. In particular, adherence to these rules avoids the spaghettilike code that is the hallmark of indiscriminate branching. Because of their benefits, these rules and other structured programming practices will be stressed frequently in subsequent parts of this book.

Figure 4.15 is a dramatic example of the contrast between unstructured and structured approaches. Both programs are designed to determine an average grade, the algorithm for which was developed earlier in this chapter. Although the programs compute identical values, their designs are in marked contrast. The purpose, logic, and organization of the unstructured program in Fig. 4.15*a* are not obvious. The user must "reverse-engineer" the code on a line-by-line basis to decipher the program.

In contrast, the major parts of Fig. 4.15*b* are immediately apparent. Title and variable-identification sections are placed at the beginning to orient the user. Program execution progresses in an orderly fashion from the top to the bottom with no major jumps. Every segment has one entrance and one exit, and each is clearly delineated by comments, spaces, and indentation. In essence, the program is distinguished by a visual clarity that is reminiscent of a well-designed flowchart. Additionally, the program is a natural evolution of the pseudocode shown in Fig. 4.12.

On the debit side, there is no question that Fig. 4.15*b* takes longer to develop than Fig. 4.15*a*. Because many pragmatic engineers are accustomed to "getting the job done" in an efficient and timely manner, the extra effort might not always be justified. Such would be the case for a short program you might develop for a one-time-only computation that you are not going to share with others. However, for larger, more important programs that are intended for repeated application by yourself or others, there is no question that a structured approach will yield long-term benefits.

Finally, we have noticed that on more than one occasion, programs that we thought were one-time-only affairs ultimately grew into important long-term tools.

---

[1]In the past few years, it should be noted that popular computer languages such as FORTRAN and BASIC have developed a few additional structures beyond the three listed here. Although they are not necessary to program algorithms, they make certain actions more convenient. Furthermore, if employed properly, they still result in well-structured code. We will introduce some of these structures when we describe these languages.

```
10 S=0                          'Average Grade Program
20 I=0
30 INPUT G                      'by S.C. Chapra
40 IF G<0 THEN 80               'May 31, 1992
50 S=S+G
60 I=I+1                        'This program determines the average
70 GOTO 30                      'value for a set of grades
80 IF I=0 THEN 110              '------------------------------------
90 A=S/I
100 GOTO 120                    'Definition of variables:
110 A=0                         'grade  = grade input by user
120 PRINT A                     'sum    = summation of grades
130 END                         'count  = number of grades
                                'averag = average grade
(a)                             '------------------------------------

                                CLS

                                grade = 0
                                sum = 0
                                count = 0

                                'accumulate grade data
                                PRINT "Grade = (terminate with negative)";
                                INPUT grade
                                DO WHILE grade >= 0
                                  sum = sum + grade
                                  count = count + 1
                                  PRINT "Grade = (terminate with negative)";
                                  INPUT grade
                                LOOP

                                'calculate average grade
                                IF count = 0 THEN
                                  average = 0
                                ELSE
                                  average = sum / count
                                END IF
                                PRINT "average grade = "; average

                                END

                                (b)
```

FIGURE 4.15
(a) An unstructured and (b) a structured version of the same program. Both versions perform identical computations but differ markedly in terms of clarity.

Therefore, we have personally tried to use structured approaches regardless of the level of application. In the long run, this is facilitated by the fact that, as you discipline yourself to program correctly, the effort involved in producing quality software decreases significantly.

4.6 FINAL COMMENTS BEFORE YOU BEGIN TO PROGRAM

In his book *Megatrends,* Robert Naisbitt states that to be truly successful in the coming years Americans "will have to be trilingual: fluent in English, Spanish and computers." Aside from its obvious commentary on our shifting demography, Naisbitt's statement is interesting in its characterization of computer use as a linguistic process.

In fact, learning to program a computer is a lot like mastering a foreign language. One insight that can be gained from this analogy is that an efficient way to learn a foreign tongue is to use it to write and speak. Similarly, the best way to learn programming is to compose computer programs. The next two parts of the book are designed to provide you with sufficient information to write problem-solving computer programs. With this knowledge as background, you can progress to the more advanced concepts covered in the subsequent chapters.

Although our primary intent is to get you on the machine as rapidly as possible, we must note one very important distinction between foreign and computer languages that will have a large bearing on your success. When speaking in a foreign language, your conversation will be with another human being. Therefore you can express yourself imperfectly and still be understood. The other person can interpret the context of the situation, your tone of voice, your gestures, and the look in your eyes; and no matter how much you mangle the language's grammar and syntax, some communication is usually possible.

In contrast, you will be conducting your computer conversations with a machine that is incapable of gazing deeply into your eyes to search out what you are really trying to say. Although this might seem almost ridiculously obvious, it is one of the most important initial insights that must be grasped by the novice programmer. That is, in order to communicate with the computer effectively, you must express yourself precisely. If you don't, your programs will simply not work. Computers and their languages are very unforgiving in this respect.

Fortunately, the rules of most high-level programming languages are easy to learn. This is due, in part, to the very fact that they are designed for communication with machines. Thus they do not involve the nuances and exceptions that represent the primary stumbling blocks to learning foreign languages. As described in the next two parts of the book, the modern dialects of BASIC and FORTRAN are particularly well designed in this regard.

PROBLEMS

4.1 List and define six steps involved in the production of high-quality engineering software.

4.2 What is an algorithm, and why is it so critical to the effective development of high-quality software?

4.3 You are at the corner of 2nd Street and Avenue C (Fig. P4.3). Write an algorithm to direct a motorist to take the shortest route to 6th Street and Avenue D. Critique your algorithm: how might it fail?

4.4 A different number is written on each of 10 index cards. Write an algorithm to search through these cards to determine the lowest number.

4.5 Using the fundamental control structures from Fig. 4.10, write a flowchart for Prob. 4.4.

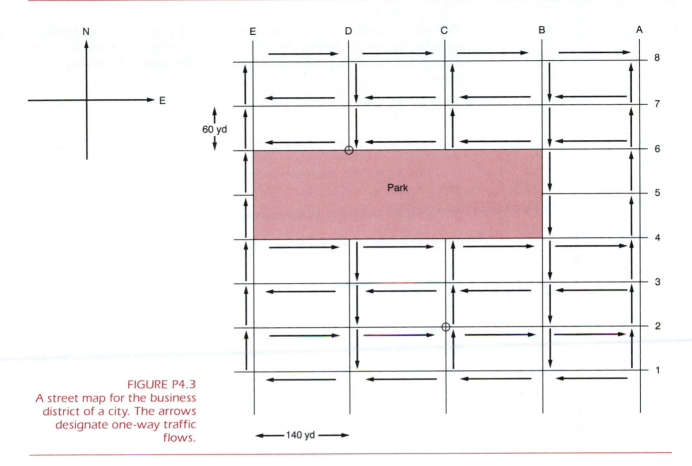

FIGURE P4.3
A street map for the business district of a city. The arrows designate one-way traffic flows.

4.6 A value for the plasticity index of soil samples is recorded on each of a set of index cards. A card marked "end of data" is placed at the end of the set. Write an algorithm to determine the sum and the average of these values.

4.7 Using the fundamental control structures from Fig. 4.10, construct a flowchart and pseudocode for Prob. 4.6.

4.8 Write an algorithm to calculate and print out the real roots of a quadratic equation

$$ax^2 + bx + c = 0$$

where a, b, and c are real coefficients. The formula used to calculate roots is the quadratic formula

$$x = \frac{-b \pm \sqrt{b^2 - 4ac}}{2a}$$

Note that if the quantity within the square root sign (that is, the determinant) is negative, the root is complex. Also note that division by zero occurs if $a = 0$. Design your algorithm so that it deals with these contingencies by printing out an error message.

4.9 Using the fundamental control structures from Fig. 4.10, construct a flowchart for Prob. 4.8.

4.10 The pseudocode in Fig. 4.14 is not "bullet-proof." For example, what if the number of digits is input as a negative number? Can you think of any other ways that a user might inadvertently cause the program to yield erroneous results? Refine this pseudocode to avoid errors of this type.

4.11 Using the fundamental control structures from Fig. 4.10, (a) construct a flowchart for an algorithm to compute the 2's complement of an 8-digit binary integer. (b) Develop pseudocode for the same algorithm.

4.12 The exponential function e^x can be represented by the following infinite series:

$$e^x = 1 + x + \frac{x^2}{2} + \frac{x^3}{3!} + \frac{x^4}{4!} + \cdots$$

Write an algorithm to implement this formula so that it computes and prints out the values of e^x as each term in the series is added. In other words, compute and print in sequence the values for

$$e^x \cong 1$$
$$e^x \cong 1 + x$$
$$e^x \cong 1 + x + \frac{x^2}{2}$$
.
.
.

up to the order term of your choosing. For each of the above, compute and print out the absolute percent relative error as

$$\% \text{ error} = \left| \frac{\text{true} - \text{series approximation}}{\text{true}} \right| \times 100\%$$

4.13 Using the fundamental control structures from Fig. 4.10, construct a flowchart for Prob. 4.12. Label the structures to identify each one by its type.

4.14 Figure P4.14 shows the reverse of a checking account statement. The bank has developed this sheet to help you balance your checkbook. If you look at it closely,

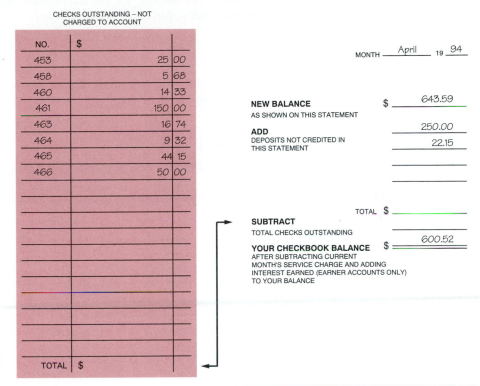

THE AREA BELOW IS PROVIDED TO HELP YOU BALANCE YOUR CHECKBOOK

FIGURE P4.14
A form used to balance a
checkbook.

you will realize that it is an algorithm. Develop a step-by-step algorithm to accomplish the same task.

4.15 Using the fundamental control structures from Fig. 4.10, construct a flowchart for Prob. 4.14.

4.16 Construct a pseudocode for Prob. 4.14.

4.17 An air conditioner both cools and removes humidity from a factory. The system is designed to turn on (1) between 7 A.M. and 6 P.M. if the temperature exceeds 75°F and the humidity exceeds 40 percent, or if the temperature exceeds 70°F and the humidity exceeds 80 percent; or (2) between 6 P.M. and 7 A.M. if the temperature exceeds 80°F and the humdidity exceeds 80 percent, or if the temperature exceeds 85°F regardless of the humidity. Write a structured flowchart that inputs temperature, humidity, and time of day and displays a message specifying whether the air conditioner is on or off.

4.18 It is relatively straightforward to compute cartesian coordinates (x, y) on the basic of polar coordinates (r, θ). The reverse process is not so simple. The radius can be computed with the following formula:

$$r = \sqrt{x^2 + y^2}$$

If the coordinates lie within the first or fourth quadrants (that is, $x > 0$), then a simple formula can be used:

$$\theta = \tan^{-1}\left(\frac{y}{x}\right)$$

The difficulty arises for all other situations. The following table summarizes the possible cases that can occur.

x	y	θ
<0	>0	$\tan^{-1}\left(\dfrac{y}{x}\right) + \pi$
<0	<0	$\tan^{-1}\left(\dfrac{y}{x}\right) - \pi$
<0	=0	π
=0	>0	$\dfrac{\pi}{2}$
=0	<0	$-\dfrac{\pi}{2}$
=0	=0	0

Write a well-structured flowchart to compute (r, θ) given (x, y). Express the final result for θ in degrees.

4.19 Develop a well-structured flowchart where the user can enter a number and one of the following units of time: second, minute, hour, and day. Design the flowchart so that it will then display the time in the remaining units.

4.20 Develop a well-structured flowchart where the user can enter the number of seconds past midnight (0 through 86,400 s). Design the flowchart so that it will then display the time in hours, minutes, and seconds. Use the 12-h format for displaying the result (that is, midnight is 12:00:00 P.M. and noon is 12:00:00 A.M.). If the user inputs a negative time or a time that is too large (>86,400 s), display an appropriate error message.

4.21 Develop a well-structured flowchart that converts Julian day (1, 2, ..., 365) into calendar day (Jan. 1, Jan. 2, ..., Dec. 31) for a nonleap year. If the user inputs a negative day or a day that is too large (>365), display an appropriate error message.

4.22 Define the following terms:
  (*a*) Structured programming
  (*b*) Modular design
  (*c*) Procedure
  (*d*) Top-down design

4.23 Compare and contrast the learning of a computer language and a foreign language.

4.24 What attributes have made modular design, top-down design, and structured programming techniques so helpful in reducing the cost of software development?

STRUCTURED BASIC

After the development of a sound and comprehensive plan, the next step in the program-development process is to compose the program using a high-level language. This part of the book shows how structured BASIC can be employed for this purpose.

Chapter 5 is devoted to the most elementary infor-

mation regarding BASIC programming. This material is designed to get you started by providing information sufficient to successfully write, execute, and document simple programs. Chapter 6 through 11 cover additional material that will allow you to develop more complicated and powerful programs.

FUNDAMENTALS OF STRUCTURED BASIC

When I show programmers the BASIC I use today, they say something like, "That looks a lot like Pascal." People who think of BASIC as the IBM PCs BASICA are surprised that it doesn't have to be a line-oriented language with little structure. BASIC has become very modern since 1983.

—Bill Gates, Founder of Microsoft, Inc.
Byte Magazine, October 1989.

Just as professional engineers and scientists have unique computing needs, so also do students. BASIC, which stands for *B*eginner's *A*ll-purpose *S*ymbolic *I*nstruction *C*ode, was developed expressly as an instructional language at Dartmouth College in the mid-1960s. Its originators, John Kemeny and Thomas Kurtz, recognized that, although FORTRAN was widely used, it had certain characteristics that posed problems for novice programmers. Kemeny and Kurtz came up with BASIC as an alternative that was easier to learn and use.

Although the ensuing years have seen the language evolve into a powerful problem-solving tool, it still retains most of the characteristics that originally made it easy to understand and use. The two most important of these characteristics are its interactive nature and its straightforward vocabulary.

The interactive nature stems from the way in which early BASIC programs were translated into machine language. BASIC was originally developed as an *interpreted language*. This means that translation from BASIC statements to machine instructions was implemented on a line-by-line basis as the program was executed. Although this slows down program execution, it allows errors to be detected and communicated to the user, and it permits the user to correct them "on the fly" during program execution. This is in contrast to a *compiled language* (such as FORTRAN and most other languages, including modern dialects of BASIC), where the entire program is translated into machine language before the program is run. As a consequence, the user must wait for the compilation phase to be completed before receiving feedback on errors. Some modern compilers are very fast, which has resolved this problem to a certain extent.

However, in many cases, the compile time delay can be extremely frustrating during program development. Of course, once the program is error-free, a compiled version is always preferable to an interpreted version because it will run quicker.

Today, although the modern forms of BASIC are now compiled efficiently, the language still maintains its interactive nature. The language developers did this by designing compiler-interpreter hybrids that act just like interpreters. These can be used during program development. Once the program is error-free, the finished version can then be formally compiled for efficient long-term application.

BASIC's second strength as a learning language relates to its simplicity. BASIC has a straightforward English-like vocabulary that is communicative and nonthreatening. Consequently, you can compose working computer programs without having to wrestle with an alien vocabulary and syntax. The experience is analogous to learning to drive a car with a standard versus an automatic transmission. Over and above all the other skills required to operate an automobile, novices learning to drive with a standard transmission must simultaneously worry about shifting gears and mastering the clutch. In contrast, novices learning to operate a car with an automatic transmission can focus their attention on the more fundamental aspects of driving such as accelerating, stopping, and avoiding solid objects. BASIC is just like driving a car with an automatic transmission. Within a very short time, you will be "on the road" to successfully developing meaningful programs for engineering problem solving.

5.1   WHICH BASIC?

During its 25-year existence, many versions of BASIC have been developed. In addition, an evolution has occurred over time as new capabilities have been incorporated into the various versions. This evolution consisted of three major phases (Fig. 5.1).

As depicted in Fig. 5.1a, the original BASICs of the late 1960s and early 1970s were primitive compared to today's language. Variable names were limited to two letters, and programs had line numbers. The very first versions did not even have INPUT statements to allow users to enter data while the program was running. Most of these limitations were imposed so that the early versions could be squeezed into the miniscule memories (4K-byte) of the microcomputers available at that time.

The first major upgrade came in 1977 when 8K-byte versions were introduced. The most noteworthy of these was Microsoft BASIC, the standard version for the popular IBM PC. As in Fig. 5.1b, it included longer variable names, INPUT statements, and more sophisticated program control structures. However, it still had major drawbacks that severely limited its professional application. These included mandatory line numbers, global variables, a 64K limitation on program size, and a 640K limit on RAM.

Today's structured BASIC programs have overcome all these limitations and added some powerful new features. Consequently, they now represent truly viable alternatives for professional use in fields such as engineering (see Fig. 5.1c). The modern dialects no longer require line numbers, include all the major control structures, and support sophisticated computing features such as record structures and recursion. They are highly modular, employing procedures with local variables and parameter passing.

```
10 REM 1964 BASIC PROGRAM
20 REM TO DETERMINE N!
30 LET N = 5
40 NF = 1
50 FOR I = 2 TO N
60 NF = NF * I
70 NEXT I
80 PRINT N, "! = "; NF
90 END
```

(a)
1964 BASIC

```
10 REM 1977 8-BYTE BASIC PROGRAM
20 REM TO DETERMINE FACTORIAL!
30 INPUT "NUMBER = "; NUMBER
40 IF NUMBER >= 0 THEN GOTO 70
50 PRINT "CANNOT COMPUTE"
60 GOTO 120
70 FACTORIAL = 1
80 FOR I = 2 TO NUMBER
90 FACTORIAL = FACTORIAL * I
100 NEXT I
110 PRINT NUMBER; "! = "; FACTORIAL
120 PRINT "WANT TO TRY ANOTHER Y/N";
130 INPUT ANS$
140 IF ANS$ = "Y" THEN GOTO 30
150 END
```

(b)
1977 BASIC

```
DECLARE FUNCTION Factorial! (number!)

'1990 Quick BASIC program to
'calculate factorials with a
'recursive function

DO
  DO
    INPUT "Number = "; number
    IF number >= 0 THEN EXIT DO
    PRINT "Number must be >= 0"
  LOOP
  PRINT number; "! = "; Factorial(number)
  INPUT "Want to try another Y/N"; ans$
LOOP UNTIL ans$ = "N" OR ans$ = "n"

END

FUNCTION Factorial (number)

  IF number = 0 THEN
    Factorial = 1
  ELSE
    Factorial = number * Factorial(number
  END IF

END FUNCTION
```

(c)
1990 Quick
BASIC

FIGURE 5.1
The historical evolution of BASIC. All three programs determine the factorial of a number.

In addition, today's BASIC programs have transcended the 64K and 640K limits and are restricted only by the size of your computer's memory.

However, over and above these significant changes to the language itself, an additional exciting development has been the evolution of the computing environments within which the language is implemented. For example, modern versions of BASIC include built-in full-screen editors with automatic syntax checking and context-sensitive help. These features greatly expedite the program-development process.

The following chapters are designed to be compatible with the Microsoft Quick-BASIC dialect. We have selected QuickBASIC because it seems to be the most widely available and best supported of the modern dialects. In particular, a subset of the language—QBASIC—comes as a standard part of MS-DOS operating system software on most personal computers. It should be noted that we have attempted to limit our discussion to general aspects of the language. Consequently, most of the following material is compatible with the other modern dialects, such as Turbo BASIC and True BASIC.

5.2   A SIMPLE BASIC PROGRAM

Figure 5.2 shows a very simple BASIC program to add two numbers. Each of the lines in this program is called a *statement*. When such a program is executed or "run," the statements are implemented in sequence. Thus, a program (which is also sometimes referred to as *source code*) is merely a series of instructions to direct the computer to perform actions.

For this program, values of 28 and 15 are assigned to the variables, a and b. Next, the addition is performed and the result stored in the variable c. Then, the value of c is displayed on the screen by the PRINT statement. Finally, the END statement signals that the program is finished.

The remainder of this chapter will be devoted to expanding Fig. 5.2 into a quality BASIC program. Because of the simple nature of the algorithm underlying Fig. 5.2, the following material does not represent a detailed description of the capabilities of BASIC. A more in-depth discussion of the language will be deferred to subsequent chapters. For the time being, we will focus on the fundamentals of BASIC programming. In addition, we will introduce some practices that will add to the quality of your finished product.

Adding Descriptive Comments.   As we will reiterate time and time again, an important attribute of a high-quality computer program is how easy it is to read and to understand. One deficiency of the program in Fig. 5.2 is that it is devoid of documentation. That is, it consists merely of a series of instructions to get a job done. Although this is not a major problem for such a simple program, inadequate documentation can detract immensely from the value of more complicated programs. If the user has to expend a great deal of time and energy to decipher your program's operation and logic, it will diminish the impact and usefulness of your work. Furthermore, if you have not used your own programs for a period of time, you will have difficulty deciphering your own poorly documented code.

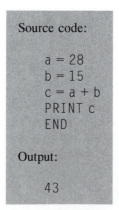

Source code:

```
a = 28
b = 15
c = a + b
PRINT c
END
```

Output:

43

FIGURE 5.2
A program to perform simple addition in BASIC. Note that the results are shown below the program source code.

One of our primary tools for the internal documentation of a program is the comment statement. The general form of this statement is

' *comment*

where *comment* is any descriptive information you would like to include. For example, the following comment statement can be included at the beginning of the program in Fig. 5.2 in order to provide a title:

' Simple Addition Program

Thus the user can immediately identify the general subject of the program.

Beyond setting forth titles, comment statements can be employed to incorporate additional documentation. Figure 5.3 is an example of how this can be done for the simple addition program. Notice that a group of comment statements has been used to form a title header at the program's beginning. We have called this version 1.1 to indicate that it is an improved version of the original source code from Fig. 5.2 (version 1.0). Along with the title, this section includes the author's name, the date the program was developed, and a concise description of what the program is to accomplish.

After the title block, comment statements can be used to interject descriptive comments as well as to make the program more readable. As seen in Fig. 5.3, this latter objective can be accomplished by using comments to delineate clearly the program's major parts. Also notice how several blank lines and a comment followed by a line of dashes are included to set off and separate some of the comments and sections. The judicious use of lines and blank spaces is one effective tool for making the program more readable.

FIGURE 5.3
A revised version of the simple addition program from Fig. 5.2. Here comments serve to document the program.

```
'Simple Addition Program (Version 1.1)
'by S.C. Chapra
'October 31, 1991
'This program adds two numbers
'and prints out the results
'-------------------------------------
'assign constants to variables
a = 28
b = 15
'perform computation
c = a + b
'display results
PRINT c
END
```

In early dialects of BASIC, a REM (short for REMark) statement was used to designate a comment, as in

```
REM Simple Addition Program
```

Although the REM statement can still be employed in modern structured BASIC, we will limit ourselves to the apostrophes because they are less obtrusive than REMs.

Aside from standing alone on its own line, a comment can also be appended to a line, as in

```
c = a + b  'this statement adds two variables
```

This is a nice way to add a comment because it stands out from the rest of the program and is easy to see. Thus, when someone reads the statement, she or he will recognize the intent of the statement. Admittedly, this is not necessary for such a simple statement. However, there are situations where such labeling can prove useful in clarifying a program line. Note that the REM statement cannot be used in this way.

It should be mentioned that several statements can be written on a single line by separating, or "delimiting," them with colons. For example, the assignment of values to a and b in Fig. 5.3 could be represented by the single line

```
a = 28: b = 15
```

This was done frequently in "Olde Style Basic" to conserve memory space. In modern BASIC, which is relatively memory rich, the indiscriminate use of colons to write several statements on a single line should be avoided because such a tactic can obscure the structure of the statements. However, it is acceptable style in certain cases such as this example, where only a few values are being assigned.

While the use of comment statements is optional in BASIC, their use is strongly encouraged. But do not overuse comments. If the action of a line or lines is obvious, do not add excess verbiage in the form of remarks. A program can quickly become cluttered with unneeded comments that actually make the program harder to read and follow. As described in the next section, careful selection of variable names helps to minimize the need for explanatory remarks.

5.3   NUMERIC CONSTANTS, VARIABLES, AND ASSIGNMENT

As an engineer, one of the primary uses you will have for a computer is to manipulate numbers. Consequently, it is important to understand how numeric quantities are stored and managed in the computer.

As illustrated by the simple program depicted in Fig. 5.3, there are two fundamental ways in which numeric data is expressed in BASIC: directly as constants and indirectly as variables. *Constants* (also called *literals* in computer jargon) are values that remain fixed during program execution, whereas *variables* are symbolic names to which a value is assigned that can be changed during execution.

TABLE 5.1   The Four Types of Numbers Used in QuickBASIC*

Type	Suffix	Range	Internal Representation
Integer (two bytes)	%	−32,768 to +32,767	
Long integer (four bytes)	&	−2,147,483,648 to +2,147,483,647	
Single-precision real (four bytes)	!	≅ −3.4E+38 to −8.4E−37 and ≅ +8.4E−37 to +3.4E+38	
Double-precision real (two bytes)	#	≅ −1.7D+308 to −4.2D−307 and ≅ +4.2D−307 to +1.7D+308	

*The ranges shown follow IEEE format. Note that some early versions of the Microsoft product used schemes that did not follow these ranges exactly.

5.3.1 Numeric Constants

Numeric constants may be either positive or negative and may be expressed in several different forms. As listed in Table 5.1, these forms relate to the way in which the numbers are stored in the computer's memory.

Integers are counting numbers with no decimal point, as in

> 508   67   −1507   883645

In QuickBASIC, there are two kinds of base 10 integers[1]: integers and long integers. The former consist of a 16-bit word and hence can range in value from −32,768 to 32,767. *Long integers* use two 16-bit words and range from −2,147,483,648 to 2,147,483,647.

The *real constants* are those with decimal points (and most usually fractional parts):

> 896.   −4.78   0.0056   −1000.4

Note that 896 is an integer and 896. is a real constant. The two forms are stored differently in the computer.

---

[1]Note that it is also possible to employ other number bases such as hexadecimal (16) and octal (8). However, we will limit the present discussion to base 10, or decimal numbers.

In addition, real constants can be expressed in a form similar to mathematical scientific notation. This form, called an *exponential constant* in BASIC, is handy for representing very large or very small numbers.

Recall that a quantity is expressed in scientific notation by converting it to a number between 1 and 10 multiplied by an appropriate integral power of 10. This can be done by moving the decimal point to the right or left until there remains a number between 1 and 10. This remaining number must then be multiplied by 10 raised to a power equal to the number of places the decimal point was moved. If the decimal has been moved right, the power is negative; and if it has been moved left, the power is positive. The general form is

$$b \times 10^n$$

where b is the decimal or integer value that remains after the number has been converted to a value between 1 and 10 (often called the *mantissa* or *significand*) and n is the integer number of places the decimal has been moved (called the *characteristic* or *exponent*). For example, the number 8,056,000 would be converted to 8.056 by moving the decimal six places to the left. Therefore in scientific notation it would be expressed as 8.056×10^6.

BASIC employs the same concept but uses a different notation:

$$b\mathrm{E}n$$

Thus in BASIC the number 8,056,000 is expressed as 8.056E6.

It should be noted that real numbers in QuickBASIC can be expressed in either single or double precision. *Single-precision numbers* are stored with about 7 significant digits, whereas *double-precision numbers* are stored with about 15. The double-precision version of the exponential form uses a D rather than an E, as in

    1.6D5   4.83702D5   15D10

Aside from the E and the D used to represent exponential numbers, four other symbols can be appended to numeric constants to specify their type. Integer or double-precision constants can be forced to be a single-precision constant by appending an exclamation point, as in

    49.7!   2000!   8.86!

In addition, if the # suffix is appended, the number will be forced to be stored in double-precision form:

    30#   1050#   3.1415926536#   2.71828183#

Short and long integers can be specified by appending the % and the & suffixes, respectively.

Note that no other letters (other than E, D) or symbols (other than !, #, &, %) such as dollar signs and commas are allowed in numeric constants. Therefore $56.23 and 1,760,453 are illegal numeric constants and would cause a syntax error in BASIC. In addition, only one minus sign may be used at the front of a constant to designate that it is negative. Minus signs cannot be used as delimiters (that is, separators), as is commonly done in social security numbers (and often with student I.D. numbers). For example, 120-44-5605 is an illegal constant. In fact, BASIC would misinterpret it as representing subtraction and, depending on the context, either yield the result, −5529, or display an error message.

You may be wondering at this point why so many different forms are needed and why we would want to force the computer to store a constant in a particular form. There are a number of reasons why we might want to go to the trouble. For example, an integer takes up less space in memory and can be processed faster during computations than a real constant. Similarly, single-precision numbers have the same advantages over double-precision constants. Thus, if we use integers rather than single precision, and single precision rather than double precision, the resulting programs will run faster and use less memory. On the other hand, double-precision constants are advantageous when a computation must be performed with a high degree of accuracy. Such situations often occur in engineering.

Although the above is true, it should be noted that in most short programs you will not be concerned with how numbers are stored on the computer. You will just write the constants naturally and let the computer determine how they should be stored. With this as background, we can now turn to the manner in which the constants are actually stored.

5.3.2 Variables and Assignment

In addition to constant quantities, it is useful to have variable quantities that can change during the execution of a program. These variable quantities are designated by symbolic *variable names.* It is common practice to refer to the variable name as the variable itself. For example, in the simple program from Fig. 5.3, we used three variables—a, b, and c.

To understand how variables work, it is useful to describe how computers store and retrieve information from primary memory. The computer identifies each location in its memory by a number called an *address.* When a language such as QuickBASIC employs a variable name, it automatically associates a particular address with the name. Consequently, when the variable name is used in a program, the computer uses the address to access the proper memory location.

It is convenient to think of this arrangement as consisting of locations or "cells" that are similar to post office boxes (Fig. 5.4). The variable is the label that the computer assigns to each of these cells. For example, in the simple addition program, each of the variables—a, b, and c—would correspond to the label of the cells. When we used the statement a = 28 from Fig. 5.3 to assign a value of 28 to a, the computer stored the value in the memory location for a (Fig. 5.4a). Then later in the program when we performed the addition with the statement c = a + b, the computer

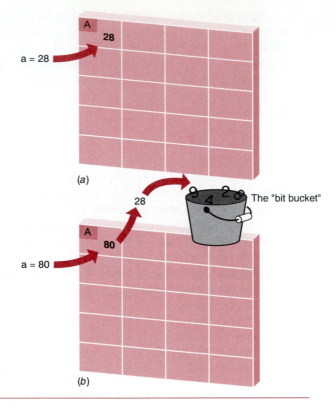

FIGURE 5.4
Analogy between computer memory locations and post office boxes. (a) The BASIC statement, a = 28, assigns the variable name, a, to one of the memory locations and stores a value of 28 in that location. (b) A subsequent assignment statement, a = 80, places a new value of 80 in the location, and the process discards the prior value of 28.

remembered that a was equal to 28 by using the address to locate the memory cell where it was stored.

Now suppose that after the addition we defined a by another assignment statement, such as

a = 80

When this line was executed, the computer would replace the old value by this new one. Using our analogy of the postal boxes, this act would be akin to placing the new number in the box. The old value would be pushed out the back and into oblivion, or into the "bit bucket" as it is sometimes called (Fig. 5.4b). This ability to change value is the reason symbolic names are called *variables.*

Because statements such as a = 80 assign a value to a variable, they are called *assignment statements.* They can be written in general form as

variable = expression

where *variable* stands for the variable name, and *expression* can be a constant, another variable, or an arithmetic expression involving constants and variables. The following statement is an example of the use of an arithmetic expression in an assignment statement:

    c = a + b

When such a statement is executed, the expression to the right of the "=" sign (that is, the *assignment operator*) is evaluated first. Then, the resulting numeric value is automatically stored in the storage location for c. For example, in Fig. 5.3, a value of 43 would be stored in c.

5.3.3 Algebraic Equations Versus Computer Assignment Statements

The equal sign in the assignment statement can be thought of as meaning "is replaced by" or "is assigned the result." This is so significant that a computer language such as Pascal does not even use the "=" sign to designate the process of assignment. For example, in Pascal, it is represented by ":=" as in

    c:=a + b;

Some computer scientists have proposed that a better representation of the process would be an arrow, as in

    c ← a + b

Both modifications are efforts to distinguish more clearly between assignment and equality. In particular, they signify that computer assignment statements differ in a very important way from standard algebraic equations. That is, assignment statements must always have a single variable on the left-hand side of the equal sign. Thus, because it violates this rule, the following legitimate algebraic equation is illegal as a BASIC assignment statement:

    a + b = c

In the same sense, although

    x = x + 1

is not a proper algebraic equation, it is a perfectly acceptable BASIC assignment statement. This is evident when the words "is replaced by" or "is assigned the result" are substituted for the equal sign. For example, if x had a prior value of 5, the new value after execution of this statement would be 6.

All the above makes eminent sense when we remember how the computer accesses memory locations to store and retrieve values. Using the postal box analogy (Fig. 5.4), if more than a single variable name occurred to the left of the assignment operator, how could the computer conceivably know where to assign the outcome? Clearly, the only way the scheme works is when a single variable name appears on the left.

5.3.4 Variable Names

The symbolic name for the variable should reflect, as simply as possible, what information is stored by the variable. Names that are short enough to type quickly and accurately are helpful, as long as the name is of sufficient length to convey the meaning or purpose of the variable.

The actual symbolic names used to designate variables must follow rules. Variables must begin with a letter, not a number. After the first letter, they may have either letters, numerical digits, or certain allowable symbols. Reserved words or keywords may not be used as variable names in BASIC. For example, PRINT always identifies output of a value and is never a variable name. However, the name may include embedded keywords; for example, PRINTOR would be acceptable even though it includes two keywords, PRINT and OR.

For QuickBASIC, variable names can contain up to 40 characters. Variable names should not contain characters other than letters (both lowercase and uppercase), numerical digits, and decimal points except for their last character, if the last character is one of the special type identifiers such as ! or #. The special type identifiers (recall Table 5.1) are used to specify the type of constant that can be stored in the variable's memory location. Note that if you omit a special type identifier from a name, the computer will assume that it is single-precision real. We will have more to say regarding variable typing below.

Variable names end when a character that is not a letter, numerical digit, decimal point, or special type identifier is encountered. The following are examples of legitimate variable names:

```
one           var1        x1938           Var!
i%            sum#        fall.velocity   PRODUCT
test.two      z2001v      I               Stress
```

All of these variable names conform to the above rules. However, not all may be ideal variable names because their meaning may not be obvious. Notice also how the decimal point is used to make names consisting of two words more readable (for example, `fall.velocity` and `test.two`).

Examples of variable names that are not acceptable for BASIC are

```
8thVariable    4567    !TEST    The1st@
INPUT          log     A-VAR    INT%1
thisvariablenameismuchtoolongtobevalidinQuickBASIC
```

The names `8thVariable`, `4567`, and `!TEST` do not begin with a letter. The name `The1st@` ends with a character, "@," that is not one of the special type identifiers.

INT%1 has one of the special characters that identify type, but it is not the last character of the name. INPUT and log are BASIC reserved words. A-VAR is actually two variable names and the subtraction operator (−). Thus, although A and VAR are legitimate names, A-VAR cannot be used to name a single variable because the computer would "think" that it signifies the expression: A minus VAR.

A few words of caution are in order when naming variables. The QuickBASIC form of BASIC ignores the distinction between uppercase and lowercase letters in variables so that Number, NUMBER, and number would all refer to the same memory location. In fact, dialects such as QuickBASIC, which have "smart" editors, will actually change the format of all other instances of a variable to the most recent way that you typed it.

The names I, I#, and I% represent three distinct variables that would label three different memory locations. However, val and val! would represent the same location because BASIC assumes that all variable without suffixes are single-precision real.

5.3.5 String Constants and Variables

Aside from numbers, computers are also designed to handle character-based information (that is, groups of letters, numerical digits, and symbols) such as names, addresses, and other miscellaneous labels. *String constants* are specific groupings of character information. In BASIC they are enclosed in quotation marks as in

```
"John Doe"  "8/5/48"  "The first result is "  "050-58-7724"
```

Note that, in contrast to numeric constants and variable names, blank spaces and symbols such as "−" or "/" can be a part of a string. Because the symbols are enclosed within the quotes, the computer does not mistake them for arithmetic operators.

String variables work in the same manner as numeric variables in the sense that they are names that designate a specific location in the computer's memory. The naming conventions are the same as for numeric variables with the exception that all string variable names must end with a "$." For example,

```
Name2$   Date$   Sum.label$   Soc.sec.no$
```

String variables can be assigned constant values in the same manner as numeric variables:

```
Name2$ = "John Doe"
Date$ = "8/5/48"
Sum.label$ = "The first result is"
Soc.sec.no$ = "050-58-7724"
```

Just as with numeric values, these statements will store each constant in a memory location which is associated with the variable name. The suffix "$" signals that the quantity stored is of the string type.

5.3.6 Variable Types in QuickBASIC

QuickBASIC has five types of variables (Table 5.2). In other words, the storage location represented by the variable can contain five types of information. These are integers, long integers, real single-precision numbers, real double-precision numbers, and characters.

Using the analogy from Fig. 5.4, the different variable types relate to different types of mailboxes for storing information. Three means are available for telling the computer the type of data that a variable will contain.

Type-declaration suffixes.   As already mentioned, the simplest means of specifying variable type is by adding a type-declaration suffix to the variable name. These suffixes are summarized in Table 5.2. QuickBASIC assumes that all variables without suffixes are single-precision real.

This is the most common way to declare a variable type. Because the convention is built into the name, the type of variable can be defined conveniently and unambiguously.

The DEFtype declaration.   A second way to inform the QuickBASIC compiler of variable type is by using the BASIC statements DEFINT, DEFLNG, DEFSNG, DEFDBL, and DEFSTR. By using these DEFtype statements, you can specify that all variables starting with a certain letter or range of letters will be of a particular type. For example:

```
DEFDBL Q, S, U
DEFLNG I-K
DEFSTR T
```

In the first example, all variables beginning with the letter Q, S, and U will be double-precision reals. In the second, any names starting with the letters I, J, and K would be long integers. The last statement declares that all variables beginning with T will be of type STRING. This means that the variable may contain string constants—that is, letters, numbers, and symbols for purposes such as labeling.

TABLE 5.2 Types of Variables Used in Microsoft QuickBASIC

Type	Declaration	Suffix	Description
Integer	INTEGER	%	Contains a 16-bit integer.
Long integer	LONG	&	Contains a 32-bit integer.
Real (single-precision)	SINGLE	!	Contains a single-precision, floating-point, or decimal number.
Real (double-precision)	DOUBLE	#	Contains a double-precision real number.
Characters	STRING	$	Contains character or alphanumeric information.

Type-declaration suffixes take precedence over DEFtype declarations. Consequently, even though the last DEFtype statement above specifies that variables starting with the character T would generally hold strings, the following statement would evoke an error message,

```
T# = "That's all folks"
```

The DEFtype statement is a convenient way to declare variable type. However, it can lead to confusion if the reader of the program does not remember which first letters correspond to which variable types.

The DIM/AS declaration. The final way to declare the type of a variable is by explicitly giving the variable type and then the variable names corresponding to the type. This is accomplished with a DIM/AS statement that is of the general form

DIM *VarName* AS *type*

where *VarName* is the name that is being typed, and *type* is chosen from the names of the different kinds of variables—INTEGER, LONG, SINGLE, DOUBLE, and STRING. For example,

```
DIM COUNT, index, Grade AS INTEGER
DIM Number.homes AS LONG
DIM sum, MEAN, one, Two, numval AS SINGLE
DIM diff AS DOUBLE
DIM StudentName AS STRING
DIM address AS STRING * 20
```

The next to the last statement declares that the variable name is a *variable-length string*. Such strings are "expandable" in that their length in memory depends on the length of the string that is assigned to them. In contrast, the last statement sets up the variable address as a *fixed-length string*. The *20 specifies that this string can contain 20 characters.

The DIM/AS statement overrides the DEFtype declarations. Aside from its expressed purpose of declaring the type of each variable, the explicit declaration has an added benefit. That is, the names and types of all variables can be located in one area of the program for easy reference by the user.

In the QuickBASIC programs in the remainder of the book, we will primarily use type-declaration suffixes to establish variable types. In almost all cases, our use will be limited to no suffixes for single-precision REAL, % for INTEGER, and $ for STRING variables.

5.3.7 CONST Statements

Certain constants are used so often that they are given special names. An example is the number 3.14159... which is commonly called π. The CONST statement provides a means to assign a name to such constants. Its general form is

$$\text{CONST } ConstName_1 = exp_1, ConstName_2 = exp_2, \ldots$$

where $ConstName_i$ is the name of the ith constant being defined and exp_i is the ith expression that is being assigned to $ConstName_i$.

An example is

```
CONST pi = 3.14159, npoints = 200
```

For this example, the names, `pi` and `npoints`, are assigned values of 3.14159 and 200, respectively.

Although it should be clear why we would want to assign a constant in the case of π, the reason for assigning a value to the other parameter, `npoints`, might not be obvious.

Oftentimes, you will employ a constant repeatedly throughout a large program. For example, `npoints` might represent the number of data that is being analyzed. This number could be employed hundreds of places in your program. If you represented it by its constant value and you wanted to change this constant, it would be extremely time-consuming to edit your program to change every possible instance where it was used. In contrast, if you employed the name `npoints` throughout your source code, you would merely have to change the single CONST statement to produce the modification.

5.4   INTRODUCTION TO INPUT-OUTPUT

Input-output or "I/O" refers to the means whereby information is transmitted to and from a program. You have already been introduced to output in the form of the PRINT command from the simple addition program. Now we will elaborate on the topic so that you can perform more elaborate input and output.

5.4.1   The CLS Statement

The CLS statement is short for *"CLear Screen."* When this command is implemented, the entire monitor screen will be cleared of all text and graphics. This statement is usually placed at the beginning of a BASIC program so that your output can be displayed on a blank space. In large programs, it is also used repeatedly to clear the screen after each major input or output action.

5.4.2   Input

As we stated at the beginning of this chapter, early versions of BASIC did not even have statements to allow users to enter information while programs were running. Today's structured BASIC has rich input capabilities. The present section focuses on the INPUT statement, which is the principle means of entering data via the keyboard during program execution.

The INPUT statement. One disadvantage of the assignment statement is that it sets a variable to a specific value. Thus, in the simple addition program (Fig. 5.3), we used assignment statements to set a equal to 28 and b equal to 15. If we wanted to add two different numbers, say a = 8 and b = 33, we would have to modify or "edit" the assignment statements in order to assign new values to the variables.

The INPUT statement provides an alternative that allows new data to be entered every time the program is run. As an example of its use, suppose that we want to assign a different value to a every time we run the program. To do this, we would replace the line

```
a = 28
```

by

```
INPUT a
```

Then when we run the program, the computer will respond by displaying a prompt character on the monitor screen. A *prompt* is a signal from the computer that you must enter information. For BASIC, the prompt is represented by a question mark. For the present example, the prompt would appear when the computer reached the INPUT statement. Suppose that you want to set a equal to 8. Then by merely typing an 8 and striking the [Enter] key you would have instructed the computer to assign 8 to the variable a.

Internally, the computer handles the operation in an identical fashion to assignment. Thus, when you strike [Enter], the computer assigns the value of 8 to a memory location labeled a.

More than one value may be entered with an INPUT statement. For example, to input values for both a and b, you could type the statement

```
INPUT a, b
```

When this statement is executed, a prompt will appear on the screen and two values separated by a comma can be typed to enter the values for a and b, as in

```
? 8, 33
```

Until the [Enter] key is struck, the [Backspace] and other editing keys can be used to alter the input data. Once the [Enter] key is pressed, BASIC examines the constants that were typed to verify that they match the variables in the input statement. If the data entered does not match the number and type of variables in the input statement, then BASIC will display an error message, discard what had been entered, and redisplay the prompt until the data is entered correctly. The message and prompt will look like

```
Redo from start
?
```

Here are some INPUT statements with acceptable entries:

INPUT Statement	Acceptable Entry
`INPUT var1`	429.07
`INPUT var1, var2`	73.6, −943
`INPUT var1#, var2$`	2D7, "Hello there"
`INPUT var1, var2, var3`	−9.06E−4, 109, 15968.8

In contrast, the following entries would cause an error message to be displayed:

INPUT Statement	Unacceptable Entry	Problem
`INPUT var1`	"Hello"	Type mismatch
`INPUT var1, var2`	73.6 −943	No commas
`INPUT var1#, var2$`	2D7, "Hi", 5	Too many data
`INPUT var1, var2, var3`	−9.06E−4, 109	Insufficient data

As shown in these examples, both numbers and strings can be entered provided that the types are consistent. It should be noted that, in most instances, strings can be input without using quotes. However, if the character string includes commas, the quotes must be used. This is because the computer "perceives" commas that are not enclosed by quotes as delimiters.[2] For example,

```
? "July 4, 1776"
```

is correct because the date includes a comma. If the quotes were not used, as in

```
? July 4, 1776
```

the computer would "perceive" that the character string constant was July 4 and that 1776 was a separate value.

Descriptive input.   One problem with the INPUT statements described to this point is that the user has been provided no clue as to what information is supposed to be input. For example,

```
INPUT a, b
```

would cause a ? to be displayed when the program is run. Unless the user knew beforehand that values for a and b were to be entered, the question mark provides no indication as to what variables are to be input. One way to remedy this situation is to employ

---

[2]A *delimiter* is a general term for a symbol that is used to separate units of information.

a PRINT statement and a string constant to display a message that alerts the user to what is required, for example:

```
PRINT "a, b = "
INPUT a, b
```

When these lines are executed, the computer will display

```
a, b =
?
```

Thus the message lets the user know what is required. The descriptive input can be improved further by adding a semicolon at the end of the first statement, as in

```
PRINT "a, b = ";
INPUT a, b
```

The semicolon acts to suppress the line advance on the computer output. Thus when the program is run, rather than being displayed on the next line, the prompt appears at the end of the message, as in

```
a, b = ?
```

There is one final improvement that can be made. It is possible to include one character string message directly in the body of an INPUT statement. For example, rather than use the PRINT/INPUT combination shown above, we could simply type

```
INPUT "a, b = "; a, b
```

When the program is run, the result would be

```
a, b = ?
```

The question mark prompt can be eliminated by using a comma in place of a semicolon

```
INPUT "a, b = ", a, b
```

with the result,

```
a, b =
```

The LINE INPUT statement. One additional capability for entering data in BASIC is LINE INPUT. This statement allows the input of a line of up to 255 characters to a single string variable, without the use of quotes. Thus, it allows you to enter long strings containing commas, quotation marks, and other special characters. For example, if the following statement were executed,

```
LINE INPUT Descartes.says$
```

the following line

```
Descartes said: "I think, therefore I am."
```

could be entered complete with the quotes and the comma. The entire line would thereafter be stored in the string variable

```
Descartes.says$.
```

5.4.3 Output

Just as it is critical to have clear input of information, it is of equal importance to carefully design a program's output. In the present section we will show how several statements are used to this end.

The PRINT statement. The PRINT statement is the most common vehicle for outputting data to the screen. Interestingly, the word "PRINT" is employed because output was traditionally directed to printers, even though today the most common output device is the display screen.

To this point, we have showed (Fig. 5.3) how it can be employed to display a single value. It can also be used to output the values of a series of variables, constants, or both, as in

```
PRINT a, b, -7075, c
```

When the values are delimited by commas, as in this example, the computer automatically spreads the values out evenly across the page. For example, if $a = 28$, $b = 15$, and $c = 43$, the output will look something like

```
28                  15                  -7075               43
```

However, if a semicolon is used as delimiter,

```
PRINT a; b; -7075; c
```

the computer will pack the numbers closely together with a single separating space, as in

```
28  15 -7075  43
```

Note that a space is also reserved for the number's sign. For positive numbers, the place for the plus sign is left blank.

You might wonder why it would ever be preferable to jam numbers together in this fashion. Actually, it is rarely desirable to do this in its own right. However, when used in conjunction with the TAB command, semicolon delimiters provide a means to exactly control the spacing of the output.

The TAB and SPC commands. Although delimiting data with commas automatically results in evenly spaced output, there may be cases where you will desire to control spacing of the output yourself. The TAB and SPC commands are available for this purpose.

The *TAB command* can be included in a PRINT statement to specify the column at which output starts. For example,

```
PRINT TAB(10); c
```

will cause the value of c to be displayed starting at column 10. Thus if c = 43, the following output results:

Note that, as mentioned above, when BASIC outputs numbers, a space is left for the sign. However, for positive numbers, the sign is not explicitly displayed. Therefore, even though the value appears to start in column 11, the field for the number actually starts in column 10.

Several TABs can be used in a single PRINT statement to space out a series of values, as in

```
PRINT a; TAB(8); b; TAB(16); c
```

For this case, the value for b would begin in column 8 and for c in column 16. Hence the TAB function works just like the tab setting on a typewriter.

The *SPC command* can be included in a PRINT statement to skip spaces between values. For example,

```
PRINT "far"; SPC(20); "out"
```

will cause the second string ("out") to be displayed 20 spaces beyond the end of the first ("far"), as in

```
far                    out
```
       ‿‿‿‿‿‿‿‿‿‿‿‿
        20 spaces

Descriptive output.   Just as input can be made more user-friendly, output can also be made more descriptive by employing character strings, semicolon delimiters, TAB functions, and blank lines. For example, in the simple addition program (Fig. 5.3) the statement

```
PRINT c
```

merely displays the total, 43. To make the output more meaningful, a message can easily be incorporated into the statement, as in

```
PRINT "Total = "; c
```

When the program is run, the result would be displayed as

```
Total = 43
```

In addition, several constants, variables, or messages can be output in a single PRINT statement. For example, the single PRINT statement

```
PRINT a; b; c
```

could be reformulated as

```
PRINT "The sum of "; a; " plus"; b; " = "; c
```

When executed, this statement would result in the output

```
The sum of 28 plus 15 = 43
```

Figure 5.5 is a final revised version of the simple addition program which includes a number of improvements. The code itself has been made clearer by adding a nomenclature section defining the variables. In addition, more expressive variable names (for example, `total` instead of `c`) have been employed.

Program I/O also has been upgraded. We have added a variety of descriptive INPUT and PRINT statements to make the program more readable and easier to use. These, along with CLS and TAB statements, create a communicative and attractive output when the program is run (Fig. 5.6).

5.5   QUALITY CONTROL

To this point, you have acquired sufficient information to compose a simple BASIC computer program. Using the information from the previous sections, you should now be capable of writing a program that will generate some output. However, the fact that a program displays information is no guarantee that these answers are correct.

```
'Simple Addition Program (Version 2.0)

'by S.C. Chapra
'October 31, 1991

'This program adds two numbers and prints
'out the results using descriptive variable
'names and enhanced I/O
'-------------------------------------------

'Definition of variables

'   first, second = variables to be added
'   total = sum of the numbers
'-------------------------------------------

'clear screen and identify program

CLS
PRINT : PRINT
PRINT TAB(25);
PRINT "Program to add two numbers"

'obtain numbers from the user

PRINT : PRINT
PRINT TAB(30); : PRINT "Input the numbers"
PRINT : PRINT TAB(29);
INPUT "First number = "; first
PRINT TAB(29);
INPUT "Second number = "; second

'perform computation

total = first + second

'display results

PRINT
PRINT TAB(32); "The sum = "; total

END
```

FIGURE 5.5
A revised version of the simple addition program that uses descriptive I/O to make the program more expressive.

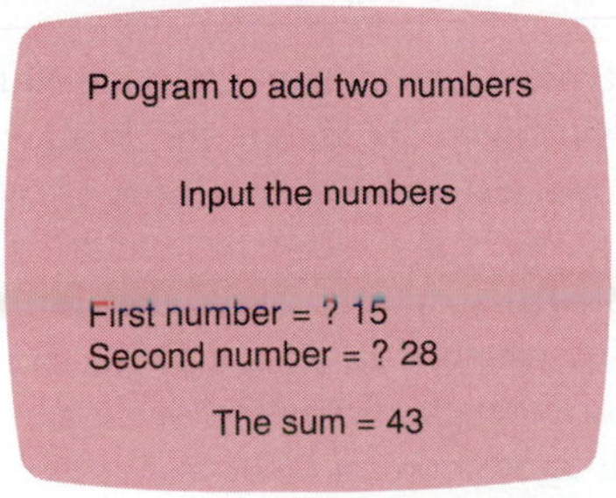

Program to add two numbers

Input the numbers

First number = ? 15
Second number = ? 28
The sum = 43

FIGURE 5.6
A run of the program from Fig. 5.5.

Consequently, we will now turn to some issues related to the reliability of your program. We are emphasizing this subject at this point to stress its importance in the program-development process. As problem-solving engineers, most of our work is used directly or indirectly by clients. For this reason, it is essential that our programs be reliable—that is, that they do exactly what they are supposed to do. At best, lack of reliability can be embarrassing; at worst, for engineering problems involving public safety, it can be tragic.

5.5.1 Errors or "Bugs"

During the course of preparing and executing a program of any size, it is likely that errors will occur. These errors are usually called *bugs* in computer jargon. This label was coined by Captain Grace Hopper, one of the pioneers of computing languages. In 1945 she was working with one of the earliest electromechanical computers when it went dead. She and her colleagues opened the machine and discovered that a moth had become lodged in one of the relays. Thus the label "bug" came to be synonymous with problems and errors associated with computer operations, and the term *debugging* came to be associated with their solution.

When developing and running a program, three types of bugs can occur:

1. *Syntax errors* violate the rules of the language, such as spelling, number formation, line numbering, and other conventions. These errors are often the result of mistakes such as typing IMPUT rather than INPUT. They usually cause the computer to display an error message. Such messages are called *diagnostics* because the computer is helping you to "diagnose" the problem.

2. *Run-time errors* are those that occur during program execution, for example, the case where there are an insufficient number of data entries for the number of variables in an INPUT statement. For such situations a diagnostic message "Redo from start" will be displayed, and you must then reenter the data properly. For other run-time errors the computer may simply terminate the run or display a message informing you at what line the error occurred. In any event, you will be cognizant of the fact that an error has taken place.

3. *Logic errors,* as the name implies, are due to faulty program logic. These are the worst types of errors because the computer may not display a diagnostic to alert you to their presence. Thus your program will appear to be working properly in the sense that it executes and generates output. However, the output will be incorrect.

All the above types of errors must be removed before a program can be employed for engineering applications. We divide the process of quality control into two categories: debugging and testing. As mentioned above, debugging involves correcting known errors until the program seems to execute as expected. In contrast, *testing* is a broader process intended to detect errors that are not obvious. Additionally, testing should also determine whether the program fills the needs for which it was designed.

As will be clear from the following discussion, debugging and testing are interrelated and are often carried out in tandem. For example, we might debug a module to remove the obvious errors. Then, after a test, we might uncover some additional bugs that require correction. This two-pronged process is repeated until we are convinced that the program is absolutely reliable.

5.5.2 Debugging

As stated above, debugging deals with correcting known errors. There are three ways in which you will become aware of errors. First, an explicit diagnostic message can provide you with the exact location and the nature of the error. Second, a diagnostic may be displayed, but the exact location of the error is unclear. Third, no diagnostics are displayed, but the program does not operate properly. This is usually due to logic errors such as infinite loops or mathematical mistakes that do not necessarily yield syntax errors. Corrections are easy for the first type of error. However, for the second and third types some detective work is required in order to identify the errors and their exact locations.

Unfortunately, the analogy with detective work is all too apt. That is, ferreting out errors is often difficult and involves a lot of "legwork," collecting clues and running down some blind alleys. In particular, as with detective work, there is no clear-cut methodology involved. However, there are some ways of going about it that are more efficient than others.

One key to finding bugs is to display intermediate results. With this approach, it is often possible to determine at what point the computation went bad. Similarly, the use of a modular approach will obviously help to localize errors in this manner. You could locate a PRINT message at the start of each module and in this way focus in on the module where the error occurs. In addition, it is always good practice to debug (and test) each module separately before integrating it into the total package.

Modern BASIC on a personal computer is an ideal tool for debugging. Because of its interactive mode of operation, the user can efficiently perform the repeated runs that are sometimes needed to zero in on the location of errors.

Structured dialects such as QuickBASIC also have very powerful debugging capabilities. These are an integral part of the environment and allow you to track the values of variables and to observe program execution on a step-by-step basis.

Although these procedures can help, there are often times when your only option will be to sit down and read the source code line by line, following the flow of logic and performing all the computations with a pencil, paper, and pocket calculator. Just as a detective sometimes has to think like the criminal in order to solve a case, there will be times when you will have to "think" like a computer to debug a program.

Finally, we cannot stress enough the fact that "an ounce of prevention is worth a pound of cure." That is, a sound preliminary program design and structured programming techniques are great ways to avoid errors in the first place. As with so many other aspects of your life, the sooner you adopt a "pay me now rather than pay me later" philosophy, the less grief you will experience as a computer programmer.

5.5.3 Testing

One of the greatest misconceptions of the novice programmer is the belief that if a program runs and displays results, it is correct. Of particular danger are those cases where the results appear to be "reasonable" but are in fact wrong. To ensure that such cases do not occur, the program must be subjected to a battery of tests for which the correct answer is known beforehand.

As mentioned in the previous section, it is good practice to debug and test modules prior to integrating them into the total program. Thus testing usually proceeds in phases. These phases consist of module tests, development tests, whole-system tests, and operational tests.

Module tests, as the name implies, deal with the reliability of individual modules. Because each is designed to perform a specific, well-defined function, the modules can be run in isolation to determine whether they are executing properly. Sample input data can be developed for which the proper output is known. This data can be used to run the module, and the outcome can be compared with the known result to verify successful performance.

Development tests are implemented as the modules are integrated into the total program package. That is, a test is performed after each module is integrated. An effective way to do this is with a sequential approach that starts with the first module and progresses down through the program in execution sequence. If a test is performed after each module is incorporated, problems can be isolated more easily, since new errors can usually be attributed to the latest module added.

After the total program has been assembled, it can be subjected to a series of *whole-system tests.* These should be designed to subject the program to (1) typical data, (2) unusual but valid data, and (3) incorrect data to check the program's error-handling capabilities.

Note that all the foregoing tests are often collectively referred to as *alpha-testing.* Once they are successful, a *beta-* or *operational testing* phase is usually initiated. These tests are designed to check how the program performs in a realistic setting. A standard way to do this is to have a number of independent subjects (including your users or customers) employ the program. Operational tests sometimes turn up bugs. In addition, they provide feedback to the developer regarding ways to improve the program so that it more closely meets the user's needs.

Although the ultimate objective of the testing process is to ensure that programs are error-free, there will be complicated cases where it is impossible to subject a program to every possible contingency. In fact, for programs of even moderate complexity, it is often practically impossible to prove that an error does not exist. At best, all that can be said is that, under rigorous testing conditions, no error could be found.

It should also be recognized that the level of testing depends on the importance and magnitude of the problem context for which the program is being developed. There will obviously be different levels of rigor applied to testing a program to determine the batting averages of your intramural softball team as compared with a program to regulate the operation of a nuclear reactor or the space shuttle. In each case, however, adequate testing must be performed, testing consistent with the liabilities connected with a particular problem context.

5.6   DOCUMENTATION

Documentation is the addition of written text to allow you or another user to implement the program more easily. (Remember, along with other people who might employ your software, you are also a user.) Although a program may seem clear and simple when you first compose it, after the passing of time the same source code may seem inscrutable. Therefore sufficient documentation must be included to allow you and other users to immediately understand your program and how it can be implemented. This task has both internal and external aspects.

5.6.1   Internal Documentation

Picture for a moment a text with just words and no paragraphs, subheadings, or chapters. Such a book would be more than a little intimidating to study. The way the pages of a text are structured—that is, all the devices that serve to break up and organize the material—serves to make the text more effective and enjoyable. In the same sense, internal documentation of computer code can greatly enhance the user's understanding of a program and how it works.

We have already broached the subject of internal documentation when we introduced the comment statement in Sec. 5.2. The following are some general suggestions for documenting programs internally:

• Include a section at the head end of the program, giving the program's title, your name, and the date. This is your signature and marks the program as your work.
• Include a second section to define each of the key variables.
• Select variable names that are reflective of the type of information the variables are to store.
• Insert spaces within statements to make them easier to read.
• Use comment statements and blank lines liberally throughout the program for purposes of labeling and clarifying. In particular, use comment statements to label clearly and set off all the sections.
• Use identation to clarify the program's structures. In particular, as will be illustrated in later chapters, indentation can be used very effectively to set off structures called selection and repetition control constructs.

5.6.2   External Documentation

External documentation refers to instructions in the form of output messages and supplementary printed matter. The messages occur when the program runs and are intended to make the output attractive and user-friendly. This involves the effective use of spaces, blank lines, and special characters to illuminate the logical sequence and structure of the program output. Attractive output simplifies the detection of errors and enhances the communication of program results.

The supplementary printed matter can range from a single sheet to a comprehensive user's manual. Figure 5.7 is an example of a simple documentation form that we recommend you prepare for every program you develop. These forms can be

PROGRAM: The Simple Addition Program (Version 2.0)

PROGRAMMER: S. C. Chapra
               Civil, Environmental, and Architectural
               Engineering
               The University of Colorado
               Boulder, CO 80309-0428

DESCRIPTION: This program adds two numbers and
               prints out the results.

SPECIAL FEATURES: The program uses descriptive
               variable names and enhances I/O.

REFERENCE: Chapra, S. C., and Canale, R. P. (1992).
               Introduction to Computing for Engineers,
               2d ed. (McGraw-Hill: New York), Chap. 5.

LISTING:

                        (insert listing here)

RUN:

                        (insert run here)

FIGURE 5.7
A simple one-page format for
program documentation. This
documentation can be stored
in a binder along with a
printout of the program.

maintained in a notebook to provide a quick reference for your program library. Note that for well-documented programs, the title information in Fig. 5.7 is unnecessary because the title section could be included directly in the program.

The user's manual of your own computer is an example of comprehensive documentation. This manual tells you how to run your computer system and disk operating programs. In addition, manuals for high-level languages and software packages also represent extensive documentation efforts. If you ever have to compose a major documentation, these manuals provide goods models of how large software packages can be effectively documented.

5.7 STORAGE AND MAINTENANCE

When the computer is shut off, the contents of its primary memory (RAM) are destroyed. In order to retain a program for later use, you must transfer it to a secondary storage device such as a hard disk or diskette before turning the machine off.

Aside from this physical act of retaining the program on a disk, storage and maintenance consist of two major tasks: (1) upgrading the program in light of experience and (2) ensuring that the program is safely stored. The former is a matter of communicating with users to gather feedback regarding suggested improvements to your design.

The question of safe storage of programs is especially critical when using diskettes. Although these devices offer a handy means for retaining your work and transferring it to others, they can be easily damaged. Table 5.3 outlines a number of tips for effectively maintaining your programs on diskettes. Follow the suggestions on this list carefully in order to avoid losing your valuable programs through carelessness.

TABLE 5.3 Suggestions and Precautions for Effective Software Maintenance

- Maintain paper printouts of all finished programs in a file or binder.
- Always maintain at least one backup copy of all diskettes.
- Handle all diskettes gently. Do not bend $5\frac{1}{4}$-in diskettes or touch their exposed surfaces.
- Affix a temporary label describing the contents to all diskettes. When working with $5\frac{1}{4}$-in disks, use only soft felt-tipped pens to write on the label once it is affixed.
- Once a diskette is complete, make it write-protected by covering (for $5\frac{1}{4}$-in diskettes) or moving (for $3\frac{1}{2}$-in diskettes) its write-protect switch.
- Keep $5\frac{1}{4}$-in diskettes in their protective envelope when not in use.
- Store diskettes upright in a diskette box or carrier. Make sure they never get too hot or too cold.
- Keep diskettes away from water and contaminants such as smoke and dust.
- Store diskettes away from magnetic devices such as televisions, tape recorders, and electric motors.
- Remove diskettes from the drive before turning the system off.
- Never remove a diskette when the red light on the drive is lit.
- Be careful not to bend the metal shutter on $3\frac{1}{2}$-in diskettes. It can get caught in the diskette drive and make it impossible to remove without damaging it or obtaining a technician's help.

PROBLEMS

5.1 Describe some of the advantages of BASIC.

5.2 Learn how to turn on your computer, format a diskette, and access BASIC on your system. Enter and run the simple addition program from Fig. 5.5.

5.3 The arithmetic operator for BASIC multiplication is the asterisk. Therefore a simple multiplication program can be developed by reformulating the assignment statement as

```
product = first * second
```

Develop a version of this program patterned after Fig. 5.5. Save the program on your diskette, and generate a paper printout or hard copy of the output. Document this program using the format specified in Fig. 5.7. Employ values of `first` = 20 and `second` = 16 for your example run.

5.4 Develop, test, and document a program to input the length, width, and height of a box. Have the program compute the volume and the surface area of the box. Also, allow the user to input the units (for example, feet) for the dimensions of the box as string constants, and display the correct units with the answers. Document this program using the format specified in Fig. 5.7. Employ values of length, width, and height equal to 5, 1.5, and 1.25 m, respectively, for your example run.

5.5 Develop, test, and document a program to input the external dimensions of a box along with the thickness of its walls. Have the program compute the internal volume, and the internal and external surface areas of the box. Also, allow the user to input the units (for example, feet) for the dimensions of the box as string constants, and display the correct units with the answers. Document this program using the format specified in Fig. 5.7. Employ values of length, width, height, and thickness equal to 5, 1.5, 1.25, and 0.1 m, respectively, for your example run.

5.6 Write a BASIC program comparable to the one given in Fig. 5.5. The program should prompt the user to input a value for the radius of a circle and should respond with values for the diameter, circumference, and area, displayed in a neat format. Use the CONST statement to set the value for π.

5.7 Write a BASIC program comparable to the one given in Fig. 5.5. The program should prompt the user for input values for a, b, and c and should respond with calculated values of y where

```
y = ac + ba + bc
```

The arithmetic operator for BASIC multiplication is the asterisk.

5.8 Write a BASIC program comparable to the one given in Fig. 5.5. The program should prompt the user for input values for the velocity of a car and the duration of a trip and should respond with the calculated distance traveled.

5.9 Repeat Prob. 5.4, except calculate the volume and surface area of a closed cylinder (that is, one with a lid), given its radius and height. Use a radius of 1 m and a height of 10 m to test your program. The BASIC formulas to compute the area and volume of a closed cylinder are

```
A = 2 * pi * r * h + 2 * pi * r * r
```

and

```
V = pi * r * r * h
```

where h = height, r = radius, and $pi = \pi$.

5.10 Repeat Prob. 5.5, except calculate the internal volume and surface area of a cylinder, given its radius, height, and wall thickness. Use a radius of 1 m, a height of 10 m, and a thickness of 0.04 m to test your program. See Prob. 5.9 for formulas to compute surface area and volume.

5.11 Repeat Prob. 5.4, except calculate the volume and surface area of a sphere, given its radius. Use a radius of 4 m to test your program. The BASIC formulas to compute the area and volume of a sphere are

```
A = 4 * pi * r * r
```

and

```
V = 4 * pi * r * r * r/3
```

where r = radius, and $pi = \pi$.

5.12 Repeat Prob. 5.5, except calculate the internal volume and surface area of a sphere, given its radius, and wall thickness. Use a radius of 4 m and a thickness of 0.04 m to test your program. See Prob. 5.11 for formulas to compute surface area and volume.

COMPUTATIONS

One of the primary engineering applications of computers is to manipulate numbers. In the previous chapter, you were introduced to some fundamental concepts that lie at the heart of this endeavor: constants, variables, and assignment. Now we will expand on these concepts and introduce some additional material that will allow you to exploit the full potential of BASIC to perform computations.

6.1 ARITHMETIC EXPRESSIONS

Obviously, in the course of engineering problem solving, you will be required to perform more complicated mathematical operations than the simple process of addition introduced in Chap. 5. One key to understanding how these operations are accomplished is to realize that a primary distinction between computer and conventional equations is that the former are constrained to a single line. In other words, a multilevel, or "built-up," formula such as

$$x = \frac{a + b}{y - z^2} + \frac{33.2}{w^3}$$

must be reexpressed so that it can be written as a single-line expression in BASIC.

The operators that are used to accomplish this are listed in Table 6.1. Notice how the two operations that are intrinsically multilevel in written algebraic expressions—that is, division and exponentiation—can be expressed in single-line form using these operators. For example, A divided by B, which in algebra is usually written

$$\frac{A}{B}$$

is expressed in BASIC as A/B. Similarly, the cube of A, which in algebra demands that

TABLE 6.1   Parentheses, Arithmetic Operators, and Their Associated Priorities

Symbol	Meaning	Priority
( )	Parentheses	1
^	Exponentiation	2
−	Negation	3
*	Multiplication	4
/	Division	4
+	Addition	5
−	Subtraction	5

the 3 be raised as a superscript, as in A^3, is expressed in BASIC as A ^ 3. Note that the power to which the number is raised can be a positive or negative integer or decimal. However, a negative number can be raised only to an integer power. In general, raising a negative number to a fractional power is not feasible.

6.1.1 Rules for Arithmetic Operators

Although the ^ and / operators allow us to write algebraic expressions on a single line, they are not sufficient to give the computer all the information that it needs to understand complicated arithmetic expressions. This is due to the fact that a computer can perform only one operation at a time.

To illustrate the difficulties that can occur because of this fact, suppose that you wanted to program the following formula:

$$x = \frac{a + b}{c + d}$$

Using the operators in Table 6.1, we might formulate the following BASIC expression:

    x = a + b / c + d

Now the computer must decide which operation to perform first. Suppose that we decided to make a rule that the computer would evaluate the operations in a left-to-right order. Thus, it would first add a to b. Next it would divide the result by c. Finally, it would add the d and store the final result in x. Unfortunately, this sequence would actually be equivalent to

$$x = \frac{a + b}{c} + d$$

which is quite different from what we intended.

We therefore need a few more rules. First, the computer allows the use of parentheses to group operations and to specify which calculations are to be executed first. In other words, parentheses are given top priority in the sense that the calculations within parentheses are implemented first. For example, the following representation would allow our example to be computed properly:

x = (a + b) / (c + d)

Thus, the additions are performed first and then the results are divided. Note that if several sets are nested within each other, the operations in the innermost parentheses are performed first.

Parentheses must always come in pairs. A good practice when writing expressions is to check that there is always an equivalent number of left and right parentheses. For example, the expression

((X + 2 ∗ (3 – B – (4 ∗ A / B) – 7))

is incorrect because there are too many left parentheses. Checking the number of left and right parentheses does not ensure that the expression is correct, but if the number of parentheses does not match, the expression is definitely erroneous.

Note that parentheses cannot be used to imply multiplication as is done in conventional arithmetic. The asterisk must always be included to indicate multiplication. For example, the algebraic multiplication

$$x = c(a + b)$$

could not be expressed as x = c (a + b) in BASIC. Rather, an asterisk would be needed as in x = c ∗ (a + b).

Although the parentheses are helpful, additional rules are needed. This is due to the fact that if they were the only criterion for establishing priority, parentheses would clutter up most computer formulas. Because this would make expressions extremely difficult to read and interpret, an additional hierarchy of priorities is needed. Such a scheme is embodied in Table 6.1.

After operations within parentheses, the next highest priority goes to exponentiation. If multiple exponentiations are required, as in x ^ y ^ z, they are implemented from left to right.[1] This is followed by negation—that is, the placement of a minus sign before an expression to connote that it is negative. (Negation is formally referred to as the *unary minus operation*.)

Next, all multiplications and divisions are performed from left to right. Finally, the additions and subtractions are performed from left to right. (Subtraction is formally referred to as the *binary minus operation*.) The use of these priorities to construct mathematical expressions is best shown by example.

---

[1] Not all formulas operate in this fashion. For example, in FORTRAN exponentiation is implemented from right to left.

EXAMPLE 6.1   Arithmetic Expressions

PROBLEM STATEMENT: Determine the proper values of the following arithmetic expressions in BASIC:

(a) $5 + 6 * 7 + 8$

(b) $(5 + 6) * 7 + 8$ (f) $5 \wedge 2 \wedge 3$

(c) $(5 + 6) * (7 + 8)$ (g) $5 \wedge (2 \wedge 3)$

(d) $-2 \wedge 4$ (h) $8/2 * 4 - 3$

(e) $(-2) \wedge 4$ (i) $(10/(2 + 3) * 5 \wedge 2) \wedge 0.5$

SOLUTION:
(a) Because multiplication has a higher priority than addition, it is performed first to yield

$$5 + 42 + 8$$

Then the additions can be performed from left to right to yield the correct answer, 55.
(b) Although multiplication has a higher priority than addition, the parentheses have even higher priority, and so the parenthetical addition is performed first:

$$11 * 7 + 8$$

followed by multiplication

$$77 + 8$$

and finally addition to give the correct answer, 85.
(c) First evaluate the additions within the parentheses,

$$11 * 15$$

and then multiply to give the correct answer, 165.
(d) Because exponentiation has a higher priority than negation, the 2 is first raised to the power 4 to give 16. Then negation is applied to give the correct answer, -16.
(e) In contrast to the expression in (d), the parentheses here force us to first transform 2 into -2. Then the -2 is raised to the power 4 to give 16.
(f) Because the order of exponentiation is from left to right, 5 is first squared to give 25, which in turn is raised to the power 3 to yield the correct answer, 15625.
(g) In contrast to the expression in (f), the normal order of exponentiation is overruled, and 2 is first cubed to give $5 \wedge 8$, which in turn is evaluated to yield the correct answer, 390625.

(*h*) Because multiplication and division have equal priority, they must be performed from left to right. Therefore, 8 is first divided by 2 to give

$$4 * 4 - 3$$

Then 4 is multiplied by 4 to yield

$$16 - 3$$

which can be evaluated for the correct answer, 13. Note that if the left-to-right rule had been disobeyed and the multiplication performed first, an erroneous result of $8/8 - 3 = -2$ would have been obtained.

(*i*) Because the parentheses in this expression are nested (that is, one pair is within the other), the innermost is evaluated first:

$$(10/5 * 5 \text{ ^ } 2) \text{ ^ } 0.5$$

Next the expression within the parentheses must be evaluated. Because it has the highest priority, the exponentiation is performed first:

$$(10/5 * 25) \text{ ^ } 0.5$$

Then the division and multiplication are performed from left to right to yield

$$(2 * 25) \text{ ^ } 0.5$$

and

$$50 \text{ ^ } 0.5$$

which can be evaluated to give the correct answer, 7.071.

When dealing with very complicated formulas, it is often good practice to break them down into component parts. Then the components can be combined to come up with the end result. This practice is especially helpful in locating and correcting typing errors that frequently occur when entering a long formula.

EXAMPLE 6.2   Complicated Arithmetic Expressions

PROBLEM STATEMENT: Write the comparable BASIC version for the algebraic expression

$$y = \frac{[(x^2 + 2x + 3)/\sqrt{5 + z}] - \sqrt{(15 - 77x^3)/(z^{3/2})}}{x^2 - 4xz - 5x^{-4/5}}$$

Treat each major term of the formula as a separate component in order to facilitate programming the expression in BASIC.

SOLUTION: The denominator and the two terms in the numerator can be programmed separately as

```
denom = x ^ 2 - 4 * x * z - 5 * x ^ (-4/5)
part1 = (x ^ 2 + 2 * x + 3)/(5 + z) ^ .5
part2 = ((15 - 77 * x ^ 3)/z ^ (3/2)) ^ .5
```

Then the total assignment statement could be formed from the component parts, as in

```
y = (part1 - part2)/denom
```

Whereas the priority rules from Table 6.1 allow you to employ less parentheses, there are cases where they can be used to enhance the clarity of a formula. For example, the following expression

$$\frac{a}{bc}$$

can be represented accurately as

```
a/b/c
```

Because of the left-to-right rule, this relationship will function properly. However, an alternative version can be developed as

```
a/(b * c)
```

It can be argued that the use of parentheses here represents an improvement because this version more faithfully follows the structure of the original equation.

While parentheses are useful for increasing the clarity of expressions, their indiscriminate use can be counterproductive. For example, the `part2` expression from Example 6.2 could have been represented as

```
part2 = ((15 - (77 * (x ^ 3)))/(z ^ (3/2))) ^ (.5)
```

Although this version will execute correctly, it is questionable whether the extra parentheses have enhanced the presentation.

Clearly, there is some judgment involved in formulating expressions. The bad news is that it is one of the many instances in engineering involving ambiguities, where no single clear-cut answer is available. The good news is that such situations allow you the freedom to express yourself and develop your own creativity and "style." In fact, it is usually these instances that will allow you to excel and distinguish yourself from your peers.

On the basis of our own programming experiences, we generally try to follow the rule of thumb: "When building expressions, always make them as simple as possible but not so simple that they are difficult to understand." Additionally, we always try to place ourselves in the position of our user or customer. In other words, we try to write formulas that *we* would be able to grasp easily if we came upon them at a later time. We have found that following these guidelines helps us to develop expressions that are generally clean, lean, and lucid.

6.1.2   Exponentiation in BASIC

Earlier in this chapter, we mentioned that negative numbers cannot be raised to noninteger powers. The reason for this lies in the manner in which BASIC performs the exponentiation operation. It is also related to how you would possibly compute a value for an expression like $2^{3.2}$ by hand or with a calculator.

For cases with an integer power, BASIC merely performs the number of multiplications specified by the exponent. Therefore, the cube of a number is evaluated as

```
2 ^ 3 = 2 * 2 * 2 = 8
```

Similarly, the cube of a negative number would be evaluated as

```
(-2) ^ 3 = (-2) * (-2) * (-2) = -8
```

This obviously poses no difficulty.

However, when a noninteger power is used, a logarithm is employed to make the evaluation, as in the following sequence of operations[2]:

$$x = 2^{3.2}$$

$$\log_{10} x = \log_{10} 2^{3.2}$$

$$\log_{10} x = 3.2 \log_{10} 2$$

$$x = 10^{3.2 \log_{10} 2} = 9.189588$$

Whereas this poses no problem when a positive number is raised to a power, it clearly creates an error when a negative is employed. This is due to the fact that it is impossible to take the logarithm of a negative number.

It should be noted that if you force an integer exponent to be treated as a real number by using a suffix, BASIC is "smart" enough to recognize that the exponent is a whole number. Therefore, the following operation will be implemented successfully as shown:

---

[2]QuickBASIC does not actually employ the base 10 or common logarithm to make this evaluation. However, because it is the most familiar of the logarithms, we used it for this example.

$$(-2)^\wedge 3! = (-2) * (-2) * (-2) = -8$$

It should be noted that this is not the case for all computer languages (for example, standard FORTRAN 77 would evaluate such an expression with the logarithmic method unless it had a "smart" optimizing compiler).

6.2 INTRINSIC, OR BUILT-IN, FUNCTIONS

In engineering practice many mathematical functions appear repeatedly. Rather than requiring the user to develop special programs for each of these functions, manufacturers include them directly as part of the language. These *built-in* (or *intrinsic* or *library*) *functions* are of the general form

  name (*argument*)

where the *name* designates the particular intrinsic function (see Table 6.2) and the *argument* is the argument that is passed to the function. An argument can be a constant, variable, another function, or an algebraic expression. (Note that there are also certain functions with more than one argument.)

TABLE 6.2   Commonly Available Built-in Functions in BASIC*

Function	Description
SQR(X)	Calculates the square root of X where X must be positive
ABS(X)	Determines the absolute value of X
LOG(X)	Calculates the natural logarithm of X where X must be positive
EXP(X)	Calculates the exponential of X
SIN(X)	Calculates the sine of X where X is in radians
COS(X)	Calculates the cosine of X where X is in radians
TAN(X)	Calculates the tangent of X where X is in radians
ATN(X)	Calculates the arctan of X where X is dimensionless, and returns a value in radians
INT(X)	Determines the greatest integer value that is less than or equal to X
FIX(X)	Determines the truncated value of X, ignoring the decimal part
SGN(X)	Determines a value dependent on the sign of X (If X is positive, +1 is returned; if X is zero, 0 is returned; if X is negative, −1 is returned.)
CINT(X)	Determines the rounded value of X
RND	Generates a random number

*The argument of the function X can be a constant, variable, another function, or an algebraic expression.

6.2.1   General Mathematical Functions

The first group of functions deals with some standard mathematical operations that have frequent application in engineering.

The square root function (SQR).   To this point we have learned one way to determine the square root by raising a number to the 1/2 power, as in

```
x ^ .5
```

Because this is such a common operation, the built-in function named SQR is included in BASIC to perform the same task, as in

```
SQR(x)
```

The absolute value function (ABS).   There are many situations where you will be interested in the value of an expression or variable but not its sign. The ABS function

```
ABS(x)
```

provides the value of x without any sign. If x is greater than or equal to 0, ABS(x) will equal x. If it is less than 0, ABS(x) will equal x times -1.

EXAMPLE 6.3   Using SQR and ABS Functions to Determine the Complex Roots of a Quadratic Equation

PROBLEM STATEMENT: Given the quadratic equation

$$f(x) = ax^2 + bx + c$$

the roots may be determined by the formula

$$x = \frac{-b \pm \sqrt{b^2 - 4ac}}{2a}$$

If $b^2 - 4ac \geq 0$, this equation can be used directly. However, if $b^2 - 4ac < 0$, then the roots are complex and a revised solution is

$$x = \text{Re} \pm i \, \text{Im}$$

where Re is the real part of the root, which can be calculated as

$$\text{Re} = \frac{-b}{2a}$$

Im is the imaginary part, which is calculated as

$$Im = \frac{\sqrt{|b^2 - 4ac|}}{2a}$$

and i represents the square root of -1. Use the ABS and SQR functions to determine the complex roots for the case where $b^2 - 4ac < 0$.

SOLUTION: The code to determine the complex roots of a quadratic equation can be written as

```
INPUT a, b, c
real = -b/(2 * a)
imaginary = SQR(ABS(b ^ 2 - 4 * a * c))/(2 * a)
PRINT real, imaginary
END
```

Notice how the ABS function is located within the argument of the SQR function. Check the nesting of the parentheses yourself.

The natural logarithm function (LOG). Logarithms have numerous applications in engineering. Recall that the general form of the logarithm is

$$y = \log_b x$$

where b is called the *base* of the logarithm. According to this equation, y is the power to which b must be raised to give x

$$x = b^y$$

The most familiar and easy-to-understand form is the *base 10*, or *common, logarithm,* which corresponds to the power to which 10 must be raised to give x; for example,

$$\log_{10} 10,000 = 4$$

because 10^4 is equivalent to 10,000.

Common logarithms define powers of 10. For example, common logs from 1 to 2 correspond to values from 10 to 100. Common logs from 2 to 3 correspond to values from 100 to 1000, and so on. Each of these intervals is called an *order of magnitude.*

Besides the base 10 logarithm, the other important form in engineering is the *base e logarithm,* where $e = 2.71828....$ The base e logarithm, which is also called the *natural, or napierian, logarithm,* is extremely important in calculus and other areas of engineering mathematics. In fact, in engineering and most areas of science, the natural logarithm is actually more prevalent than the common logarithm. For this reason, **BASIC** includes the *LOG* function to compute the natural logarithm, as in

```
LOG(x)
```

Unfortunately, BASIC does not include a similar function to compute the common logarithm. However, it is simple to convert the base e to the base 10 version using the following formula:

$$\log_b x = \frac{\ln x}{\ln b}$$

where, as you have already learned in calculus, $\ln x$ is the standard nomenclature for $\log_e x$. Thus, the base e logarithm can be converted to base 10 by

$$\log_{10} x = \frac{\ln x}{\ln 10} = \frac{\ln x}{2.3025}$$

or in BASIC

```
log10x = LOG(x)/LOG(10)
```

The exponential function (EXP). An additional mathematical function that has widespread engineering applications is the exponential function, or e^x, where e is the base of the natural logarithm. Just as 10^x is the inverse of $\log_{10} x$, so also e^x is the inverse of $\ln x$. Therefore

$$x = e^{\ln x}$$

The exponential formula is often used to characterize processes that grow or decay. For example, the radioactive decay of carbon 14(^{14}C) can be computed by the formula

$$A = A_0 e^{-0.00012t}$$

where A is the amount of carbon 14 present after t years, and A_0 is the initial quantity of carbon 14. In BASIC, the EXP function would be employed to compute this formula as

```
A = A0 * EXP(-0.00012 * t)
```

Aside from its direct use, the exponential function provides a means to compute other values, such as the hyperbolic trigonometric functions as described in the next section.

6.2.2 Trigonometric Functions

Other mathematical functions that have great relevance to engineering are the trigonometric functions. In BASIC the sine (SIN), cosine (COS), tangent (TAN), and

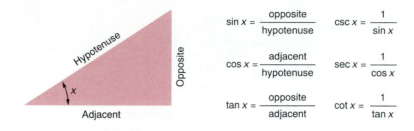

$$\sin x = \frac{\text{opposite}}{\text{hypotenuse}} \qquad \csc x = \frac{1}{\sin x}$$

$$\cos x = \frac{\text{adjacent}}{\text{hypotenuse}} \qquad \sec x = \frac{1}{\cos x}$$

$$\tan x = \frac{\text{opposite}}{\text{adjacent}} \qquad \cot x = \frac{1}{\tan x}$$

FIGURE 6.1 Definitions of the trigonometric functions.

inverse or arctangent (ATN) functions are commonly available. As outlined in Fig. 6.1, other trigonometric functions can be derived by reciprocals of the first three.

For all these functions, the angle must be expressed in radians (Fig. 6.2). A *radian* is the measure of an angle at the center of a circle that corresponds to an arc length that is equal to the circle's radius (Fig. 6.2b). The conversion from degrees to radians is (Fig. 6.2c)

$$\pi \text{ radians} = 180°$$

Therefore, if x is expressed originally in degrees, the sine of x can be computed in BASIC as

```
SIN(x * 3.14159/180)
```

FIGURE 6.2
Relationship of radians to degrees: (a) The angle *x* of the right triangle placed in a standard position at the center of a circle; (b) the definition of a radian; (c) π radians = 180°.

Aside from the functions listed in Table 6.2 and shown in Fig. 6.1, other trigonometric functions can be determined with intrinsic functions. For example, hyperbolic trigonometric functions can be computed with the EXP function by employing the following relationships:

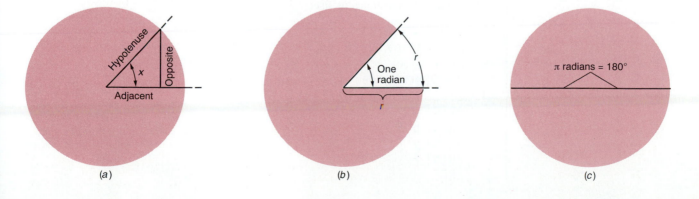

(a)   (b)   (c)

$$\sinh x = \tfrac{1}{2}(e^x - e^{-x})$$

$$\cosh x = \tfrac{1}{2}(e^x + e^{-x})$$

$$\tanh x = \frac{\sinh x}{\cosh x}$$

A group of additional trigonometric functions that can be evaluated on the basis of intrinsic functions is listed in Appendix A.

EXAMPLE 6.4

How to Determine π Using the ATN Function

PROBLEM STATEMENT: π is one of the most commonly used constants employed in engineering-oriented programs. Although we all probably remember that it is approximately equal to 3.14, it is more difficult to remember that it is equal to 3.141593 in the 7 significant digits used for single-precision real variables in the IEEE standard. Furthermore, it would be impractical for most of us to try to remember that in double precision, it is 3.141592653589793. Develop an assignment statement so that your computer determines π internally.

SOLUTION: Recall that the tangent of 45° is equal to 1. Using radians (45° $= \pi/4$), this identity can be expressed as

$$\tan\left(\frac{\pi}{4}\right) = 1$$

By taking the arctangent of this function we get

$$\frac{\pi}{4} = \tan^{-1}(1)$$

Multiplying both sides by 4 yields

$$\pi = 4\tan^{-1}(1)$$

This relationship can now be added to any computer program to compute π. The following code demonstrates its use:

```
pi = 4 * ATN(1)
pi# = 4# * ATN(1#)
angle = 30
y = SIN(angle)
z = SIN(angle * pi/180)
PRINT pi
PRINT pi#
PRINT y
PRINT z
END
```

If this program is run, the following results would be printed:

```
3.141593
3.141592653589793
-.9880316
.5
```

Observe how the program computes the sine of 30 radians (or 1719°) as -0.9880316. In contrast, when the radian correction is employed, the anticipated result for $\sin 30°$ is obtained: 0.5.

6.2.3 Miscellaneous Functions: INT, FIX, SGN, CINT, and RND

Aside from general mathematical and trigonometric functions, there are other built-in functions that are available and have utility to engineers.

Functions to determine integer values (INT and FIX). The INT gives the greatest integer (that is, whole number) that is less than or equal to the argument. Because of this definition, it is sometimes referred to as the *greatest integer function*. For example,

```
x = 3.9
y = INT(x)
```

will give y equal to 3. Also, it is important to realize that

```
x = -2.7
y = INT(x)
```

will give y equal to -3, because it is the greatest integer that is less than -2.7.
The FIX function discards the decimal part of a number and returns the integer part. For positive numbers, this gives the same result as does the INT function. However, for negative numbers, it is decidedly different. For example,

```
x = -2.7
y = INT(x)
```

will give y equal to -2, because the FIX simply truncates, or "chops off," the 0.7 to attain its result.

EXAMPLE 6.5   *Using the INT Function to Truncate and Round Positive Numbers*

PROBLEM STATEMENT: It is often necessary to round numbers to a specified number of decimal places. One very practical situation occurs when performing calculations pertaining to money. Whereas an economic formula might yield a result of $467.9587, in reality we know that such a result must be rounded to the nearest cent.

According to rounding rules, because the first discarded digit is greater than 5 (that is, it is 8), the last retained digit would be rounded up, giving a result of $467.96. The INT function provides a way to accomplish this same manipulation with the computer.

SOLUTION: Before developing the rounding formula, we will first derive a formula to truncate, or "chop off," decimal places. First, we can multiply the value by 10 raised to the number of decimal places we desire to retain. For example, in the present situation we desire to retain two decimal places; therefore, we multiply the value by 10^2, or 100, as in

$$467.9587 \times 100 = 46795.87 \tag{E6.5.1}$$

Next we can apply the `INT` function to this result to chop off the decimal portion:

```
INT(46795.87) = 46795
```

Finally we can divide this result by 100 to give

$$46795/100 = 467.95$$

Thus we have retained two decimal places and truncated the rest. The entire process can be represented by the BASIC formula

```
INT(x * 10 ^ D)/10 ^ D
```
$\tag{E6.5.2}$

where x is the number to be truncated and D is the number of decimal places to be retained.

Now it is a very simple matter to modify this relationship slightly so that it will round. This can be done by looking more closely at the effect of the multiplication in Eq. (E6.5.2) on the last retained and the first discarded digit in the number being rounded (Fig. 6.3). Multiplying the number by 10^D essentially places the last retained digit in the unit position and the first discarded digit in the first decimal place. For example, as shown in Eq. (E6.5.1), the multiplication moves the decimal point between the last retained digit (5) and the first discarded digit (8). Now it is straightforward to see that if the first discarded digit is above 5, adding 0.5 to the quantity will increase the last retained digit by 1. For example, for Eq. (E6.5.1)

FIGURE 6.3
Illustration of the retained and discarded digits for a number rounded to two decimal places.

$$46795.87 + 0.5 = 46796.37 \qquad \text{(E6.5.3)}$$

If the first discarded digit were less than 5, adding 0.5 would have no effect on the last retained digit. For instance, suppose the number were 46795.49. Adding 0.5 yields 46795.99, and therefore the last retained digit (5) is unchanged.

To complete the process, the INT function is applied to Eq. (E6.5.3):

$$\texttt{INT(46796.37)} = 46796$$

The result can be divided by 100 to give the correct answer, 467.96. The entire process is represented by

$$\texttt{INT(x * 10 ^ D + .5)/10 ^ D} \qquad \text{(E6.5.4)}$$

It should be noted that numbers can be rounded to an integer value by setting $D = 0$, as in

$$\texttt{INT(467.9587 * 10 ^ 0 + .5)/10 ^ 0} = 468$$

or to the tens, hundreds, or higher places by setting D to negative numbers. For example, to round to the 10's place, $D = -1$ and Eq. (E6.5.4) becomes

$$\texttt{INT(467.9587 * 10 ^ (-1)+.5)/10 ^ (-1)} = 470$$

The sign function (SGN). There are occasions where you will need to know the sign of a quantity without regard for its magnitude. The sign function fills this need. If the quantity being evaluated is positive, the SGN function will be $+1$. If it is zero, SGN will be 0. If it is negative, SGN will be -1. For example,

```
x = -3.14159
y = SGN(x)
```

will yield a value of -1 for y.

EXAMPLE 6.6

Using the FIX and SGN Functions to Truncate and Round Both Positive and Negative Numbers

PROBLEM STATEMENT: Notice that we did not truncate or round negative numbers in Example 6.5. Develop relationships for truncating and rounding that hold for both positive and negative numbers.

SOLUTION: Because of the way that the INT function operates on negative values, Eq. (E6.5.2) is inadequate for truncating negative numbers. For example, -467.9587 truncated to two decimal places would be [according to Eq. (E6.5.2)]

$$x = INT(-467.9587 * 10 \wedge 2)/10 \wedge 2 = -467.96$$

which is erroneous. A more general truncation formula that works for either positive or negative values can be developed by substituting the FIX for the INT function in Eq. (E6.5.2) to give

$$FIX(x * 10 \wedge D)/10 \wedge D \tag{E6.6.1}$$

For example, using the same values as before,

$$FIX(-467.9587 * 10 \wedge 2)/10 \wedge 2 = -467.95$$

which is correct. Because FIX and INT yield identical results for positive values, Eq. (E6.6.1) is valid for truncating both positive and negative numbers.

For rounding, we might try to use the FIX function in place of the INT in Eq. (E6.5.4):

$$FIX(x * 10 \wedge D + .5)/10 \wedge D$$

However, this does not work. For example, -467.9587 is rounded to two decimal places, as in

$$FIX(-467.9587 * 10 \wedge 2 + .5)/10 \wedge 2 = -467.95$$

which is incorrect. A slight adjustment of the formula using the SGN function can be made to correct this problem:

$$FIX(x * 10 \wedge D + SGN(x) * .5)/10 \wedge D \tag{E6.6.2}$$

For the example,

$$FIX(-467.9587 * 10 \wedge 2 + SGN(-467.9587) * .5)/10 \wedge 2$$

which when evaluated yields the correct result of -467.96.

It should be noted that Eq. (E6.5.4) also yields the correct result, as in

$$INT(-467.9587 * 10 \wedge 2 + .5)/10 \wedge 2 = -467.96$$

Whereas the INT function proved inadequate for the truncation of negative numbers, it performs correctly for rounding. This is directly attributable to the fact that it returns the greatest integer that is less than or equal to the argument.

The previous examples have been expressly designed to illustrate the workings of the INT, FIX, and SGN functions. In fact, as described next, there is an intrinsic function ready made for rounding.

The rounding function (CINT). The CINT function returns a rounded integer. This function can be employed to round either positive of negative numbers. For example, the QuickBASIC statement

```
PRINT CINT(4.3), CINT(4.8), CINT(-4.3), CINT(-4.8)
```

would yield the output

```
4      5      -4      -5
```

The random-number function (RND). The subject of statistics is of great significance in all areas of engineering. One important aspect of this subject is the random number. To understand what this means, suppose that you toss an honest coin over and over again and record whether it comes up heads (H) or tails (T). Your result for 20 tosses might look like

HTTHHTHHHHTHTTTHTHTT

This is called a *randomly generated sequence* because each letter represents the result of an independent toss of the coin. If we repeated the experiment, we would undoubtedly come up with a different sequence because, for an honest coin, the outcome of each toss is random. That is, there is an equal likelihood that each toss will yield either a head or a tail, and there is no way to predict the outcome of an individual toss beforehand.

The same sort of experiment could be performed with a die (that is, one of a pair of dice). For this case, every toss would result in either 1, 2, 3, 4, 5, or 6, and the sequence of 20 tosses might look like

53122341626444153566

Random-number sequences are very important in engineering because many of the systems that you will study have aspects that vary in a random fashion. Consequently, it is useful to know that most dialects of BASIC have a random-number function represented by

```
y = RND
```

Every time this function is implemented, a random decimal fraction somewhere between 0 and 1 is returned.

EXAMPLE 6.7   *Simulating a Random Coin Toss with the Computer*

PROBLEM STATEMENT: Use the RND function to simulate the random toss of a coin.

SOLUTION: The following program fragment can be used to simulate the random toss of a coin:

```
x = RND
y = INT(x + .5)
PRINT x, y
```

The first line generates a random number between 0 and 1. Then the second line rounds the number to an integer. If the random number (x) is less than 0.5, the rounded value will be 0 and we will consider the toss a tail. If it is greater than or equal to 0.5, the rounded value will be 1 and we will consider the toss a head. A run yields

```
.7055475        1
```

Therefore, a head has been tossed. To obtain additional values, the statements can be repeated. For example,

```
x = RND
y = INT(x + .5)
PRINT x, y
x = RND
y = INT(x + .5)
PRINT x, y
.
.
.
```

If these lines were repeated five times, the result would be

```
.7055475        1
.533424         1
.5795186        1
.2895625        0
.301948         0
```

Thus for this case, the first three tosses would be heads and the last two would be tails.

Every time the program at the end of Example 6.7 is run on a particular computer, the same series of random numbers will be generated. This is because the computer generates the random number on the basis of a starting value called a *seed*. Unless the seed is changed, a particular computer will generate the same sequence of random numbers every time it is run. The seed can be changed with a RANDOMIZE statement.

This statement can be written in two ways. One way is to include any integer number between $-32,768$ and $32,767$, as in

```
RANDOMIZE 26
```

This changes the seed to 26, and if this line were included at the beginning of the program at the end of Example 6.7, a different sequence of random numbers would result. A second way to change the seed is to merely write

```
RANDOMIZE
```

When the statement is executed, the computer will query you, as in

```
Random Number Seed (-32768 to 32767)?
```

whereupon you could enter 26 or any other integer number within the limits.

6.2.4   Other Library Functions

Aside from the standard library functions described in the previous sections, each computer has its own special functions. For instance, the following example illustrates a built-in function called TIMER. These and other specialized QuickBASIC functions are listed in the Appendix B.

EXAMPLE 6.8   *Developing a "Random" Random Number Generator*

PROBLEM STATEMENT: One problem with the RND function is that you have to invoke the RANDOMIZE command if you want to generate a different sequence of random numbers each time you run the program. Develop a scheme so that the computer does this automatically.

SOLUTION: One way to solve this problem relies on the intrinsic function TIMER. This is one of several functions that capitalizes on the fact that a personal computer has a built-in clock. The TIMER function returns the number of seconds that have elapsed since midnight. Consequently, any time you run your program, the TIMER will likely return a different value (unless you have the improbable habit of performing runs at exactly the same number of seconds after midnight). Therefore, the following statement would have a high probability of generating a different sequence of random numbers every time you ran a program:

```
RANDOMIZE TIMER
```

Note that the number of seconds elapsed since midnight can clearly exceed the 32,767 limit for the seed number. However, the RANDOMIZE function handles this situation without an error message.

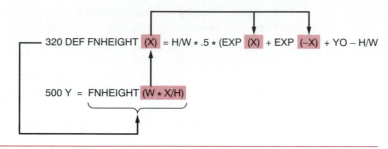

FIGURE 6.4
An example of a DEF
statement.

6.3   USER-DEFINED STATEMENT (ONE-LINE) FUNCTIONS

Aside from built-in functions, you might be interested in defining functions of your own. The DEF, short for *def*inition, statement function is used for this purpose and has the general form

DEF FN*variable (ArgList) = expression*

where the FN is followed by any legal *variable* name, the *ArgList* is the comma-delimited list of arguments that are passed to the function, and the *expression* is the actual arithmetic expression. For example, we can define a function to add two numbers as

DEF FNadd (x, y) = x + y

Then a line to perform the addition could be written as

total = FNadd (first, second)

For modern BASIC, several arguments may be specified for a statement function. Other variables in the statement function have the same value as they had in the main program at the point the user-defined function was invoked (Fig. 6.4).

A user-defined function must always be placed before the point in the program at which the function is to be used. A good stylistic practice is to place all statement functions together at the beginning of the program.

EXAMPLE 6.9   A User-Defined Statement Function

PROBLEM STATEMENT: The height of a hanging cable is given by the formula (Fig. 6.5)

$$y = \frac{H}{w} \cosh\left(\frac{w}{H}x\right) + y_0 - \frac{H}{w}$$

(E6.9.1)

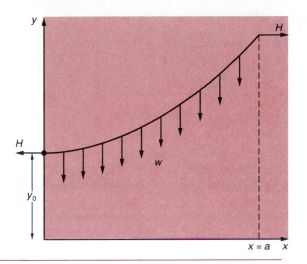

FIGURE 6.5
A section of a hanging cable.

where H is the horizontal force at $x = a$, y_0 is the height at $x = 0$, and w is the cable's weight per unit length. In Sec. 6.2.2, we noted that the hyperbolic cosine can be computed by

$$\cosh x = \tfrac{1}{2}(e^x + e^{-x}) \tag{E6.9.2}$$

Compose a computer program to calculate the height of a cable at equally spaced values of x from $x = 0$ to a maximum x, given w and H. Employ a user-defined statement function to compute the height. Test the program for the following case: $w = 10$, $H = 2000$, and $y_0 = 10$ for $x = 0$ to 40. Print out values of y at four equally spaced points from $x = 0$ to 40.

SOLUTION: Figure 6.6 shows a program to compute the height of a hanging cable. Notice how a statement has been set up to calculate the height as a user-defined function. Several other statements access this function to compute the cable height. Notice how the argument of `FNHeight` in these lines is an arithmetic expression, `w * x/H`, that is passed to the `DEF` statement. There its value is ascribed to the dummy variable `x`, which is used to evaluate the height. This result is then passed back and substituted for `FNHeight (w * x/H)` in the four lines where the function is invoked. The `x` is called a *dummy variable* because its use as the argument does not affect its value in the rest of the program. The value of `x` in the user-defined function is actually equal to the value of `w * x/H`.

Before moving on, we would like to explain a little more clearly how the argument list for the function operates. When a variable appears in a function's argument list, it performs two tasks. Using Fig. 6.4 as an example, it first points back to line 500

Listing:

```
'Hanging Cable Program

'by S.C. Chapra
'November 3, 1991

'This program computes the height of a
'hanging cable at four equally-spaced
'points along the x dimension
'-----------------------------------------

'Definition of variables

'w = weight per length
'H = horizontal force
'y0 = initial height
'xmax= maximum length
'xinc = increment of x where y is computed
'-----------------------------------------

'function to compute cable height
DEF FNHeight (x) = H / w * .5 * (EXP(x) + EXP(-x)) + y0 - H / w

'clear screen and identify program
CLS
PRINT "Program to compute height of hanging cable"
PRINT

'obtain numbers from user
INPUT "weight per length = "; w
INPUT "horizontal force = "; H
INPUT "initial height = "; y0
INPUT "maximum length = "; xmax

'perform computation and display results

PRINT : PRINT "Distance         Height": PRINT

PRINT x, y0                    'print initial values

xinc = xmax / 4
x = x + xinc
y = FNHeight(w * x / H)
PRINT x, y
x = x + xinc
y = FNHeight(w * x / H)
PRINT x, y
x = x + xinc
y = FNHeight(w * x / H)
PRINT x, y
x = x + xinc
y = FNHeight(w * x / H)
PRINT x, y

END
```

Run:

```
Program to compute height of hanging cable

weight per length = ? 10
horizontal force = ? 2000
initial height = ? 10
maximum length = ? 40

Distance          Height

0                 10
10                10.25005
20                11.00083
30                12.25422
40                14.01335
```

FIGURE 6.6
A program to determine the
height of a hanging cable.

in order to fetch the value stored in the memory location for the expression $W * X/H$. Second, it indicates how this value is employed on the right side of the function assignment operator. To reinforce the notion that the two X's represent different entities, you should recognize that line 320 could have just as easily been programmed as

```
DEF FNHEIGHT(Z) = H/W * .5 * (EXP(Z) + EXP(-Z)) + YO - H/W
```

We will explore this topic in more detail when we discuss procedures in Chap. 9.

Although Example 6.9 shows how a user-defined function is employed, it does not really illustrate all of its advantages. These advantages become significant when the function is long and is used numerous times throughout a program. In these instances, the concise representation provided by the user-defined function is a potent asset. In addition, when the function is extremely long, BASIC provides an even more powerful capability called a multiline user-defined function, or *FUNCTION procedure*. We will also discuss these in Chap. 9.

PROBLEMS

6.1 Which of the following are valid assignment statements in BASIC? State the probable errors for those that are incorrect.

(a) r1 = −4 ^ 1.5

(b) a + b = c + d

(c) x − y = 7

(d) c = c + 22

(e) 4 = x + y

6.2 If r = 3, s = 5, and t = −2, evaluate the following. Perform the evaluation step by step, following the priority order of the individual operations. Show all your work.

(a) −s * r + t

(b) s + r/s * t

(c) s + t ^ t * t ^ r

(d) r * (r + (r + 1) + 1

(e) t + 10 * s − s ^ t ^ r

6.3 Write BASIC statements for the following algebraic equations.

(a) $x = \dfrac{a}{c} + \dfrac{c - d^2}{g} \dfrac{r^5}{2f^2 - 4}$

(b) $r = \dfrac{cb^3}{r^4}(3d + 8x) - \tfrac{4}{9}x$

(c) $y = \dfrac{a - b}{x^2 \sin x}$

6.4 In civil engineering the vertical stress under the center of a loaded circular area is given by the following formula. Write this equation as a BASIC assignment statement.

$$s = q\left[1 - \frac{1}{[(r^2/z^2) + 1]^{3/2}}\right]$$

6.5 In engineering economics the annual payment, A, on a debt that increases linearly with time at a rate G for n years is given by the following equation. Write this equation as a BASIC assignment statement.

$$A = G\left[\frac{1}{n} - \frac{n}{(1 + i)^n - 1}\right]$$

6.6 In environmental engineering the following formula is employed to calculate the dissolved oxygen concentration in a river at a location x downstream from a sewage treatment plant effluent. Write this equation as a BASIC assignment statement.

$$c = \frac{k_1 D_0}{k_1 - k_2}\left(e^{-k_2(x/u)} - e^{-k_1(x/u)}\right)$$

6.7 In chemical engineering van der Waals equation gives pressure of a gas as

$$p = \frac{RT}{(v/n) - b} - a\left(\frac{n}{V}\right)^2$$

Write this equation as a BASIC assignment statement.

6.8 In electrical engineering the electric field at a distance x from the center of a ring-shaped conductor is given by

$$E = \frac{1}{4\pi\epsilon_0}\frac{Qx}{(x^2 + a^2)^{3/2}}$$

Write this equation as a BASIC assignment statement.

6.9 In fluid mechanics the pressure of a gas at the stagnation point is given by

$$p = p_0\left(1 + m^2\frac{k - 1}{2}\right)^{k/(k-1)}$$

Write this equation as a BASIC assignment statement.

6.10 The saturation value of dissolved oxygen in fresh water is calculated by the equation

$$c_{sf} = 14.652 - 0.41022T_c + 7.9910 \times 10^{-3}T_c^2 - 7.7774 \times 10^{-5}T_c^3$$

where T_c is temperature in degrees Celsius. Saltwater values can be approximated by multiplying the freshwater result by

$$c_{ss} = [1 - (9 \times 10^{-6})n]c_{sf}$$

where n is salinity in milligrams per liter. Finally, temperature in degrees Celsius is related to temperature in degrees Fahrenheit, T_F, by

$$T_c = \tfrac{5}{9}(T_F - 32)$$

Develop a BASIC computer program to calculate c_{ss}, given n and T_F. Test the program for $n = 18{,}000$ and $T_F = 70$. Note that both c_{sf} and c_{ss} are in milligrams per liter.

6.11 A projectile is fired with a velocity v_0, at an angle of θ above the horizon, off the top of a cliff of height y_0 (Fig. P6.11). The following formulas can be used to compute how far horizontally (x) the projectile will carry prior to impacting on the plane below.

$$(v_y)_0 = v_0 \sin \theta$$

$$(v_x)_0 = v_0 \cos \theta$$

$$t = \frac{-(v_y)_0 - \sqrt{(v_y)_0^2 + 2gy_0}}{g}$$

$$x = (v_x)_0 t$$

where $(v_y)_0$ and $(v_x)_0$ are the vertical and horizontal components of the initial velocity, t is the time of flight, and g is the gravitational constant (or -9.81 m/s^2, where the negative sign designates that we are assigning a negative value to downward acceleration). Prepare a program to compute x, given v_0, θ, and y_0. If $v_0 = 180$ m/s, $\theta = 30°$, and $y_0 = -150$ m, x will be equal to 3100 m. Use this result to test your program. Then employ the program to determine what value of θ results in the greatest value of x. Use trial and error to determine this optimal value of x.

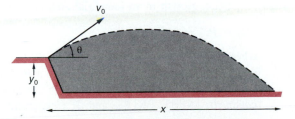

FIGURE P6.11

6.12 Develop a simple program that will produce a single random number over a given range. Employ the RANDOMIZE and the TIMER functions so that the program is likely to generate a different number in the range any time it is run. Test it by generating a random number between 3 and 9. Run the program 10 times to verify that it is operating properly.

6.13 Given X and Y, the coordinates of a point on the circumference of a circle centered at the origin, write a user-defined function that calculates the area.

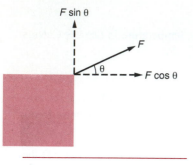

$F \sin \theta$

F

θ

$F \cos \theta$

FIGURE P6.14

6.14 Figure P6.14 shows a force F acting on a block at an angle θ. The horizontal and vertical components of the force are given by $F \cos \theta$ and $F \sin \theta$, respectively. Write a program similar to the one shown in Fig. 6.6 to compute the horizontal and vertical components of F for given values of F and θ. Use the following values for F and θ:

Force, lb	Angle, degrees
1000	30
50	65
1200	130

Your program should have at least one application of each of the following BASIC statements: *comment*, INPUT, PRINT, and DEF.

6.15 Write a computer program that requests numerical values to be rounded and the number of places to the right of the decimal point to be retained. Employ a DEF statement to compute the rounded number. Use the following data to test your program:

Number to Be Rounded	Number of Places to the Right of the Decimal Point
−3.14159265	3
86,400.761	1
0.6666666	4

6.16 Population growth in an industrial area is given by the following formula:

$$p(t) = P(1 - e^{-at})$$

where $p(t) =$ population at any time ≥ 0, $P =$ maximum population, $a =$ constant [yr], and $t =$ time [yr]. Write a BASIC computer program to compute the population at $t = 5$ years with $P = 100,000$ for $a = 0.05$, 0.2, and 0.8. Use *comment*, INPUT, PRINT, and DEF statements in your program.

6.17 The maximum velocity of a rocket $\bar{v}$ is given by

$$\bar{v} = a \ln \left(\frac{m_0}{m_0 - m_1} \right)$$

where $a =$ relative velocity of exhaust gases, $m_0 =$ total mass of rocket plus fuel at start of flight, and $m_1 =$ mass of fuel. Write a BASIC computer program to compute $\bar{v}$ for $m_1 = 25,000$, 30,000, and 35,000, given $m_0 = 40,000$ and $a = 25,000$. Use *comment*, INPUT, PRINT, and DEF statements in your program.

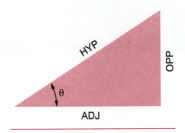

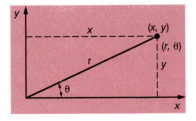

FIGURE P6.19
The relationship of cartesian
and polar coordinates.

6.18 A triangle is defined by sides ADJ, OPP, and HYP (Fig. P6.18). Write a BASIC computer program to calculate HYP and θ using two DEF statements. Employ the following data to test your program:

ADJ	OPP
10	10
5	0
2	6

6.19 Sometimes it is useful to make transformations between polar and cartesian coordinates. The following equations are employed for this purpose (Fig. P6.19):

$$x = r \cos \theta$$

and

$$y = r \sin \theta$$

Write a program that reads polar coordinates for a point and displays the corresponding cartesian coordinates. Employ user-defined functions to make the conversions.

6.20 Examples 6.5 and 6.6 were not intended as a definitive treatment of rounding. Rather, they were intended to illustrate the use of the INT, FIX, and SGN intrinsic functions. In fact, it should be noted that the rounding schemes suggested in Example 6.6 are not strictly correct. The difficulty relates to the special case where the first discarded digit is a 5 and all other discarded digits are 0's. For this situation, rounding up is just as incorrect as rounding down! Either choice is arbitrary. A better approach would be to adopt a rule that would give a 50 percent chance of either rounding up or down. One way to do this is to adopt the following rule:

• When the first discarded digit is exactly 5 and all other discarded digits are 0, round off so as to leave the last retained digit an even number.

If this rule is followed faithfully, errors due to rounding tend to compensate for one another.

Perform some numerical experiments to determine how both Eqs. (E6.5.4) and (E6.6.2) round numbers of this sort. For example, you might try rounding values such as 467.05, 467.15, 467.25, and so on, to one decimal place. Determine whether they round up or round down. Also investigate how the negatives of these numbers are rounded. Discuss your results.

CHAPTER 7

SELECTION

To this point we have written programs that perform instructions sequentially. That is, the program statements are executed line by line starting at the top of the program and moving down to the end. However, a strict sequence is only one of the ways in which a program can be executed. The flow of logic can branch (selection) or loop (repetition) depending on certain conditions. The present chapter deals with the statements required to branch, or to reroute the flow of logic to a line or statement other than the next in sequence.

Although selection can greatly enhance the power and flexibility of your programs, its indiscriminate and undisciplined application can introduce enormous complexity to a program. Because this greatly detracts from program clarity, computer scientists have developed a number of conventions to impose order on selection. Some of these actually involve special BASIC statements that provide very clear and coherent ways to perform selection. These are called *control-flow structures.*

Figure 7.1 shows the two principal control-flow structures that are used for selection. Unfortunately, such structures were not available in early versions of BASIC. However, because of their great utility, they have been incorporated into all modern dialects. We will stress the use of these structures in this chapter.

The coherence of a program also derives from the way in which the selection statements are applied; that is, their clarity involves the "style" with which the programmer chooses to construct the code. Nowhere is this more manifest than in the use of the GOTO statement.

7.1   THE GOTO STATEMENT

The GOTO is the simplest of all transfer statements. It allows you to directly override the sequence in which the program is executed. The general form is

GOTO *line-identifier*

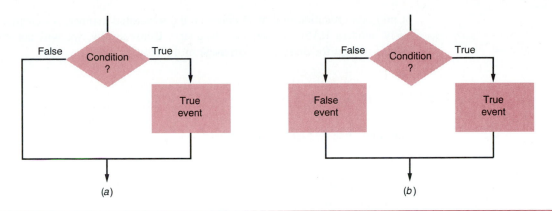

FIGURE 7.1
The two fundamental control constructs used for selection: the (a) single- and (b) double-alternative decision patterns. These are the fundamental building blocks for all types of selection. The combination of these two patterns allows any possible selection process to be implemented.

where *line-identifier* denotes the line that is to be executed next. Because early forms of BASIC employed line numbers, they were used for this purpose, as in the following program fragment:

```
100 GOTO 130
110 INPUT a,b
120 PRINT a,b
130 c = d + e
```

Under normal circumstances, the program would advance to line 110 after executing line 100. However, because of the GOTO statement, the program jumps to line 130.

Although line numbers are no longer required for modern BASIC, they can still be employed as line identifiers. In addition, the newer dialects also allow alphanumeric line labels. These may be any combination of from 1 to 40 letters or digits, starting with a letter and ending with a colon. Using line labels, our example can be rewritten as

```
GOTO sum
INPUT a,b
PRINT a,b
sum:c = d + e
```

As in this case, line labels are preferable for GOTO transfers because they are more descriptive than numbers.

By allowing transfer to any other line, the GOTO greatly increases the program's flexibility. However, a word of caution is in order. All the jumping around that results from the indiscriminate use of GOTO statements can lead to programs that are extremely difficult to understand, debug, and maintain. For this reason, structured programming considerations mandate that they be employed only when absolutely necessary.

Negative feelings about GOTO statements are so strong that in some programming languages (Pascal, for one) the GOTO construct is discouraged and is rarely seen. Languages that shun the GOTO statement usually have other constructs that control the flow of program execution in a more coherent and structured manner. As mentioned previously, modern BASIC is one such language. Consequently, we will not employ the GOTO (except for illustrative purposes) in the ensuing pages.

7.2   THE SINGLE-LINE IF/THEN STATEMENT

The single-line IF/THEN statement represents the simplest form of selection. Its general format is

IF *LogicalExpression* THEN *TrueStatement* ELSE *FalseStatement*

where *LogicalExpression* represents an expression that is either true or false, *TrueStatement* is a BASIC statement that is to be implemented if the expression is true, and *FalseStatement* is a BASIC statement that is to be implemented if the expression is false.

An example of the single-line IF/THEN statement is

IF number > 0 THEN PRINT number ELSE PRINT -number

For this example, if the value of the variable number is greater than zero, then the variable will be displayed. However, if its value is less than or equal to zero, its negative or a zero will be displayed. After the action is implemented, control will transfer to the next line. Note that this form of selection corresponds to Fig. 7.1b.

There are instances when you will merely want to implement an action if an expression is true (Fig. 7.1a). For these cases, the ELSE part of the statement can be omitted, as in

IF number > 0 THEN average = sum/number

For this example, if number is greater than zero, the statement average = sum/number will be executed and then control will transfer to the following line. If it is equal to or less than zero, the statement will be ignored and control will transfer immediately to the next line.

It should be noted that early forms of BASIC made widespread use of this statement to transfer control to lines other than the succeeding line. This type of "logical GOTO" can still be invoked, as in

IF number > 0 THEN GOTO 200 ELSE GOTO 1000

For this example, if number is greater than zero, the program would jump to line 200. Otherwise, it would transfer to line 1000. This type of statement was so commonly employed that the GOTO could be omitted, as in

IF number > 0 THEN 200 ELSE 1000

We must stress that the use of the single-line IF/THEN statement for this purpose is to be avoided. Along with the simple GOTO, it was one of the principal reasons why early BASIC programs were sometimes so difficult to understand.

7.3 FORMULATING LOGICAL EXPRESSIONS

The logical expression in an IF/THEN statement usually represents the relation between two values. Therefore a more detailed representation is

value$_1$ *relation value*$_2$

where the *values* can be variables, constants, or arithmetic expressions. The *relation* is any one of the relational operators in Table 7.1. These include equality, inequality, and testing whether one value is greater or less than another.

The distinction between the equal sign as a relational operator versus its use in assignment now becomes important. The `var = 0` portion of the first example in Table 7.1 does not cause a value of zero to be assigned to `var`. Rather, it merely tests whether `var` is currently equal to zero. In contrast, the statement `var = 99` following the `THEN` causes `var` to take on the value of 99 if it has previously been equal to zero. If `var` is not equal to zero when the IF/THEN statement is reached, it will still maintain its previous value after the statement is executed.

The difference between the relational "=" and the assignment "=" is so important that some languages use different symbols for each. For example, Pascal uses the equal sign for logical expressions but specifies assignment with symbols ":=" as in

```
IF var = 0 THEN var := -var;
```

In contrast, FORTRAN employs the equal sign for assignment but specifies relational equality with ".EQ." as in

```
IF (VAR .EQ. = 0.) VAR = -VAR
```

TABLE 7.1 Relational Operators

Operator	Relationship	Example
=	Equal	`IF var = 0 THEN var = 99`
<>	Not equal	`IF c$<>"9" THEN PRINT c$`
<	Less than	`IF var < 0 THEN var = -var`
>	Greater than	`IF x% > y THEN z = y/x%`
<=	Less than or equal to	`IF 3.9<=x/3 THEN INPUT y`
>=	Greater than or equal to	`IF var >=0 THEN CLS`

Unfortunately, BASIC uses the same symbol "=" for both operations. However, once the distinction is understood, programmers rarely confuse the two uses.

Proceeding with the other relational operators, the inequality tests whether the expression on the right side of the <> is different from that on the left. For the example in Table 7.1, the value of c$ will be displayed only if c$ is not equivalent to the character string "9".

Testing whether one value is less than another is accomplished with the < operator. Likewise, the > determines if the value on the left side of the > is greater than that on the right side. The operators >= allow the left-hand value to be either exactly equal to or greater than the right-hand value. Similarly, the operators <= will be true if the two are equal or if the left-hand value is less than the right-hand value. Remember the rules for working with inequality operators, which require the operator to change if the values on the left and right side of inequality are exchanged. X > Y is the same test as Y < X.

The order of evaluation of the IF/THEN statement is important to note. First the condition is tested. If the condition is true, then the statements following the THEN are carried out. In the first example in Table 7.1, var is tested to determine whether it is equal to zero prior to performing the assignment following THEN. If the IF/THEN did not perform the evaluation in this fashion, var would never equal zero. It would always equal 99.

Arithmetic and relational operators can be combined in a condition being tested. The arithmetic operations are always implemented before testing the relationship. In the fifth example of Table 7.1, x is first divided by 3, and then 3.9 is tested to see if it is equal to or less than the quotient.

Strings and character data can also be tested for their relationship to other string and character data. As in the second example in Table 7.1, character data cannot be directly compared to numeric data. That is why the 9 has to be enclosed in quotation marks. Enclosing the 9 in quotation marks causes it to be treated as a string character and overrides its numeric value. If the 9 is not set in quotations, BASIC will stop processing the line and report an error. Just as string variables cannot have numbers assigned to them or vice versa, relational operations require that both sides of the relation have either numeric data or string data, but not both in one relational operation.

Be cautious when testing for the equality or inequality of two real expressions. Remember that real numbers are stored in the computer as series of binary digits and an exponent. If the value to be tested cannot be exactly represented by the significant digits of a single-precision real variable, then the test may fail, even though it is, in fact, "almost" true. For example, the following condition would test as false:

```
1/3 = 0.333
```

Even though 0.333 is approximately equal to $\frac{1}{3}$, the computer must sense an exact equality for the condition to be true. More subtle differences can occur, as illustrated in the following example.

EXAMPLE 7.1 *Problems with Testing Equalities*

PROBLEM STATEMENT: The following code was written to investigate the testing of equalities between real numbers:

```
a = 1/82
b = a * 82
messag1$ = "equality holds"
messag2$ = "equality does not hold"
IF b = 1 THEN PRINT messag1$ ELSE PRINT messag2$
PRINT "b = "; b
END
```

Run:

```
equality does not hold
b = .9999999
```

Although we would expect that 1/82 multiplied by 82 should be exactly equal to 1, such is not the case for this code. Because of the approximate way that the computer stores real numbers, the equality does not hold.

One way to circumvent this problem is to reformulate it. Rather than testing whether the two numbers are equal, you can test whether their difference is small. For example, the IF/THEN could be reformulated as

```
IF ABS(b - 1) < .00001 THEN PRINT messag1$ ELSE PRINT messag2$
```

This statement tests whether the absolute value of the difference between b and 1 is less than 0.00001—that is, less than 0.001 percent. Formulated in this way, the IF/THEN would test true, as desired. This is due to the fact that the discrepancy introduced by the internal binary representation is much smaller than the tolerance used for the comparison. In fact the difference occurs in the seventh significant digit, which corresponds to about a 0.00001 percent difference.

Although you should be aware of the problem outlined in the previous example, it is usually limited to equality. Testing whether one value is less than or greater than another can usually be performed without running into this difficulty.

7.4 COMPOUND IF/THEN STATEMENTS AND LOGICAL OPERATORS

Modern BASIC allows the testing of more than one logical condition in the same IF/THEN statement. These *compound IF/THEN statements* employ the logical operators AND, OR, and NOT to simultaneously test conditions (Table 7.2).

TABLE 7.2 Commonly Used Logical Operators

Operator	Meaning	Example
NOT	*The logical complement.* This negates a logical expression. In the example, if x <= 0 is false, the expression is true (and vice versa).	`IF NOT x <= 0 THEN PRINT "positive"`
AND	*Conjunction.* Both logical expressions must be true; otherwise the compound expression is considered false.	`IF x > 0 AND x < 5 THEN PRINT "0 to 5"`
OR	*Disjunction (inclusive "or").* If either or both of the expressions are true, the compound expression is true.	`IF a$ = "y" OR a$ = "Y" THEN PRINT "Yes"`

7.4.1 Conjunction

The AND is employed when both conditions must be true, as in the general form

> IF *expression*$_1$ AND *expression*$_2$ THEN *TrueStatement*

If both conditions are true, then this statement will execute the *TrueStatement*. If either or both *expressions* are false, the program will advance to the next line.

7.4.2 Disjunction (the Inclusive "OR")

In contrast, the OR is employed when either one or the other or both of the conditions are true, as in

> IF *expression*$_1$ OR *expression*$_2$ THEN *TrueStatement*

If either or both conditions are true, the *TrueStatement* will be performed.

7.4.3 Logical Complement

The NOT is used for the case where the *TrueStatement* is completed if a condition is false, as in

> IF NOT *expression* THEN *TrueStatement*

If the *expression* is false, the *TrueStatement* will be executed.

When a logical expression is false, its logical complement is true, and vice versa. For example, the complement of $x <= 0$ is $x > 0$. Thus, the first example from Table 7.2 could be rewritten as

> IF x > 0 THEN PRINT "positive"

The complements for the relational operators are summarized in Table 7.3. Note that in most cases, it is clearer and more convenient to use the complement of a relational operator than to employ the NOT. On the other hand, there will be other instances where the NOT will be desirable.

Compound expressions can contain more than one of the logical operators, for example,

> IF a > 0 OR b < 0 AND c = 0 THEN CLS

However, just as with arithmetic expressions, the order in which the operations are evaluated must follow rules. The priorities for the operators (along with all the other priorities established to this point) are specified in Table 7.4. Thus, all other things being equal, the NOT would be implemented prior to the AND and the AND prior to the OR.

TABLE 7.3   Complements for the Relational Operators

Operator	Complement	Identities
=	<>	a = b is equivalent to NOT (a <> b)
<	>=	a < b is equivalent to NOT (a >= b)
>	<=	a > b is equivalent to NOT (a <= b)
<=	>	a <= b is equivalent to NOT (a > b)
>=	<	a >= b is equivalent to NOT (a < b)
<>	=	a <>b is equivalent to NOT (a = b)

TABLE 7.4   The Priorities of Operators Used in BASIC

Type	Operator	Priority
Arithmetic	Parentheses, () Exponentiation, ^ Negation, − Multiplication, ∗, and Division, / Addition, +, and Subtraction, −	Highest
Relational	=, <, >, <=, >=, <>	
Logical	NOT AND OR	Lowest

For complicated logical expressions, parentheses and position also dictate the order. Thus operations within parentheses are implemented first. Then the priorities in Table 7.4 are used to determine the next step. Finally, within the same priority class, operations are executed from left to right.

EXAMPLE 7.2   *Priorities in Compound Logical Expressions*

PROBLEM STATEMENT: Determine whether the following compound logical expressions are true or false:

(*a*) a * b > 0 AND b = 2 AND x > 7 OR NOT y$ > "d"
(*b*) a * b > 0 AND b = 2 AND (x > 7 OR NOT y$ > "d")

Note that a = -1, b = 2, x = 1, and y$ = "b".

SOLUTION: (*a*) Using the priorities from Table 7.4, we can determine that the expression is true by the following sequence of evaluations:

a * b > 0 AND b = 2 AND x > 7 OR NOT y$ > "d"

-2 > 0 AND 2 = 2 AND 1 > 7 OR NOT "b" > "d"

F   AND   T   AND F   OR NOT   F

                                              True

          False

                  False

                  True

First, all the relational operators are evaluated because they have higher priority than the logical operators. Of the logical operators, the NOT comparison is implemented first because it has a higher priority than AND and OR. It negates the last false and converts it to a true. The leftmost AND is implemented next because of the left-to-right rule. It is false because both expressions must be true for an AND to be true. Next, the conjunction of two falses leads to a false. Finally, the OR comparison is true because one of the conditions being compared is true.
(*b*) Because of the introduction of the parentheses, we are forced to evaluate the OR comparison before the ANDs. As a result, the entire operation yields a false result, as in

```
a * b > 0 AND b = 2 AND (x > 7 OR NOT y$ > "d")

-2  > 0   AND 2 = 2 AND (1 > 7 OR NOT "b" > "d")

 F  AND  T   AND   (F  OR NOT    F)
                                ‿‿‿
                                 True
                         ‿‿‿‿‿‿‿‿‿
                            True
        ‿‿‿‿‿‿‿
        False
        ‿‿‿‿‿‿‿‿‿‿‿‿‿‿‿‿‿
                 False
```

Modern BASIC also employs several other operators (Table 7.5). The most commonly used is the XOR operator that tests true when one or the other is true but not when both of the conditions are true. It is called the *exclusive "OR."* Note that the operators in Table 7.5 have lower priority than the NOT, AND, and OR.

In summary, there is a rich array of logical operators in the BASIC language. Table 7.6 summarizes their behavior in the form of a truth table.

TABLE 7.5   Additional Logical Operators in BASIC*

Operator	Name	Meaning
XOR	Exclusive "or"	Tests true when one or the other but not both of the conditions are true.
EQV	Equivalence	Tests true when either both are true or both are false.
IMP	Implication	Tests false when the first expression is true and the second is false. All other combinations are true.

*Note that these are listed in descending priority order. Thus, the XOR will be implemented prior to the EQV, and the EQV will be executed prior to the IMP.

TABLE 7.6   A Truth Table Summarizing the Behavior of the Logical Operators Available in QuickBASIC

x	y	NOT x	x AND y	x OR y	x XOR y	x EQV y	x IMP y
T	T	F	T	T	F	T	T
T	F	F	F	T	T	F	F
F	T	T	F	T	T	F	T
F	F	T	F	F	F	T	T

7.5   STRUCTURED PROGRAMMING OF SELECTION

The statements introduced in the previous sections can be orchestrated to introduce all sorts of complicated logic into your computer programs. However, as cautioned at the beginning of this chapter, their indiscriminate application can lead to code that is so complex that it is practically indecipherable (so-called spaghetti code).

7.5.1   Improper Selection with the IF/THEN Statement and GOTOs

There are several ways that the IF/THEN construct can be employed to program a double-alternative decision pattern. Figure 7.2 shows a flowchart that will serve to illustrate this point. Figure 7.2*a* through *d* presents four sets of BASIC code that all implement the algorithm in the flowchart. Figure 7.2*a* and *b* both employ the form of the condition shown in the flowchart; that is, they test whether A is less than zero. They differ in that Fig. 7.2*a* has the false alternative first, whereas Fig. 7.2*b* has the true alternative first. Figure 7.2*c* and *d* are alternatives that result from utilizing the complement of the original condition; that is, they test whether A is greater than or equal to zero. They also differ regarding the order of the alternatives. Notice how in all cases, GOTO statements are required to transfer control properly.

The fact that four separate versions can be developed to perform the same double-alternative decision is one example of how selection can introduce flexibility and an associated complexity into a program. The result is that the logic of such statements is often very difficult to decipher. In order to reduce this potential confusion, computer scientists have constrained selection by developing the block-structured IF/THEN/ELSE construct.

7.5.2   The Block Structured IF/THEN/ELSE Construct

The block-structured IF/THEN/ELSE construct can be represented generally as

```
IF LogicalExpression THEN
    True Statements
    .
    .
    .
ELSE
    FalseStatements
    .
    .
    .
END IF
```

If the *LogicalExpression* is true, control moves to the *TrueStatements*. After these are executed, control jumps immediately to the line following the END IF statement. If the expression is false, control jumps to the *FalseStatements* following the ELSE. The for-

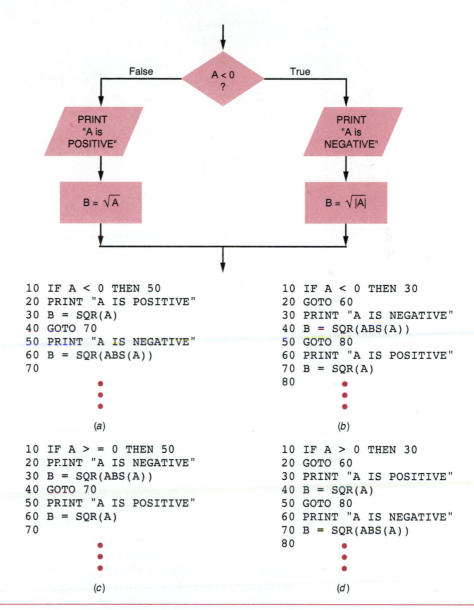

```
10  IF A < 0 THEN 50
20  PRINT "A IS POSITIVE"
30  B = SQR(A)
40  GOTO 70
50  PRINT "A IS NEGATIVE"
60  B = SQR(ABS(A))
70
            •
            •
            •
           (a)
```

```
10  IF A < 0 THEN 30
20  GOTO 60
30  PRINT "A IS NEGATIVE"
40  B = SQR(ABS(A))
50  GOTO 80
60  PRINT "A IS POSITIVE"
70  B = SQR(A)
80      •
        •
        •
       (b)
```

```
10  IF A > = 0 THEN 50
20  PRINT "A IS NEGATIVE"
30  B = SQR(ABS(A))
40  GOTO 70
50  PRINT "A IS POSITIVE"
60  B = SQR(A)
70
            •
            •
            •
           (c)
```

```
10  IF A > 0 THEN 30
20  GOTO 60
30  PRINT "A IS POSITIVE"
40  B = SQR(A)
50  GOTO 80
60  PRINT "A IS NEGATIVE"
70  B = SQR(ABS(A))
80      •
        •
        •
       (d)
```

FIGURE 7.2
Four sets of early BASIC code that implement the algorithm specified by the flowchart.

mat is reiterated in Fig. 7.3, which shows a BASIC example along with the flowchart and pseudocode representations of the construct. Note that as depicted in Fig. 7.4, a version is available for the single-decision alternative.

The formats shown in Figs. 7.3 and 7.4 are referred to as *block-structured IF/ THEN/ELSE* constructs. The benefits of using these constructs are great. Each IF/

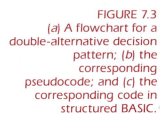

FIGURE 7.3
(a) A flowchart for a double-alternative decision pattern; (b) the corresponding pseudocode; and (c) the corresponding code in structured BASIC.

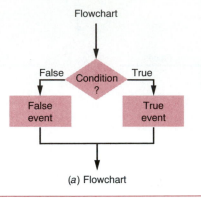

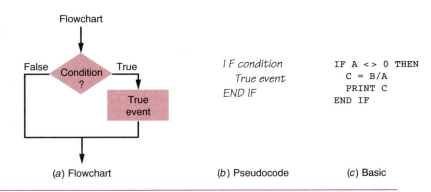

(a) Flowchart

IF condition
    True event
ELSE
    False event
END IF

(b) Pseudocode

```
IF A < 0 THEN
    PRINT "A IS NEGATIVE"
    B = SQR (ABS(A))
ELSE
    PRINT "A IS POSITIVE"
    B = SQR(A)
END IF
```

(c) Basic

FIGURE 7.4
(a) A flowchart for a single-alternative decision pattern; (b) the corresponding pseudocode; and (c) the corresponding code in structured BASIC.

(a) Flowchart

IF condition
    True event
END IF

(b) Pseudocode

```
IF A <> 0 THEN
    C = B/A
    PRINT C
END IF
```

(c) Basic

THEN/ELSE has only one entry point and one exit. The BASIC statements that are a part of the IF/THEN/ELSE construct are immediately apparent, and the true and false alternatives are clearly delineated. This is reinforced by indentation. The primary objective here is to make the resulting code coherent and easily understandable to someone other than the original author of the program. These conventions help regulate the writing of programs so that others may more readily grasp the organization of an unfamiliar program. They also serve to standardize across different languages.

All forms of selection (with the exception of the single-line versions from Sec. 7.2) should be constructed as block structures for easier readability. If there is only one BASIC statement associated with the THEN and the ELSE conditions, a one-line IF/THEN statement is reasonable; however, if there is more than one statement in either, then a block-structured IF/THEN/ELSE construct with proper indentation is much easier to understand. Figure 7.5, which redisplays Figs. 7.2a and 7.3c, should make this evident. Even though this is a very simple example, the superiority of the structured approach is obvious. For more complicated code, the benefits become even more pronounced.

```
10 IF A < 0 THEN 50            IF A < 0 THEN
20 PRINT "A IS POSITIVE"          PRINT "A IS NEGATIVE"
30 B = SQR(A)                     B = SQR(ABS(A))
40 GOTO 70                     ELSE
50 PRINT "A IS NEGATIVE"          PRINT "A IS POSITIVE"
60 B = SQR(ABS(A))                B = SQR(A)
70 REM CONTINUE                END IF
```
          (a)   (b)

FIGURE 7.5 A side-by-side comparison of Figs. 7.2a and 7.3c.
Both sets of code accomplish the same objective, but the structured
version in (b) is much easier to read and decipher.

7.5.3   Nesting of Block IF/THEN/ELSE Constructs

An IF/THEN/ELSE construct can have another IF/THEN/ELSE statement as part of its THEN or ELSE statements. The inclusion of a construct within a construct is called *nesting*. For the case of the block IF/THEN/ELSE construct, nesting provides an alternative to having involved conditional expressions with many AND, OR, and NOT clauses. By using simpler conditional expressions it may be easier to follow the logic of the decisions.

The more complex nested IF/THEN/ELSE structures are much easier to understand when proper rules for indentation are followed. The rules for forming and indenting the nested IF/THEN/ELSE statements are essentially the same as those for a single IF/THEN/ELSE structure. Just remember that indentation of the conditional statement ELSE and the END IF is always determined by the IF/THEN/ELSE statement to which they refer. An example employing nested IF/THEN/ELSE statements will demonstrate the use of nesting and proper indentation.

EXAMPLE 7.3   Nesting of IF/THEN/ELSE Constructs

PROBLEM STATEMENT: Develop a program to determine the real and complex roots of the quadratic equation.

SOLUTION: A structured flowchart for this problem is shown in Fig. 7.6. Note that there are two nested selections. The first selection is designed to avoid division by zero in the quadratic equation for the case where $a = 0$. The second selection branches to compute the real or complex roots, depending on whether the term within the square root (called the "discriminant") is positive or negative.

A structured code to implement the flowchart is listed in Fig. 7.7. Indentation is used to show clearly the true and false options of each IF/THEN/ELSE. Problem 7.10 at the end of the chapter deals with refining this program.

At this point you may wonder why all BASIC programs have not been written in a structured form. What are the disadvantages of using this type of program organization?

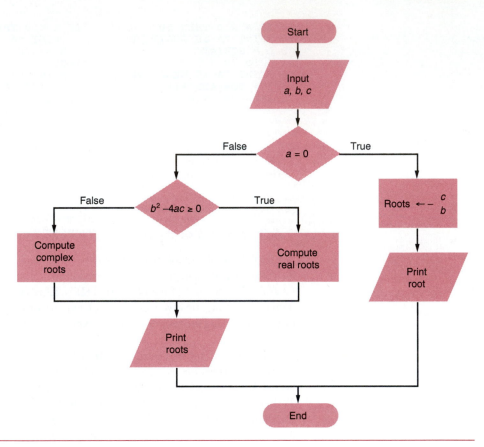

FIGURE 7.6
A structured flowchart to
determine the real and
complex roots of a quadratic
equation. Note that the first
selection prevents division by
zero in the equation if $a = 0$.

For one thing, the amount of typing is increased in the structured code. In addition, the speed of execution of the structured code may be slightly slower than that of a nonstructured version. Thus there is a trade-off here. The structured program is easier to understand and change later, but more time may be consumed developing it initially. However, in the overall scheme of program development and use, the time and effort are almost always a worthwhile investment. Additionally, as you employ structured programming constructs on a regular basis, they will become second nature to you. As your experience grows, you will realize that the structured constructs are actually simpler to program initially and unquestionably easier to modify in the future.

7.5.4 The CASE Construct

The CASE construct is a special type of nested IF/THEN/ELSE block. As depicted in the flowchart and pseudocode in Fig. 7.8, the CASE construct selects one path from a group of alternative paths based on the values of a single variable. This is a very useful construction in situations where the program must perform several independent tasks depending on an "option" variable.

```basic
'Quadratic Roots Problem

'by S.C. Chapra
'July 27, 1992

'This program determines the real or
'complex roots of a quadratic equation
'----------------------------------------

'Definition of variables:

'a = second-order term
'b = first-order term
'c = zero-order term
'root = single root
'real1 = real part of first root
'imag1 = imaginary part of first root
'real2 = real part of second root
'imag2 = imaginary part of second root
'discr = discriminant

'input data and print headings
CLS
INPUT "a, b, c = ", a, b, c
PRINT : PRINT "Results:": PRINT

IF a = 0 THEN
  root = -c / b
  PRINT "single root = "; root
ELSE
  discr = b ^ 2 - 4 * a * c
  IF discr > 0 THEN
    'compute real roots
    real1 = (-b + SQR(discr)) / (2 * a)
    real2 = (-b - SQR(discr)) / (2 * a)
  ELSE
    'compute complex roots
    real1 = -b / (2 * a)
    real2 = real1
    imag1 = SQR(ABS(discr)) / (2 * a)
    imag2 = -imag1
  END IF
  'output results
  PRINT "root1 = "; real1; imag1; "i"
  PRINT "root2 = "; real2; imag2; "i"
END IF

END
```

FIGURE 7.7
Structured BASIC program to implement the alorithm from Fig. 7.6. Notice how indentation clearly sets off the program's structure.

The general format of the CASE construct in QuickBASIC is

```
SELECT CASE TestExpression
    CASE List₁
        Block₁
    CASE List₂
        Block₂
            .
            .
            .
    CASE ELSE
        ElseBlock
END SELECT
```

where *TestExpression* is any numeric or string expression, *Block$_i$* is the group of statements corresponding to the *i*th case, and *List$_i$* is the value(s) or range(s) of the *TestExpression* coresonding to the *i*th case.

FIGURE 7.8
(*a*) Flowchart and
(*b*) pseudocode for the CASE
construct.

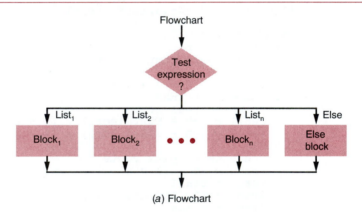

(*a*) Flowchart

```
SELECT CASE Test Expression
    CASE List₁
        Block₁
    CASE List₂
        Block₂

            ●
            ●
            ●

    CASE Listₙ
        Blockₙ
    CASE ELSE
        Else Block
END CASE
```

(*b*) Pseudocode

Note that the *List* can express individual values,

```
CASE 5,8,-1
```

ranges of values,

```
CASE 15 TO 20
```

relational values,

```
CASE IS < 0
```

or combinations of the above,

```
CASE -4 TO -2, 12, 15, IS <> 20
```

EXAMPLE 7.4   The CASE Construct

PROBLEM STATEMENT: Suppose that each class at a university is to go to a different room to register for courses. The classroom assignments are as follows:

Class	Identification Code	Classroom
Freshmen	1	CR2–3
Sophomores	2	CE1–4
Juniors	3	ME1–6
Seniors	4	CR2–6
All others		CH2–8

The identification code is a number that signifies the student's class. Design a program that uses the CASE construct to display the proper classroom number in response to the student's identification code.

SOLUTION: The BASIC program to accomplish the objective is listed in Fig. 7.9. Notice that the CASE construct is used to display the correct message on the basis of the student's code.

As shown in Fig. 7.10, the CASE construct can be programmed using IF/THEN/ELSEIF statements. As programmed in Fig. 7.10, the two versions would execute in precisely the same manner. The CASE would be preferred because it is more descriptive of the underlying structure. Although the CASE is preferred for situations where the branching depends on a single variable, the IF/THEN/ELSEIF must be employed when the branching depends on several variables or more complicated conditions.

FIGURE 7.9
A program fragment
illustrating the CASE
construct in QuickBASIC.

```
'Program fragment to illustrate
'the SELECT CASE construct

CLS
'create a menu
PRINT : PRINT : PRINT
PRINT "Select an option": PRINT
PRINT TAB(10); "1.   Freshman"
PRINT TAB(10); "2.   Sophomore"
PRINT TAB(10); "3.   Junior"
PRINT TAB(10); "4.   Senior"
PRINT
PRINT "Enter the number corresponding"
PRINT "to your status";
INPUT class
PRINT
CLS
SELECT CASE class
  CASE 1
    PRINT "Proceed to room CR2-3"
  CASE 2
    PRINT "Proceed to room CE1-4"
  CASE 3
    PRINT "Proceed to room ME1-6"
  CASE 4
    PRINT "Proceed to room CR2-6"
  CASE ELSE
    PRINT "Proceed to room CH2-8"
END SELECT

END
```

Run:
```
Select an option

          1.    Freshman
          2.    Sophomore
          3.    Junior
          4.    Senior

Enter the number corresponding
to your status 4

Proceed to room CR2-6
```

FIGURE 7.10
A CASE construct
programmed with
IF/THEN/ELSEIF statements.
This version behaves similarly
to the SELECT CASE in Fig. 7.9.

```
IF class = 1 THEN
   PRINT "Proceed to room CR2-3"
ELSEIF class = 2 THEN
   PRINT "Proceed to room CE1-4"
ELSEIF class = 3 THEN
   PRINT "Proceed to room ME1-6"
ELSEIF class = 4 THEN
   PRINT "Proceed to room CR2-6"
ELSE
   PRINT "Proceed to room CH2-8"
END IF
```

PROBLEMS

7.1 The following code is designed to display a message to a student according to his or her final exam grade (`Final`) and total points accrued in a course (`Points`).

```
IF Final <= 60 THEN PRINT "Failing Grade"
IF Final > 60 THEN PRINT "Passing Grade"
IF Final > 60 AND Points < 200 THEN PRINT "Poor"
IF Final > 60 AND Points >= 200 AND Points < 400 THEN PRINT "GOOD"
IF Final > 60 AND Points > 400 THEN PRINT "Excellent"
```

Reformulate this code using block-structured IF/THEN/ELSE constructs. Test the program for the following cases:

```
Final = 50, Points = 250
Final = 70, Points = 321
Final = 90, Points = 561
```

7.2 As an industrial engineer you must develop a program to accept or reject a rod-shaped machine part according to the following criteria: its length must be not less than 7.75 cm or greater than 7.85 cm; its diameter must be not less than 0.335 cm or greater than 0.346 cm. In addition, under no circumstances may its mass exceed 5.6 g. Note that the mass is equal to the volume (cross-sectional area times length) multiplied by the rod's density (7.8 g/cm^3). Write a structured computer program to input the length and diameter of a rod, and then display whether it is accepted or rejected. Test it for the following cases. If it is rejected, display the reason (or reasons) for rejection.
(*a*) Length = 7.71, diameter = 0.338
(*b*) Length = 7.81, diameter = 0.341
(*c*) Length = 7.83, diameter = 0.343
(*d*) Length = 7.86, diameter = 0.344
(*e*) Length = 7.63, diameter = 0.351

7.3 An air conditioner both cools and removes humidity from a factory. The system is designed to turn on (1) between 7 A.M. and 6 P.M. if the temperature exceeds 75°F and the humidity exceeds 40 percent, or if the temperature exceeds 70°F and the humidity exceeds 80 percent; or (2) between 6 P.M. and 7 A.M. if the temperature exceeds 80°F and the humidity exceeds 80 percent, or if the temperature exceeds 85°F, regardless of the humidity.

Write a structured computer program that inputs temperature, humidity, and time of day and displays a message specifying whether the air conditioner is on or off. Test the program with the following data:
(*a*) Time = 7:40 P.M., temperature = 81°F, humidity = 68 percent
(*b*) Time = 1:30 P.M., temperature = 72°F, humidity = 79 percent

(c) Time = 8:30 A.M., temperature = 77°F, humidity = 50 percent

(d) Time = 2:45 A.M., temperature = 88°F, humidity = 28 percent

7.4 Develop a structured program that uses the IF/THEN/ELSE construct to mimic the SGN intrinsic function.

7.5 Develop a structured program that uses the CASE construct to mimic the SGN intrinsic function.

7.6 The American Association of State Highway and Transportation Officials provides the following criteria for classifying soils in accordance with their suitability for use in highway subgrades and embankment construction:

Grain Size, mm	Classification
>75	Boulder
2–75	Gravel
0.05–2	Sand
0.002–0.05	Silt
<0.002	Clay

Develop a structured program that uses the CASE construct to classify a soil on the basis of grain size. Test the program with the following data:

Soil Sample	Grain Size, mm
1	2×10^{-4}
2	10
3	0.6
4	120
5	0.01

7.7 Develop a structured program that uses the CASE construct to assign a letter grade to students on the basis of the following scheme: A (90–100), B (80–90), C (70–80), D (60–70), and F (<60). For the situation where the student's grade falls on the boundary (for example, 80), have the program give the student the higher letter grade (for example, a B).

7.8 Write a structured BASIC program that reads values of X and outputs absolute values of X without using the ABS function.

7.9 Determine whether the following are true or false. Show all the steps in the evaluation of the condition in a manner similar to Example 7.2. (*Note:* X = 10, Y = 20, and Z = 30.)

(a) X <= 10 OR Y = 20

(b) X <= 20 OR Y >= 50

(c) (Y = 20 AND Z = 30) AND (X > 10 OR Y > 30)
(d) (Y = 20 OR Z = 50) AND X >= 10 AND Y <= 20
(e) (Y = 50 AND X = 10) OR (Z < 40 OR Y > 30)

7.10 Develop a program to determine the real and complex roots of the quadratic equation. Base your code on Fig. 7.7, but:
(a) Modify the code so that it covers every possible numerical input. (*Note:* The version in Fig. 7.7 does not cover every possible case.)
(b) Improve the output. For example, as programmed in Fig. 7.7, the output for the case where a, b, c = 1, 4, 2 looks like

```
root1 = -.5857865 0 i
root2 = -3.414214 0 i
```

Reprogram the code so that the imaginary part is omitted when it is equal to zero. Similarly, for the case where there are complex roots, display the sign between the real and the imaginary components of the root. Use selection constructs to accomplish these enhancements.

7.11 As previously reviewed in Prob. 6.19, it is relatively straightforward to compute cartesian coordinates (x, y) on the basic of polar coordinates (r, θ). The reverse process is not so simple. The radius can be computed with the following formula:

$$r = \sqrt{x^2 + y^2}$$

If the coordinates lie within the first or fourth quadrants (that is, $x > 0$), then a simple formula can be used:

$$\theta = \tan^{-1}\left(\frac{y}{x}\right)$$

The difficulty arises for all other situations. The following table summarizes the possible cases that can occur

x	y	θ
<0	>0	$\tan^{-1}\left(\dfrac{y}{x}\right) + \pi$
<0	<0	$\tan^{-1}\left(\dfrac{y}{x}\right) - \pi$
<0	=0	π
=0	>0	$\dfrac{\pi}{2}$
=0	<0	$-\dfrac{\pi}{2}$
=0	=0	0

Write a well-structured program to compute (r, θ) given (x, y). Express the final result for θ in degrees. Test the program for the following combinations of x and y:

x	y
0	1
−1	0
−1	1
0	−1
−1	−1
0	0

7.12 Develop a well-structured program where the user can enter a number and one of the following units of time: second, minute, hour, and day. Design the program so that it will then display the time in the remaining units. Employ a menu (similar to the one in Fig. 7.9) to allow the user to specify the unit.

7.13 Develop a well-structured program where the user can enter the number of seconds past midnight (0 through 86,400 s). Design the program so that it will then display the time in hours, minutes, and seconds. Use the 12-h format for displaying the result (that is, midnight is 12:00:00 P.M. and noon is 12:00:00 A.M.). If the user inputs a negative time or a time that is too large (>86,400 s), display an appropriate error message.

7.14 Develop a well-structured program that converts Julian day $(1, 2, \ldots, 365)$ into calendar day (Jan. 1, Jan. 2, ..., Dec. 31) for a nonleap year. If the user inputs a negative day or a day that is too large (>365), display an appropriate error message.

7.15 The following CASE statement can be written in QuickBASIC:

```
SELECT CASE i
  CASE 1
    PRINT "The first = "; i
  CASE 2, 3, 9
    PRINT "The second, third and ninth ="; i
  CASE 4 TO 8
    PRINT "The fourth through eighth ="; i
  CASE ELSE
    PRINT "All others"; i
END SELECT
```

Rewrite this code using IF/THEN/ELSEIF statements.

REPETITION

Computers excel at performing tasks that are repetitive, boring, and time-consuming. Even very simple programs can accomplish huge tasks by executing the same chore over and over again. One of the earliest applications of mechanical calculation was to tally and categorize the U.S. population for the 1890 census, a very repetitive task. Even today, one of the primary applications of computers is to maintain, count, and organize many pieces of information into smaller, more meaningful collections.

Nowhere is the repetitive power of the computer more valuable than in engineering. Many of our applications hinge on the ability of computers to implement portions of programs repetitively. This is true in many areas of our discipline but is especially relevant to computationally intensive or *"number-crunching"* calculations. Before discussing the actual statements and constructs used for repetition, we will present a simple example to demonstrate the major concept underlying repetition—the loop.

8.1 GOTO, OR INFINITE, LOOPS

Suppose that you want to perform a statement or a group of statements many times. One way of accomplishing this is merely to write the set of statements over and over again. For example, a repetitive version of the simple addition program is

```
INPUT a, b
PRINT a + b
INPUT a, b
PRINT a + b
INPUT a, b
PRINT a + b
        .
        .
        .
```

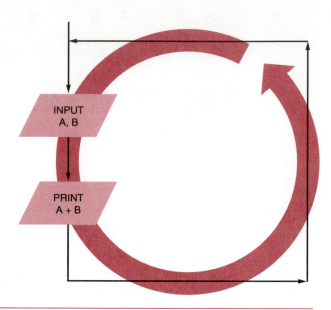

FIGURE 8.1
A GOTO loop. Once it starts,
it never stops.

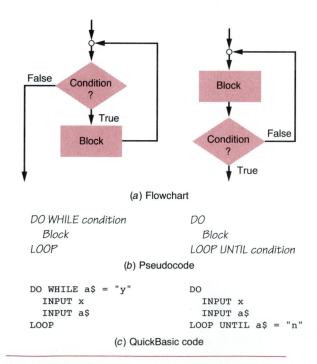

(a) Flowchart

```
DO WHILE condition            DO
    Block                         Block
LOOP                          LOOP UNTIL condition
```

(b) Pseudocode

```
DO WHILE a$ = "y"             DO
   INPUT x                       INPUT x
   INPUT a$                      INPUT a$
LOOP                          LOOP UNTIL a$ = "n"
```

(c) QuickBasic code

FIGURE 8.2 Comparison of the DO WHILE and the
DO UNTIL loop structures.

Such a sequential approach is obviously inefficient. A much more concise alternative can be developed using a GOTO statement, as in

```
repeat: INPUT a, b
        PRINT a + b
        GO TO repeat
```

As depicted in Fig. 8.1, the use of the GOTO statement directs the program to automatically circle or "loop" back and repeat the INPUT and PRINT statements. Such repetitive execution of a statement or a set of statements is called a *loop*. Because one of the primary strengths of computers is their ability to perform large numbers of repetitive calculations, loops are among the most important structures in programming.

A flaw of Fig. 8.1, called a *GOTO loop*, is that it is a *closed*, or *infinite, loop*. Once it starts, it never stops. Consequently, provisions must be made so that the loop is exited after the repetitive computation is performed satisfactorily.

Two different approaches are employed to terminate loops. *Decision loops* are terminated on the basis of the state of a logical expression. In this sense they are related to the selection constructs discussed in the previous chapter. Because they are based on a decision, they may repeat a different number of times on every occasion that they are executed. In contrast, *counter loops* repeat a prespecified number of times.

8.2   DECISION LOOPS

There are two fundamental types of decision loop constructs—the DO WHILE and the DO UNTIL structures. As depicted in Fig. 8.2*a*, the DO WHILE is used to create a loop that repeats as long as a condition is true. Because the condition precedes the event, it is possible that the action will not be implemented if the condition is false on the first pass.

In contrast, as shown in Fig. 8.2*b*, the DO UNTIL is designed to have the event execute repeatedly until the condition is true. Then control is passed outside the loop. A key feature of this type of loop is that the event is always executed at least once.

8.2.1   The DO WHILE (or PreTest) Loop

The fundamental syntax of the BASIC DO WHILE loop is

```
DO WHILE LogicalExpression

    StatementBlock

LOOP
```

The loop will execute only if the *LogicalExpression* is true. When it becomes false, control transfers to the next statement following the loop. Because the test is at the beginning of the loop, this construct is also called a *PreTest loop*.

As with all decision loops, something must happen within the loop to make the *LogicalExpression* false. Otherwise, the loop would never terminate. This important point is illustrated in the following example.

EXAMPLE 8.1   *Using a DO WHILE Loop to Input, Count, and Total Values*

PROBLEM STATEMENT: A common application of repetition is to enter data from the keyboard. There are occasions when the user will have prior knowledge of the number of values that are to be input to a program. However, there are just as many instances when the user either will not know or will not desire to count the data beforehand. This is typically the case when large quantities of information are to be entered. In these situations, it is preferable to have the program count the data. Employ the DO WHILE loop to input, count, and total a set of values. Use the results to compute the average.

SOLUTION: The important issue that needs to be resolved in order to develop this program is how to signal when the last item is entered. A common way to do this is to have the user enter an otherwise impossible value to signal that data entry is complete. Such impossible data values are sometimes called *sentinel,* or *signal, values.* The DO WHILE statement can then be used to test whether the sentinel value has been entered and whether to transfer control out of the loop. The sentinel value can be a negative number if all data values are known to be positive. A large number, say 999, might be employed if all valid data was small in magnitude. In the following code the number −999 is used as the sentinel value:

```
CLS

count = 0
total = 0

INPUT "Value = (-999 to terminate)"; value
DOWHILE value <> -999
  count = count + 1
  total = total + value
  INPUT "Value = (-999 to terminate)"; value
LOOP

IF count > 0 THEN
  average = total/count
  PRINT:PRINT "number of values ="; count
  PRINT "average  = " average
ELSE
  PRINT
  PRINT "no data entered; average not computed"
END IF
```

Run:

```
Value = (-999 to terminate)? 20
Value = (-999 to terminate)? 100
Value = (-999 to terminate)? 60
Value = (-999 to terminate)? 40
Value = (-999 to terminate)? 80
Value = (-999 to terminate)? -999

number of values = 5
average = 50
```

Before the loop is implemented, `count` and `total` are set to zero. This is called *initialization.* On each pass through the loop, the values are summed by accumulating them in the variable `total`. For the initial pass through the loop, the first value is added to `total` and the result assigned to `total` on the left side of the assignment operator. Therefore, after the first pass, `total` will contain the first value. On the next pass the second value is added, and `total` will then contain the sum of the first and second values. As the iterations continue, additional values are input and added to `total`. Thus, after the loop is terminated, `total` will contain the sum of all the values. The variable `count` operates in a similar fashion but is incremented by 1 on each pass. Finally, both `total` and `count` are then used to compute the average in the IF/THEN/ELSE block.

The first INPUT statement, which is called a *priming input,* is required so that the user has the option of entering no values at all. If the first entry were −999, then the body of the loop would not be implemented and `count` would remain at zero. In the same vein, the computation of the average value is conditional in order to avoid division by zero for this case.

Also, observe how the input of data is the last statement in the loop. This ordering prevents the sentinel value from being incorporated into `total` or tallied in `count`.

Note that most modern BASICs have an alternative form to implement this construct, called the WHILE/WEND loop, that was actually developed before the DO WHILE. Its general structure is,

WHILE *LogicalExpression*

　　StatementBlock

WEND

This construct operates in a fashion identical with the DO WHILE loop and is interchangeable with it. However, because of the introduction of the DO WHILE statement, the WHILE/WEND is falling out of favor.

8.2.2 The DO UNTIL (or PostTest) Loop

The fundamental syntax of the BASIC DO UNTIL loop is

```
DO

    StatementBlock

LOOP UNTIL LogicalExpression
```

In contrast to the DO WHILE construct, this loop will cycle at least one time. This is a consequence of the test residing at the end of the loop. Hence, this construct is sometimes called the *PostTest loop*.[1]

    Also, note that it will continue to execute only if the *LogicalExpression* is false. When it becomes true, control transfers to the next statement following the loop; thus, the reason for using the word UNTIL to describe the loop should be clear.

EXAMPLE 8.2   Employing a DO UNTIL Loop to Check the Validity of Data

PROBLEM STATEMENT: One situation where a DO UNTIL loop would be of use is when ensuring that a valid entry is made from the keyboard. For example, suppose that you developed a program that required nonzero data. Such would be the case where the data might serve as the divisor in a formula, and hence division by zero had to be avoided. It would be advisable to have the program check for zero values upon input. Use the DO UNTIL loop to accomplish this objective.

SOLUTION: The following program fragment inputs a value and checks to ensure that it is greater than zero:

```
DO
   INPUT "Value ="; value
LOOP UNTIL value > 0
```

Clearly, the user would not be able to proceed beyond this set of statements unless a positive `value` was entered.

    It should be noted that a DO WHILE statement could have been employed for the same purpose, as in

```
value = -1
DO WHILE value <= 0
   INPUT "Value ="; value
LOOP
```

Notice that more lines are required for this version because of the need for the priming

---

[1]Note that it is called the *repeat until* structure in Pascal.

assignment. From the perspective of economy of effort, the DO UNTIL would be preferable. Also, note that the *LogicalExpression* in this version is the complement of the one used for the DO WHILE. This is due to the fact that the DO WHILE terminates on a false, whereas the DO UNTIL terminates on a true.

8.2.3 The EXIT DO (Break or MidTest) Loop

The fundamental syntax of the EXIT DO loop is (Fig. 8.3)

```
DO

    StatementBlock

    IF LogicalExpression THEN EXIT DO

    StatementBlock

LOOP
```

Thus, this construct repeats until a *LogicalExpression* located in the body of the loop tests true. At this point, control transfers to the statement following the loop. For this reason, it is sometimes referred to as a *break loop* because it involves "breaking out" in the middle of a loop. It is also called a *MidTest loop* because of the position of the logical test in the body of the loop rather than at either end.

It should be noted that the break loop is not a standard control construct in some other high-level languages. However, it is well-designed because the EXIT clause always

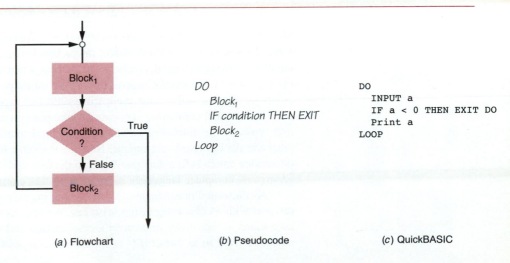

FIGURE 8.3
The EXIT DO or break loop.

(a) Flowchart

(b) Pseudocode

(c) QuickBASIC

transfers control to the line immediately after the loop. Consequently, when used properly, it does not lead to spaghetti code.

In addition, for newer dialects such as FORTRAN 90, it represents a significant structure. As in the following example, there are clearly cases where it is necessary.

EXAMPLE 8.3   *Using a Break Loop to Check the Validity of Data*

PROBLEM STATEMENT: One problem with the program fragment in Example 8.2 is that no guidance was given to the user regarding what she or he had done wrong. This could be remedied by including additional information in the body of the loop, as in

```
DO
   INPUT "Value = (must be positive)"; value
LOOP UNTIL value > 0
```

One problem with this approach is that the guidance is embedded in the query. In some cases, you might want to take a different approach and provide feedback only when the user errs. This mechanism is referred to as an *error trap* or *diagnostic*. Employ a break loop to design such a program fragment.

SOLUTION: The break loop can be employed to separate the input action from the error message, as in

```
DO
   INPUT "Value = "; value
   IF value > 0 THEN EXIT DO
   PRINT "Value must be positive, try again"
LOOP UNTIL value > 0
```

8.2.4   Logical Loops and "What If?" Computations

Most early mainframe computers (prior to the 1970s) operated as *batch-processing systems*. This mode allowed little direct interaction between the user and the machine. Often there were significant delays between the time a batch job was submitted and the time the output was received. Consequently, these batch programs were usually written in computer languages that were compatible with this type of environment.

Today, most computing applications take place in an *interactive environment*. The user typically sits in front of a display monitor and communicates directly with the computer via the keyboard or a mouse. Because response from the computer is direct, this interaction amounts to a dialogue between the user and the machine. Contemporary programs and computer languages must adapt to this environment.

As illustrated in Example 8.3, one facet of the user-machine interaction is providing the user with feedback regarding mistakes. Another example involves the case where a user wants to repeatedly implement a computation and view the result. These are commonly referred to as *"what if?"* or interactive computations. The following example is designed to illustrate some features of loops and other aspects of QuickBASIC that can facilitate such computations.

EXAMPLE 8.4   *Designing an Interactive Version of the Simple Addition Program*

PROBLEM STATEMENT: In Chap. 5, we presented the simple addition program as an example of a very elementary BASIC program. Develop a more sophisticated version employing loops and other facets of QuickBASIC such as selection constructs. Design the program so that it is interactive. In particular, provide the user with feedback and allow him or her to repeat the program as many times as desired.

SOLUTION: Before designing the program, we will first explore how loops can be employed to allow the user to perform multiple runs. Because most programs would usually perform the action once before questioning the user, this is clearly a case where the DO UNTIL loop would be used. One way to set up the algorithm would be

```
DO
    .
    .
    .
    action
    .
    .
    .
    INPUT "Do you want to repeat the action (y or n)"; q$
LOOP UNTIL q$ = "n"
```

This loop will cycle until the user enters an "n" to signify a negative response to the query. However, a problem with this scheme is that the user might enter an uppercase "N". If this is done, the loop test will not work as expected because to the computer "N" ≠ "n". Therefore, the user might want to terminate the action but be forced to perform it again because of the computer's distinction between uppercase and lowercase.

An alternative involves reformulating the UNTIL statement as

```
LOOP UNTIL q$ = "n" OR q$ = "N"
```

For this version, if the user enters either "n" or "N", the test is satisfied.

However, a problem still exists. What happens if the user wants to type an "n" but mistakenly strikes one of the adjacent keys (for example, an "m" or a "b")? Close inspection of the loop indicates that because the struck key was not an "n", the loop will cycle again as though the user had typed a "y"! Clearly, this is unacceptable.

A way to avoid this is to ensure that the user enters only an "n" or a "y". This can be accomplished with a break loop as previously described in Example 8.3. One way to do this is shown here:

```
'Simple Addition Program (Interactive Version)

'by S.C. Chapra
'July 31, 1992

'This program queries the user for two
'numbers, asks for the summation, and
'then provides feedback on the validity
'of the user's response.
'-------------------------------------------

'Definition of variables

'first, second = variables to be added
'total = sum of the variables
'answer = user's response for the sum
'd$ = dummy variable
'q$ = query variable
'-------------------------------------------

DO
  CLS
  INPUT "First number = "; first
  INPUT "Second number = "; second
  total = first + second
  PRINT
  INPUT "Please enter the answer = "; answer
  PRINT
  IF answer = total THEN
    PRINT "Good job!"
  ELSE
    PRINT "Sorry, the correct answer is "; total
    PRINT
    INPUT "Strike [Enter] key to continue", d$
    PRINT
    PRINT "Remember, if you get stuck,"
    PRINT "use those old fingers and toes!"
  END IF
  DO
    PRINT
    INPUT "another run (y or n) "; q$
    IF q$ = "y" OR q$ = "Y" OR q$ = "n" OR q$ = "N" THEN EXIT DO
    PRINT : PRINT "You did not enter a y or an n.  Try again"
  LOOP
LOOP UNTIL q$ = "n" OR q$ = "N"
CLS
END
```

FIGURE 8.4
An interactive version of the simple addition program. This version illustrates how selection and repetition constructs are employed to provide feedback and allow the user to perform multiple runs. See Example 8.4 for further explanation of the program.

```
DO
    .
    .
    .
    action
    .
    .
    .
    DO
        INPUT "Do you want to repeat the action"; q$
        IF q$ = "y" OR q$ = "Y" OR q$ = "n" OR q$ = "N" THEN EXIT DO
        PRINT "You did not enter a y or n. Try again."
    LOOP
LOOP UNTIL q$ = "n" OR q$ = "N"
```

Now we can use this structure to refine the simple addition program. Code to solve this problem is listed in Fig. 8.4, and a sample session is depicted in Fig. 8.5. Observe how an IF/THEN/ELSE structure is employed to give the user feedback regarding the correct answer for the summation. In particular, if the user enters the incorrect answer, we provide a tip to help him or her on the next try.

Also, observe how nested logical loops are employed to query the user to determine

---

FIGURE 8.5
Input/output from a session
using the program from
Fig. 8.4.

```
First number = ? 1
Second number = ? 2

Please enter the answer = ? 4

Sorry, the correct answer is 3

Strike [Enter] key to continue

Remember, if you get stuck,
use those old fingers and toes!

another run (y or n) ? y


First number = ? 2
Second number = ? 3

Please enter the answer = ? 5

Good job!

another run (y or n) ? b

You did not enter a y or an n. Try again

another run (y or n) ? N
```

whether another run is desired. As described above, the inner break loop ensures that the user has provided either a "y" or an "n" signifying "yes" and "no," respectively, in answer to the query. Then, if the response is "y", the outer DO UNTIL loop repeats. If it is "n", the loop terminates.

8.2.5 The INKEY$ Function

The program in Fig. 8.4 pauses at several junctures and proceeds only after the user responds to a query. In all cases, the queries are triggered by an INPUT statement involving the entry of a character or a numerical quantity. In most instances, the query flowed naturally from the dialogue. However, there was one instance where the pause and response was somewhat artificial,

```
INPUT "Strike [Enter] key to continue"; dummy$
```

As displayed, this code employs a dummy INPUT statement to allow the user to signal that she or he is ready to continue by pressing the [Enter] key. Although this represents an adequate mechanism, it has some deficiencies. First, it is somewhat artificial in that the INPUT variable, aptly named dummy$, has no purpose other than to allow the user to proceed. Second, the user must locate the [Enter] key. Third, it is conceivable that a non-character key, such as the [End] or [Home] keys, could be struck inadvertently. In this case, an error message could result.

An alternative, which permits any key to be struck, is provided by the INKEY$ function. The *INKEY$* function returns a string constant related to a character read from the standard input device (usually the keyboard). This function is employed for a number of different input manipulations. In the present context, its relevant behavior relates to the fact that if you strike any key on the keyboard, the INKEY$ will return a character string (whose value depends on which key happened to be struck). In contrast, if a key is not struck, it returns a null string (that is, " ").

On the basis of this knowledge, a simple way can be devised to have the computer pause until the user strikes a key. One way to do this is via the following statements:

```
PRINT "Strike any key to continue"
DO
  key$ = INKEY$
LOOP UNTIL key$ <> ""
```

As long as you do not strike any key, the DO UNTIL loop will continue to execute because INKEY$ will be equal to the null string (" "). Once you press a key, a character string will be returned and the loop will terminate.

It should be noted that there are a variety of formats for this operation. For instance, the DO UNTIL version can be written as a single line, as in

```
DO:LOOP UNTIL INKEY$ <> ""
```

This version has the disadvantage that its structure has been obscured by writing it as a single line. However, because the whole line amounts to a single action, this is one example of where the use of the colon is acceptable.

In addition, it should be noted that the DO WHILE can be employed to perform the identical action

```
DO WHILE INKEY$ ="":LOOP
```

For this case, the complement of the condition is used because of the different termination conditions for the WHILE and the UNTIL constructs.

Finally, the EXIT DO can also be employed,

```
DO:IF INKEY$ <> "" THEN EXIT DO:LOOP
```

The reason that all three logical loop constructs can be employed relates to the fact that for this case all that matters is that the loop terminates, not where it terminates.

8.3 COUNTER (FOR/NEXT) LOOPS

Suppose that you wanted to perform a specified number of repetitions, or iterations, of a loop. One way, employing a DO WHILE construct, is depicted in the flowchart and program fragment in Fig. 8.6. The loop is designed to repeat 10 times. The variable c is a

FIGURE 8.6
A (a) flowchart and (b) QuickBASIC code for a counter loop constructed with a DO WHILE loop.

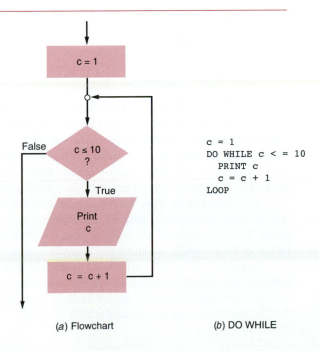

```
c = 1
DO WHILE c < = 10
    PRINT c
    c = c + 1
LOOP
```

(a) Flowchart   (b) DO WHILE

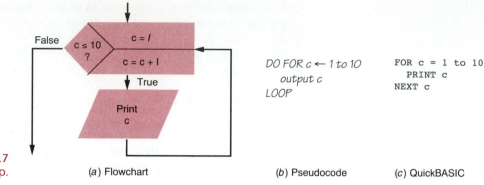

FIGURE 8.7
The BASIC FOR/NEXT loop.   (a) Flowchart   (b) Pseudocode   (c) QuickBASIC

counter that keeps track of the number of iterations. If c is less than or equal to 10, an iteration is performed. On each pass, the value of c is displayed and c is incremented by 1. After the tenth iteration, c has a value of 11. Therefore the test will fail, and control is transferred out of the loop.

Although the DO WHILE loop is certainly a feasible option for performing a specified number of iterations, the operation is so common that a special set of statements is available in BASIC for accomplishing the same objective in a more efficient manner. Called the *FOR/NEXT* loop, its general format is (Fig. 8.7)

FOR *index = start* TO *finish* STEP *increment*

   StatementBlock

NEXT *index*

The FOR/NEXT loop works as follows. The *index* is a variable that is set at an initial value (*start*). The program then executes the body of the loop, moves to the NEXT statement, and then loops back to the FOR statement. Every time the NEXT statement is encountered, the *index* is automatically increased by the *increment*. Thus the *index* acts as a counter. Then, when the *index* is greater than the final value (*finish*), the computer automatically transfers control out of the loop to the line following the NEXT statement.

An example of the FOR/NEXT loop is provided in Fig. 8.7c. Notice how this version is much more concise than the DO WHILE loop in Fig. 8.6. Also notice that the terms "STEP *increment*" are omitted from the FOR statement. If STEP terms are not included, the computer automatically assumes a default value of +1 for the *increment*.

As shown in Fig. 8.7c, all BASIC statements within a FOR/NEXT loop should be indented two spaces from that of the FOR statement. The NEXT statement should be indented the same number of spaces as its associated FOR statement. Any number of BASIC statements can be included within the loop.

EXAMPLE 8.5   Using a FOR/NEXT Loop to Input and Total Values

PROBLEM STATEMENT: As in Example 8.1, a common application of repetition is to enter data from the keyboard. The FOR/NEXT loop is particularly well suited for cases where the user has foreknowledge of the number of values that are to be entered. Employ the FOR/NEXT loop to input a set number of values to a program. Also, design the code so that the values are totaled.

SOLUTION: A program fragment to perform the above task is

```
CLS

count = 0
total = 0

INPUT "Number of values ="; count
FOR i = 1 TO count
   INPUT "Value =": value
   total = total + value
NEXT i
```

The number of data values is input and stored in the variable count. Thereafter, count serves as the *finish* parameter for the loop. Because the *increment* is not specified, a step size of +1 is assumed. Therefore the loop repeats count times while entering one value per iteration.

The size of the *increment* can be changed from the default of 1 to any other value. The *increment* does not have to be an integer, nor does it even have to be positive. If a negative value is used, then the logic for stopping the loop is reversed. With a negative *increment*, the loop terminates when the value of the *index* is less than that of the *finish*. Step sizes such as -5, -1, 0.5, or 60 would all be possible.

Note that the *index* does not have to be included with the NEXT statement. However, it greatly improves the readability of the program because it reinforces, along with indentation, the connection between associated FOR and NEXT statements. Therefore, it should be used. Remember that FOR and NEXT statements must be used in pairs and that the FOR statement must always come before its NEXT statement.

EXAMPLE 8.6   Examples of Using FOR/NEXT Loops

PROBLEM STATEMENT: Illustrate the use of FOR/NEXT loops with (*a*) an increment greater than 1 and (*b*) a negative increment.

SOLUTION:
(*a*) The following program displays numbers from 2 to 22 with an increment of 4:

```
FOR count = 2 TO 22 STEP 4
   PRINT count;
NEXT count
PRINT
PRINT "After exiting loop, count ="; count
```

Run:

```
2 6 10 14 18 22
After exiting loop, count = 26
```

Notice that upon exit from the loop, count has been incremented by 4 beyond the *finish* limit of the loop. If you do not recognize this, you might run into problems later on in your code. For example, you might misuse count in a computation.

(*b*) The following program displays numbers from 20 down to −5, using an increment of −6:

```
FOR i = 20 TO -5 step -6
   PRINT i;
NEXT i
PRINT:PRINT i
```

Run:

```
20 14 8 2 -4
-10
```

Observe how the semicolon suppresses the carriage control for the PRINT statement within the loop. The additional PRINT outside the loop triggers a carriage control so that the final PRINT occurs on a separate line.

8.4   NESTING CONTROL STRUCTURES

Repetition and selection structures may be enclosed completely within other repetition and selection structures. This arrangement, called *nesting,* allows algorithms of great power and complexity to be designed. In the present section, we will examine several examples of how this can be done.

8.4.1   Nesting Counter Loops

Counter loops may be enclosed completely within other counter loops. This arrangement is valid only if it follows the rule:

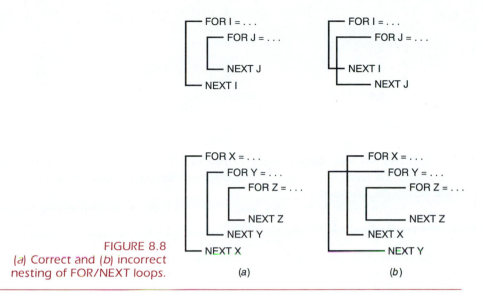

FIGURE 8.8
(a) Correct and (b) incorrect
nesting of FOR/NEXT loops.

- If a FOR/NEXT loop contains either the FOR or the NEXT parts of another loop, it must contain both of them.

Figure 8.8 presents some correct and incorrect versions. The incorrect versions will yield error messages. Matching the *index* between the FOR and NEXT statements is a good way to ensure that improper nesting does not occur. This is an added reason for including the loop *index* in the NEXT statement. Also, notice how indentation is used to delineate clearly the extent of the loops. This also helps to avoid improper nesting.

EXAMPLE 8.7   Example of Nested FOR/NEXT Loops

PROBLEM STATEMENT: Demonstrate the nesting of FOR/NEXT loops.

SOLUTION: The following program illustrates both the nesting of loops and the associated indentation:

```
FOR i = 1 TO 3
   FOR j = 1 TO 5
      PRINT i * j;
   NEXT j
   PRINT
NEXT i
PRINT "After termination of loops:"
PRINT "i ="; i
PRINT "j ="; j
```

Run:

```
1  2  3   4   5
2  4  6   8   10
3  6  9   12  15
After  termination  of  loops:
i = 4
j = 6
```

A PRINT has been placed between the two NEXT statements so that a carriage return occurs. Therefore, each line of output represents the output from the PRINT statement within the innermost loop.

For the first pass of the outer loop, $i = 1$ and the inner loop cycles though $j = 1$, 2, 3, 4, 5. Therefore, the product $i * j$ is displayed as 1 2 3 4 5. On the second pass, $i = 2$ and the inner loop again cycles though $j = 1$, 2, 3, 4, 5. Therefore, the product is displayed as 2 4 6 8 10. On the last pass of the outer loop, $i = 3$ and the inner loop again cycles though $j = 1$, 2, 3, 4, 5. Therefore, the product is displayed as 3 6 9 12 15.

8.4.2   Nesting Decision and Counter Loops

It should be mentioned that beyond counter loops, all the loops described in the present chapter can be nested. Such structure will be illustrated in the following example.

EXAMPLE 8.8   *Nested Decision and Counter Loops*

PROBLEM STATEMENT: Loops are commonly used to increment and print out values in a computation. For example, suppose that you wanted to print out a value for the beginning of a calculation (time = 0) and then print out values for subsequent time increments. A FOR/NEXT loop can be employed to accomplish this as in

```
last = 2
t = 0
PRINT "time ="; t
FOR i = 1 TO last
   t = t + 1
     PRINT "time =", t
NEXT i
```

Output:

```
time = 0
time = 1
time = 2
```

Now suppose that you want to print out intermediate times within a unit increment. If you did this using increments that are evenly divisible into 1 (for example, 0.1, 0.2, 0.25), it would be relatively simple to modify the program fragment to accomplish this action.

However, suppose that you wanted to use a fractional increment that was not an even factor of 1, say 0.4. Design code to accomplish this objective.

SOLUTION: One way to do this involves nesting a decision loop within the FOR/NEXT loop as delineated in the flowchart in Fig. 8.9. The corresponding QuickBASIC code is

```
last = 2
tprnt = 0.4
t = 0
PRINT "time = ", t
FOR i = 1 TO last
   tend = t + 1
   tstep = tprnt
   DO
      IF tstep > tend -t THEN tstep = tend - t
      t = t + tstep
      PRINT "time = ", t
   LOOP UNTIL t >  = tend
NEXT i
END
```

Output:

```
time = 0
time = 0.4
time = 0.8
time = 1
time = 1.4
time = 1.8
time = 2
```

Notice how the statement

```
IF tstep > tend -t THEN tstep = tend -t
```

is used to "shorten up" the increment tstep, when using tprnt would overshoot tend.

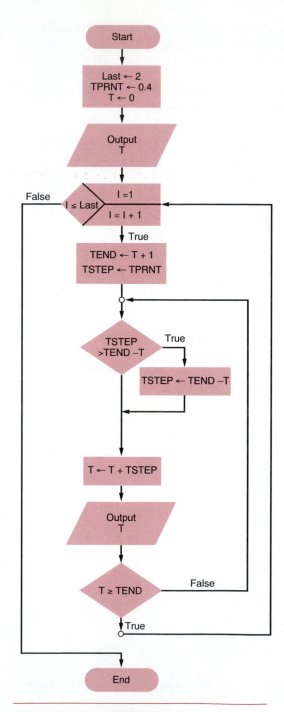

FIGURE 8.9 A flowchart showing nested counter and DO UNTIL constructs to solve Example 8.8.

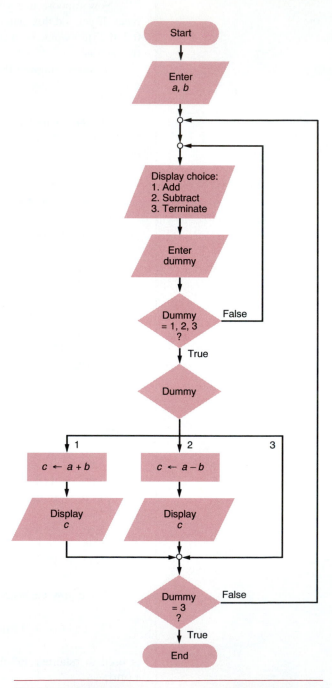

FIGURE 8.10 A flowchart for an interactive addition/subtraction program. This code includes a menu that provides the user with a number of options.

8.4.3 Nesting Repetition and Selection Structures

Repetition and selection constructs can also be nested. As with loops, each internal structure is indented. As a rule of thumb, we usually indent two spaces. The following example provides a simple illustration of how this is done.

EXAMPLE 8.9   *Design of an Interactive Menu*

PROBLEM STATEMENT: Everyone is familiar with ordering from a restaurant menu. Similarly, we might want the user to choose among a variety of options in a computer program. Use selection and repetition structures to design a menu for a simple addition/subtraction program.

SOLUTION: One way to do this is outlined in Figs. 8.10 and 8.11. After entering two numbers, the program moves into a group of nested structures.

The user is first presented with a menu consisting of three choices which are selected by entering a number: (1) add, (2) subtract, or (3) terminate. The menu is enclosed within a DO UNTIL loop that is designed to ensure that the user makes one of these choices before proceeding.

After making one of the three choices, the flowchart moves on to a CASE structure that either adds or subtracts and displays the result. Note that because of the simplicity of this example, a double-selection structure (IF/THEN/ELSE) would have been adequate. However, the CASE is used here because menus often have many options.

FIGURE 8.11
Computer code
corresponding to the
algorithm depicted in
Fig. 8.10.

```
CLS
INPUT "Enter two numbers = ", first, second
DO
  DO
    CLS
    PRINT "Make a menu selection"
    PRINT
    PRINT "1.  Add"
    PRINT "2.  Subtract"
    PRINT "3.  Terminate"
    PRINT : INPUT "Enter a number"; dummy
  LOOP UNTIL dummy = 1 OR dummy = 2 OR dummy = 3
  CLS
  SELECT CASE dummy
    CASE 1
      total = first + second
      PRINT "sum = "; total
    CASE 2
      total = first - second
      PRINT "difference = "; total
  END SELECT
  PRINT : PRINT "Strike any key to proceed"
  DO: LOOP UNTIL INKEY$ <> ""
LOOP UNTIL dummy = 3
CLS
END
```

The user is then allowed to inspect the result. After striking any key, the outer DO UNTIL loop tests whether the user had made menu selection 3 (terminate). If so, the flow drops out of the loop and the program ends. Otherwise, the loop cycles back and reprints the menu so that the user can make an additional selection.

As illustrated by the previous example, nested repetition and selection constructs can be employed to develop a user-friendly interactive program. Although the present example was simple, such nesting provides a powerful capability that can become logically complex. Close adherence to structured programming conventions such as indentation, as manifested by Figs. 8.10 and 8.11, is essential to keeping such algorithms lucid and maintainable.

8.4.4   The EXIT FOR

Just as the EXIT DO allows you to terminate a logical loop from its interior, the FOR/NEXT loop can be exited from its interior using the EXIT FOR statement. When this clause is executed, control is transferred to the statement immediately following the NEXT. This is useful at times when you do not wish to execute all the iterations, as in the following code:

```
INPUT n
FOR i = 0 TO n
  PRINT i;
  IF i > 9 THEN EXIT FOR
NEXT i
```

If the user enters any value of n that is less than or equal to 9, the loop will print out the values for i up to the entered value. For example, if n is input as 3, the program will display

```
0  1  2  3
```

If the user enters a value greater than or equal to 9, the program will display only 9 digits. For example, if n is entered as 25, the program will display

```
0  1  2  3  4  5  6  7  8  9
```

Although it might not be clear where such a capability could come in handy, there are occasions where it will be convenient to exit a FOR/NEXT loop in this manner. It should be noted that, as with the break loop, the EXIT FOR is not a standard control construct in some other high-level languages. However, it is well-designed because the EXIT clause always transfers control to the line immediately after the loop. Consequently, when used properly, it does not lead to spaghetti code.

Nevertheless, it should be mentioned that when many loops are nested, the use of EXIT statements can lead to some jumps that start to resemble GOTOs. For example, suppose that a DO UNTIL loop is nested within a FOR NEXT. If the EXIT FOR is placed within the DO UNTIL loop, it will cause transfer out of both loops to the statement following the NEXT. Although there will be times when this might be useful and straightforward, there could also be occasions when it would make your code difficult to follow. Hence, the EXIT statements, in general, should be employed with some discretion.

PROBLEMS

8.1 What will be displayed when the following loops are implemented?

(a)
```
a = 2
DO
   a = a * 3
   PRINT 2 * a
LOOP UNTIL a > 18
```

(b)
```
i = 5
FOR i = 2 TO 4 STEP 0.8
   PRINT i ^2;
NEXT i
PRINT i
```

(c)
```
m = 4: n = -10
FOR i = 1 TO m STEP 2
   FOR k = 9 TO n STEP -6
      PRINT i * k;
   NEXT k
   PRINT
NEXT i
```

(d)
```
FOR i = 1 TO 10 STEP 4
   j = i
   DO
      j = j - 3
      PRINT i; j
   LOOP UNTIL j <= 0
   PRINT
NEXT i
```

8.2 The exponential function can be represented by the following infinite series:

$$e^x = 1 + x + \frac{x^2}{2!} + \frac{x^3}{3!} + \frac{x^4}{4!} + \ldots$$

Write a program to implement this formula so that it computes the value of e^x as each term in the series is added. In other words, compute and display in sequence

$$e^x \cong 1$$

$$e^x \cong 1 + x$$

$$e^x \cong 1 + x + \frac{x^2}{2}$$

up to the order term of your choosing. For each of the above, compute the absolute percent relative error as

$$\% \text{ error} = \left| \frac{\text{true} - \text{series approximation}}{\text{true}} \right| \times 100\%$$

Use the library function for e^x in your computer to determine the "true value." Have the program display the series approximations and the error at each step along the way. Employ loops to perform the analysis. As a test case, use the program to evaluate $e^{1.5}$ up to the term $x^{10}/10!$.

8.3 Repeat Prob. 8.2, but make the order of the series expansion dependent on a user-supplied stopping criterion. In other words, have the user specify that the series should be expanded until the percent relative error falls below a prescribed level. Estimate the percent relative error with the following formula:

$$\% \text{ error} = \left| \frac{\text{present} - \text{previous}}{\text{present}} \right| \times 100\%$$

where the present estimate is the latest value of the series approximation and the previous estimate is the value calculated with one fewer term. Test your program for the same case as Prob. 8.2, employing a stopping criterion of 0.1 percent. Have the program display the series approximations and the error at each step along the way.

8.4 Repeat Prob. 8.2 for the hyperbolic cosine which can be approximated by

$$\cosh x = 1 + \frac{x^2}{2!} + \frac{x^4}{4!} + \ldots$$

As a test case, employ the program to evaluate $\cosh 78°$ up to the term $x^{10}/10!$. Remember that both the cosh and the above formula use radians. Also, recall that the cosh can be computed exactly with the exponential function,

$$\cosh x = \frac{e^x + e^{-x}}{2}$$

8.5 Repeat Prob. 8.4, but using the stopping criterion as specified in Prob. 8.3.

8.6 Repeat Prob. 8.2 for the sine, which can be approximated by

$$\sin x = x - \frac{x^3}{3!} + \frac{x^5}{5!} - \frac{x^7}{7!} + \ldots$$

As a test case, employ the program to evaluate $\sin 30°$. Remember, both the sine function and the above formula use radians.

8.7 Write a computer program to evaluate the following series:

$$sum = \sum_{i=0}^{n} \frac{1}{2^i} \quad \text{for } n = 20$$

Display *sum* and *n* during each iteration. Does the series converge to a constant value or diverge to infinity?

8.8 Repeat Prob. 8.7, except with the following formula ($n = 20$):

$$sum = \sum_{i=1}^{n} \frac{1}{i^2}$$

8.9 Repeat Prob. 8.7, except with the following formula ($n = 20$):

$$\text{sum} = \sum_{i=0}^{n} \left(\frac{2}{5}\right)^i$$

8.10 Consider the following formula:

$$\frac{1}{1 - x} = 1 + x + x^2 + \dots$$

Write a computer program to test the validity of this formula for values of 0.6, -0.3, 2, and $-3/2$.

8.11 The series

$$f(x) = \frac{4}{\pi}\left(\sin x + \frac{\sin 3x}{3} + \frac{\sin 5x}{5} + \dots\right)$$

is an approximation of $f(x)$ as shown in Fig. P8.11. Write a computer program to evaluate the accuracy of the series as a function of the number of terms at $x = \pi/2$. The program should display the error in the approximation and the number of terms in the series.

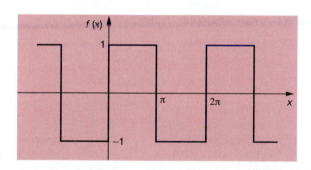

FIGURE P8.11

8.12 Economic formulas are employed extensively in engineering projects. For example, the following formula allows you to calculate the annual payment, A, needed for a loan:

$$A = \frac{i(1 + i)^n}{(1 + i)^n - 1}P$$

where P is the loan value, i is the annual interest rate, and n is the number of years for the loan.

Every year part of your payment is used to cover the interest. This amount is equal to the interest rate multiplied by the principal (that is, the amount left to pay at any particular time). The rest of your annual payment goes to reduce the principal.

Suppose that you are involved in the construction of a $450,000 storage tank for a gas processing facility. You can obtain a 10-year loan ($n = 10$) at an annual interest rate of 12 percent ($i = 0.12$). Develop a program to display a table showing the cumulative payment, interest, and principal (that is, the amount left to pay) for each year of the loan.

8.13 Develop a well-structured QuickBASIC program to display a table of temperatures in degrees Celsius and Fahrenheit. Query the user for the lower bound, upper bound, and increment for the Celsius temperatures. Use error traps to ensure that (1) the lower bound is less than the upper bound and (2) the increment is less than the range from the lower to the upper bound. Test the program for the case where the lower and upper bounds are 0 and 100, respectively, and the increment is 15°C. Also set up tests to ensure that your error traps are working properly.

8.14 In many cultures, certain numbers are ascribed qualities. For example, we are all familiar with "lucky 7" and "Friday the 13th." Ancient Greeks called the following number the "golden ratio":

$$\frac{\sqrt{5} - 1}{2} = 0.61803\ldots$$

This ratio was employed for a number of purposes, including the development of the rectangle in Fig. P8.14. These proportions were considered aesthetically pleasing to the Greeks. Among other things, many of their temples followed this shape.

The golden ratio is related to an important mathematical sequence known as the Fibonacci numbers, which are

0, 1, 1, 2, 3, 5, 8, 13, 21, 34, ...

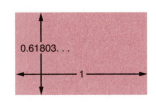

FIGURE P8.14
The golden ratio can be used to generate the rectangle shown above.

Thus, each number after the first two represents the sum of the preceding two. This sequence pops up in many diverse areas of science and engineering. In the context of the present discussion, an interesting property of the Fibonacci sequence relates to the ratio of consecutive numbers in the sequence; that is, $0/1 = 0$, $1/1 = 1$, $1/2 = 0.5$, $2/3 = 0.667$, $3/5 = 0.6$, $5/8 = 0.625$, and so on. If the series is carrier out far enough, the ratio approaches the golden ratio!

Develop a well-structured QuickBASIC program to compute and display the Fibonacci sequence and the corresponding ratios of its terms. Allow the user to enter the number of terms in the sequence. Employ error traps along with a diagnostic message to prevent the user from entering numbers less than 1. Also, in the event that the user requests more than 10 terms, limit the display to 10 terms at a time and display subsequent pages at the user's discretion. Finally, design the program so that the entire computation can be repeated as many times as the user desires.

8.15 The following equation is a simple model of the velocity of a falling object as a function of time:

$$v = \frac{gm}{c}[1 - e^{-(c/m)t}] \tag{P8.15.1}$$

where v = velocity [m/s], c = a drag coefficient [kg/s], m = mass [kg], g =

gravitational acceleration = 9.8 m/s², and t = time [s]. Suppose that you had to determine the drag coefficient required in order that a 68.1-kg object would reach a v = 44.87 m/s in 10 s. A problem with making such an estimate is that it is mathematically impossible to solve Eq. (P8.15.1) for c explicitly.

An alternative is to solve the problem interactively. In other words, the equation can be programmed to compute v repeatedly as the user makes guesses until the correct result for c is obtained. Design the program as a "what if?" type of calculation. Also, devise it so that the key input parameters are within the following bounds:

$0 < m < 100$ kg

$0 < t < 50$ s

$0 < c < 20$ kg/s

8.16 Develop an interactive program that allows the user to perform addition, subtraction, multiplication, and division of two numbers. Employ ideas from Figs. 8.4 and 8.11 when designing your code. In particular, employ a menu to allow the user to select from among addition, subtraction, multiplication, and division. Try to be as creative as possible in providing feedback and help to the user.

8.17 Write structured QuickBASIC code for the flowchart shown in Fig. P8.17.

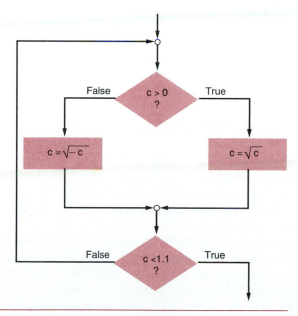

FIGURE P8.17

8.18 Write structured QuickBASIC code to input, count, and sum a collection of positive numbers. Use an appropriate sentinel value to signal that all the numbers have been entered. Also, keep track of the maximum and the minimum values that are entered. After performing this task, print out the number of values, the average, and the minimum and maximum.

CHAPTER 9

MODULAR PROGRAMMING

It is often desirable to execute a particular set of statements several times, in different parts of a program. It would be inconvenient and time-consuming to rewrite these statements separately every time they are required. Fortunately, it is possible in BASIC to write such sections once and then have them available at different locations within the program to use as many times as necessary. These "miniprograms," which are referred to as *procedures,* are the key to modular programming in BASIC. In the present chapter, we will introduce the two types available in modern BASIC: the SUB and FUNCTION procedures.[1] However, before doing this, we would like to review how early forms of BASIC attempted to incorporate miniprograms.

9.1   THE GOSUB/RETURN

As outlined in Fig. 9.1, the GOSUB and RETURN statements were an attempt to introduce modularity to early dialects of BASIC. The GOSUB transfers control to a line in a fashion identical with a GOTO statement. Thereafter execution proceeds normally until the RETURN statement is encountered, whereupon control is transferred back to the line immediately following the GOSUB.

The GOSUB/RETURN combination is just like having a GOTO in the main program that branches to another area of the program and a GOTO that then branches back to the statement immediately following the calling GOTO. The advantage of using GOSUB/RETURN instead of two GOTO statements is that the former conveniently allows the subroutine to be called from several locations. The RETURN ensures that control is automatically transferred back to the correct line in the main program; whereas the

---

[1] In other high-level languages, the terminology for the SUB and FUNCTION procedures differs from QuickBASIC. For example, in FORTRAN, a procedure is generally referred to as a *subprogram*. The SUB procedure is called a *subroutine subprogram*, and the FUNCTION procedure is called a *function subprogram*. In Pascal, they are simply called procedures and functions, respectively.

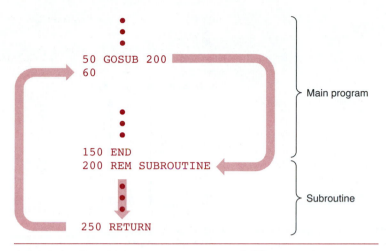

```
          •
          •
          •
   50 GOSUB 200
   60
                                        Main program
          •
          •
          •
   150 END
   200 REM SUBROUTINE
          •
          •
          •                             Subroutine
   250 RETURN
```

FIGURE 9.1 The GOSUB was the only available option for modular programming in the older dialects of BASIC. It is now obsolete because of the development of procedures.

use of a GOTO in the subprogram would require additional programming to return to the proper location.

Although the GOSUB/RETURN was a convenient first step toward modular programming, it had several disadvantages. For example, it was just like a GOTO in that it encouraged the "jumping around" that tended to make unstructured programs disorganized and confusing. Also, it was not truly modular in the sense that the GOSUB/RETURN was not a separate entity but rather an integral part of the program.

Today, the problems associated with the GOSUB are a moot point. Because of the development of the SUB procedure, the GOSUB is now obsolete and should never be used.

9.2   SUB PROCEDURES

The SUB procedure differs from the GOSUB in that it is truly an autonomous module or miniprogram. It is invoked by a CALL statement which is of the general form

CALL *name(arg$_1$, arg$_2$,..., arg$_n$)*

where *name* is a variable name that can be up to 40 characters long and *arg$_i$* is the *i*th argument passed to the procedure. These arguments can be either constants, variables, or expressions.

The SUB procedure itself has the general form

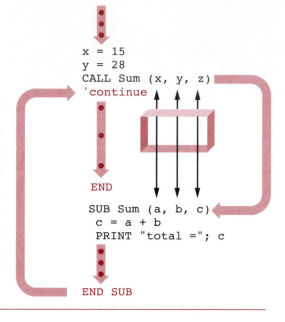

SUB *name (arg₁, arg₂,. . . , argₙ)*

.

.

.

Statements

.

.

.

END SUB

where *name* is the same *name* as was used in the CALL statement and each of the arguments corresponds to the arguments listed in the CALL. In the SUB statement all the arguments must be variable names. These arguments are not required to have the same names used in the CALL statement. Only their position determines which values get passed to which variables. Although the names may differ, the variable types must match.

As in the old GOSUB, the CALL statement transfers control to the procedure, where execution proceeds normally until the END SUB is encountered. Thereupon, control is transferred back to the statement immediately following the CALL. The process is depicted graphically in Fig. 9.2.

EXAMPLE 9.1   SUB Procedure to Compute Volume and Surface Area of a Rectangular Prism

PROBLEM STATEMENT: Figure 9.3*a* summarizes formulas for the volume and surface area of a rectangular prism. Develop a procedure to compute these values.

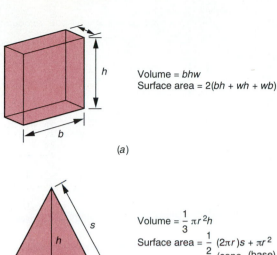

Volume = *bhw*
Surface area = 2(*bh* + *wh* + *wb*)

(a)

Volume = $\frac{1}{3} \pi r^2 h$

Surface area = $\frac{1}{2} \underset{\substack{\text{(cone} \\ \text{wall)}}}{(2\pi r)s} + \underset{\substack{\text{(base)}}}{\pi r^2}$

(b)

where $s = \sqrt{r^2 + h^2}$

Volume = $\pi r^2 h$
Surface area = $\underset{\substack{\text{(cylinder} \\ \text{wall)}}}{2\pi rh} + \underset{\substack{\text{(bases)}}}{2\pi r^2}$

(c)

FIGURE 9.3
Formulas to compute the
volume and the surface areas
of a number of geometric
solids: (a) rectangular prism,
(b) cone, (c) cylinder, and
(d) sphere.

Volume = $\frac{4\pi r^3}{3}$

Surface area = $4\pi r^2$

(d)

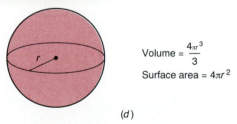

SOLUTION: Figure 9.4 shows BASIC code to solve this problem. The main program is organized into three procedure CALLs to a preprocessor (inputs the data), a processor (performs the computation), and a postprocessor (displays the results). Note that all the data needed by each SUB procedure is included as arguments.

Main Program:

```
DECLARE SUB PreProcessor (l!, w!, h!)
DECLARE SUB Processor (l!, w!, h!, v!, a!)
DECLARE SUB PostProcessor (v!, a!)

'Rectangular Prism Program

'by S.C. Chapra
'May 31, 1992

'This program calculates the volume and
'surface area of a rectangular prism
'------------------------------------

'Definition of variables

'p.len = prism base length
'p.wid = prism base width
'p.heit = prism height
'p.vol = prism volume
'p.area = prism surface area
'------------------------------------

CALL PreProcessor(p.len, p.wid, p.heit)
CALL Processor(p.len, p.wid, p.heit, p.vol, p.area)
CALL PostProcessor(p.vol, p.area)

END
```

Procedures:

```
SUB PreProcessor (l, w, h)

   'inputs dimensions of rectangular prism
   '------------------------------------

   'Definition of variables

   'l = prism base length
   'w = prism base width
   'h = prism height

      '------------------------------------

   CLS
   PRINT "Enter dimensions for a rectangular prism:"
   PRINT
   INPUT "base length = "; l
   INPUT "base width = "; w
   INPUT "height = "; h

END SUB
```

```
SUB Processor (l, w, h, v, a)

   'computes the volume and area for a
   'rectangular prism
   '------------------------------------

   'Definition of variables

   'l = prism base length
   'w = prism base width
   'h = prism height
   'v = prism volume
   'a = prism surface area
   '------------------------------------

   v = l + h + w
   a = 2 * (l * h + w * h + w * l)

END SUB
```

```
SUB PostProcessor (v, a)

   'displays results
   '------------------------------------

   'Definition of variables

   'v = prism volume
   'a = prism surface area
   '------------------------------------

   CLS
   PRINT "volume = "; v
   PRINT "area = "; a

END SUB
```

Output:

```
Enter dimensions for a rectangular prism:

base length = ? 1
base width = ? 2
height = ? 3

volume =  6
area =  22
```

FIGURE 9.4 A program using SUB procedures to compute the volume and surface area of a rectangular prism.

In the procedures themselves, observe that abbreviated variable names are used for the arguments. Although the same names as were used in the main program could have been chosen, different names are employed to highlight the modular nature of the SUB procedure and to reinforce the fact that the two sets of variables can be given different names.

Finally, we have used comments to label modules and define variables. Although this is not really necessary for such a simple program, it is absolutely essential for large and complex programs. Even though it takes extra effort, it is recommended that you document all your programs in this manner.

9.3 PASSING PARAMETERS TO PROCEDURES

Note that the procedure's arguments are collectively referred to as its *parameter list,* and each argument is called a *parameter.* A *formal parameter* is the name by which the argument is known within the procedure. The *actual parameter* is the specific variable, expression, or constant passed to the procedure when it is called.

In contrast to the open-ended nature of the GOSUB, the parameter list provides a means to control the flow of information between the calling program and the SUB procedure. This quality is illustrated by the following example.

EXAMPLE 9.2   The Behavior of the Parameter List

PROBLEM STATEMENT: Investigate how the parameter list controls the flow of information by examining the output of the following program:

```
CLS
a = 1: b = 2: c = 3
PRINT a; b; c; d
CALL Test(a, b, c)       Main program
PRINT a; b; c; d
END

SUB Test(b, a, d)
  PRINT a; b; c; d
  d = 4                  Procedure
  c = b
END SUB
```

SOLUTION: The first PRINT statement yields

```
1 2 3 0
```

Thus, a, b, and c have their original assigned values of 1, 2, and 3, respectively. The variable d is equal to 0 at this point because it has yet to be assigned a value.

Next, the values a, b, and c are passed to the SUB procedure, where they are temporarily associated with the variable names b, a, and d. Consequently, when the second PRINT statement is implemented, the result is

```
2  1  0  3
```

Notice the "local" nature of variable c in the SUB. The labels that are passed as parameters b, a, and d initially have values related to the variables in the main program a, b, and c. Consequently, a label such as c has a value of zero when it is first printed in the SUB. It is then assigned a value of 1 by the statement c = b. However, because c is not "passed" back to the main program, this assignment holds only in the SUB and has no effect on the value of c in the main program.

After the two assignment statements in the SUB procedure are implemented, control passes back to the main program, where the final PRINT yields

```
1  2  4  0
```

Observe how the value of 4 is assigned to the variable c because of its position in the parameter list. Similarly, the SUB procedure assignment statement, d = 4, has no effect on the variable d in the main program. This is due to the fact that d is not included in the CALL parameter list. Therefore, the two d's are distinct.

Although the foregoing example is not very realistic, it serves to illustrate how the parameter list operates. In particular, it demonstrates how actions and variables in a SUB procedure can be either isolated from or connected to other parts of a program. This quality is essential for implementation of modular programming. Before exploring how this is done, we will first discuss how the *ParameterList* actually works within the computer.

As described earilier, when a variable appears in the main program, a location is defined in the computer's memory. When an action is taken to assign a value to the variable, this is where the value resides. Now, when a variable name is passed to a procedure via the *ParameterList,* a new memory location is *not* created. Rather, the computer knows to go back to the main program to the original memory location in order to obtain the value. It "knows" this because of the presence of the variable in the parameter list. This is referred to as passing the parameter *by reference* or *by address.*

In contrast, suppose that the argument in the parameter list is a constant. For example, if you wanted to pass a value of 2 as the first argument in the SUB procedure from Example 9.2, you could merely include the statement

```
CALL Test(2, b, c)
```

For this case, the "2" is copied to a temporary memory location, and then the address of this location is passed to the procedure. This is referred to as passing the parameter *by value.*

This operation has special significance because variables can also be passed by value. This is done by enclosing the variable in parentheses. For example, suppose that we wanted to pass the variable a by value. This could be accomplished by

```
CALL Test((a), b, c)
```

The following example provides a further illustration of this feature and its behavior.

EXAMPLE 9.3   *Passing by Value and by Reference*

PROBLEM STATEMENT: Develop a short program that employs passing by value and by reference in the same procedure.

SOLUTION: The following program can be written:

```
CLS
x = 1: y = 0
PRINT "before SUB: x="; x; "y="; y
CALL Add((x), y)
PRINT "after SUB: x="; s; "y="; y
END

SUB Add (x, y)
   y = x + 1
   x = 0
   PRINT "in SUB: x="; x; "y="; y
END SUB
```

The output for this program would be

```
before SUB: x = 1 y = 0
in SUB:     x = 0 y = 2
after SUB:  x = 1 y = 2
```

Notice that because x is passed by value, even though it is modified in the procedure, its original value is maintained in the main program. The same is not true for y, which is passed by reference.

Another example of passing by value is when an expression is passed to a procedure. For example,

```
CALL Test(a + 5, b, c)
```

Just as with the constant, the computer first evaluates the expression and places the result in a temporary memory location. Then the address of this location is passed to the procedure. Now it should be clear why the parentheses are used to pass a variable by value. In essence, the parentheses "trick" the computer into thinking that you are passing an expression!

Before proceeding, we should point out why you might want to pass a variable by value. There will be occasions where you may want to perform operations on a variable in a procedure that change its value. At the same time, you might want to retain the original value for some other purpose. On such occasions, passing by value provides a handy means to perform the desired operation while maintaining the integrity of the original variable.

9.4   FUNCTION PROCEDURES

Aside from the SUB, the FUNCTION procedure represents the other major vehicle for modular programming in BASIC. Recall that in Sec. 6.3, we introduced the user-defined function (or DEF FN) to program an expression that was used repeatedly in a program. Although these one-line functions[2] are useful, there are often occasions when several lines are needed to compute the desired quantity. The *FUNCTION procedure*, sometimes called a function subprogram, provides a convenient way to do this.

In contrast to the one-line DEF FN, the FUNCTION procedure is a miniprogram whose general format is

```
FUNCTION name (arg₁, arg₂,..., argₙ)
          .
          .
          .

name = expression
          .
          .
          .

END FUNCTION
```

where *name* is the name of the FUNCTION procedure, arg_i is the ith argument passed to the procedure, and *expression* is the return value that is assigned to the function.

Notice that somewhere in the body of the function, the *expression* must be assigned to the function's *name*. As illustrated by the following example, this is the way that the result is returned to the place where the function was invoked.

EXAMPLE 9.4   A Simple Function Procedure to Calculate a Factorial

PROBLEM STATEMENT: Employ a FUNCTION procedure to determine the factorial. Recall that the factorial is computed as

---

[2] Note that in QuickBASIC, it is possible to program multiline DEF FN functions. However, because the FUNCTION procedures described in the present section offer greater flexibility and control, the multiline DEF FN function will not be described here and its use is not recommended.

$$n! = 1 * 2 * 3 * \cdots * (n - 2) * (n - 1) * n$$

For example, 4 factorial would be calculated as

$$4! = 1 * 2 * 3 * 4 = 24$$

Note that be definition, $0! = 1$.

SOLUTION: The following program illustrates how the factorial can be computed as a **FUNCTION** procedure:

```
CLS

FOR i = 0 TO 5
  PRINT i; "! = "; Factorial(i)
NEXT i
END

FUNCTION Factorial (n)
  x = 1
  FOR i = 1 TO n
    x = x * i
  NEXT i
  Factorial = x
END FUNCTION
```

Output:

```
0! = 1
1! = 1
2! = 2
3! = 6
4! = 24
5! = 120
```

Observe how the function is invoked by name in the PRINT statement. If you wanted to save the result in a memory location, you could add an assignment statement such as

```
f3 = Factorial (3)
```

In addition, you can use the function as part of an expression as in

```
sin.approx = x - x ^ 3/Factorial (3) + x ^ 5/Factorial (5)
```

9.4.1 Comparison between SUB and FUNCTION Procedures

Note that FUNCTION and SUB procedures are similar in many respects. For example, information can be passed to a function by either reference or value with a parameter list.

Although they behave similarly in many respects, SUB and FUNCTION procedures differ in some significant ways. Figure 9.5 delineates the differences among SUB and FUNCTION procedures, and the other types of functions that are available in QuickBASIC. The major difference between a SUB and a FUNCTION procedure is that the latter returns a result that can be used directly, just as if it were a variable.

To this point, we have covered two other types of statements that behave in a similar fashion—the intrinsic functions and the statement functions discussed in Chap. 6. Recall how an intrinsic function such as SIN(X) can be used as the part of, say, a mathematical expression. The FUNCTION procedure can be employed in a similar

---

FIGURE 9.5
Summary and comparison of the SUB and FUNCTION procedures, and the other types of functions available in QuickBASIC.

INTRINSIC OR LIBRARY FUNCTIONS: Built-in functions that perform commonly employed mathematical or trigonometric operations and that generate a single value that can be used just like a variable—e.g., as part of a mathematical expression, in a PRINT statement, etc.

```
Definition:        Defined by system
Application:       y = SIN(X)
```

STATEMENT FUNCTIONS: User-defined functions that return a single value. The function is defined by a one-line statement, which must be placed in the main program.

```
Definition:        FNround (x) = INT(x + 0.5)
Application:       y = FNround(x)
```

FUNCTION PROCEDURES: User-defined miniprograms that consist of several lines. The FUNCTION procedure generates a single value that can be invoked by the procedure name.

```
Application:       y = Area(radius)
Definition:        FUNCTION Area (r)
                      Area = 3.14159 * r ^ 2
                   END FUNCTION
```

SUB PROCEDURES: User-defined miniprograms that consist of several lines. Several results may be generated and passed back to the main program. The results are passed between the SUB procedure and the main program via the SUB's arguments or via SHARED statements.

```
Application:       CALL Sum(x, y, z)
Definition:        SUB Sum (a, b, c)
                      c = a + b
                      PRINT c
                   END SUB
```

fashion. However, they differ from the intrinsic and the one-line user-defined statement functions in some fundamental ways. In contrast to the intrinsic function, which is supplied by the system, the FUNCTION procedure is concocted by the user. In contrast to the user-defined statement function, which is limited to a single line, the function subprogram can consist of many statements. Thus, as seen in Fig. 9.5, the FUNCTION procedure lies midway between statement functions and SUB procedures.

9.5 LOCAL AND GLOBAL VARIABLES

As described in the previous sections, two-way communication is required between a procedure and the main program. The procedure needs data to act on and must return its results for utilization in the main program.

When the GOSUB was used in early versions of BASIC, this was not an issue because all variables were in common or *global*. Every part of the program had access to all variables.

In modern BASIC, the parameter list provides a controlled means for transferring information. In other words, the information that is passed in and out of the procedure must be specified explicitly by the programmer. Because both the main program and the procedure have direct access to them, the variables in the list can be thought of as being global to that procedure specifically. In contrast, variables that reside exclusively in the procedure are called *local variables*.

It should be mentioned at this point that when a local variable appears in a procedure, a location is defined in the computer's memory. When an action is taken to assign a value to the variable, this is where the value for this variable resides. This is in contrast to parameter-list variables which, as noted previously, do not have independent memory locations.

The ability to have local variables in a procedure can be a great asset. In essence, they serve to make each procedure a truly autonomous entity. For example, changing a local variable will have no effect on variables outside the procedure. In fact, true local variables can even have the same name as variables in other parts of the program and still be distinct entities. One advantage of this is that it decreases the likelihood of "side effects" that occur when the value of a global variable is changed in a subprogram and is then used erroneously elsewhere in another part of the program. In addition, the existence of local variables also facilitates the debugging of a program. The subprograms can be debugged individually without fear that a working procedure will cease functioning because of an error in another procedure.

9.5.1 SHARED Statement

Although local variables are a great benefit, there will be occasions when you will want your subprograms to have direct access to certain variables in the calling program. To this point, we have shown one way in which this can be done: the *ParameterList*.

A variable may also be made available for use by a subprogram with the SHARED statement. The SHARED statement directs the compiler to set aside a certain area in the computer's memory which the main program and the procedure may access to obtain values of the variable. SHARED statements have the general form

SHARED *VariableList*

where the *VariableList* are the comma-delimited variables that are to be shared.

The SHARED statement can appear only in a SUB or FUNCTION procedure. When it appears in a particular procedure, it causes the variables to be shared between that procedure and the calling module. For example, in the program from Fig. 9.4, the procedure PostProcessor could have been represented as

```
SUB PostProcessor
  SHARED p.vol, p.area
  CLS
  PRINT "volume = "; p.vol
  PRINT "area = "; p.area
END SUB
```

The CALL in the main program would be modified to

```
CALL PostProcessor
```

Notice how the new version of the procedure must now employ variable names from the main program.

Aside from the SHARED statement, there is also a SHARED attribute.[3] This is used when you desire variables to be globally accessible everywhere in your program. Its general form is

COMMON SHARED *VariableList*

When this statement appears in the main program, it makes the *VariableList* global. For example, in the case of the program from Fig. 9.4, the following statement could have been included in the main program:

```
COMMON SHARED p.len, p.wid, p.heit, p.vol, p.area
```

If this had been done, all the procedure parameter lists could have been omitted. However, it would also mean that all the procedures would have to be written in terms of the main-program variable names. For instance, the formula for the volume in the *Processor* procedure would have to be written as

```
p.vol = p.len * p.heit * p.wid
```

---

[3] An *attribute* is a keyword that defines a more general characteristic of the program than a statement.

9.5.2 The STATIC Statement and Attribute

The variable values within a procedure are automatically reset each time the procedure is accessed.[4] There will be occasions when you might want to override this normal state of affairs and have procedure variables retain their value between CALLs. The STATIC statement, which is used for this purpose, has the general form

```
STATIC VariableList
```

When this statement appears in the procedure, it makes the *VariableList* static between CALLs.

It should also be noted that there will be times when you might want to freeze all the procedure variables between CALLs. In this case, the STATIC attribute can be used. This is done by appending the word "STATIC" to the first line of the procedure. The way this is done and the use of the SHARED and STATIC statements are illustrated in the following example.

EXAMPLE 9.5   Use of SHARED and STATIC in a Procedure

PROBLEM STATEMENT: Show how SHARED and STATIC invoke different behavior in procedure variables.

SOLUTION: The following program fragment can be written:

```
CLS
FOR i = 1 TO 2
  CALL Adder
NEXT i
PRINT: PRINT "in main program:"
PRINT "x = "; x; "y = "; y
END

SUB Adder
  STATIC x
  x = x + 1
  y = y + 2
  PRINT "in sub:"
  PRINT "x = "; x; "y = "; y
END SUB
```

The output for this program would be

```
in sub:
x = 1 y = 2
x = 2 y = 2
```

---

[4] Numeric variables are reset to zero and strings to the null character, which is the string equivalent of zero (that is, "").

```
in main program:
x = 0 y = 0
```

Observe how the STATIC statement causes the procedure to retain the value for the variable x from the first to the second call. In addition, see how none of the information is transferred to the main program.

The action of the STATIC attribute and the SHARED statement can be illustrated by editing the procedure to give

```
SUB Adder STATIC
   SHARED x
   x = x + 1
   y = y + 2
   PRINT "in sub:"
   PRINT "x = "; x; "y = "; y
END SUB
```

which results in the output

```
in sub:
x = 1 y = 2
x = 2 y = 4
in main program:
x = 2 y = 0
```

Now, both x and y are held static, so both sums are implemented. However, because only x is included in the SHARED statement, the value of y never changes in the main program. Thus, the y in the main program is clearly shown to be a distinct variable from the y in the procedure.

At this point, you might be wondering why there are seemingly identical means for sharing information between the main program and procedures: the *ParameterList,* and the SHARED statement or attribute. Although they might seem identical, there are some subtle differences that might make one preferable to another in a particular problem context. For example, the *ParameterList* is limited in size. In cases where you must transfer many variables to and from a procedure, numerous SHARED statements can allow many more values to be transferred.

Another example relates to the fact that the SHARED attribute is clearly different from both the *ParameterList* and the SHARED statement. It provides a concise means of making variables truly global. Thus, it should be used only for those variables that you know will be needed in many places throughout your program. It should be noted that when this is done, you should take care to give these variables unique and recognizable names to avoid side effects.

A further distinction relates to the fact that the *ParameterList* allows you to pass an argument to a procedure "by value." Neither the SHARED attribute nor the SHARED statement can be employed in this fashion.

Although these and other examples can be developed, there is one particular distinction between the approaches that has general significance. This relates to the fact that SHARED is ineffective for procedures that are designed to be used by many different programming applications. This is because the SHARED statement and attribute specify fixed names for the variables. In contrast, the *ParameterList* is generic. It allows you to write a procedure in terms of its own variable names. Then, whenever it is called, the *ParameterList* provides a means to employ other names for the variables via the CALL statement. This capability is at the heart of modular programming, which we will discuss in more detail next.

9.6 MODULAR PROGRAMMING

Just as the chapters are the building blocks of a novel, procedures are the building blocks of a computer program. Each procedure can have all the elements of a whole program, but in general, it is smaller and more specialized. Some computer experts suggest that a reasonable size limit for a procedure should be no more than about 50 to 100 lines. Because this size will fit on several monitor screens, it allows quick viewing of the whole procedure. It is a reasonable guideline because it facilitates visualizing the purpose and the composition of the procedure.

In addition to allowing you to view the procedure easily, the 50- to 100-line limit will also encourage you to restrict the procedure to one or two well-defined tasks. It is this characteristic of procedures that makes them the ideal vehicle for expressing a modular design in program form.

Procedures should always be laid out clearly, using blank spaces and lines. Every procedure should begin with a short description of the purpose of the procedure, and definitions of local variables (that is, those used exclusively in the procedure) should be included after the title and before the body of the procedure. Some programmers include the author, date, and an edition number (for example, 1.0, 1.1, 2.0) to keep track of the version.

The statements in the body of the procedure should be organized in a fashion similar to that of the main program. Structures should be identified by indentations, and remarks should be included as needed.

EXAMPLE 9.6   Menu-Driven Program to Access Procedures

PROBLEM STATEMENT: Modify the program from Fig. 9.4 so that it provides the user with a choice of determining the area and volume of a cone or a cylinder.

SOLUTION: A well-structured modular program to solve this problem is displayed in Fig. 9.6. A sample run is shown in Fig. 9.7.

Several important features of the program bear mentioning:

Main Program:

```
DECLARE SUB Cone ()
DECLARE SUB Cylinder ()
DECLARE SUB Display (solid$, area!, volume!)

'Volume/Area Program
'by S.C. Chapra
'June 3, 1992

'This program calculates the volume and
'surface area of a cone and a cylinder
'-------------------------------------
'Definition of variables:
'd = dummy variable
'-------------------------------------

CONST pi = 3.14159

DO
  DO                      'create a menu
    CLS
    PRINT "Select one of the following options"
    PRINT "to determine the volume and area for:"
    PRINT
    PRINT "1. Cone"
    PRINT "2. Cylinder"
    PRINT "3. Terminate the program"
    PRINT : INPUT "enter a number"; d
  LOOP UNTIL d = 1 OR d = 2 OR d = 3
  SELECT CASE d          'call SUBs according to
    CASE 1               'menu selection
      CALL Cone
    CASE 2
      CALL Cylinder
  END SELECT
LOOP UNTIL d = 3

END
```

Procedures:

```
SUB Display (solid$, area, volume)

  'Displays results
  '-------------------------------------
  'Definition of variables:
  'solid$ = type of solid
  'area = solid surface area
  'volume = solid volume
  '-------------------------------------

  CLS
  PRINT "For the "; solid$
  PRINT "area = "; area;
  PRINT " and volume = "; volume

  'loop until user strikes
  DO
  LOOP UNTIL INKEY$ <> ""

END SUB
```

```
SUB Cone

  'Inputs dimensions and computes volume
  'and area for a cone
  '-------------------------------------
  'Definition of variables:
  'r = cone radius
  'h = cone height
  'v = cone volume
  'a = cone surface area
  's = cone side length
  '-------------------------------------

  'input dimensions
  CLS
  PRINT "Enter dimensions for a cone:"
  PRINT
  INPUT "radius = "; r
  INPUT "height = "; h

  'perform computations
  s = SQR(r ^ 2 + h ^ 2)
  v = pi * r ^ 2 * h / 3
  a = 2 * pi * r * s / 2 + pi * r ^ 2

  'display results
  CALL Display("cone", a, v)

END SUB
```

```
SUB Cylinder

  'Inputs dimensions and computes volume
  'and area for a cylinder
  '-------------------------------------
  'Definition of variables:
  'r = cylinder radius
  'h = cylinder height
  'v = cylinder volume
  'a = cylinder surface area
  '-------------------------------------

  'input dimensions
  CLS
  PRINT "Enter dimensions for a cylinder:"
  PRINT
  INPUT "radius = "; r
  INPUT "height = "; h

  'perform computations
  v = pi * r ^ 2 * h
  a = 2 * pi * r * h + 2 * pi * r ^ 2

  'display results
  CALL Display("cylinder", a, v)

END SUB
```

FIGURE 9.6 A program using procedures to compute the volume and surface area of a cone and a sphere.

1. A CONST statement is used to define π in the main program. This value is then globally available for use throughout the rest of the program. Specifically, it is employed in the procedures to calculate the volumes and surface areas according to the formulas from Fig. 9.3.

2. DO UNTIL loops and a SELECT CASE are used in tandem to present and implement a menu in the main program. The interior DO loop cycles until the user makes an appropriate menu selection (1, 2, or 3). Then the SELECT CASE employs

---

FIGURE 9.7
Output generated by the program from Fig. 9.6. Notice that values input by the user are highlighted.

Screen 1:
```
Select one of the following options
to determine the volume and area for:

1. Cone
2. Cylinder
3. Terminate the program

enter a number  1
```

Screen 2:
```
Enter dimensions for a cone:

radius =   1
height =   2
```

Screen 3:
```
For the cone
area =   10.1664   and volume =   2.094393
```

Screen 4:
```
Select one of the following options
to determine the volume and area for:

1. Cone
2. Cylinder
3. Terminate the program

enter a number  2
```

Screen 5:
```
Enter dimensions for a cylinder:

radius =   1
height =   2
```

Screen 6:
```
For the cylinder
area =   18.84954   and volume =   6.28318
```

Screen 7:
```
Select one of the following options
to determine the volume and area for:

1. Cone
2. Cylinder
3. Terminate the program

enter a number  3
```

the entered value to CALL the appropriate procedure. When the user enters a value of 3, the outer DO loop is exited and the program terminates.

3. Separate procedures are employed for the cone and the cylinder. Observe that neither of these procedures has a parameter list or a SHARED. These were omitted because there is no need to pass any information between these procedures and the main program.

4. Each of the procedures, Cone and Cylinder, performs input and calculations. However, because their output is very similar, a separate procedure, Display, is employed to print out results. Notice that this procedure does have a parameter list in order to receive the computed area and volume for each case. Also, see that we have passed a string, solid$, so that Display can output a customized label for each case.

As in Fig. 9.7, the program's output consists of a series of screens that allow the user to compute areas and volumes for the two types of solids interactively. Of course the program could be refined in many ways (see Prob. 9.8). However, it serves to illustrate how procedures can be employed to create a modular program.

Procedures facilitate the process of program development for two major reasons. First, they are discrete, task-oriented modules that are, therefore, easier to understand, write, and debug. Second, they can be used over and over both within a program and among programs. Testing a small procedure to prove it works properly is far easier than testing an entire program. Once a procedure is thoroughly debugged and tested, it can be incorporated into other programs. Then, if errors occur, the procedure is unlikely to be the culprit, and the debugging effort can be focused on other parts of the program.

Many professional programmers accumulate a collection, or library, of useful procedures that they access as needed. Thus instead of programming many lines of code, they combine and manipulate larger, more powerful pieces of code—that is, procedures. Programming with these custom modules saves overall effort and, more important, improves program reliability. The modules in a library can be thoroughly debugged under a number of different conditions to ensure proper function. Such thorough testing can rarely be afforded for sections of code that are used only once.

It should be mentioned that QuickBASIC is an ideal language for this approach. In fact, the dialect permits you to build a "superprogram" by loading several separate BASIC files into memory simultaneously. This is accomplished by creating a file with the extension .MAK (called a *MAK file*) that lists the program files that you want to include in the superprogram. Once a superprogram has been loaded in this way, each of the separate program files is formally referred to as a *module*. Any of an individual module's procedures can then be called by any of the other modules. Although this is an advanced technique that is beyond the scope of the present book, you should be aware of this powerful capability. You can consult the QuickBASIC user's manual or many of the fine books available on the language to learn more concerning this capability.

In the remainder of the book we will be developing a number of programs that have general applicability in engineering. You should view these programs as the beginning of your own procedure library, with an eye to their utility in your future academic and professional efforts.

9.7   RECURSIVE PROCEDURES

Recursion is a powerful problem-solving tool that is available in certain computer languages such as modern BASIC. A *recursive process* is one that calls itself (see Fig. 9.8). The factorial calculation provides a good example for the use of recursion. This can be seen by realizing that it can be represented by the sequence

```
5! = 5 · 4!
4! = 4 · 3!
3! = 3 · 2!
2! = 2 · 1!
1! = 1 · 0!
```

The following example illustrates how this pattern can be concisely expressed by exploiting the recursive ability of modern BASIC.

FIGURE 9.8
A visual example of recursion.

EXAMPLE 9.7   A Recursive Function Procedure to Calculate a Factorial

PROBLEM STATEMENT: Employ recursion to determine the factorial.

SOLUTION: The following program illustrates how the factorial can be computed using a recursive FUNCTION procedure:

```
CLS

FOR i = 0 TO 5
  PRINT i; "! = "; Factorial(i)
NEXT i
END

FUNCTION Factorial(n)
  IF n > 0 THEN
    Factorial = n * Factorial(n - 1)
  ELSE
    Factorial = 1
  END IF
END FUNCTION
```

The output for this program would be identical with that from Example 9.4.

Although Example 9.7 employed a FUNCTION, SUB procedures can also call themselves recursively. Note that a recursive process must always have a terminating condition. Otherwise, it would go on forever. In the case of the factorial, the condition arises naturally from the fact that $0! = 1$. Consequently, when $0!$ is called, the IF statement shifts to an assignment statement, Factorial = 1. At this point, because the function is no longer calling itself, the process terminates.

PROBLEMS

9.1   Write a program to determine the roots of a quadratic equation. Pattern the computation after Fig. 7.7. However, use separate procedures to (1) input the coefficients, (2) compute the roots, and (3) print out the results.

9.2   Reprogram Fig. 6.6 so that it employs procedures.

9.3   Write a procedure that accomplishes the same operation as the SGN function (recall Table 6.2).

9.4   Write a program to determine the values of all the trigonometric functions in Fig. 6.1 given an angle. Use a procedure to input the angle and transform it to radians if it is entered in degrees. Employ another procedure to determine the values of the trigonometric functions and print out the results.

9.5   Write a SUB procedure to input and convert temperatures from degrees Fahrenheit to degrees Celsius and kelvins by the formulas

$$T_c = \tfrac{5}{9}(T_F - 32)$$

and

$$T_K = T_C + 273.15$$

Test it with the following data: $T_F = 20$, 203, and 98.

9.6   Write a SUB procedure to input and convert temperatures from degrees Celsius to degrees Fahrenheit and Rankine by the formulas

$$T_F = \tfrac{9}{5}T_c + 32$$

and

$$T_R = T_F + 460$$

Test it with the following data: $T_C = -10$, 27, and 90.

9.7   Develop the procedures in Probs. 9.5 and 9.6. Then merge them into a menu-driven program to perform general temperature conversions. The main program menu should include such choices as (1) to convert from Fahrenheit to Celsius and kelvin, (2) to convert from Celsius to Fahrenheit and Rankine, and (3) to terminate the program. Test the program with the following data: 10°C, 25°F, 120°C, and 500°F.

9.8   Using Fig. 9.6 as your starting point, develop a well-structured program to compute the volumes and surface areas for all the geometric solids from Fig. 9.6. Enhance your program in any way you see fit. For example, you might include error traps to prevent the user from entering negative values or zeros for the dimensions. Another suggestion might be to allow the user to specify units which could then be displayed with the results. You should also try to develop ideas of your own to make the program as user-friendly and reliable as possible.

9.9   Develop a well-structured, modular, and menu-driven program to convert between polar and rectangular coordinates, and vice versa. Consult Probs. 6.19 and 7.11 for further information regarding the conversions.

9.10 Write a program to determine the range of a projectile fired from the top of a cliff as described in Prob. 6.11. Employ a modular approach; and use three procedures to input the data, perform the computation, and output the results. Design both the main program and the input procedure so that they are menu-driven. The main program menu should include such choices as whether to input or modify data, to perform the computation, to output results, or to terminate the program. Figure 9.6 can serve as the model for how this can be done. The input procedure menu should include such choices as whether to input or modify (1) the initial velocity, (2) the angle, and (3) the cliff height, or to (4) return to the main menu. Use a CASE construct as the basis for the input procedure. Finally, include DO UNTIL loops in the main program to allow the user to repeat the computation, if so desired.

9.11 Write a FUNCTION procedure subprogram to linearly interpolate between two points: x_1, y_1, and x_2, y_2. Place the interpolated value as the value of the function. Specifically, the user-defined function will have five values in its argument list: x_1, y_1, x_2, y_2, and the value of x at which the interpolation is to be performed. Note also that your function should check whether the value of x is within the interval from x_1 to x_2 and should print out an error message if not.

9.12 Write a FUNCTION procedure that checks whether the first value in the argument list is equal to one of the four remaining values. If the first value is identical with one of the arguments, assign a value of 1 to the function's name; otherwise assign it a value of 0.

Test your main program and function by assigning different values to the arguments; after the function is executed, have the program write a message stating whether a match was found.

9.13 Although it is already available as an operator (^), integer exponentiation is an excellent example of a recursive function. Exponentiation can be expressed recursively as

$$x^0 = 1$$

$$x^1 = x \cdot x^0$$

$$x^2 = x \cdot x^1$$

.

.

.

$$x^n = x \cdot x^{n-1}$$

Develop a recursive FUNCTION procedure to implement this algorithm. Call the procedure xpower. Specify the power as a short integer, n%.

9.14 As described previously in Prob. 8.14, the "golden ratio" is related to an important mathematical sequence known as the Fibonacci numbers, which are

0, 1, 1, 2, 3, 5, 8, 13, 21, 34, . . .

Thus, each number after the first two represents the sum of the preceding two. An interesting property of the Fibonacci sequence relates to the ratio of consecutive numbers in the sequence; that is, $0/1 = 0$, $1/1 = 1$, $1/2 = 0.5$, $2/3 = 0.667$, $3/5 = 0.6$, $5/8 = 0.625$, and so on. If the series is carried out far enough, the ratio approaches the golden ratio

$$\frac{\sqrt{5} - 1}{2} = 0.61803\ldots$$

The Fibonacci sequence can be expressed recursively as in

$f(0) = 0$

$f(1) = 1$

$f(2) = f(1) + f(0)$

$f(3) = f(2) + f(1)$

.

.

.

$f(n) = f(n - 1) + f(n - 2)$

Develop a recursive FUNCTION procedure to implement this algorithm. Call the procedure `fibon`. Specify the argument as a short integer, `n%`. Ensure that the user cannot enter a negative value or a noninteger value for the argument. Test it by computing the golden ratio.

9.15 Compound interest is computed by the following sequence:

$A_1 = (1 + i)A_0$

$A_2 = (1 + i)A_1$

.

.

.

$A_n = (1 + i)A_{n-1}$

Close inspection shows that the process can be expressed concisely by the general formula

$$A_n = (1 + i)^n A_0 \qquad\qquad\qquad \text{(P9.15.1)}$$

where A_0 is the original amount that was borrowed, A_n is the amount owed after n periods, and i is the fractional rate of interest. Use a recursive function to program the process. Employ Eq. (P9.15.1) to verify that your procedure yields correct results.

9.16 Write a recursive function to compute the sum of the integers

`1, 2, 3, . . . , n`

Specify the argument as a short integer, `n%`. Ensure that the user cannot enter a negative value or a noninteger value for the argument.

9.17 Write individual QuickBASIC SUB procedures to perform complex addition, subtraction, multiplication, and division. Integrate them into a modular program that allows a user to enter two complex numbers. Then have the user select the desired operation from a menu and display the result. Allow the user to repeat the process as many times as he or she desires. Note that if the two complex numbers are $z_1 = x_1 + iy_1$ and $z_2 = x_2 + iy_2$, the operations are defined as

$$z_1 + z_2 = (x_1 + x_2) + i(y_1 + y_2)$$

$$z_1 - z_2 = (x_1 - x_2) + i(y_1 - y_2)$$

$$z_1 z_2 = x_1 x_2 - y_1 y_2 + i(x_1 y_2 + x_2 y_1)$$

$$\frac{z_1}{z_2} = \frac{x_1 x_2 + y_1 y_2 + i(x_2 y_1 - x_1 y_2)}{x_2^2 + y_2^2}$$

DATA STRUCTURE

In earlier chapters, we have seen how algorithm structure makes computer code much easier to use and to modify. In a similar fashion, data can also be organized to make its use more efficient. Such organization is referred to as *data structuring*.

The present chapter focuses on three features that foster more efficient and succinct organization of information in BASIC programs. First, READ/DATA statements allow constants to be assigned to variables in a very concise manner. Second, arrays permit the use of subscripted variables. These permit us to access a group of values with a single variable name. The subscript provides a handy way to distinguish among the individual values. Finally, records allow several types of information to be accessed with a single name. Along with their other advantages, these features can be employed to input information to a program in a neat and effective manner.

10.1   READ AND DATA STATEMENTS

Data can be entered into a BASIC program in a variety of ways. For example, the INPUT statement allows the user to enter different values for a variable on each run without modifying the program itself. Although this capability has its advantages, there will be times when it will be inconvenient to enter all or a part of the data from the keyboard. This is particularly true when entering large amounts of data.

In addition to large data sets, there will also be times when you will need the same information to be entered every time a program is run. For example, a table of physical properties for a group of materials might be required in programs used for engineering design. Again, it is inefficient to make the user input such information one value at a time with an INPUT statement.

One remedy might be to assign values to the variables within the body of the program with assignment statements. Although this circumvents the problem of retyping the data for each run, it can add quite a few lines to your program. The combination of DATA and READ statements provides a more concise and efficient alternative for incorporating large amounts of information into the program code. It allows the data to be kept together, apart from the body of the program but closely attached, so that the resulting code is coherent and uncluttered. Having data in a separate place may also facilitate program maintenance, especially if the data is to be changed.

The general forms of these statements are

```
READ var₁, var₂, . . .
```

and

```
DATA const₁, const₂, . . .
```

where $const_i$ is the ith numeric or string constant that is being entered and var_i is the ith numeric or string variable name to which $const_i$ is being assigned.

The DATA statement contains the actual information, whereas the READ statement assigns the information to a particular variable. Functionally, the READ and DATA statements

```
READ number
DATA 7
```

are equivalent to the assignment statement

```
number = 7
```

When a READ statement is executed, the values are read in the order they appear in the DATA statement. Figure 10.1 presents a visual depiction of this sequential process. Conceptually it is akin to having an imaginary pointer that moves along the DATA statement as each READ statement is implemented.

FIGURE 10.1
Illustration of how information is input with READ and DATA statements. A pointer indicates how the data is entered in sequence as each READ is implemented. Note that line numbers are employed here merely to illustrate how the READ/DATA statements work.

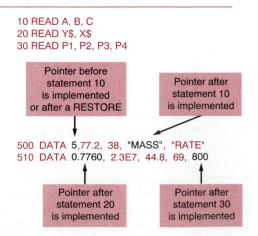

It is not necessary to read all the DATA values; for example, the value of 800 at the end of the second DATA statement in Fig. 10.1 is not used. However, if the variables in the READs exceed the DATA items, an error message such as "Out of DATA" will appear on your monitor, and the run will be terminated.

EXAMPLE 10.1 Using READ and DATA Statements

PROBLEM STATEMENT: Employ READ and DATA statements to enter the first four values of flow from Table 10.1. Then determine the average flow. Finally, display the average along with the years and flows in tabular form.

SOLUTION:

```
CLS
READ Y1$, F1, Y2$, F2, Y3$, F3
READ Y4$, F4
sum = F1 + F2 + F3 + F4
average = sum/4
PRINT "year      flow"
PRINT Y1$, F1
PRINT Y2$, F2
PRINT Y3$, F3
PRINT Y4$, F4
PRINT "average = "; average
DATA "1954", 125, "1955", 102
DATA "1956", 147, "1957", 76
DATA "1958", 95, "1959", 119
END
```

Output:

```
year        flow
1954        125
1955        102
1956        147
1957        76
average = 112.5
```

In this example, observe that extra items are included in the DATA statements. Because the READ statements never reach them, they are never accessed.

In some programs, you may wish to READ the DATA information more than once. A RESTORE statement of the form

```
RESTORE
```

TABLE 10.1   Thirty Years of Water-Flow Data for a River

Year	Flow*	Year	Flow*
1954	125	1969	63
1955	102	1970	115
1956	147	1971	71
1957	76	1972	106
1958	95	1973	53
1959	119	1974	99
1960	62	1975	153
1961	41	1976	112
1962	104	1977	82
1963	81	1978	132
1964	128	1979	96
1965	110	1980	102
1966	83	1981	90
1967	97	1982	122
1968	79	1983	138

*In billions of gallons per year.

can be added to your code to restore the pointer to the first data item (see Fig. 10.1). In addition, a line identifier may be typed after RESTORE to bring the pointer back to the beginning of a particular line. For example, the following statements can be added to the program from Fig. 10.1

```
RESTORE 510
READ X1
```

These will cause a value of 0.7760 to be assigned to X1. This option is one case where a line number (or better yet, a descriptive line label) can be beneficial.

The READ statement must be placed before the lines where the data is used. The DATA statement can appear anywhere in the main program. Two methods are recommended for placing DATA statements. One option, which is preferable in most instances, is to position them just before the END statement for the main program. This alternative is advantageous when dealing with large amounts of data or when several READs access the same DATA statement. A second alternative, which is applicable only when the READ statements reside in the main program, is to place the DATA statement immediately following the associated READ.

You can include as many DATA lines as necessary. Each statement may include as many values as can fit on a line (usually up to 255 characters). However, in most cases, you will limit yourself to the 80 columns that constitute conventional personal computer displays.

Although READ/DATA statements provide a concise means to enter data, even they prove inadequate when dealing with large quantities of information. For such cases, data files represent the preferred alternative. These will be described in Chap. 11.

In summary, READ/DATA statements provide an excellent vehicle for neatly organizing and assigning a reasonable amount of information within a program. Now, we will move on to another means for structuring information—the array.

10.2   ARRAYS

To this point, we have learned that a variable name conforms to a storage location in primary memory. Whenever the variable name is encountered in a program, the computer can access the information stored in the location. This is an adequate arrangement for many small programs. However, when dealing with large amounts of information, it proves too limiting.

For example, suppose that in the course of a water-resources engineering project you were using stream-flow measurements to determine how much water you might be able to store in a projected reservoir. Because some years are wet and some are dry, natural stream flow varies from year to year in a seemingly random fashion.[1] Thus you might require about 25 to 50 years of measurements to estimate the long-term average flow accurately. Data of this type is listed in Table 10.1.

A computer would come in handy in such an analysis. However, it would obviously be inconvenient to concoct a different variable name for each year's flow. Similar problems arise continually in engineering and other computer applications. For this reason, subscripted variables, or arrays, have been developed.

An *array* is an ordered collection of data. That is, it consists of a first item, second item, third item, and so on. For the stream-flow example, an array would consist of the first year's flow, the second year's flow, the third year's flow, and the like.

In programming, all the individual items, or *elements,* of the array are referenced by the same variable name. In order to distinguish among the elements, each is given a subscript. This is a standard practice in mathematics. For example, if we assign flow the variable name F, an index or subscript can be attached to distinguish each year's flow, as in

$$F_1 \quad F_2 \quad F_3 \quad F_4 \quad F_5$$

where F_i refers to the flow in year i. These are referred to as *subscripted variables.* In BASIC a similar arrangement is employed; but because subscript symbols are not allowed, the index is placed in parentheses, as in

```
F(1)  F(2)  F(3)  F(4)  F(5)
```

and the parentheses notation is still called a "subscript."

Note that for BASIC, the parentheses are part of the array variable name. This makes these elements different from F, F1, F2, and the like. In fact, F(1), F, and F1

---

[1]Natural phenomena like stream flows are actually not as random as they might first appear. The field of *time-series analysis* is used to detect the underlying structure or trends exhibited by such data.

FIGURE 10.2
Visual depiction of memory
locations for unsubscripted
variables and subscripted
variables. The program on
the left assigns values to F1,
F, and F(1) through F(4),
with the resulting constants
being stored in the memory
locations depicted on
the right.

```
F1 = 12
F = 15
FOR i = 1 TO 4
  F(i) = i * i
NEXT i
```

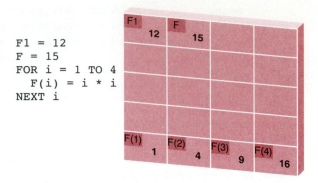

are considered to be different variables and can be used in the same program. Each would be assigned a different storage location, as depicted in Fig. 10.2.

Assignment statements can be employed to assign values to the elements of an array in a fashion similar to assigning them to nonarray variables. For example, using the data in Table 10.1,

```
F(1) = 125
F(2) = 102
F(3) = 147
       .
       .
       .
```

INPUT statements can also be employed, as in

```
INPUT F(1)
INPUT F(2)
INPUT F(3)
       .
       .
       .
```

Although individual statements can be employed, loops can be used to perform the same operation more efficiently. The following example illustrates the advantages of utilizing arrays and loops when dealing with large quantities of data.

EXAMPLE 10.2   Inputting Data to an Array with Loops

PROBLEM STATEMENT: Solve the same problem as in Example 10.1, but utilize arrays and loops to make the code more well structured and concise.

SOLUTION:

```
CLS
n = 4
sum = 0
FOR i = 1 TO n
  READ year$(i), flow(i)
  sum = sum + flow(i)
NEXT i
average = sum/n
PRINT "year       flow"
FOR i = 1 TO n
  PRINT year$(i), flow(i)
NEXT i
PRINT "average = "; average
DATA "1954", 125, "1955", 102
DATA "1956", 147, "1957", 76
DATA "1958", 95, "1959", 119
END
```

The output for this program would be identical with that of Example 10.1. Notice how we have used more descriptive names, year$ and flow, for the arrays. In addition, notice how the indices (i) of the loops are used as subscripts for the arrays. Thus, on the initial pass through the first loop, i is 1, and the program will read the values year$(1) and flow(1). Then, on subsequent passes, i will be incremented in order to read year$(2), flow(2), year$(3), flow(3), and so on.

10.2.1 Dimensioning Arrays

BASIC automatically sets aside 11 memory locations for any subscripted variable it encounters in a program. These correspond to the subscripts 0 through 10.

If a negative subscript or a subscript greater than 10 is needed, a DIM statement is required. The term "DIM" is short for "DIMensioning" or sizing an array. Its general form is

$$DIM \; var_1(n_1), \; var_2(n_2), \; . \; . \; .$$

where var_1 and var_2 are the variable names of different arrays and n_1 and n_2 are the upper limits on the number of elements in each array. For example, if we wanted to perform an analysis on all the data in Table 10.1, we would have to include the following DIM statement in our program:

```
DIM year$(30), flow(30)
```

This would set aside 31 storage locations (subscript 0 through 30), which would be adequate to hold the 30 years and flows in Table 10.1.[2]

Although there are only 30 values in Table 10.1, we could just as well have set aside more locations, as in

```
DIM year$(100), flow(100)
```

This might be done if we anticipated that at a later time additional data (say, from the years after 1983) would be obtained. Thus we do not have to specify the exact size of an array in the DIM statement; we merely have to make sure that there are enough elements available to store our data. If not enough elements have been set aside, an error will occur.

The DIM statement must be placed before the first line where the subscripted variable is utilized. Otherwise the variable is automatically dimensioned to 10 when it is first used. Redimensioning after this point would result in an error message. Therefore, it is good general practice to place DIM statements at the beginning of a program and outside of loops.

Note that lower limits other than 0 can be specified for the dimension. For example, the following DIM statement could be used to make the subscript of flow equal to the corresponding range of years from Table 10.1:

```
DIM flow(1954 TO 1983)
```

In a similar fashion, negative subscripts could be established, as in

```
DIM displacement(-20 TO 20)
```

Although arrays allow storage of large numbers of elements, their size is not unlimited. Therefore, for programs involving very large arrays, it is advisable to estimate their sizes realistically so that you do not exceed the available memory capacity. Dynamic dimensioning provides an alternative means to accomplish this objective.

As described above, the use of a constant value for the subscript limit is called *static dimensioning*. In this case, the memory locations for the array are allocated when the program is compiled.

Partly because BASIC was originally interpreted, most dialects also allow you to use numerical expressions other than constants to specify the size of arrays. This is called *dynamic dimensioning*. For example, you might input the number of data points as a variable before dimensioning, as in

```
INPUT "Number of years of flow = "; n
DIM flow(n), year$(n)
```

---

[2]For this case, a dimension of 29 would actually have been adequate to hold the 30 flow values. However, many individuals feel more comfortable starting to count at 1 rather than 0. Therefore, we have disregarded the zero element and begin entering the flows with flow(1).

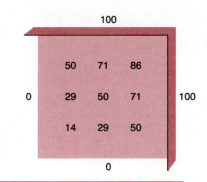

FIGURE 10.3
Temperature measurements
on a heated plate.

FIGURE 10.4
A two-dimensional array, or
matrix, used to store the
temperatures from Fig. 10.3.
Note that the first and second
subscripts designate the row
and the column of the
elements, respectively.

Columns

Rows

A(1, 1) 100	A(1, 2) 100	A(1, 3) 100	A(1, 4) 100	A(1, 5) 100
A(2, 1) 0	A(2, 2) 50	A(2, 3) 71	A(2, 4) 86	A(2, 5) 100
A(3, 1) 0	A(3, 2) 29	A(3, 3) 50	A(3, 4) 71	A(3, 5) 100
A(4, 1) 0	A(4, 2) 14	A(4, 3) 29	A(4, 4) 50	A(4, 5) 100
A(5, 1) 0	A(5, 2) 0	A(5, 3) 0	A(5, 4) 0	A(5, 5) 0

In this case, the memory locations for the array are allocated when the program is run. This capability is extremely useful when you do not have a lot of available memory because it allows you to specify exactly how much memory you require and no more.

10.2.2 Two-Dimensional Arrays

There are many engineering problems where data is arranged in tabular or rectangular form. For example, temperature measurements on a heated plate (Fig. 10.3) are often taken at equally spaced horizontal and vertical intervals. Such a two-dimensional arrangement is formally referred to as a *matrix*. It should be noted that in mathematics the one-dimensional array is often called a *vector*.

As shown in Figure 10.4, the horizontal sets of elements are called *rows*, whereas the vertical sets are called *columns*.[3] The first subscript designates the number of the

---

[3]A handy way to keep the notions of a row and a column straight is to remember that the rows of a theater are horizontal and the columns of a temple are vertical.

Program:

```
'Heated Plate Program

'by S.C. Chapra
'Aug. 5, 1992

'This program reads temperatures for a
'heated plate from data statements.
'It then converts the temperatures from
'Celcius to Fahrenheit.  The results are
'displayed as rounded integers.
'-----------------------------------

'Definition of variables:
'tempC() = temperature in degrees C
'tempF() = temperature in degrees F
'nr = number of rows of temperatures
'nc = number of columns of temperatures
'-----------------------------------

DIM tempC(20, 20), tempF(20, 20)

DEF FNCtoF (T) = (9 / 5) * T + 32
DEF FNround (x) = FIX(x + SGN(x) * .5)

'input plate temperatures
'-----------------------------------
CLS
PRINT "Temperature (F):": PRINT
READ nr, nc
FOR i = 1 TO nr
  FOR j = 1 TO nc
    READ tempC(i, j)
    tempF(i, j) = FNCtoF(tempC(i, j))
    PRINT FNround(tempF(i, j));
  NEXT j
  PRINT
NEXT i

DATA 5,5
DATA 100, 100, 100, 100, 100
DATA 0, 50, 71, 86, 100
DATA 0, 29, 50, 71, 100
DATA 0, 14, 29, 50, 100
DATA 0, 0, 0, 0, 100

END
```

Output:

```
Temperature (F):

212   212   212   212   212
32    122   160   187   212
32    84    122   160   212
32    57    84    122   212
32    32    32    32    212
```

FIGURE 10.5
Program to input, convert,
and output a
two-dimensional array for the
temperatures from Fig. 10.3.

row in which the element lies. The second subscript designates the column. For example, element A(2, 3) is in row 2 and column 3.

Matrices are so common in mathematically oriented fields like engineering that BASIC has the capability of storing data in *two-dimensional arrays*. These are extremely useful when dealing with information such as the temperatures from Fig. 10.3.

EXAMPLE 10.3 *Procedures and Two-Dimensional Arrays*

PROBLEM STATEMENT: Write a structured computer program to input the temperature data from Fig. 10.4. Then convert it to degrees Fahrenheit by the formula

$$T_F = \tfrac{9}{5}T_c + 32$$

Display the results as integers.

SOLUTION: Figure 10.5 shows the program. Notice how user-defined statement functions are employed to perform the temperature conversion (`FNCtoF`) and to round the results (`FNround`). Recall that the rounding function was derived previously in Example 6.6.

Also note that the rounding is performed when the results are being displayed. This was done so that the values stored in the array `tempF` are unaffected by the rounding process.

In modern BASIC, arrays with more than two dimensions can be employed. These are also useful in engineering. For example, if a third dimension of information were available for the plate in Fig. 10.3 (suppose temperatures were available at several levels into the plate), a three-dimensional array would come in handy. If the plate's temperatures were measured at several different times, a fourth dimension could be added to specify each case. Other problem contexts might require additional dimensions. In fact, QuickBASIC allows up to 255 distinct dimensions, more than would ever be needed in practice.

10.2.3   How Arrays Are Stored in Memory

Although it is valid for illustrating the two-dimensional structure of a matrix, Fig. 10.4 is somewhat misleading regarding the way in which the computer actually stores such information in its memory. In the present section, we would like to elaborate on this topic. We do this because there are instances where an accurate understanding of this process can be helpful. In particular, it has relevance to how the information in an array is "passed" to a procedure.

If you think about it, there are two simple ways to keep track of the values of an array in a computer's memory. One way would be to assign labels to individual memory locations, as in Fig. 10.4. Although this would certainly work, keeping track of the two-dimensional layout in this way would take up additional memory.

An alternative, which requires less memory, would be to design the language so that the elements are stored sequentially. Then, the location (or address) of the first element would be all that would be required to find the array. An individual element could then be easily located by determining how far down it was located from the first element. This downward distance is dubbed the element's *displacement*.

A hypothetical example might help clarify this alternative. Suppose that the first five flow values from Table 10.1 are stored with the following program fragment:

```
FOR i = 1 TO 5
  READ flow(i)
NEXT i
DATA 125, 102, 147, 76, 95
```

The way that these values are stored in the computer can be represented by a *storage table* of the form

Name	Value	Address[4]
flow(0)	0	0304
flow(1)	125	0305
flow(2)	102	0306
flow(3)	147	0307
flow(4)	76	0308
flow(5)	95	0309

This table shows how the array would be stored in a series of sequential memory locations. The locations are represented by their *address,* which is the code employed by the computer to label a specific memory location. When a computer interprets or compiles code, it replaces variable names with an assigned memory address.

Observe that we have included the zero element, flow(0), even though we stored the first value in flow(1). We have done this because, in the default case, the zero element is the initial location that marks the beginning of the sequence of addresses holding the array. Hence, the computer knows how to locate this zero element by means of its address 0304. Consequently, it is called the array's *base address.* Also, note that this initial element is considered to have a displacement of 0. If we wanted to find the address of the fourth element, we could merely calculate it by the simple formula

$$a_i = a_0 + d_i$$

where a_0 is the base address, a_i is the address of the ith element, and d_i is the displacement of the ith element. For this simple case, the displacement is simply the subscript,

---

[4] These are fictitious decimal addresses which we have concocted for this example (as well as for Fig. 10.6). In fact, the memory addresses in the computer would be labeled with binary codes.

$$d_i = i$$

Consequently, the address of the element, `flow(4)`, could be calculated as

$$a_4 = 0304 + 4 = 0308$$

Such a system becomes a little more complicated when storing a two-dimensional array. Because the computer's memory addresses are accessed sequentially, the information must still be stored in sequence. This means that a decision has to be made regarding how this is done: by columns or by rows? As shown in Fig. 10.6, storage is conventionally done by columns.

Next, a formula for determining displacement must be devised. Suppose that we are dealing with an array with $n + 1$ rows and $m + 1$ columns. Our intent is to come up with a displacement to substitute into the following formula:

$$a_{i,j} = a_{0,0} + d_{i,j} \tag{10.1}$$

For column zero (second subscript, $j = 0$), the displacement is simply the first subscript, i. For column one ($j = 1$), the displacement would amount to the number of elements in the zero column ($n + 1$) plus the first subscript, i. For column two ($j = 2$), the displacement would be equal to the number of elements in the zero and one columns $[j \cdot (n + 1)]$ plus the first subscript, i. Generalizing this scheme, a simple formula can be developed as

a) Program

```
n = 3; m = 2
DIM a (3, 2)
FOR i = 1 TO n
    FOR j = 1 TO m
        READ a (i, j)
    NEXT j
NEXT i
DATA 4, -1
DATA 7, 3
DATA 2, 5
```

b) Matrix

	(j = 0)	(j = 1)	(j = 2)
(i = 0)	0	0	0
(i = 1)	0	4	-1
(i = 2)	0	7	3
(i = 3)	0	2	5

c) Storage table

Element	Value	Displacement	Address
a (0, 0)	0	0	5063
a (1, 0)	0	1	5064
a (2, 0)	0	2	5065
a (3, 0)	0	3	5066
a (0, 1)	0	4	5067
a (1, 1)	4	5	5068
a (2, 1)	7	6	5069
a (3, 1)	2	7	5070
a (0, 2)	0	8	5071
a (1, 2)	-1	9	5072
a (2, 2)	3	10	5073
a (3, 2)	5	11	5074

$$d_{i,j} = j \cdot (n + 1) + i \qquad\qquad (10.2)$$

The application of this formula will be illustrated in the following example.

EXAMPLE 10.4   *Displacement for Two-Dimensional Arrays*

PROBLEM STATEMENT: Test Eqs. (10.1) and (10.2) for the matrix shown in Fig. 10.6 by verifying that element `a(3, 1)` has an address of 5070.

SOLUTION:
For this case, $a_{0,0} = 5063$, $n = 3$, $m = 2$, $i = 3$, and $j = 1$. Substituting these values into the displacement equation (10.2) yields

$$d_{3,1} = 1 \cdot (3 + 1) + 3 = 7$$

This result can be introduced into the address equation (10.1) to give

$$a_{3,1} = 5063 + 7 = 5070$$

Substitution of other values from the table will confirm that the formula is generally valid.

We have reviewed the foregoing material to acquaint you with how arrays are handled within the computer. As described next, such information has relevance to the manner in which arrays are passed to procedures.

10.2.4   Arrays in Procedures

Arrays can be passed to procedures via argument or by using the SHARED statement or attribute. Furthermore, arrays can also exist in subprograms as local variables. In this section, we will briefly review these features.

Arguments.   Arrays are passed to procedures as arguments in a similar fashion to the way that nonsubscripted variables are handled. The only difference is that they are followed by a ( ) when entered into the parameter list. Suppose that we were transferring a one-dimensional array `Temp` of length n to a SUB procedure named `Processor`. The CALL statement might look like

```
CALL Processor(Temp(), n)
```

The corresponding SUB statement could be written as

```
SUB Processor(T(), n)
```

Notice that just as with unsubscripted variables, different variable names can be used in the main program and the procedure. Also, note that the array does not have to be re-dimensioned in the procedure.

When an array is transferred to a procedure as an argument, it is *passed by reference* (recall Sec. 9.3). That is, a new memory location is not set up to store the value. Rather, the procedure uses the array's base address and displacement to locate the proper values in the memory locations for the variable *in the main program*. The presence of the variable name in the procedure's argument list provides a path that the computer follows to determine the location of the original memory location. The ( ) informs the computer that it is dealing with an array. It, therefore, knows that it has to deal with base addresses and displacements to locate the values it needs.

Aside from passing entire arrays, individual elements can be passed to procedures. For example,

```
CALL Adder (x(1), x(2), sum)
```

could be used in conjunction with the corresponding SUB procedure

```
SUB Adder (a, b, c)
   c = a + b
END
```

Notice that, when individual elements are passed, the SUB must use nonsubscripted variables to represent the values. In this example, the SUB would go to the memory locations for x(1) and x(2) in order to determine the values of a and b used in the procedure. If the values of a or b are modified in the procedure, this will result in x(1) and x(2) being changed in the calling program.

Finally, it should be mentioned that arrays cannot be passed by value (recall Sec. 9.3). This is important if you desire to retain the original values for the case where the procedure modifies array elements. In these situations, an option is to save the original values in another array before you invoke the procedure.

SHARED.   Arrays can also be passed to procedures using SHARED statements and attributes. For example, the following *SHARED statement* would give the procedure access to the temperatures:

```
DIM Temp (20)
CALL Processor (n)
     .
     .
     .

SUB Processor (n)
SHARED Temp()
```

The *SHARED attribute* is handled a little differently in that a DIM SHARED statement would be placed in the main program, as in

```
DIM SHARED Temp (20)
CALL Processor (n)
        .
        .
        .
SUB Processor (n)
```

The first statement makes the array available globally throughout the program. Notice that this statement combines the functions of the DIM statement and the SHARED attribute into a single entity. Consequently, if you choose to use the DIM SHARED attribute for an array, it is unnecessary to dimension it with an independent DIM statement. In fact, if you try to use both a DIM and a DIM SHARED for the same variable, an error message will be displayed.

Arrays as local variables in procedures.   Finally, it should be noted that arrays can also act as local variables in procedures. For these cases, a DIM statement must be placed at the beginning of the procedure. This and other aspects of arrays and procedures are explored in the following example.

EXAMPLE 10.5   **Procedures and Two-Dimensional Arrays**

PROBLEM STATEMENT: Write a structured and modular computer program to input the temperature data from Fig. 10.3. Then convert it to degrees Rankine by the formulas

$$T_F = \tfrac{9}{5} T_c + 32$$

and

$$T_R = T_F + 460$$

Display the results as integers.

SOLUTION: Figure 10.7 shows the program. Notice how user-defined functions are employed to make part of the temperature conversion (FNCtoF) and to round the results (FNround). Recall that the rounding function was derived previously in Example 6.6. Also note that the rounding is performed when the results are being displayed. This was done so that the values stored in the array tempR are unaffected by the rounding process.

Also notice how the array TF is a local variable in the procedure Processor. This was done because TF is not needed in any other part of the program. Notice that because it is used locally, TF is dimensioned at the beginning of the procedure.

Main Program:

```
DECLARE SUB PreProcessor (TC!(), nr!, nc!)
DECLARE SUB Processor (TC!(), nr!, nc!, TR!())
DECLARE SUB PostProcessor (TC!(), TR!(), nr!, nc!)

'Heated Plate Program

'by S.C. Chapra
'June 2, 1992

'This program reads temperatures for a
'heated plate from data statements.
'It then converts the temperatures from
'Celcius to Rankin.  The results are
'displayed as integers.
'---------------------------------------

'Definition of variables:
'tempC() = temperature in degrees C
'tempR() = temperature in degrees R
'nr = number of rows of temperatures
'nc = number of columns of temperatures
'---------------------------------------

DIM tempC(20, 20), tempR(20, 20)

DEF FNCtoF (T) = (9 / 5) * T + 32
DEF FNround (x) = FIX(x + SGN(x) * .5)

CALL PreProcessor(tempC(), nr, nc)
CALL Processor(tempC(), nr, nc, tempR())
CALL PostProcessor(tempC(), tempR(), nr, nc)

DATA 5,5
DATA 100, 100, 100, 100, 100
DATA 0, 50, 71, 86, 100
DATA 0, 29, 50, 71, 100
DATA 0, 14, 29, 50, 100
DATA 0, 0, 0, 0, 100

END
```

Procedures:

```
SUB PreProcessor (TC(), nr, nc)

   'inputs plate temperatures
   '-----------------------------------

   READ nr, nc
   FOR i = 1 TO nr
     FOR j = 1 TO nc
       READ TC(i, j)
     NEXT j
   NEXT i

END SUB
```

```
SUB Processor (TC(), nr, nc, TR())

   'Converts temperatures for the heated
   'plate from Celcius to Fahrenheit
   '-----------------------------------

   'Definition of local variables:
   'TF() = temperature in degrees F
   '-----------------------------------

   DIM TF(nr, nc)

   FOR i = 1 TO nr
     FOR j = 1 TO nc
       TF(i, j) = FNCtoF(TC(i, j))
       TR(i, j) = TF(i, j) + 460
     NEXT j
   NEXT i

END SUB
```

```
SUB PostProcessor (TC(), TR(), nr, nc)

   'Display results as integers
   '-----------------------------------

   CLS

   PRINT "Temperature (C):": PRINT
   FOR i = 1 TO nr
     FOR j = 1 TO nc
       PRINT FNround(TC(i, j));
     NEXT j
     PRINT
   NEXT i
   PRINT
   PRINT "Temperature (R):": PRINT
   FOR i = 1 TO nr
     FOR j = 1 TO nc
       PRINT FNround(TR(i, j));
     NEXT j
     PRINT
   NEXT i

END SUB
```

Output:

```
Temperature (C):

100  100  100  100  100
0   50   71   86   100
0   29   50   71   100
0   14   29   50   100
0    0    0    0   100

Temperature (R):

672  672  672  672  672
492  582  620  647  672
492  544  582  620  672
492  517  544  582  672
492  492  492  492  672
```

FIGURE 10.7 Modular program to input, convert, and output a two-dimensional array for the temperatures from Fig. 10.3.

10.3 RECORDS

The arrays discussed to this point represent an example of data structuring. They have allowed us to access a group of numbers using a single name. Furthermore, the two-dimensional arrangement of information in a matrix embodies information over and above the actual values of the numbers themselves. For example, the way the array was used to store the temperatures in the program from Fig. 10.7 corresponds to the spatial distribution shown in Fig. 10.3.

As described to this point, an array has the limitation that all of its information must be of the same type. Data does not always come in this form. Often it consists of different types of information. For example, think of the type of data that is associated with you as an individual. Aside from your name, you also have several identification numbers such as your social security and driver's license number. Furthermore, you could be described by your sex, height, weight, and eye and hair color. Such information could be compiled in a database.

As seen in Fig. 10.8, each individual piece of information on a file is called an *item*. Examples might be a name, a social security number, a flow rate, or a price. A *record* is a group of items that relate to the same object or individual. For example, in

FIGURE 10.8
The various components associated with data files. The figure uses the analogy between an office file system and a file system employed on a computer.

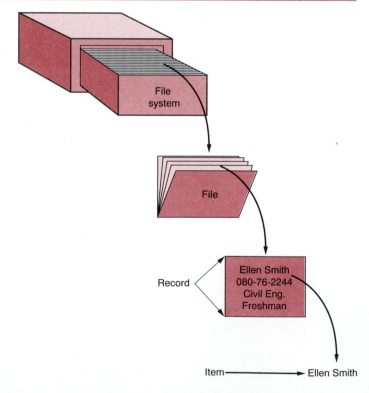

your university's central computer there is undoubtedly a record that refers to your academic standing. It probably consists of a number of items, including your name, social security or identification number, class, department, grades in individual courses, credit hours, and so on. This record is in turn a part of the *file* which contains the academic records of all the students at your school. Finally, this file is one of many files that are maintained in the *database* or *file system* of the university's mainframe computer.

In the following chapter, we will show how BASIC will allow you to generate a file of information. However, beyond formally developing a file, QuickBASIC also allows us to make an array into a file of sorts. That is, it allows us to view the individual rows of a matrix as a record. The following general representation allows us to do this:

```
TYPE UserType
    ElementName AS TypeName
    ElementName AS TypeName
        .
        .
        .
END TYPE
```

These statements allow programmers to set up their own customized type. The *UserType* represents the name given to this type. Each element of the record is then given an identifier (*ElementName*) and a type (*TypeName*). Note that the type names are as follows:

Type	TypeName
Integer	INTEGER
Long integer	LONG
Single-precision real	SINGLE
Double-precision real	DOUBLE
Character	STRING*num

The parameter *num* establishes the longest string that you want to store in a particular element name. This is referred to as a *fixed-length string*.

After establishing the user type, the array can be dimensioned as

```
DIM var(n) AS UserType
```

where *var* is the array name and *n* is the number of records to be stored in the array.

Finally, the *i*th element of the array can be invoked in the program through the variable name *var.ElementName* (i). The following example shows how these statements are employed in a program.

EXAMPLE 10.6   Using a Record in Conjunction with an
Equation of State

PROBLEM STATEMENT: The ideal gas law models the behavior of a single mole of
gas as

$$P = \frac{RT}{V}$$

where P = pressure [atm], V = volume [L], R = the universal gas constant
[=0.08205 (atm · L)/(mole · K)], and T = temperature [K].

A modification of this formula, which can handle nonideal conditions, is the
Redlich and Kwong equation of state:

$$P = \frac{RT}{V - b} - \frac{a}{\sqrt{T}V(V + b)}$$

where

$$a = \frac{0.4278R^2T_c^{2.5}}{P_c}$$

$$b = \frac{0.0867RT_c}{P_c}$$

where P_c = the critical pressure of the substance [atm] and T_c = the critical temperature [K]. These parameters have been compiled for numerous gases. For example,

Compound	T_c	P_c
Methane	190.6	45.4
Ethylene	282.4	49.7
Nitrogen	126.2	33.5
Water	647.1	217.6

Develop a program to compute pressure using both formulas for each of these
gases for a particular temperature and volume. Test it for T = 400 K and V = 100 L.
Employ READ/DATA statements to input the critical temperatures and pressures into
record arrays.

SOLUTION:

```
CLS

TYPE ChemData
  Compound AS STRING *15
  Tc AS SINGLE
  Pc AS SINGLE
END TYPE

DIM Chem(20) AS ChemData

CONST R = .08205     'gas constant

FOR i = 1 TO 4
  READ Chem(i).Compound, Chem(i).Tc, Chem(i).Pc
NEXT i

INPUT "Temperature (K) = "; T
INPUT "Volume (L) = "; V
PRINT
PRINT "Compound          P(ideal)     P(R&K)"
P.ideal = R * T/V
FOR i = 1 TO 4
  a = .4278 * R ^ 2 * Chem(i).Tc ^ 2.5/Chem(i).Pc
  b = .0867 * R * Chem(i).Tc/Chem(i).Pc
  P.rk = R * T/(V - b) - a/(SQR(T) * V * (V + b))
  PRINT Chem(i).Compound, P.ideal, P.rk
NEXT i

DATA "Methane", 190.6, 45.4
DATA "Ethylene", 282.4, 49.7
DATA "Nitrogen", 126.2, 33.5
DATA "Water", 647.1, 217.6

END
```

Output:

```
Temperature (K) = 400
Volume (L) = 100
Compound            P(ideal)        P(R&K)
Methane             .3282           .328139
Ethylene            .3282           .3279446
Nitrogen            .3282           .3282111
Water               .3282           .3275647
```

The syntax for passing a record to a procedure is similar to an array with one notable modification: the record's type must be specified in the procedure's argument list. For example, suppose that the record used in the previous example were being passed to a function to compute the pressure. The FOR/NEXT loop from the example could be written as

```
FOR i = 1 TO 4
  P.rk = Pressure(R, T, V, i, Chem())
  PRINT Chem(i).Compound, P.ideal, P.rk
NEXT i
```

and the **FUNCTION** procedure as

```
FUNCTION Pressure (R, T, V, i, Chem() AS ChemData)
  a = .4278 * R ^ 2 * Chem(i).Tc ^ 2.5/Chem(i).Pc
  b = .0867 * R * Chem(i).Tc/Chem(i).Pc
  Pressure = R * T/(V - b) - a/(SQR(T) * V * (V + b))
END FUNCTION
```

Thus, the clause AS ChemData informs the procedure that Chem() is a record.

PROBLEMS

10.1   Given the following code, what is the output?

```
FOR i = 1 TO 3
  FOR j = 1 TO 3
    a(i, j) = i * j
  NEXT j
NEXT i
FOR i = 1 TO 3
  FOR j = 1 TO 3
    PRINT a(j, i)
  NEXT j
NEXT i
```

10.2   Write QuickBASIC statements to display (that is, PRINT) the contents of an array, a. Note that the array begins with the element a(1) and ends with the element a(150). Write the statements so that the array is displayed in reverse order, (*a*) placing each element on a line by itself and (*b*) placing two elements on each line.

10.3   Write a FOR/NEXT loop to fill each element of a one-dimensional array of 100 elements with the element's subscript.

10.4 (*a*) What is the largest INTEGER array you can use on your computer (to the nearest kilobyte)? That is, how big a dimension?
(*b*) How big is the largest array of single-precision real numbers you can use on your computer (to the nearest kilobyte)?
(*c*) How big is the largest n x n array of single-precision real numbers you can use on your computer (to the nearest kilobyte)?
(*d*) Contrast the results of (*a*) and (*b*), and draw conclusions related to the use of integer and real arrays in computer programs.

10.5 Write QuickBASIC statements to display the contents of a 5×3 two-dimensional array, y. Design the code so that it prints out the results in a single column, in a similar fashion to the way in which the array is stored in the computer (recall Fig. 10.6 and accompanying discussion).

10.6 Write a well-structured modular program to input a series of numbers and compute their average and standard deviation. Whereas the average (or *arithmetic mean*) provides an estimate of the "location" or center of a group of data, the standard deviation is a statistic that quantifies its "spread about the mean." A computationally efficient formula for the *standard deviation* is

$$s_x = \sqrt{\frac{\sum x_i^2 - [(\sum x_i)^2/n]}{n-1}} \tag{P10.6.1}$$

where all summations are from $i = 1, n$. Employ the following algorithm as the basis for the program:

- *Step 1* Use a loop to input and count the numbers. Example 8.1 might serve as a model for inputting and counting. Employ an array to store the values as subscripted variables.
- *Step 2* Compute the arithmetic mean as the sum divided by the total number of values, and the standard deviation as in Eq. (P10.6.1)
- *Step 3* Display the individual values.
- *Step 4* Display the arithmetic mean and standard deviation.

Test the program by determining the average flow for the data in Table 10.1.

10.7 Repeat Prob. 10.6, but rather than inputting the data via the keyboard, use READ/DATA statements to incorporate the data into the program. Employ subscripted variables to store the values for the flows.

10.8 Repeat Prob. 10.7, but use an array that has subscripts equal to the years at which the flows were measured.

10.9 Suppose that you have an array a of dimension n which contains values that are increasing. An example of such data, for the case where n = 11, is

```
a(0) = -10        a(4) = 6.5        a(8) = 21
a(1) = -6         a(5) = 9.3        a(9) = 22
a(2) = 0          a(6) = 12         a(10) = 33
a(3) = 2          a(7) = 15         a(11) = 49
```

Develop a modular program that employs a procedure to enter this array using READ statements. Then, set up a query in the main program that prompts the user for a number, x. Pass this number to another procedure that determines a value k such that x is between $a(k)$ and $a(k + 1)$. Return values of -1 or n if x is outside the range. Display this number for the user. Employ a DO UNTIL structure to allow the user to repeat the process as many times as desired.

10.10 Repeat Prob. 10.9, but design it to work for arrays that contain values that are monotonic (that is, either increasing *or* decreasing).

10.11 Suppose that QuickBASIC employed default arrays that started with a 1 instead of a 0 subscript. Rederive Eq. (10.2) for this case.

10.12 Write nested FOR/NEXT loops to fill each element of a three-dimensional array with the product of its subscripts. Test your code for a $4 \times 5 \times 3$ matrix.

10.13 Employ arrays and READ/DATA statements to develop a modular computer program to determine your grade for this course.

10.14 Employ arrays, READ/DATA statements, and records to create a modular program to determine the statistics for a sports team.

10.15 Select 10 of your favorite issues from the New York Stock Exchange. Write, debug, and document a BASIC program that reads the name of the stock, its selling price today, and its 52-week low. Use a record to hold this information. The program should then display the name of the stock, its selling price, and the percentage gain over the 52-week low.

10.16 Employ READ/DATA statements and records to enter the following Student Aptitude Test (SAT) data into a computer program:

ID Number	Name	Math	Verbal
551556681	Reckhow, Sarah	736	710
337653878	Chen, David	650	680
263748294	Morales, Juan	690	550
876770183	Hornacek, Heather	620	640
627839874	Kelly, Carlos	720	480
863454637	Jones, Isaac	780	715

Include a procedure in your program which determines and displays the average and standard deviation (see Prob. 10.2) for the math and verbal scores. Include another procedure that displays the name, ID number, and scores for the student with the highest total score (that is, math plus verbal).

10.17 Employ READ/DATA statements and records to enter the following information into a computer program:

Name	Team	AB (At Bats)	H (Hits)
Kruk	Phil	355	121
Van Slyke	Pitt	378	126
Sheffield	SD	390	127
Grace	Chi	380	121
Butler	LA	376	119
DeShields	Mon	408	128

Include a procedure in your program which determines the batting average (=H/AB) for each player. Include another procedure that displays the name, team, and batting average for all the players on the list. Note that batting averages are typically expressed to three decimal places (for example, 0.305).

10.18 Members of a sales force receive commissions of 5 percent for the first $1000, 7 percent for the next $1000, and 10 percent for anything over $2000. Compute and display the commissions and social security numbers for the following salespeople: John Smith, who has $5325.76 in sales and a social security number of 080-55-6642; Bubba Holottashakingoinon, who has $8050.63 in sales and whose number is 306-77-5520; and Bubbette Schumacker, who has $6643.96 in sales and a social security number of 707-00-1122. Use READ/DATA statements and records to input the information.

10.19 When dealing with numbers, it is often necessary to sort them in ascending order. There are many methods for accomplishing this objective. A simple approach, called a *selection sort,* is probably very close to the commonsense approach you might use to solve the problem. First you would scan the set of numbers and pick the smallest. Next you would bring this value to the head of the list. Then you would repeat the procedure on the remaining numbers and bring the next smallest value to the second place on the list. This procedure would be continued until the numbers were ordered. Develop a well-structured modular program to implement the selection sort. Test it on the data from Table 10.1.

10.20 Another simple sorting algorithm, called the *bubble sort,* consists of starting at the beginning of the array and comparing adjacent values. If the first value is smaller than the second, nothing is done. However, if the first is larger than the second, the two values are switched. Next the second and third values are compared, and if they are not in the correct order (that is, smallest first), they are also switched. This process is repeated until the end of the array is reached. This entire sequence is called a *pass.*

After the first pass, the process returns to the top of the list and another pass is implemented. However, on this pass, you only have to go down to the next-to-the-last element. The procedure is repeated until a pass with no switches occurs. The name "bubble sort" comes from the fact that the smallest value tends to rise to the top of the list like a bubble rising through a liquid. Develop a well-structured modular program to implement the bubble sort. Test it on the data from Table 10.1.

10.21 Aside from sorting, another important operation performed on data is searching—that is, locating an individual item in a group of data. One common method for accomplishing this objective is the *binary search*. Suppose that you are given an array $A(i)$ of length n and must determine whether the value V is included in this array. After sorting the array in ascending order (see Probs. 10.19 and 10.20), you employ the following algorithm to conduct a binary search: Compare V with the middle term $A(m)$. Three results can occur: (*a*) If $V = A(m)$, the search is successful; (*b*) if $V < A(m)$, V is in the first half; and (*c*) if $V > A(m)$, V is in the second half. In the case of (*b*) or (*c*), the search is repeated in the half in which V is known to lie. Then V is again compared with $A(m)$, where m is now redefined as the middle value of the smaller list. The process is repeated until V is located or the array is exhausted.

Develop a SUB procedure to implement a binary search. Use structured programming techniques and a modular design so that the SUB can be easily integrated into other programs. Document the SUB both internally and externally.

10.22 Write a program to input your friends' and family's names and birthdays into an array of records.

10.23 Modify the program from Prob. 10.22 so that when the user enters today's date, the record containing the next birthday is displayed.

ADVANCED INPUT-OUTPUT

Because it represents the interface between the user and the program, input-output (or *I/O*) is a critical aspect of software design. In the early days of computing, input was via punched cards and output was on paper produced by large line printers. As a result, the first computer languages had limited capabilities for input-output. Today, we enter data into computer programs via a keyboard, with a mouse, or directly from measuring instruments, and display our results on a monitor, laser printer, or pen plotter. Consequently, most modern languages have well-developed capabilities for implementing I/O. Modern dialects, such as QuickBASIC, provide the capability to design a machine-user interface that is expressive, efficient, and friendly.

11.1   FORMATTED OUTPUT

There are many cases in engineering where you would want to control the format of a program's output. This is particularly important when the program is to be used for report generation. You have already been introduced to several ways to do this using the PRINT, TAB, and SPC commands. Now we will delve more deeply into the topic by describing the PRINT USING command, which allows the explicit formatting of output.

11.1.1   PRINT USING

The PRINT USING statement has the general form,

PRINT USING *FormatString*; *item*$_1$; *item*$_2$; *etc.*

where the *FormatString* is either a string constant or a variable that specifies the format of the items that are being displayed. Notice that semicolons are employed to delimit the items.

The PRINT USING statement can be employed to format either numeric or string data. Separate rules apply to each case.

Numeric data.   An example of a PRINT USING statement to output numbers is

```
PRINT USING "#####"; 8; 99
```

The string constant "#####" defines the *field*—that is, the number of spaces on the displayed line reserved for an item. The symbol # is employed to reserve a space for a digit. Thus, for this example, a 5-digit length field is set aside for each number that is displayed. Note that the statement could also be written using a string variable:

```
format$ = "#####"
PRINT USING format$; 8; 99
```

If the items do not fill the field, each is displayed right-justified (that is, placed at the far-right end of the field) and the unused portion filled with blanks. For our example, the result would be[1]

```
____8___99
↑
└─────────── position 1
```

Thus the 8 is displayed in position 5 and the 99 in positions 9 and 10. Spaces between numbers can be interjected by merely leaving blanks in the *FormatString,* as in

```
PRINT USING "##### "; 8; 99
                  ↑
                  └─────────── space
```

with the resulting output looking like:

```
____8____99
```

The 8 is still displayed in position 5, but the 99 has moved over to positions 10 and 11 because of the extra space that has been inserted between the values.

In addition to controlling the spacing, the PRINT USING also allows decimal points to be placed in the output. The decimal is merely located in the desired position in the format string, for example,

```
PRINT USING "###.## "; 3.56; .85; 20.697; 15050
```

with the resulting output looking like:

```
__3.56___0.85__20.70_%15050.00
```

Notice that if the format string specifies that at least 1 digit is to precede the decimal point, it is always displayed (even as a 0, as in the case of 0.85). Also observe that

---

[1]Note that in all this chapter's example output, we will use underscores to clearly indicate blank spaces.

because 20.697 has one too many fractional digits, it is automatically rounded to 20.70 when it is displayed. Finally, recognize that 15050 is too big for the field. For this case, BASIC displays the number in the proper format (two decimal places) but affixes a percent sign in front of the number to signify in a conspicuous manner that the field has been exceeded.

Some additional symbols that are employed in PRINT USING statements are described in Table 11.1. The most important from the perspective of engineering is the exponential format. For this case, carets (^^^^) are used to specify the part of the field taken up by the exponential term,

```
PRINT USING "##.## ^^^^ "; 45678; .0015638; 3.568E11
```

with the resulting output looking like:

```
_4.57E+04_ _1.56E-03_ _3.57E+11
```

Note that you may also use five carets (^^^^^) to allow numbers with large exponents (for example 1.0E+103) to be displayed.

EXAMPLE 11.1   Displaying a Table with the PRINT USING Statements

PROBLEM STATEMENT: The students in a class have obtained the following grades:

Student Name	Average Quiz	Average Homework	Final Exam
Smith, Jane	82.00	83.90	100
Apicella, Fred	96.55	92.00	77
Horvath, Sarah	97.30	89.67	87

Their final grade can be computed with the following weighted average:

$$FG = 0.5AQ + 0.2AH + 0.3FE$$

where FG = final grade, AQ = average quiz, AH = average homework, and FE = final exam grade. Develop a computer program to compute the final grade for each student. Display the students' grades in a tabular format as shown above, with the final grade added as the last column.

SOLUTION: The code to determine the final grade and to display the table is depicted in Fig. 11.1. Notice that all columns are aligned properly and all numbers have the same number of decimal places. It is this kind of control that makes the PRINT USING so valuable when preparing tables for reports and other engineeering documents.

TABLE 11.1   Additional Symbols Employed in Numeric Fields of the PRINT USING Statement*

Symbol	Description	Example
^^^^	Positioned at the end of the field. You can also use five carets to allow E+xxx to be printed for very large numbers.	PRINT USING "#.## ^^^^"; -256178 -.26E+06 PRINT USING "#.## ^^^^^"; -256178 -.26E+006
+	Positioned at the beginning or end of a format string, it causes the sign of the number to be printed before or after the number.	PRINT USING "+##.#"; -29; 5.25 -29.0 +5.3 PRINT USING "##.#+"; -29; 5.25 29.0- 5.3+
−	Positioned at the end of a format string, it causes negative numbers to be printed with a trailing negative sign.	PRINT USING "##.#-"; -29; 5.25 29.0- 5.3
,	Positioned anywhere to the left of the decimal point, it causes a comma to be printed to the left of every third digit to the left of the decimal. Positioned at the end of a format string after the decimal, the comma is printed as part of the string.	PRINT USING "#####,."; -15245 -15,245

*Note that as in the text, we will use underscores to clearly indicate blank spaces in the examples shown here.

String data.   When the **PRINT USING** statement is employed to display strings, three characters are used to control formatting: !, \, and &. The exclamation point is utilized if only the first character of the string is to be output, as in

```
PRINT USING "!"; "Moe"; "Larry"; "Curly"
MLC
```

Main Program:

```
DECLARE SUB PreProcessor (n!)
DECLARE SUB Processor (n!, grade!())
DECLARE SUB PostProcessor (n!, grade!())

'Grade Calculation Program

'by S.C. Chapra
'August 7, 1992

'This program determines the final
'grades for three students
'-------------------------------------

'Definition of variables:
'n = number of students
'nam$() = student names
'quiz() = average quiz
'hmwk() = average homework
'exam() = final exam
'grade() = final grade
'------------------------------- -------

CLS

n = 3

DIM SHARED nam$(n), quiz(n), hmwk(n), exam(n)
DIM grade(n)

CALL PreProcessor(n)
CALL Processor(n, grade())
CALL PostProcessor(n, grade())

DATA "Smith, Jane", 82, 83.9, 100
DATA  "Apicella, Fred", 96.55, 92, 77
DATA "Horvath, Sarah", 97.3, 89.67, 87

END
```

Procedures:

```
SUB PreProcessor (n)

  'inputs names and grades
  '-------------------------------------

  FOR i = 1 TO n
     READ nam$(i), quiz(i), hmwk(i), exam(i)
  NEXT i

END SUB


SUB Processor (n, grade())

  'computes final grade
  '-------------------------------------

  FOR i = 1 TO n
     grade(i) = .5 * quiz(i) + .2 * hmwk(i) + .3 * exam(i)
  NEXT i

END SUB


SUB PostProcessor (n, grade())

  'displays table of grades
  '-------------------------------------

  PRINT "Student"; TAB(25); "Average";
  PRINT TAB(35); "Average"; TAB(45);
  PRINT "Final"; TAB(55); "Final"
  PRINT "Name"; TAB(25); "Quiz";
  PRINT TAB(35); "Homework"; TAB(45);
  PRINT "Exam"; TAB(55); "Grade"
  PRINT

  FOR i = 1 TO n
     PRINT USING "\                    \"; nam$(i);
     PRINT USING "###.##    "; quiz(i); hmwk(i);
     PRINT USING "###.##    "; exam(i); grade(i)
  NEXT i

END SUB
```

Output:

Student Name	Average Quiz	Average Homework	Final Exam	Final Grade
Smith, Jane	82.00	83.90	100.00	87.78
Apicella, Fred	96.55	92.00	77.00	89.78
Horvath, Sarah	97.30	89.67	87.00	92.68

FIGURE 11.1 A program and output to determine the grades for three students (see Example 11.1).

Spaces can be inserted by placing spaces within the quotes specifying the format,

```
PRINT USING "! "; "Moe"; "Larry"; "Curly"
M_L_C
```

The backslash is used to specify that a set number of characters are to be output. For example, \\ specifies that two characters are to be displayed and \ \ (one blank between the slashes) specifies three. If the string values to be displayed are greater than the field specified by the format string, then the extra characters are ignored. If the string values to be displayed are less than the field, then the string is left-justified and blanks are used to fill out the field, for example,

```
PRINT USING "\ \"; "Moe"; "Larry"; "Curly"
Moe_LarrCurl
```

The final symbol, &, is employed when the entire string value is to be output exactly as it is stored:

```
PRINT USING "&"; "Moe"; "Larry"; "Curly"
MoeLarryCurly
```

Spaces can be added within the quotes to leave room between the string values when they are displayed:

```
PRINT USING " &"; "Moe"; "Larry"; "Curly"
__Moe__Larry__Curly
```

Numbers and strings. A final feature of the PRINT USING statement bears mention. That is, it is possible to mix strings and numbers in the *FormatString*. For example, suppose that you would like to output the answer to a computation along with a label and units. The following program fragment illustrates how the PRINT USING statement can be employed for this purpose:

```
pi = 4 * ATN(1)
r = 10
area = pi * r ^ 2
PRINT USING "area = ###.## square meters"; area
```

When this fragment is executed, the resulting output is

```
area = 314.16 square meters
```

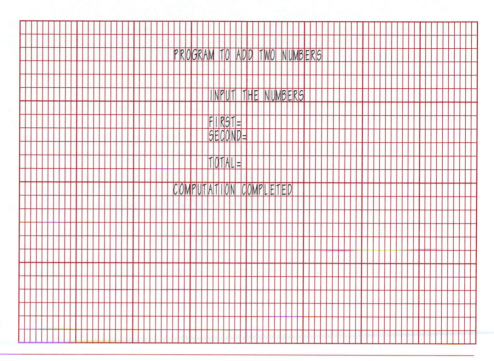

FIGURE 11.2
Layout form.

Such capabilities allow you to make your output clear and self-explanatory. They can be particularly effective when combined with modern BASIC's ability to manipulate strings (Sec. 11.2).

11.1.2 Screen Design

An exciting aspect of personal computing is the interaction that can be developed between the user and the machine. By effectively designing input and output, the programmer can cause the computer to become a more efficient and enjoyable tool for engineering problem solving.

To this point, you have been introduced to a number of statements that can be employed to design an effective user-software interface. These are screen clearing (CLS), line and space skipping (PRINT, TAB, and SPC), data formatting (delimiters, rounding, and PRINT USING), and string messages. With these tools at your disposal, the screen can be conceived of as an artist's canvas, where you can position your input and output in an effective and aesthetically pleasing fashion.

An additional tool is the screen layout form shown in Fig. 11.2. This particular form is 80 columns wide and 24 rows deep.[2] This layout form provides a vehicle for

---

[2]Actually, in text mode, the displays are 25 rows deep, but the twenty-fifth row is used by QuickBASIC to display messages such as "Press any key to continue."

visualizing how information will appear on a monitor screen. As shown in Fig. 11.2, you can sketch your desired output on the form and then use it as a guide for developing the corresponding BASIC code.

11.1.3 The LOCATE Statement

One additional type of statement can facilitate screen design. To this point, you have the capability to move from left to right (TAB) and from top to bottom (PRINT) to position data on the screen. This is a somewhat limited state of affairs. The situation would be similar to that of an artist who had to paint a canvas starting at the upper left corner and then proceeding from left to right and top to bottom.

For Microsoft BASIC, the LOCATE statement allows you to position the cursor at any of the 25x80 cells on the screen. Its general form is

LOCATE *RowNumber, ColumnNumber*

where *RowNumber* specifies the cell's row (1 to 25) and *ColumnNumber* specifies its column (1 to 80). For example, the following statements would move the cursor to the middle of the screen and display an asterisk:

```
LOCATE 12,40
PRINT "*"
```

EXAMPLE 11.2   Screen Design Using LOCATE

PROBLEM STATEMENT: Develop QuickBASIC code for a simple addition program. Employ LOCATE statements to design the display so that it follows the layout of Fig. 11.2

SOLUTION: Statements to accomplish the objective are:

```
CLS
LOCATE 3, 24
PRINT "PROGRAM TO ADD TWO NUMBERS"
LOCATE 6, 29
PRINT "Input the numbers"
LOCATE 8, 29
INPUT "First = "; first
LOCATE 9, 29
INPUT "Second = "; second
LOCATE 11, 29
PRINT "Total = "; first + second
LOCATE 13, 24
PRINT "computation completed"
END
```

The concept of screen design is closely related to the topic of computer graphics. Several problems at this chapter's end explore the idea in more detail.

11.2   STRING MANIPULATIONS

In the same way that numbers are manipulated by arithmetic operations and numeric intrinsic functions, strings can also be manipulated. This is a particularly strong feature of the BASIC language.

11.2.1   Concatenation

The process of combining strings is called *concatenation*. For example, the following program fragment concatenates several string variables and constants:

```
a$ = "Go"
b$ = "away"
c$ = a$ + " " + b$ + "!"
PRINT c$
```

with the resulting output:

```
Go away!
```

11.2.2   String Manipulation Functions

The BASIC language has a rich array of functions to manipulate strings.

LCASE$ and UCASE$ functions.   *UCASE$* returns a string expression with all letters in uppercase. *LCASE$* returns a string expression with all letters in lowercase. The following code illustrates how these statements work:

```
a$ = "Joe Romig"
PRINT a$
a$ = UCASE$(a$)
PRINT a$
a$ = LCASE$(a$)
PRINT a$
```

with the resulting output:

```
Joe Romig
JOE ROMIG
joe romig
```

EXAMPLE 11.3   *Designing a Robust Query*

PROBLEM STATEMENT: Today, more and more programs are being designed in an interactive fashion. In Example 8.4, you designed an interactive loop that performs an action and then queries the user as to whether he or she wants to repeat the action. The control structure had the general format:

```
DO
    .
    .
    .
    action
    .
    .
    .
    DO
        INPUT "Do you want to repeat the action"; q$
        IF q$ = "y" OR q$ = Y" OR q$ = "n" OR q$ = "N" THEN EXIT DO
        PRINT "You did not enter a y or an n. Try again."
    LOOP
LOOP UNTIL q$ = "n" OR q$ = "N"
```

Use string manipulation to make this scheme more concise.

SOLUTION: Rather than testing whether both lowercase and uppercase responses have been made, the LCASE$ function can be employed to make the tests more concise:

```
DO
    .
    .
    .
    action
    .
    .
    .
    DO
        INPUT "Do you want to repeat the action"; q$
        IF LCASE$(q$) = "n" or LCASE$(q$) = "y" THEN EXIT DO
        PRINT "Enter either y or n, try again"
    LOOP
LOOP UNTIL LCASE$(q$) = "n"
```

Thus, if an uppercase "N" or "Y" is entered, the function will convert it to lowercase. Therefore, both uppercase and lowercase responses are treated identically.

LTRIM$ and RTRIM$ functions. The *LTRIM$* statement returns the right portion of a string after removing any blanks at the beginning. Its general format is

```
LTRIM$(StringExpression)
```

The *RTRIM$* statement is similar but returns the left portion and deletes blanks at the end. Its general format is

```
RTRIM$(StringExpression)
```

The following code illustrates how these statements work:

```
a$ = "   Yo    "
b$ = "   Bubba!    "
PRINT LTRIM$(a$) + LTRIM$(b$)
PRINT RTRIM$(a$) + RTRIM$(b$)
c$ = LTRIM$(RTRIM$(a$)) + " " + LTRIM$(RTRIM$(b$))
PRINT c$
```

with the resulting output:

```
Yo___Bubba!
___Yo___Bubba!
Yo Bubba!
```

MID$ function. The *MID$* function returns a string consisting of a subset of another string. Its general format is

```
MID$(StringExpression, start, [length])
```

where *StringExpression* is the expression from which the subset is taken, *start* is an integer expression indicating the position in *StringExpression* where the substring starts, and *length* is the number of characters to extract. Note that the argument *length* is optional. If it is omitted, the MID$ function returns all the characters to the right of the *start* character.

The following code illustrates the MID$ statement:

```
a$ = "Introductory Circuit Analysis"
PRINT a$
PRINT MID$(a$, 14)
PRINT MID$(a$, 14, 7)
```

with the resulting output:

```
Introductory Circuit Analysis
Circuit Analysis
Circuit
```

LEFT$ and RIGHT$ functions.   The *LEFT$* statement returns a string consisting of the leftmost *n* characters of a string. Its general format is

```
LEFT$(StringExpression, n)
```

The *RIGHT$* statement returns a string consisting of the rightmost *n* characters of a string. Its general format is

```
RIGHT$(StringExpression, n)
```

The following code illustrates how these statements work:

```
a$ = "Bridget Turner"
PRINT LEFT$(a$, 7)
PRINT RIGHT$(a$, 6)
```

with the resulting output:

```
Bridget
Turner
```

INSTR and LEN functions.   The *INSTR* statement returns the position of the first occurrence of a string in another string. If no match is found, a 0 is returned. Its general format is

```
INSTR([start,]string1, string2)
```

where *string1* is the string being searched, *string2* is the string you are trying to detect, and *start* is an integer specifying the position at which you want to start the search. Note that the argument *start* is optional. If it is omitted, the search is begun at the start of *string1*.

The following code illustrates how the INSTR function works:

```
BillSays$ = "Ack!"
x = INSTR(BillSays$, "!")
PRINT x
```

with the resulting output:

```
4
```

The *LEN* function returns the length (that is, the number of characters and blanks) of a string. The following code illustrates how this statement works:

```
a$ = "Whizzer White"
PRINT LEN(a$)
```

with the resulting output:

13

EXAMPLE 11.4 Program to Divide a String

PROBLEM STATEMENT: Develop some code to divide a string into its component parts. Test it on the string "Bill Coleman."

SOLUTION: The code to solve this problem can be written as

```
a$ = "Bill Coleman"
PRINT "The name to be divided is"; a$
y = LEN$(a$)
x = INSTR(a$, " ")
PRINT "The name's length is "; y
PRINT "The blank is in the position"; x
PRINT "The first name is "; LEFT$(A$, x)
PRINT "The surname is "; RIGHT#(a$, y - x)
```

When the program is run, the following output would be generated:

```
The name to be divided is Bill Coleman
The name's length is 12
The blank is in position 5
The first name is Bill
The surname is Coleman
```

The foregoing commands can be used to manipulate strings in a variety of ways. The following example provides another illustration of how they can be employed.

EXAMPLE 11.5 Program to Change a Number into Dollars

PROBLEM STATEMENT: Develop some code to take a double-precision number and convert it to a string with commas inserted after every third place to the left of the decimal point. Also add a dollar sign to the head end of the string and show the cents at its tail end.

SOLUTION: The code to solve this problem is depicted in Fig. 11.3 along with a sample run. An EXIT DO loop is included in the main program to prevent the user from entering a negative number.

A function is employed to create the string. First, the number is transformed into a string and its decimal point is located with the INSTR function. If the number does not include a decimal point, its decimal position is set at the right end of the string. Then, if the decimal place is greater than four (meaning that the number is greater than or equal to 1000), a loop is employed to enter the commas into the string. After this is accomplished, a dollar sign is concatenated to the front end of the string. This result is then assigned to the FUNCTION name, `Delimit$`, and the result returned to the main program where is is displayed.

Aside from the functions described above, there are several others that allow powerful string manipulation in QuickBASIC. Together, all these functions can be marshaled to exert great control over input and output. Several problems at the end of the chapter are designed to help you start to explore this capability.

FIGURE 11.3
A program to convert a numeric value to a string. For this case, a dollar sign and commas are employed to express the string as a dollar amount.

Program:

```
DECLARE FUNCTION Delimit$ (n#)
CLS

'Enters positive number
DO
  INPUT "Enter your salary"; number#
  IF number# >= 0 THEN EXIT DO
  PRINT "Salary must be a positive number"
  PRINT "Remember, you're an engineer now!"
LOOP

'Displays salary with commas, dollar sign,
'and cents
number$ = Delimit$(number#)
PRINT "Your salary is "; number$

END


FUNCTION Delimit$ (n#)
  n$ = STR$(n#)
  decimal% = INSTR(n$, ".")
  IF decimal% = 0 THEN
    decimal% = LEN(n$) + 1
  END IF
  IF decimal% > 4 THEN
    FOR i% = decimal% - 3 TO 3 STEP -3
      n$ = LEFT$(n$, i% - 1) + "," + MID$(n$, i%)
    NEXT i%
  END IF
  Delimit$ = "$" + LTRIM$(n$)
END FUNCTION
```

Sample Run:

```
Enter your salary? -555555555
Salary must be a positive number
Remember, you're an engineer now!
Enter your salary? 555555555.25
Your salary is $555,555,555.25
```

11.3   FILES

To this point, we have covered three methods for entering data into a program (see Table 11.2). Notice in the table that all of them have disadvantages related to large quantities of data and data for which frequent change might be required. In addition, all the methods deal exclusively with the information requirements of a single program. If the same data is to be used for another program, it must be retyped.

Another option is to store information on an external memory device such as a disk. This is conceptually similar to the way we have been saving our programs on disks so that they are not destroyed when the computer is shut off. Along with our programs, we can store data in external memory files just as office information is stored in file cabinets for later use.

A number of advantages result from storing data on files. First, data entry and modification become easier. In addition, files make it possible to run several programs using the data from the same file and, alternatively, to run the same program from several different files. Before proceeding to the specifics of manipulating files, we will introduce some new terminology.

11.3.1   File Types

In its most general sense, a file can be defined as a section of external storage that has a name. There are two types of files used on personal computers: sequential and direct-access files. *Sequential files* are those in which items are entered and accessed in sequence. That is, there is a first item, a second item, and so on. The key concept is that when we want to access one of these items, we have to start at the beginning of the file and make our way through each item until we come across the one we want. Using an analogy from musical recording devices, sequential files are very much like cassette tapes. If you want to listen to a tune at the end of a tape, you have to fast-forward through the music you want to skip.

In contrast, for *direct-access,* or *random-access, files,* we can access a record directly without having to make our way through all previous records. Continuing the musical analogy, random-access files are like a compact disk player. For CDs, you punch in the selection you want and the player moves directly to that selection. Random access works in a similar fashion by allowing direct movement to a specific location in the file.

Direct-access files have a number of advantages, not the least of which is the fact that they retrieve data more quickly and more efficiently than sequential files. They are especially attractive for large-scale business databases such as a bank's billing or an airline's ticketing program.

Although sequential files are inefficient, they have their place in data processing. They are especially well suited for data that is to be processed sequentially anyway, such as mailing labels. In engineering, data is often in sequential form; for example, data is often listed in temporal order. Consequently, many engineering applications are adequately handled by sequential files. Finally, and most important in the present context, they are easier to explain and are more uniform than random-access files. Because of the scope of this text, we have chosen to devote this chapter to sequential files. You can

TABLE 11.2 Advantages and Disadvantages of Four Ways in Which Information Is Entered into BASIC Programs

Method	Advantages	Disadvantages	Primary Applications
Assignment (and CONST) statements	Are simple and direct.	Takes up space in the program.	Appropriate for some simple programs.
	Values are "wired" into the program and do not have to be entered on each run.	If a new value is to be assigned, the user must edit the assignment statement.	Good for assigning values to true constants (such as π or the acceleration of gravity) that are used repeatedly throughout a program and never change from run to run.
INPUT statement	Permits different values to be assigned to variables on each run.	Values must be entered by user on each run. This can become tedious.	Appropriate for those variables, when few in number, that take on different values on each run.
	Allows interaction between user and the program.		
READ/DATA statements	Provide a concise way to assign values.	If a new value is to be assigned, the user must edit the DATA statement. Also, for large quantities of information, these can take up a lot of line space.	Same as for the assignment statement but for cases where many (but not hundreds of) constants are to be assigned.
	The DATA statement can be accessed more than once by different READ statements throughout the program.		
Files	Provide a concise way to handle large numbers of input data.	Are somewhat more complicated than the other input options.	Appropriate for large quantities of data and where data must be shared between programs.
	Allow data to be transferred between programs.		

consult your computer's user's manual in the event that you desire or need to employ direct-access files.

11.3.2 BASIC File Statements

Before describing how sequential files are used, it must be noted that they are among the least standardized elements of BASIC. Fortunately, most of the differences relate to nomenclature—that is, the specific statements that are used to designate the fundamental file manipulations. From a functional perspective, the actual operations themselves are fairly similar from system to system. Therefore, in the following discussion we have tried to keep the description of the procedures as general as possible. Where we do use specific nomenclature, we have opted to use statements compatible with the QuickBASIC system. It is possible that some of the specific names that we employ may differ somewhat from those used on your system. You may therefore have to obtain details from your dialect's user's manual in order to implement the material effectively.

In the following pages, we will first introduce the fundamental statements that are used. Then we will show how they are employed to execute a number of standard operations.

The following types of statements are fairly standard on sequential file systems:

- Opening a file—that is, establishing commmunication between a program and a file
- Transferring data to a file from a program
- Transferring data from a file to a program
- Closing a file—that is, terminating communication between a program and a file

Before describing these statements, we must first introduce the conventions for identifying or labeling a file in a program.

Identifying a file.   The following material relates strictly to files created for MS-DOS (that is, the Microsoft disk operating system that is presently prevalent on personal computers). Other operating systems (notably for Apple computers like Macintosh) may have somewhat different schemes for file identification.

The file specification of *FileSpec* is used to identify the file in auxiliary storage. Its general form is

> *DeviceName:FileName*

where the *DeviceName* specifies the storage device on which the file is located and the *FileName* is the actual name by which the file is identified. The file name itself consists of a *name* and an *extension*. Therefore a more complete representation of the file specification is

> *DeviceName:name.extension*

If the *DeviceName* is omitted (as will be the case for all the following material), BASIC assumes that the device is the default disk drive for your computer. Also note that directory paths can be included in the file specifications, as in

$DeviceName: \backslash dir_1 \backslash \; . \; . \; . \; \backslash dir_n \backslash name.extension$

The *name* of the file consists of up to eight characters followed (optionally) by a period and an *extension* of up to three characters. The *name* should describe the contents of the file. It must start with a letter followed by letters, numbers, or select special symbols. (@, \$, &, #, %, ^ , !, and – are among the acceptable symbols; check your user's manual for a complete list.) Note that because of its employment as a delimiter between the *name* and the *extension,* a period cannot be used. In addition, blank spaces cannot be used. For this reason, many programmers use an underscore to represent file names that consist of two names; for example, `FILE_RED` would be a valid name.

The *extension* usually serves to identify the file type. For example, BAS is used to designate a BASIC file. If you do not supply an extension, the system will assume an extension of BAS when you store or retrieve a program from the disk in QuickBASIC. The extensions COM and EXE are used by the disk operating system to designate files that are in machine language. Aside from these (and a few other standard extensions), you can concoct your own extensions to identify the contents of a file. For example, the extension DAT could be used to designate that data is stored in a file, whereas TXT could be used if a word-processing test is stored. Extensions are particularly handy to discriminate between files with the same name.

In a later part of the present chapter, we will create a file to store the stream-flow data from Table 10.1. An appropriate name for such a file could be FLOW.DAT.

Opening a file.   The general form of the statement to open a file is

`OPEN` *FileSpec* `FOR` *mode* `AS` #*FileNumber*

where *mode* designates the mode in which the file is to be opened. The term "`OUTPUT`" is used if information is to be output from your program to the file, "`INPUT`" is used if information is to be input to your program from the file, and "`APPEND`" is used if new information is to be appended to an existing file. The *FileNumber* is a positive integer between 1 and 255 that is employed to identify the file in input-output statements within the BASIC program. This number is required because it is common to utilize several files in a program. For such cases, it is good practice to number them in ascending order, starting at 1.

It is important to realize that the *FileNumber* identifies the file *within the program,* whereas the *FileSpec* identifies it *on the external storage device.* Therefore, if a file is used in two different programs, it could have different file numbers, but the same *FileSpec* would be required in both instances. For example, for the first program, the file might be opened for output of data from the program to the disk with

`OPEN "GRADE1.BAS" FOR OUTPUT AS #1`

whereas for the second, it could be opened for input of data from the disk to the program with

`OPEN "GRADE1.BAS" FOR INPUT AS #2`

Transferring data from a file.   Although it would probably be more logical to first explain how data is transferred to a file, there is a good reason for discussing the reverse process first. This is because the data must be in a particular form in the file in order for it to be transferred properly from the file to the program. By first understanding how information is transferred from the file, we can then appreciate some of the problems associated with the reverse process.

The general form of the statement to transfer data from a file to the program is

```
INPUT #FileNumber, VariableList
```

where the *FileNumber* is the same one designated in the OPEN statement and the *VariableList* is the list of values that are being transferred from the file. An example of such a statement, along with an appropriate OPEN statement, is

```
OPEN "FLOW.DAT" FOR INPUT AS #1
INPUT #1, year, flow
```

The INPUT# statement works just like the INPUT statement that is used to accept data from the keyboard. Thus, if a number of items are to be input in a row, they must be delimited by commas. Similarly, string values must be enclosed in quotations. The presence or absence of these characters on the file is related to the way in which they were originally output to the file, as described next.

Transferring data to a file.   The PRINT#, PRINT# USING, and WRITE# statements can be employed to transfer data to a file. They differ in the way the data is formatted as it is output to the file. In general, the PRINT# and the PRINT# USING statements are employed to create reports and text files for later displaying, whereas the WRITE# is most commonly used to create files for subsequent processing by other programs.

The general forms of the PRINT# and the PRINT# USING statements are

```
PRINT #FileNumber, VariableList
```

and

```
PRINT #FileNumber, USING FormatString; VariableList
```

where the *FileNumber* corresponds to the one defined by the corresponding OPEN statement and the *VariableList* are the values being transferred to the file.

The PRINT# and the PRINT# USING statements output data to the file in a fashion identical with the way the corresponding PRINT and PRINT USING statements will display it on the monitor screen. Thus comma and semicolon delimiters work the same way in PRINT# statements as they do in PRINT statements. Similarly, the *FormatString* of the PRINT# USING controls the format to the file in a fashion identical with the one by which the PRINT USING controls the format to the screen.

Although the PRINT# and the PRINT# USING statements are suited to store a report or text file, they are less convenient when creating a file that is to be input to another program. This is due to the fact that, as described in the preceding section, the statements used to input the data from the file to another program require that the data be delimited by commas and that string values be enclosed in quotation marks. Although this can be done with either the PRINT# or the PRINT# USING statement, it cannot be accomplished very conveniently.

The WRITE# statement is designed to avoid these inconveniences. Its general form is

WRITE #*FileNumber*, *VariableList*

For this case, the *VariableList* can be simply delimited with commas, as in

WRITE #1, name1$, number

The WRITE# statement automatically sets such data in the proper format for subsequent input. That is, as the data is output to the file, commas are automatically inserted between items and string values are enclosed in quotation marks.

Closing a file.   Before terminating a run, it is good practice (and sometimes mandatory) to close all files. Among other things, this will prevent the computer from writing other information onto the files and possibly destroying valuable data. The command to close a file is

CLOSE #*FileNumber*

11.3.3   Standard File Manipulations

Now that we have introduced some of the BASIC file commands, we can show how they are used to perform some of the standard file manipulations that are needed for routine engineering applications. These are

- Creating a file
- Displaying a file
- Adding new records to a file
- Modifying or deleting records that are already on a file

Creating a file.   Figure 11.4 shows a simple program to transfer the data from Table 10.1 into a file. Notice that the file name is input as the string variable, filnam$. This generalized format allows the program to be used to store other information consisting of pairs of string [year$(i)] and numeric [flow(i)] constants. For example, if we had up to 50 years of flow data from another stream, we could use this program to store it in a file. However, to accomplish this correctly, we would have to give the file a name other than "FLOW.DAT." Otherwise the new data would be transferred onto the file and supplant (and hence destroy) the old data (unless APPEND is used).

FIGURE 11.4
A program fragment to
output data from a program
to a file.

```
'Create a Sequential File
INPUT "Create a file named"; filnam$
OPEN filnam$ FOR OUTPUT AS #1
   DO
      INPUT "year = (enter -999 to terminate)"; year
      IF year = -999 THEN EXIT DO
      INPUT "flow = "; flow
      WRITE #1, year, flow
   LOOP
CLOSE #1
```

FIGURE 11.5
A program fragment to
transfer data from the disk
back to the program.

```
'Print Out File Contents
INPUT "List a file named"; filnam$
CLS
PRINT " year    flow"
PRINT
OPEN filnam$ FOR INPUT AS #2
   DO WHILE NOT EOF(2)
      INPUT #2, year, flow
      PRINT USING "#####  "; year;
      PRINT USING "####.# "; flow
   LOOP
CLOSE #2
```

Displaying the contents of a file.   Figure 11.5 shows a program to transfer the flow data back from the file to the program. This version also displays the data so that we can check that the data transfer has been performed as expected. Notice that in this example, we use a file number of 2 instead of 1 as in the previous example. As long as we are consistent within a program, either number could be used. However, also notice that both examples would employ the name "FLOW.DAT," which is how the file would be referenced on the disk.

As the data is being input, we must develop a way to detect when there is no more data to be read from the file. Otherwise, if we try to read data that is not present, we will obtain an error and the program will stop.

There are several ways to detect that we are at the end of a file. One way is with the end-of-file function. When a file is created, an end-of-file mark is placed after the last item of data. The end-of-file function can detect this mark. The function can then be used in the DO WHILE loop, as seen in Fig. 11.5. As long as we have not read all the data from the file, EOF will be false and control will transfer to the next line. However, when the last item is read, the EOF becomes true and control transfers out of the loop. Thus, if we are looping to read the data, this statement allows us to transfer control out of the loop before we attempt to read a nonexistent record.

Appending record to a file.   Just as we have used INPUT and OUTPUT to designate input and output modes, APPEND can be used to append or add data to the end of a file. Figure 11.6 shows a program to append additional years of flow data to the file created for Table 10.1.

```
'Adds Data to the End of a File
INPUT "Add data to a file named"; filnam$
CLS
PRINT " year    flow"
PRINT
OPEN filnam$ FOR APPEND AS #1
  DO
    INPUT "year = (enter -999 to terminate)"; year
    IF year = -999 THEN EXIT DO
    INPUT "flow = "; flow
    WRITE #1, year, flow
  LOOP
CLOSE #1
```

FIGURE 11.6
A program fragment to append additional data to a file.

```
'Correct a Value from a Sequential File
INPUT "File to be edited"; filnam$
OPEN filnam$ FOR INPUT AS #1
OPEN "dumm.dat" FOR OUTPUT AS #2
INPUT "Year to be corrected"; year.corr
DO WHILE NOT EOF(1)
  INPUT #1, year, flow
  IF year.corr <> year THEN
    WRITE #2, year, flow
  ELSE
    PRINT "current value = "; flow
    INPUT "change to "; flow
    WRITE #2, year, flow
  END IF
LOOP
CLOSE #1
CLOSE #2
KILL filnam$
NAME "dumm.dat" AS filnam$
```

FIGURE 11.7 A program fragment to modify an existing record on a file.

The same procedure can be used to insert data into the middle of a file. For this case, we would read all the records up to the point where the insertion is to be made and write these onto a new file. Then we could add the new information onto the new file. Finally, the remaining records could be read onto the new file.

Modifying or deleting records from a file. Once we have stored data on a file, there will be occasions when we will have to return and modify or delete some of the records. Both these operations can be performed in a fashion simlar to the previous method for inserting new data into an existing file.

Figure 11.7 shows a program to modify the flow for a particular year. Notice that a dummy file, dumm.dat, is created to hold the modified file as it is being created. After the modification is accomplished, the old file is destroyed with a KILL statement, which has the general form

KILL *FileSpec*

This statement deletes the *FileSpec* from the disk storage device. Then, the dummy file is given the original name of the file with the NAME statement, which has the general form

NAME *FileSpec_{old}* AS *FileSpec_{new}*

PROBLEMS

11.1 The program in Fig. 11.3 is not "bulletproof." For example, you can enter a number with more than two decimal places and it will display the final result with these extraneous numbers. Modify the program so that it circumvents this problem and any other deficiencies you can determine.

11.2 Develop and test a QuickBASIC program that enters a string and then displays it in reverse order.

11.3 Develop a QuickBASIC program that enters a string that includes blank spaces and commas. Use a LINE INPUT statement (recall Chap. 5) to enter the string. Have the program display the string minus the spaces.

Test the program by entering

Everything should be simple as possible, but no simpler

and displaying

Everythingshouldbesimpleaspossiblebutnosimpler

11.4 Store the months of the year in a single string, month$, as in

month$ = "janfebmaraprmayjunjulaugsepoctnovdec"

Have the user enter a date in the following format:

08/05/48

Design a program that uses month$ to display the date as

Aug. 5, 1948

11.5 Members of a sales force receive commissions of 5 percent for the first $1000, 7 percent for the next $1000, and 10 percent for anything over $2000. Compute and display the commissions and social security numbers for the following salespeople: John Smith, who has $5325.76 in sales and a social security number of 080-55-6642; Priscilla Olkowski, who has $8050.63 in sales and whose number is 306-77-5520; and Bubba Schumacker, who has $6643.96 in sales and a social security number of 707-00-1112. Output the results in a neat tabular format.

11.6 Write, debug, and document a computer program to determine statistics for your favorite sport. Pick anything from softball to bowling to basketball. Design the program so that it is user-friendly and provides valuable and interesting information to anyone (for example, a coach or player) who might use it to evaluate athletic performance. In particular, display the results in a neat tabular format.

11.7 Expand the program in Example 11.1 so that it also includes a class participation grade (PG). The final grade would be recalculated as

$$FG = 0.5AQ + 0.2AH + 0.25FE + 0.05PG$$

Give Sarah and Jane PGs equal to 95, and give Fred a 75. Include this new information on the table along with their identification numbers (Jane: 073-93-3336; Fred: 544-01-9969; and Sarah: 139-93-4446).

11.8 Select several of your favorite issues from the New York Stock Exchange. Write, debug, and document a BASIC program that reads the name of the stock, its selling price today, and its 52-week low. Use READ/DATA and arrays to incorporate the data into the program. The program should then display the name of the stock, its selling price, and the percentage gain over the 52-week low. Output the results in a neat tabular format.

11.9 Repeat Prob. 11.8, but employ a sequential file to store the information on a disk. Employ code similar to the program fragments from Figs. 11.4 through 11.7. Utilize subscripted variables to store the data.

11.10 Using Fig. 11.7 as a model, develop a program fragment to delete a record from a file.

11.11 Develop a modular computer program to determine your grade for this course. Store all relevant data in a sequential file.

11.12 Develop a modular, well-structured, menu-driven program to generate and edit a sequential file.

11.13 Employ the PRINT, CLS, LOCATE, and other QuickBASIC statements to form a design on the screen. The design can be your initials, your school's block letter, your fraternity's or sorority's Greek letters, or any other symbol, image, or logo. Use a screen layout form like the one in Fig. 11.2 to design your display.

11.14 Computer graphics are based on a simple idea that relates back to the very definition of plotting itself. Plotting consists of locating points on a two-dimensional space. Computer output media, such as printer paper or monitor screens, also represent two-dimensional spaces. Figure 11.2 shows how the screen of a typical computer monitor can be visualized as a grid of cells. Such a screen is said to be in the *text mode*.

Computer graphics in text mode are built up from the fundamental operation of placing a character into one of these cells. For example, a character such as an asterisk may be placed in one of the cells. If several asterisks are placed in a row of cells, a line can be formed. Similarly, one vertical column and one horizontal row may be positioned to form a set of axes.

Develop your own text-mode plot program. Express the program in procedure form so that you can use it to add a plotting capability to other program. Have the following data serve as input to the procedure: one-dimensional arrays, `x` and `y`, holding the `x`, `y` coordinates of the points; `n`, the total number of points; and the minimum and maximum limits of the `x` and `y` axes—`xmin`, `xmax`, `ymin`, `ymax`.

STRUCTURED FORTRAN

After the development of a sound and comprehensive plan, the next step of the program-development process is to compose the program using a high-level language. This part of the book shows how structured FORTRAN is employed for this purpose.

Chapter 12 is devoted to the most elementary information regarding FORTRAN programming. This material is designed to get you started by providing information sufficient to successfully write, execute, and document simple programs. *Chapters 13* through *18* cover additional material that will allow you to develop more complicated and powerful programs.

FUNDAMENTALS OF STRUCTURED FORTRAN

I don't know what the language of the year 2000 will look like but I know it will be called FORTRAN.

— C.A.R. "Tony" Hoare
Annals of the History of Computing, January 1984

At the start of the electronic computing era, business and scientific programmers had related but primarily distinct computing needs. Business-oriented work usually involved sophisticated input-output, large amounts of data, and relatively simple computations. In contrast, scientific and engineering computing entailed simple input-output, smaller quantities of data, and fairly intricate calculations. Unfortunately, the assembly language used by the early programmers had few arithmetic operators, and programming of computations was difficult and often unreliable. For this reason, there was a need in the 1950s for a high-level language expressly designed to accommodate technical computing.

FORTRAN, which stands for *FOR*mula *TRAN*slation, was introduced commercially by IBM in 1957 to meet this need. As the name implies, one of its distinctive features is that it uses a notation that facilitates the writing of mathematical formulas. Because of such features, FORTRAN is the native language of most engineers and scientists who attained computer fluency on large mainframe machines in the 1960s and 1970s.

12.1   WHY FORTRAN?

Today, although other languages are available for implementing computations, FORTRAN is still very popular. One reason for this popularity relates to the fact that FORTRAN has special calculation capabilities that are still not available in most other languages. For example, FORTRAN allows computations involving complex numbers to be performed very conveniently.

Beyond having a broad range of computational capabilities, FORTRAN also makes them very efficient and accurate. Because it is a compiled language, FORTRAN programs execute quite rapidly in comparison with programs written in interpreted languages such as old-style BASIC. In addition, because FORTRAN can be implemented on mainframe computers, it can exploit the high precision that is available on such machines. Finally, because it is so computationally oriented, FORTRAN compilers are capable of optimization of algorithms. Thus, beyond translating your program into machine language, FORTRAN compilers can be used to eliminate unnecessary instructions and to increase execution speed by streamlining your algorithm wherever possible.

Aside from its utility for computations, the availability of FORTRAN also contributed to its popularity. Because of IBM's dominance in the computer marketplace, other manufacturers developed FORTRAN compilers to remain competitive with IBM. Thus FORTRAN compilers have been readily available for all types and sizes of computers. Even personal computers, which were originally too limited to handle a full FORTRAN compiler, have evolved to the point that today most popular machines can accommodate FORTRAN.

As a result of the widespread use of FORTRAN, a wealth of programs have been written in the language. These programs are so ubiquitous that it behooves every engineer to learn the language in order to tap this body of knowledge.

12.2   WHICH FORTRAN?

Interestingly, the movement of FORTRAN to new and different computer systems did not spawn a lot of FORTRAN dialects. Most versions adhere to a common set of statements and syntax. This is probably a result of two factors: the role of a dominant company like IBM in its creation and development, and the efforts of the *American National Standards Institute (ANSI)* to promote uniform dialects.

The dialect originally introduced in 1957 came to be known as FORTRAN I. It was followed by a number of improved versions that culminated in 1962 in FORTRAN IV. In about 1966 a standard FORTRAN was adopted by ANSI. This dialect, which has come to be known as FORTRAN 66, essentially corresponds to FORTRAN IV. Even today, some programs still use this version.

In 1978 an ANSI standard called FORTRAN 77 was adopted. Much of the significant engineering software written since then has been composed in this dialect. Consequently, modern engineers should be fluent in FORTRAN 77.

The compiler used to develop the programs in this book is the Microsoft FORTRAN 5.1 product which is designed for personal computers. Although it has its own extended features, it is compatible with FORTRAN 77. In addition, there are other compilers that run on other types of machines. For example, the VS FORTRAN[1] compiler runs on IBM mainframes and compatible equipment and is also compatible with FORTRAN 77. Other computer manufacturers such as Digital Equipment Corporation (DEC) offer their own FORTRAN 77 compilers. In each case, the individual implementations have a few system-specific features.

All the programs contained here will employ standard FORTRAN 77 with no

---

[1]VS is short for "virtual storage" and refers to the way that programs are stored in internal memory.

dialect-specific features. Thus, we will adopt a "classic" FORTRAN 77 that should be compatible with many types of computing systems. Although this means that we will not be able to exploit some neat features of some of the versions, we believe that compatibility is more important. This is due to the fact that, more than some other computer languages, FORTRAN programs must often be run on different types of machines.

Recently, a version known as FORTRAN 90 has been developed. This new standard includes extensive refinements and improvements to the language. At present, ANSI has decided that the United States will employ both FORTRAN 77 and 90. Its global counterpart (the International Standards Organization, or the ISO) has decided to switch exclusively to FORTRAN 90.

Wherever possible, we will illustrate some of the new features of FORTRAN 90. However, when we discuss these features, we will take pains to distinguish them from those available on FORTRAN 77.

12.3 "TALKING" TO THE COMPUTER IN FORTRAN

How you actually go about interacting with the computer to enter and run a FORTRAN program depends on the system you employ. In the early years of computing, most machines operated as *batch-processing systems.* This mode allowed little or no direct interaction between the user and the machine. Each program was typed by the user on punched cards (Fig. 12.1) with a card-punch machine called a *keypunch.* Together with the program, data and job-control commands were also typed on cards. The entire "card

FIGURE 12.1
A punch card.

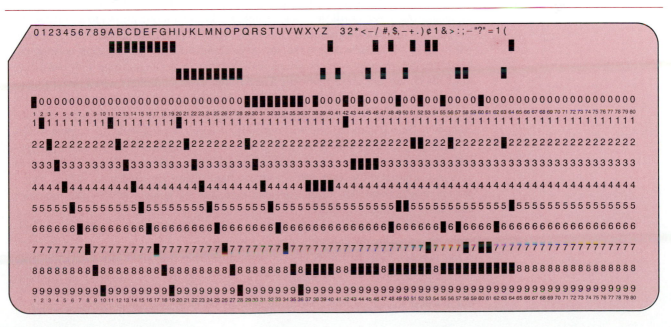

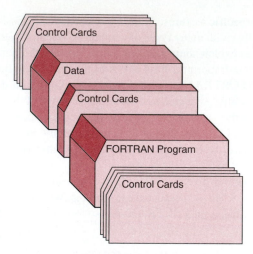

FIGURE 12.2
A FORTRAN package,
or "job."

deck" was assembled into what was called a FORTRAN package, or "job" (Fig. 12.2). Each package would usually be submitted to a computer operator who would run it and then return the job along with printed output.

The key to this mode of operation was the separation between the user and the execution of the program. Often there was a significant delay between the time that a batch job was submitted and the time that the output was received. If an error was detected, the user would not know it until she or he received the output. In that event, the program would have to be corrected and the job resubmitted.

Although batch systems are still used today, an alternative mode called a *conversational system* is also available. As the name implies, this mode allows more interaction between the user and the machine. Usually the system operates as a *timesharing* system, where many individuals access the computer simultaneously. Information is often entered into the computer via a *terminal*. This is typically a video display unit consisting of a cathode-ray tube and a keyboard. The terminal is either directly wired into the computer or connected by a telecommunications hookup. A set sign-on procedure, which varies from system to system, is usually required to establish communication with the computer. During sign-on, you will typically be required to provide the computer with information regarding your needs for the session (for example, the language and peripheral equipment you require). In addition, you must usually type in an identification code and a password to gain access to the system. These are needed for billing and to prevent unauthorized individuals from gaining access and tampering with your programs and data.

Once you have gained access to the computer, you are ready to type in your program. In contrast to the punching of cards, the typing of a program is a much more interactive process. Most terminals have editing capabilities that allow you to modify lines in an efficient manner. Once the program is to your liking, it is submitted to the computer by typing appropriate commands. We will discuss these commands in further detail in a later section. For the time being, it is sufficient to understand that the running of the program for a compiled language such as FORTRAN proceeds in stages. Whereas in batch mode you lose contact once the process begins, conversational systems allow you to maintain some control. Thus as each phase is completed, you can be apprised of the progress. If a problem occurs in the first phase, you can be alerted immediately so that you can correct the mistakes and run the program again. It is this level of control that usually makes conversational systems preferable to batch-mode systems.

The operation of a FORTRAN program on a personal computer or workstation is quite similar to operation on a timesharing system in most respects. As with the time-sharing system, the user employs an *editor* to type and correct the program until it is acceptable. A number of commands are then employed to take the program through the stages needed to come up with the final result.

Now that you know something about the manner in which you will interact with the computer, we will take a closer look at the FORTRAN program itself. As mentioned previously, a program is a set of instructions to tell the computer what to do. In FORTRAN each instruction is usually written as a single line, and each instruction is typically aligned in specific columns. The conventions regarding the alignment stem from the use of punch cards in the early years of computing. Although cards are employed infrequently today, the spacing conventions have usually been maintained. It should be noted that some systems do not employ these conventions and use numbered lines in an interactive mode that is very similar to BASIC. Others do not require the punch card format to be observed. Obviously, if your system employs such schemes, you will not need to observe the spacing conventions. However, it is not a bad idea to acquaint yourself with the following practices because they are still employed in many versions of FORTRAN. Consequently, if you observe the conventions, it will be easier to run your programs on many different computers.

The 80 columns of a punch card are called the card's *field*. Parts of the field are set aside for specific purposes. These are illustrated on the coding form shown in Fig. 12.3. A coding form is a piece of paper on which the program can be written and checked for mistakes prior to being entered into the computer. Note that the form has an 80-column field just like the punch card.

Aside from representing an example of a coding form, Fig. 12.3 describes how the specific parts of the field are used for particular purposes. You should examine this figure carefully to acquaint yourself with these rules.

As shown in Fig. 12.3, FORTRAN accepts programs in lines of 80 characters or less. Usually all characters beyond column 72 are ignored. The last eight columns were traditionally set aside for labeling each card, usually by numbering them in ascending order. This makes sense when one thinks in terms of cards rather than terminal lines. When using cards, there was always the distinct possibility that you would drop a large deck on the floor. Because card-sorting machines were available, the numbers in the last columns could serve as the basis for putting a shuffled deck back into correct order.

Notice that some program lines are numbered using columns 1 through 5. The number can start in any of the five columns but cannot extend into the sixth. Numbers can therefore range from 1 to 99999 and do not have to be in any particular order. However, each line number can be used only once. Thus its sole function is to provide a distinct identity to a particular line within the program.

```
DATA PROCESSING CENTER
Texas A & M University
PROGRAM     INTRODUCTION TO COMPUTING FOR ENGINEERS      PUNCHING INSTRUCTIONS   GRAPHIC   PUNCH   PAGE  OF
PROGRAMMER  S. C. CHAPRA                      DATE
Statement: Name | Description | Operand | Comments                                      Identification square
C    THIS CODING FORM IS USED TO DEMONSTRATE THE PURPOSE OF THE VARIOUS      LINE    1
C    FIELDS ON A PUNCHED CARD                                                LINE    2
C    COMMENT CARDS, OF WHICH THE PRESENT LINE IS AN EXAMPLE, ARE             LINE    3
C    DESIGNATED BY A "C" IN COLUMN 1                                         LINE    4
C    OTHERWISE, COLUMNS 1 THROUGH 5 ARE RESERVED FOR STATEMENT NUMBERS.      LINE    5
C    FOR EXAMPLE, THE FOLLOWING LINES ARE STATEMENTS 1, 83, AND 2006.        LINE    6

    1 FORMAT(I8)                                                             LINE    7
   83 Y=A+(B-X)**2.-5.*C/D                                                   LINE    8
 2006   Y = A + (B-X)**2. - 5.*C/D                                           LINE    9

C    COLUMNS 7 THROUGH 72 ARE RESERVED FOR THE ACTUAL STATEMENTS.            LINE   10
C    THE COMPUTER IGNORES BLANK SPACES WITHIN THIS FIELD AND, THEREFORE,     LINE   11
C    SPACING CAN BE USED TO FACILITATE READING.  FOR INSTANCE, STATEMENTS    LINE   12
C    83 AND 2006 ACCOMPLISH THE SAME ALGEBRAIC MANIPULATIONS BUT             LINE   13
C    STATEMENT 2006 IS MORE READABLE DUE TO THE INCLUSION OF BLANKS          LINE   14
C    IF A STATEMENT IS TOO LONG FOR ONE CARD, IT CAN BE CONTINUED ON         LINE   15
C    THE FOLLOWING CARD BY PLACING ANY NUMBER OTHER THAN ZERO IN COLUMN 6    LINE   16

        CONC = CONC + (THETA -XMAX) / (THETA - XMIN) * POPUL - (THETA -      LINE   17
     1  YMAX) / (THETA -YMIN) * POPUL                                        LINE   18

C    COLUMNS 73 THROUGH 80 CAN BE USED TO NUMBER THE LINES IN A PROGRAM      LINE   19
C    OR FOR ANY OTHER LABELING PURPOSE                                       LINE   20
```

FIGURE 12.3
A coding form; the particular form shown outlines the conventions governing the use of the columns in a FORTRAN instruction that originated from the days when all jobs were prepared on punched cards.

The reasons for numbering lines will become clearer as we learn more about programming. A first example is provided in Fig. 12.4. A computer executes a program by starting at the first line and then progressing from one line to the next unless instructed to do otherwise. Therefore, one reason for numbering the lines in columns 1 to 5 is to override normal sequence. To do this, you could type a statement directing the computer to jump or "go to" the line of your choice.[2] The only way that the computer would know where it was supposed to go would be if the destination statement had a line number (Fig. 12.4).

Such jumping around is not done in the simple program in Fig. 12.5. For this example, each statement is executed in a straightforward sequence. First, values are assigned to the variables A and B. The following line sums these variables and assigns the answer to C. Next the computer is instructed to display the answer. When the program is executed, this will make the computer display the result on an output device. On personal computers and timesharing systems, the PRINT statement is usually designed to direct

[2]Note that this sort of "jumping around" is becoming passé in modern programming. We will have a lot more to say about this important topic as we get further into FORTRAN.

the output to the monitor screen. On batch systems, the output can also be directed to a screen, but in many cases it is directed to a printer.

The asterisk in the PRINT statement tells the computer the manner, or *format*, in which the answer is to be displayed. The asterisk instructs the computer that we want the answer to be *list-directed*. That is, we will not try to control exactly how the result is output and will leave it up to the computer to display it in an acceptable fashion. Further on in this chapter, we will learn a little more about how this can be done in a more refined manner. In a later chapter, we will take an extensive look at how to control the formatting of both input and output. For the time being, the simple approach embodied by the asterisk will suffice.

FIGURE 12.4
A FORTRAN program starts at the top line and progresses downward unless instructed to do otherwise. One way to circumvent the normal sequence is with a GO TO statement that directs the computer to jump or "go to" the line of your choice. One reason for numbering a FORTRAN line is to let the computer know where it is "going to."

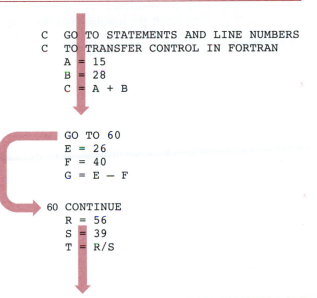

```
C   GO TO STATEMENTS AND LINE NUMBERS
C   TO TRANSFER CONTROL IN FORTRAN
    A = 15
    B = 28
    C = A + B


    GO TO 60
    E = 26
    F = 40
    G = E − F

60  CONTINUE
    R = 56
    S = 39
    T = R/S
```

FIGURE 12.5
A simple FORTRAN program to add two numbers.

Listing:

```
      PROGRAM ADDER
C Simple Addition Program
      A = 28.
      B = 15.
      C = A + B
      PRINT *, C
      END
      ↑
      |_____ col. 7
```

Output:

```
43.000000
```

Before moving on to learn how to run a program, we should mention a number of points:

- Notice that all the letters in Fig. 12.5 (except the comment) are capitalized. Many FORTRAN compilers require that programs be typed in uppercase.
- Certain FORTRAN statements are called *executable* because they represent instructions to the computer to carry out tasks during the program's execution. All others are called *nonexecutable*. In Fig. 12.5, the PROGRAM statement, the comment, and the END statement are nonexecutable.
- Observe that the last line in the program is an END. The END is a nonexecutable statement that identifies the last physical line in a FORTRAN program or subprogram. There is only one END statement in a program or subprogram. It informs the compiler where it is to terminate the translation of the program into machine language. Consequently, every program or subprogram must have END as its last line.

12.4   HOW TO PREPARE AND EXECUTE FORTRAN PROGRAMS

Every computer system has its own protocol for preparing and executing a FORTRAN program. Therefore we are somewhat limited in the guidance we can provide regarding these tasks. However, although the details differ, the general steps are the same. In the present section, we will describe these general steps as background information that you should find useful when you learn the specific procedures for your system.

FIGURE 12.6
Schematic of the four steps involved in running a FORTRAN program. The boxes, indicate the type of program generated by each of the system programs, which are designated as ovals. External arrows indicate other information required or generated by the system programs.

The four general steps for preparing and running a FORTRAN program on any interactive system are outlined in Fig. 12.6. (A somewhat different, but closely related, process is employed for batch systems.) The following sections describe these steps in detail.

12.4.1   Creating and Editing Programs

After the program has been composed, the first step in implementing it is to type it and save it in the computer. This involves two phases: allocating the storage space for the

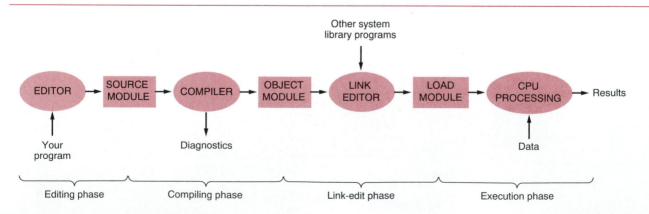

program and then typing the program into this space. Allocation of storage is often automatically accomplished by simply naming the program. However, in some cases the amount of storage must be specified.

For an interactive system, the actual typing of the program is performed on the terminal or computer keyboard. The typing process employs an *editor,* which is designed to facilitate the creation of files. Some editors are line-oriented, and others are organized around pages. Most will have capabilities for inserting or deleting characters, words, or lines anywhere in a program. Most will also have the ability to move or copy one section of a program to any other location.

All the above capabilities should allow you to create a file containing your FORTRAN program. When you are satisfied that the program is correct, the file is saved. At this point, it is referred to as the *source module* or the *source code.*

12.4.2 Compiling Programs

In the compilation step, the FORTRAN version of the program stored as the source module is translated into a machine-usable form—the *object module* or the *object code.* If the computer detects errors in your program, *diagnostics,* or *error messages,* will be displayed. The errors can be corrected by returning to the editing step and then resubmitting the corrected source module.

Once the translation is successfully completed, the compiled object module can be used over and over again without repetition of the translation process. This is why compiled programs are much faster to run than interpreted programs, which must be translated line by line every time they are executed.

12.4.3 Linking Programs

Before they can be run, almost all programs will require other special programs from the system library. For example, most computers cannot compute the square root directly, and so a special library program is needed for this purpose. Other programs may contain the routines to display on the terminal screen or on a hard-copy device such as a printer. These system library programs are combined with your object module in the *link* step. This process may occur automatically, or it may require special instructions on your part.

In addition to merging programs, linking also specifies the *memory map* for your program. That is, the link program takes care of all details necessary to place your program in the computer's memory so that the CPU can process it.

12.4.4 Executing or Running Programs

After the object module is linked to the system library programs, it is called a *load module.* At this point, it is ready to be executed. This is usually accomplished by a simple command. On some systems this entails merely typing the program's name and then striking the [Return] key on the terminal. In other cases, you may have to type the word "RUN" followed by the program name. Whatever the procedure, the effect is the same. For the simple addition program of Fig. 12.5, the result of the addition—43.000000—would be displayed on the output device.

12.5   A SIMPLE FORTRAN PROGRAM

Now that you have been introduced to how a program is executed, we can begin to discuss the actual mechanics of developing a FORTRAN computer program. Figure 12.5 shows a very simple FORTRAN program to add two numbers. Each of the lines in this program is called a *statement*. When such a program is executed or "run," the statements are implemented in sequence. Thus, a program (that is, the *source code*) is merely a series of instructions to direct the computer to perform actions.

For this program, values of 28 and 15 are assigned to the variables, A and B. Next, the addition is performed and the result stored in the variable C. Then, the value of C is displayed on the screen by the PRINT statement. Finally, the END statement signals that the program is finished.

The remainder of this chapter will be devoted to expanding Fig. 12.5 into a quality FORTRAN program. Because of the simple nature of the algorithm underlying Fig. 12.5, the following material does not represent a detailed description of the capabilities of FORTRAN. A more in-depth discussion of the language will be deferred to subsequent chapters. For the time being, we will focus on the fundamentals of FORTRAN programming. In addition, we will introduce some practices that will contribute to the quality of your finished product.

12.5.1   Adding Descriptive Comments: The Comment Statement

As we will reiterate time and time again, an important attribute of a high-quality computer program is how easy it is to read and understand. One deficiency of the program in Fig. 12.5 is that it is almost devoid of documentation. In other words, it consists merely of a series of instructions to get a job done. Although this is not a major problem for such a simple program, inadequate documentation can detract immensely from the value of larger and more complicated programs. If the user has to expend a great deal of time and energy to decipher your program's operation and logic, it will diminish the impact and usefulness of your work. Furthermore, if you have not used your own programs for a period of time, you will have difficulty deciphering your own poorly documented code.

One of our primary tools for the internal documentation of a program is the comment statement. The general form of this statement is

C *comment*

where the letter C is in the first column[3] and the *comment* is any descriptive information you would like to include. For example, the *comment* included at the beginning of Fig. 12.5 provides a title for the program, and thus the user can immediately identify the general subject of the program.

Beyond setting forth titles, comments can be employed to incorporate additional documentation. Figure 12.7 is an example of how this can be done for the simple addition program. Notice that a group of comments has been used to form a title section or

---

[3]An asterisk may also be used in place of the C to specify a comment. We prefer the C and will employ it consistently throughout this text.

```
      PROGRAM ADDER
C Simple Addition Program (Version 1.1)
C
C by S.C. Chapra
C June 10, 1992
C
C This program adds two numbers
C and prints out the results
C-----------------------------------
C
C assign constants to variables
C
      A = 28.
      B = 15.
C
C perform computation
C
      C = A + B
C
C display results
C
      PRINT *, C

      END
```

FIGURE 12.7
A revised version of the simple addition program from Fig. 12.5, including comments that serve to document the program.

"header" at the program's beginning. We have called this version 1.1 to indicate that it is an improved version of the original code in Fig. 12.5 (version 1.0). Along with the title, this section includes the author's name and a concise description of what the program is to accomplish.

After the title section, comments can be used to interject descriptive statements as well as to make the program more readable. As seen in Fig. 12.7, this latter objective can be accomplished by using comments to clearly delineate the program's major sections. Also notice how several blank lines and a dashed line are included to set off and separate some of the comments and modules. The judicious use of lines and blank spaces is one effective tool for making the program more readable.

While the use of comments is optional in FORTRAN, it is strongly encouraged. But do not overuse comment statements. If the action of a line or lines is obvious, do not add excess verbiage in the form of remarks. A program can quickly become cluttered with unneeded comments that actually make it harder to read and follow. As described in the next section, careful selection of variable names helps to minimize the need for explanatory comments.

12.6 NUMERIC CONSTANTS, VARIABLES, AND ASSIGNMENT

As an engineer, one of the primary uses you will have for a computer is to manipulate numbers. Consequently, it is important to understand how numeric quantities are stored and manipulated on the computer.

There are two fundamental ways in which numeric data is expressed in FORTRAN: directly as constants and indirectly as variables. *Constants,* also called *literals* in computer jargon, are values that remain fixed during program execution; whereas *variables* are symbolic names to which a value is assigned that can be changed during execution.

12.6.1 Numeric Constants

Numeric constants are numbers that remain fixed throughout program execution. They may be either positive or negative and may be expressed in integer or real form. *Integers* have no decimal point, as in

896   67   −1507   883

where the standard integer values for a 32-bit word size can range from −2,147,483,648 to 2,147,483,647.

The *real variables* are those with decimal points and, most often, fractional parts:

896.   −4.78   0.0056   −100.4

Note that 896 is an integer and 896. is a real constant. The two forms are stored differently in the computer.

In addition, real constants can be expressed in exponential notation. *Exponential constants* are similar in spirit to scientific notation and are handy for expressing very large or very small numbers. To review, a number is expressed in scientific notation by converting it to a number between 1 and 10 multiplied by an appropriate integral power of 10. This can be done by moving the decimal point to the right or left until there remains a number between 1 and 10. This remaining number must then be multiplied by 10 raised to a power equal to the number of places the decimal point was moved. If the decimal has been moved right, the power is negative, and if it has been moved left, the power is positive. The general form is

$$b \times 10^n$$

where b is the value that remains after the number has been converted to a value between 1 and 10 (commonly called the *mantissa*) and n is the integer number of places the decimal was moved (referred to as the *exponent*). For example, the number 8,056,000 would be converted to 8.056 by moving the decimal six places to the left. Therefore in scientific notation it would be expressed as 8.056×10^6.

FORTRAN employs the same concept, using the exponential notation

$$b\text{E}n$$

Thus in FORTRAN the number 8,056,000 is expressed as 8.056E6. As we will discuss shortly, a double-precision exponential form is also available, one using a D in place of the E.

Letters (other than E and D) and symbols such as dollar signs and commas are not allowed in numeric constants. Therefore $56.23 and 1,760,453 are illegal numeric constants and would constitute a syntax error in FORTRAN. In addition, only one minus sign may be used at the front of a constant to designate that it is negative. Minus signs cannot be used as delimiters (that is, separators), as is commonly done in social security numbers. For example, 120-44-5605 is an illegal constant in FORTRAN. In fact, FORTRAN would misinterpret it as representing subtraction and compute a result of -5529. However, a minus sign can be used to designate negative exponents. Consequently, $-3.2E-4$ is a legal FORTRAN representation of -0.00032.

It should be noted that real numbers in FORTRAN can be expressed in either single or double precision. *Single-precision numbers* are usually stored with about 7 significant digits, whereas *double-precision numbers* are stored with about 15 (different machines range from 14 to 18). If a number is written with more than 7 digits, it will automatically be truncated when stored unless it is written in the double-precision version of the exponential form. This form uses a D rather than an E, as in

$$3.1415926536D0 \qquad 2.56001753D-6 \qquad -2.050000768D9$$

You may be wondering at this point why so many different forms are needed and why we would want to force the computer to store a constant in a particular form. There are a number of reasons why we might want to go to the trouble. For example, an integer takes up less space in memory and can be processed faster during computations than a real constant. Single-precision numbers have similar advantages relative to double-precision constants. Thus, if we use integers rather than single precision, and single precision rather than double precision, the resulting programs will run faster and take up less memory. On the other hand, double-precision constants are advantageous when a computation must be performed with a high degree of precision.

12.6.2 Variables and Assignment

In addition to constant quantities, it is necessary to have variable quantities that can change during the execution of a program. These quantities are designated by symbolic variable names. For example, in the simple addition programs from Figs. 12.5 and 12.7, we used three variables—A, B, and C. (It is common practice to refer to the variable name as the variable itself.)

To understand how variables work, it is convenient to think of the computer memory as consisting of a number of cells that are similar to post office boxes (Fig. 12.8). The variable is the name that the computer assigns to each of these cells. For example, in the simple addition problem, each of the variables—A, B, and C—would correspond to the label of the cells. When we use the line A = 28 to assign a value of 28 to A, the computer stores the value in the memory location for A (Fig. 12.8a). Then later in the program when we perform the addition C = A + B, the computer "remembers" that A is equal to 28 by referring to the memory cell where it is stored.

Now suppose that after the addition we redefine A by another assignment statement, such as A = 80. When this line is executed, the computer replaces the old value with this

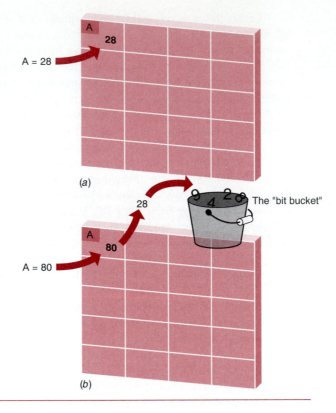

FIGURE 12.8
Analogy between computer memory locations and post office boxes. (*a*) The FORTRAN statement $A = 28$ assigns the variable name A to one of the memory locations and stores a value of 28 in that location. (*b*) A subsequent assignment statement, $A = 80$, places a new value of 80 in the location, and the process destroys the prior value of 28.

new one. Using our analogy of the postal boxes, this act would be akin to placing the new number in the box and pushing the old one out the back and into oblivion, or into the "bit bucket" as it is sometimes called (Fig. 12.8*b*). This ability to change value is the reason symbolic names are called "variables."

Because lines such as $A = 28$ and $B = 15$ assign constant values to variables, they are called *assignment statements*. They can be written in general form as

variable = *expression*

where *variable* stands for the variable name and *expression* can be a constant, another variable, or an arithmetic expression involving constants and variables. An example of the use of an arithmetic expression in an assignment statement is the line

C = A + B

When such a statement is executed, the expression to the right of the "=" sign is evaluated first. Then the resulting numeric value is automatically stored in the storage location for C. For example, in the program from Fig. 12.7, a value of 43 would be stored in C.

Note that, aside from addition, variables can also be subtracted, multiplied, and divided, as in

```
C = A - B
C = A * B
C = A/B
```

As shown in Fig. 12.3, all the above assignment statements must lie within columns 7 through 72.

In general, most compilers ignore blank spaces between terms in assignment statements. For example, the addition statement could be written as

```
C  =  A  +  B
```

12.6.3   Algebraic Versus Computer Assignment Statements

The equal sign (that is, the *assignment operator*) in the above statements can be thought of as meaning "is replaced by" or "is assigned the result." This is so significant that a computer language such as Pascal does not even use the "=" sign to designate the process of assignment. For example, in Pascal, it is represented by ":=" as in

```
c := a + b
```

Some computer scientists have proposed that a better representation of the process would be an arrow, as in

```
c ← a + b
```

Both modifications are efforts to distinguish more clearly between assignment and equality. In particular, they signify that computer assignment statements differ in a very important way from standard algebraic equations. That is, assignment statements must always have a single variable on the left-hand side of the equal sign. Thus, because it violates this rule, the following legitimate algebraic equation is illegal as a FORTRAN assignment statement:

```
A + B = C
```

In the same sense, although

```
X = X + 1
```

is not a proper algebraic equation, it is a perfectly acceptable FORTRAN assignment statement. This is evident when the words "is replaced by" or "is assigned the result" are substituted for the equal sign. For example, if X had a prior value of 5, the new value, after execution of $X = X + 1$, would be 6.

All the above makes eminent sense when we remember how the computer accesses memory locations to store and retrieve values. Using the postal box analogy (Fig. 12.8), if more than a single variable name occurred to the left of the assignment operator, how could the computer conceivably know where to assign the outcome? Clearly, the only way the scheme works is when a single variable name appears on the left.

12.6.4 Variable Names

The symbolic name for the variable should reflect, as simply as possible, what information is stored by the variable. Mnemonic names that are easy to remember are best. Names that are short enough to type quickly and accurately are helpful, as long as the name is long enough to convey the meaning or purpose of the variable. While all the following are valid FORTRAN variable names, those in the first row are inferior because their meanings are more obscure than the meanings for those in the second row:

```
SXEQW      Z19943     ABC4DE     I1949J     S
PROD23     FLUX       TOTAL      NAME1      SUM
```

Aside from being communicative, the actual symbolic names used to designate variables must follow rules (Table 12.1). Names must begin with a letter followed by up to five additional letters and numbers. Therefore, names cannot contain symbols, nor may they be longer than six characters. The following are some examples of unusable FORTRAN names:

```
100001     99VAR     A+VAR
```

The names 100001 and 99VAR do not begin with a letter. The name A+VAR provides a good example why names cannot include symbols. Because the symbol "+" connotes addition, the computer would not recognize A+VAR as a single variable name but would perceive it as the sum of two separate variables, A plus VAR.

FORTRAN reserved words or keywords should not be used as variable names. For example, the reserved word END has a specific function in FORTRAN and should not be used as a variable name.

TABLE 12.1   The Fundamental Rules for Naming Variables in FORTRAN 77

Rule	Correct	Incorrect
Every name must begin with a letter of the alphabet.	FUEL VELOC1	1MASS 2001
Names must consist of only letters and numbers.	AREA3 TEMP1	AREA#3 TEMP.1
Names must consist of six or fewer characters.	SUM3 RESULT	VELOCITY1 TOOLONG

12.6.5 String Constants and Variables

Aside from numbers, computers are also designed to handle alphanumeric information (that is, groups of letters, numbers, and symbols) such as names, addresses, and other miscellaneous labels. *Character* or *string constants* are specific groupings of character-based information. In FORTRAN they are enclosed in single quotation marks, or apostrophes, as in

```
'John Doe'    'July 4, 1976'    'The first result is'
```

Note that, in contrast to numeric constants and variable names, blank spaces and symbols such as "−" or "/" can be a part of a string. Because the symbols are enclosed within the apostrophes, the computer does not mistake the symbols for arithmetic operators.

String variables work in the same way as numeric variables in the sense that they are names that designate a specific location in the computer's memory. As will be described in the following section, a CHARACTER statement must be used to declare these variables.

String variables can be given constant values in the same manner as numeric variables. One way is via an assignment statement, as in

```
NAME = 'John Doe'
DATE = 'July 4, 1976'
SUM1 = 'The first result is'
```

Just as with numeric values, these statements will store each constant in a memory location which is associated with the variable name.

12.6.6 Variable Types

FORTRAN has six types of variables (Table 12.2). In other words, the storage location represented by the variable can contain six types of information. These are integers, real single-precision numbers, real double-precision numbers, complex numbers with an imaginary portion, character(s), and logical data that is either true or false.[4]

Using the analogy from Fig. 12.8, the different variable types relate to different types of mailboxes for storing information. Three means are available for telling the computer the type of data that a variable will contain:

Naming conventions. The simplest means of specifying variable type is by the first letter in the variable name. Unless instructed otherwise, FORTRAN assumes that all names starting with the letters I, J, K, L, M, or N are integers; whereas names beginning with any other letter of the alphabet are real single-precision variables. For example,

---

[4]Note that some compilers allow the specification of several different types of variables within a particular class. For example, Microsoft FORTRAN allows the specification of 1-, 2-, and 4-byte integers, as well as different versions of the other types.

TABLE 12.2   Types of Variables Used in FORTRAN 77

Type	Declaration	Description
Integer	INTEGER	Contains an integer; unless otherwise declared, variables starting with the letters I through N are assumed to be of this type.
Real (single-precision)	REAL	Contains a single-precision, floating-point, or decimal number; unless otherwise declared, variables starting with the letters A through H and O through Z are assumed to be of this type.
Real (double-precision)	DOUBLE PRECISION	Contains a double-precision real number.
Complex number	COMPLEX	Contains a complex number with an imaginary portion.
Characters	CHARACTER	Contains character or alphanumeric information.
Logical	LOGICAL	Contains values that are either true or false.

X, SUM, PRODUC, ONE, TWO, and THREE would represent real variables; and N, I, ICOUNT, LAST, MEAN, and MAX would represent integers.

This is the most common way to declare a variable type. Because the convention is built into the language, the type of variable can be defined conveniently and unambiguously. However, there are two shortcomings to this approach. First, sometimes the best name for one type of variable may start with a letter corresponding to the other type. For example, the maximum value of a group of numbers would be nicely named as MAXVAL. However, this label automatically limits the value to integers. Second, the naming convention does not extend to the last four variable types from Table 12.2.

The IMPLICIT declaration.   A second way to inform the FORTRAN compiler of variable type is by using the IMPLICIT statement, which is of the general form

IMPLICIT *type* (*range*), . . . , *type* (*range*)

where *type* is chosen from the names of the different kinds of variables—INTEGER, REAL, DOUBLE PRECISION, COMPLEX, CHARACTER, or LOGICAL—and *range* is a list of the first letters that you want to designate the corresponding type, for example:

```
IMPLICIT REAL (A,B,C), INTEGER (D-F), LOGICAL (L)
IMPLICIT DOUBLE PRECISION (Q,S,U)
IMPLICIT CHARACTER (M-P)
IMPLICIT CHARACTER*80 (T)
```

The IMPLICIT statement overrides the FORTRAN naming conventions so that, in the first example, all variables beginning with the letter L will be logical variables and not integer variables (as would be the case by convention). If the type of variable is not declared by an IMPLICIT statement, then the original FORTRAN naming conventions hold. For example, the letters I and J are not specified by the above IMPLICIT statements, and so names beginning with these letters would remain integers.

The last IMPLICIT statement declares that all variables beginning with T be of type CHARACTER. This means that the variable may contain string constants—that is, letters, numbers, and symbols for purposes such as labeling. The *80 signifies that the string constant may be up to 80 characters long. If the length is not specified (as in the third example), the variable may include a string of only one character.

The IMPLICIT statement is a convenient way to declare variable type. However, it can lead to confusion if the reader of the program does not remember which first letters correspond to which variable types.

The explicit declaration.   The final way to declare the type of a variable is by explicitly giving the variable type and then the variable names corresponding to the type. This is accomplished with an explicit statement that is of the general form

$$type \; variable_1, \; variable_2, \; \ldots, \; variable_n$$

where the *type* is as in the IMPLICIT statement (that is, REAL, INTEGER, and so on) and *variable_i* is the name of the *i*th variable of a particular *type*. Explicit declaration should always be done early in the program and prior to the first executable statement. Some examples of explicit declaration are

```
REAL SUM, NUMVAL, ONE, TWO, MEAN
DOUBLE PRECISION DIFF
INTEGER ICOUNT, YVAR, GRADE
CHARACTER LETTER
CHARACTER*80 STRING
LOGICAL OK, DONE
```

An explicit statement of variable type overrides both the naming conventions and any IMPLICIT statement declarations. Aside from its expressed purpose of declaring the type of each variable, the explicit declaration has an added benefit. That is, the names and types of all variables can be located in one area of the program for easy reference by the user.

In the FORTRAN programs in the remainder of the book, we will primarily use explicit declarations to establish variable types. The declarations will be located at the beginning of each program, immediately following the portion of each module where the

```
      PROGRAM ADDER
C Simple Addition Program (Version 1.2)
C
C by S.C. Chapra
C June 10, 1992
C
C This program adds two numbers
C and prints out the results
C-------------------------------------
C
C Definition of variables
C
C FIRST, SECOND = variables to be added
C TOTAL = sum of the numbers
C
      REAL FIRST, SECOND, TOTAL
C
C assign constants to variables
C
      FIRST = 28.
      SECOND = 15.
C
C perform computation
C
      TOTAL = FIRST + SECOND
C
C display results
C
      PRINT *, TOTAL
C
      END
```

FIGURE 12.9
A revision of the simple addition program from Fig. 12.7. A section has been included to define the variables used in the program. Notice that more-descriptive variable names are employed and that a REAL declaration statement is included.

use of the variables is described. For certain short examples involving numeric variables, we may also employ naming conventions for convenience.

Figure 12.9 is a revised version of the simple addition program illustrating a more communicative use of variable names. Notice how a module has been incorporated to define the meaning of the variables at a single location at the program's beginning and that more descriptive names have been used to specify the variables. FIRST, SECOND, and TOTAL are much more suggestive of the function of the variables than the original names—A, B, and C.

12.6.7 PROGRAM and PARAMETER Statements

Before proceeding, we must briefly discuss two additional statements.

PROGRAM statement.   You have probably noticed that each of the codes displayed to this point begins with a PROGRAM statement of the general form

```
PROGRAM name
```

where *name* is a legal FORTRAN variable name. This name should be distinct from any other names in the program. The PROGRAM statement has utility in identifying the program and distinguishing it from other program units (such as the subprograms that we will be describing in Chap. 16). Although it is optional, we will employ the PROGRAM statement in all our codes as a matter of good style.

PARAMETER statement. Certain constants are used so often that they are given special names. An example is the number $3.14159\ldots$, which is commonly called π. The PARAMETER statement provides a means to assign a name to such constants. Its general form is

```
PARAMETER ( param₁ = const₁ , param₂ = const₂ , . . . )
```

where $param_i$ is the ith parameter name and $const_i$ is the ith constant value.

An example is

```
PARAMETER (PI = 3.14159, NPOINT = 200)
```

For this example, the names, PI and NPOINT, are assigned values of 3.14159 and 200, respectively.

Although it should be clear why we would want to assign a constant in the case of π, the reason for assigning a value to the other parameter, NPOINT, might not be obvious. Oftentimes, you will use a constant repeatedly throughout a large program. For example, NPOINT might represent the number of data that is being analyzed. This number could be employed hundreds of places in your program. If you represented it by its constant value and you wanted to change this constant, it would be extremely time-consuming to edit your program to change every possible instance where it was used. In contrast, if you employed the name NPOINT throughout your code, you would merely have to change the single PARAMETER statement to produce the modification. We will provide examples of this usage in later chapters.

12.7 INTRODUCTION TO INPUT-OUTPUT

"Input-output" refers to the means whereby information is transmitted to and from a program. You have already been introduced to output with the PRINT command in the simple addition program. Now we will elaborate on the topic.

12.7.1 The List-Directed READ Statement

One disadvantage of the assignment statement is that it assigns a specific value to a variable. Thus, in the simple addition example program, we use assignment statements to set FIRST equal to 28 and SECOND equal to 15. If we want to add two different numbers, say, FIRST = 8 and SECOND = 33, we have to retype the two lines in order to assign new values to the variables.

The list-directed READ statement is an alternative that allows new data to be entered every time the program is run. Its general form is

READ *, var_1 , . . . , var_n

where var_i is the name of the ith variable to which the information is being assigned.

For example, rather than assign a value to FIRST in the simple addition program, we could alternatively input a value with

READ *, FIRST

When this statement is executed, it instructs the computer to expect that a value for FIRST will be entered via the computer's *default* [5] input device. When using the list-directed READ, this is usually the keyboard.

Just as with the PRINT statement, the asterisk instructs the computer that the value will be entered in the list-directed or *free format*. For input, this means that the data does not have to follow a special format (for example, a specific number of decimal places). If one value is to be input, it is merely typed on the keyboard and entered.

For example, the computer would execute the simple addition program line by line until it encountered the statement: READ *, FIRST. Upon reaching this statement, the computer would stop and wait for you to enter a value for FIRST. Suppose that you want to set FIRST equal to 15. Then by merely typing in a 15 and striking the [Enter] key, you would have instructed the computer that FIRST = 15. It would then continue executing, reading in a value for SECOND (say, 28) and then displaying the correct answer, TOTAL = 15 + 28 = 43.

More than one value may be entered with a READ statement. For example, to input values for both FIRST and SECOND, you could type

READ *, FIRST, SECOND

Then, when the computer came to this line, you would merely type the two values separated by a comma, or by one or more spaces. For example,

15, 28

and strike the [Enter] key to enter values for FIRST and SECOND. Thus a value of 15 would be stored in FIRST and 28 would be stored in SECOND; the program would sum the two values and output the result, 43.000000.

Some examples of READ statements along with acceptable entries are shown in Fig. 12.10. In all but the second of these examples, the number of data entered is consistent with the READ statement. If more entries are included on an input line than required by the READ statement, the extra values are ignored. The values 99 and 17 would be

---

[5]The term "default" indicates that this is the device that the computer will access if it is not directed to do otherwise. In later chapters, we will show how you can override this default and direct the computer to use other devices for input-output.

Program:

```
C
C Program to demonstrate READ statements
C
      REAL ONE, TWO, THREE
      REAL A, B, C, GRADE
      INTEGER N
      CHARACTER*30 FINAME, LANAME
C
      READ *, ONE, TWO, THREE
      READ *, N
      READ *, FINAME, LANAME, GRADE
      READ *, A, B, C
      PRINT *, ONE, TWO, THREE
      PRINT *, N
      PRINT *, FINAME, LANAME, GRADE
      PRINT *, A, B, C

      END
```

Data:

```
1.02E2,19.,33.29
89 99 17
'Joe','Smith' 74.5
17.9
21
0.11234e-3
```

FIGURE 12.10
A short program showing some READ statements along with the accompanying data.

ignored by READ *, N and any subsequent READ statements. If the [Enter] key is pressed without entering enough data items (as is the case in Fig. 12.10), the READ will just keep looking for data until all the variables have been assigned a value.

The data entered must match the variable type specified in the READ statement. Therefore FINAME and LANAME would have to have been previously defined as character variables using, for example,

```
CHARACTER*30 LANAME, FINAME
```

If this has not been done, an error message results and execution is terminated when the computer attempts to read the character data.

12.7.2 The List-Directed PRINT Statement

The list-directed PRINT has the general form

```
PRINT *, exp₁,..., expₙ
```

where exp_i is the ith *expression* that is to be output.

As already described when discussing Fig. 12.5, the list-directed PRINT directs output to the computer's default display, which on interactive systems is typically the

Program:

```
C
C Program to demonstrate PRINT statement
C
      REAL SUM
      INTEGER I, N
      CHARACTER*20 NAME
C
      SUM = 199.5
      NAME = 'Tom'
      I = 2
      N = 7
C
      PRINT *, 'The resulting output is:'
      PRINT *
      PRINT *, 10, SUM, NAME
      PRINT *, I + N
      PRINT *, 'Total ='
C
      END
```

Output:

FIGURE 12.11
A short program showing
some PRINT statements along
with the resulting output.

```
The resulting output is:

        10        199.500000Tom
         9
Total =
```

monitor. In addition, the asterisk instructs the computer to display the results in free format. This means that integer numbers will be displayed without a decimal, and real numbers will be displayed in either decimal or scientific notation.

Some examples of PRINT statements along with the resulting output are shown in Fig. 12.11. The *expression* following the PRINT statement can be either a constant, a variable, or a mathematical expression. The data enclosed between apostrophe marks will be displayed in the output without the marks but otherwise unchanged. The following statement causes a blank line to be output:

```
PRINT *
```

12.7.3 Descriptive Input

The following material on descriptive input is relevant only to those computer systems where a FORTRAN program can be run interactively. That is, when the program is executed, intermediate results are output to the user's terminal. For batch systems employing input media such as punched cards, the following material does not apply. However, the subsequent section on descriptive output has relevance to both interactive and noninteractive systems.

One problem with the list-directed READ statements described to this point is that the user has been provided no clue as to what information is supposed to be input. For example,

```
READ *, FIRST
```

would cause the computer to stop and wait for you to enter a value. In most cases, the terminal screen would provide no explicit indication that the program required data. Unless the user knew beforehand that a value of FIRST was to be entered, the system provides no clue as to what variables are to be input. One way to remedy this situation is to use a PRINT statement and string constants to display a message that alerts the user to what is required, for example,

```
PRINT *, 'First = '
READ *, FIRST
```

When these lines are executed, the computer will display

```
First =
_
```

where the message 'First = ' signifies that the computer is prompting you for information. Thus the PRINT statement lets the user know what is required.

12.7.4 Descriptive Output

Just as input can be made more user-friendly, output can also be made more descriptive using character strings. For example, in the simple addition program, the statement

```
PRINT TOTAL
```

merely results in the answer, 43.000000, being displayed. A message can easily be incorporated into the statement, as in

```
PRINT *, 'Total = ', TOTAL
```

When the program is run, this results in

```
Total =     43.000000
```

Figure 12.12 is a final revised version of the simple addition program. We have added a variety of descriptive PRINT statements to make the program more readable and easier to use. A run of the program is shown in Fig. 12.13. We will return to the topic of input and output in more detail in Chap. 18.

```
      PROGRAM ADDER
C Simple Addition Program (Version 2.0)
C
C by S.C. Chapra
C June 10, 1992
C
C This program adds two numbers
C and prints out the results
C-------------------------------------
C
C Definition of variables
C
C FIRST, SECOND = variables to be added
C TOTAL = sum of the numbers
C
      REAL FIRST, SECOND, TOTAL
C
C identify program
C
      PRINT *, 'Program to add two numbers'
      PRINT *
C
C obtain numbers from user
C
      PRINT *, 'First = '
      READ *, FIRST
      PRINT *, 'Second = '
      READ *, SECOND
C
C perform computation
C
      TOTAL = FIRST + SECOND
C
C display results
C
      PRINT*
      PRINT *, 'Sum of ', FIRST, ' plus ', SECOND
      PRINT *, 'is equal to ', TOTAL
C
      END
```

FIGURE 12.12
A final revision of the simple addition program. Descriptive output are incorporated in this version.

FIGURE 12.13
A run of the program from Fig. 12.12.

```
Program to add two numbers

First =
28
Second =
15

Sum of         28.000000 plus         15.000000
is equal to         43.000000
```

12.8 QUALITY CONTROL

To this point, you have acquired sufficient information to compose a simple FORTRAN computer program. Using the information from the previous sections, you should now be capable of writing a program that will generate some output. However, the fact that a program displays information is no guarantee that these answers are correct. Consequently, we will now turn to some issues related to the reliability of your program. We are emphasizing this subject at this point to stress its importance in the program-development process. As problem-solving engineers, most of our work is used directly or indirectly by clients. For this reason, it is essential that our programs be reliable; that is, they must do exactly what they are supposed to do. At best, lack of reliability can be embarrassing; at worst, for engineering problems involving public safety, it can be tragic.

12.8.1 Errors or "Bugs"

During the course of preparing and executing a program of any size, it is likely that errors will occur. These errors are usually called "*bugs*" in computer jargon. This label was coined by Captain Grace Hopper, one of the pioneers of computing languages. In 1945 she was working with one of the earliest electromechanical computers when it went dead. She and her colleagues opened the machine and discovered that a moth had become lodged in one of the relays. Thus the label "bug" came to be synonymous with problems and errors associated with computer operations, and the term "debugging" came to be associated with their solution.

When developing and running a program, three types of bugs can occur:

* *Syntax errors* violate the rules of the language, such as spelling, number formation, line numbering, and other conventions. These errors are often the result of mistakes such as typing IMPUT rather than INPUT. They usually cause the computer to display an error message. Such messages are called *diagnostics* because the computer is helping you to "diagnose" the problem.
* *Run-time errors* are those that occur during program execution, such as the case where you input the incorrect type of constant. For example, if in response to the following READ statement

```
READ *, X
```

you enter a string constant[6]

```
'Bob'
```

a run-time error diagnostic such as "invalid number is input" will be displayed.

For other run-time errors the computer may simply terminate the run or display a message informing you at what line the error occurred. In any event, you will be cognizant of the fact that an error has taken place.

---

[6]Assuming of course that you have not declared X as a CHARACTER variable.

• *Logic errors,* as the name implies, are due to faulty program logic. These are the worst types of errors because the computer may not display a diagnostic to alert you of their presence. Thus your program will appear to be working properly in the sense that it executes and generates output. However, the output will be incorrect.

All of the above types of errors must be removed before a program can be employed for engineering applications. We divide the process of quality control into two categories: debugging and testing. As mentioned above, *debugging* involves correcting known errors until the program seems to execute as expected. In contrast, *testing* is a broader process intended to detect errors that are not obvious. Additionally, testing should also determine whether the program fills the needs for which it was designed.

As will be clear from the following discussion, debugging and testing are interrelated and are often carried out in tandem. For example, we might debug a module to remove the obvious errors. Then, after a test, we might uncover some additional bugs that require correction. This two-pronged process is repeated until we are convinced that the program is sufficiently reliable.

12.8.2 Debugging

As stated above, debugging deals with correcting known errors. There are three ways in which you will become aware of errors. First, an explicit diagnostic message can provide you with the exact location and the nature of the error. Second, a diagnostic may be displayed, but the exact location of the error is unclear. Third, diagnostics are not displayed, but the program does not operate properly. This is usually due to logic errors such as infinite loops or mathematical mistakes that do not necessarily yield syntax errors. Corrections are easy for the first type of error. However, for the second and third types some detective work is required in order to identify the errors and their exact locations.

Unfortunately, the analogy with detective work is all too apt. That is, ferreting out errors is often difficult and involves a lot of "legwork," collecting clues and running down some blind alleys. In particular, as with detective work, there is no clear-cut methodology involved. However, there are some ways of going about it that are more efficient than others.

One key to finding bugs is to display intermediate results. With this approach, it is often possible to determine at what point the computation went bad. Similarly, the use of a modular approach will obviously help to localize errors in this manner. You could locate a PRINT message at the start of each module and in this way focus in on the module where the error occurs. In addition, it is always good practice to debug (and test) each module separately before integrating it into the total package.

Modern FORTRAN on a personal computer is an ideal tool for debugging. Because of its interactive mode of operation, the user can efficiently perform the repeated runs that are sometimes needed to zero in on the location of errors.

It should be noted that some FORTRAN implementations (such as the Microsoft version) have very powerful debugging capabilities. These are an integral part of the environment and allow you to track the values of variables and to observe program execution on a step-by-step basis.

Although these procedures can help, there are often times when your only option will be to sit down and read the code line by line, following the flow of logic and performing all the computations with a pencil, paper, and pocket calculator. Just as a detective sometimes has to think like a criminal in order to solve a case, there will be times when you will have to "think" like a computer to debug a program.

12.8.3 Testing

One of the greatest misconceptions of the novice programmer is the belief that if a program runs and displays results, it is correct. Of particular danger are those cases where the results appear to be reasonable but are in fact wrong. To ensure that such cases do not occur, the program must be subjected to a battery of tests for which the correct answer is known beforehand.

As mentioned in the previous section, it is good practice to debug and test modules prior to integrating them into the total program. Thus testing usually proceeds in phases. These phases consist of module tests, development tests, whole-system tests, and operational tests.

Module tests, as the name implies, deal with the reliability of individual modules or sections of the program. Because each is designed to perform a specific, well-defined function, the modules can be run in isolation to determine whether they are executing properly. Sample input data can be developed for which the proper output is known. This data can be used to run the module, and the outcome can be compared with the known result to verify successful performance.

Development tests are implemented as the modules are integrated into the total program package. That is, a test is performed after each module is integrated. An effective way to do this is with a sequential approach that starts with the first module and progresses down through the program in execution sequence. If a test is performed after each module is incorporated, problems can be isolated more easily, since new errors can usually be attributed to the latest module added.

After the total program has been assembled, it can be subjected to a series of *whole-system tests.* These should be designed to subject the program to (1) typical data, (2) unusual but valid data, and (3) incorrect data to check the program's error-handling capabilities.

Note that all the foregoing tests are often collectively referred to as *alpha-testing.* Once they are successfully implemented, a *beta-* or *operational testing* phase is usually initiated. These tests are designed to check how the program performs in a realistic setting. A standard way to do this is to have a number of independent subjects (including your users or customers) employ the program. Operational tests sometimes turn up bugs. In addition, they provide feedback to the developer regarding ways to improve the program so that it more closely meets the user's needs.

Although the ultimate objective of the testing process is to ensure that programs are error-free, there will be complicated cases where it is impossible to subject a program to every possible contingency. In fact, for programs of even moderate complexity, it is often practically impossible to prove that an error does not exist. At best, all that can be said is that, under rigorous testing conditions, no error could be found.

It should also be recognized that the level of testing depends on the importance and magnitude of the problem context for which the program is being developed. There will obviously be different levels of rigor applied to testing a program to determine the batting averages of your intramural softball team as compared with a program to regulate the operation of a nuclear reactor or the space shuttle. In each case, however, adequate testing must be performed, testing consistent with the liabilities connected with a particular problem context.

Finally, we cannot stress enough the fact that "an ounce of prevention is worth a pound of cure." That is, a sound preliminary program design and structured programming techniques are great ways to avoid errors in the first place. As with so many other aspects of your life, the sooner you adopt a "pay me now rather than pay me later" philosophy, the less grief you will experience as a computer programmer.

12.9   DOCUMENTATION

Documentation is the addition of English-language descriptions to allow you or another user to implement the program more easily. (Remember, along with other people who might employ your software, you are also a user.) Although a program may seem clear and simple when you first compose it, after the passing of time the same code may seem inscrutable. Therefore sufficient documentation must be included to allow you and other users to immediately understand your program and how it can be implemented. This task has both internal and external aspects.

12.9.1   Internal Documentation

Picture for a moment a text with just words and no paragraphs, subheadings, or chapters. Such a book would be more than a little intimidating to study. The way the pages of a text are structured—that is, all the devices that serve to break up and organize the material—serves to make the text more effective and enjoyable. In the same sense, internal documentation of computer code can greatly enhance the user's understanding of a program and how it works.

We have already broached the subject of internal documentation when we introduced the comment statement in Sec. 12.5.1. The following are some general suggestions for documenting programs internally:

- Include a section at the head end of the program, giving the program's title, your name, and the data. This is your signature and marks the program as your work.
- Include a second section to define each of the key variables.
- Select variable names that are reflective of the type of information the variables are to store.
- Insert spaces within statements to make them easier to read.
- Use comment statements and blank lines liberally throughout to program for purposes of labeling and clarification.
- In particular, use comment statements to label clearly and set off all the sections.
- Use indentation to elucidate the program's structures. In particular, as will be il-

lustrated in a later chapter, indentation can be used very effectively to set off selection and repetition.

12.9.2 External Documentation

External documentation refers to instructions in the form of output messages and supplementary printed matter. The messages occur when the program runs and are intended to make the output attractive and user-friendly. This involves the effective use of spaces, blank lines, and special characters to illuminate the logical sequence and structure of the program output. Attractive output simplifies the detection of errors and enhances the communication of program results.

The supplementary printed matter can range from a single sheet to a comprehensive user's manual. Figure 12.14 is an example of a simple documentation form that we rec-

---

FIGURE 12.14
A simple one-page format for program documentation. This documentation can be stored in a binder along with a printout of the program.

PROGRAM: The Simple Addition Program (Version 2.0)

PROGRAMMER: S. C. Chapra
Civil, Environmental and Architectural Engineering
The University of Colorado
Boulder, CO 80309-0428

DESCRIPTION: This program adds two numbers and displays the results.

SPECIAL FEATURES: The program uses descriptive variable names and enhanced I/O.

REFERENCE: Chapra, S. C., and Canale, R. P. (1993). *Introduction to Computing for Engineers,* 2d Ed. (McGraw-Hill: New York), Chap. 12.

LISTING:

(insert listing here)

RUN:

(insert run here)

ommend you prepare for every program you develop. These forms can be maintained in a notebook to provide a quick reference for your program library. Note that for well-documented programs, the title information in Fig. 12.14 is unnecessary because the title section could be included directly in the program.

The user's manual for your own computer is an example of comprehensive documentation. This manual tells you how to run your computer system and disk operating programs. In addition, manuals for high-level languages and software packages also represent extensive documentation efforts. If you ever have to compose a major documentation, these manuals provide good models of how large software packages can be effectively documented.

12.10   STORAGE AND MAINTENANCE

When the computer is shut off, the contents of primary memory (RAM) are destroyed. In order to retain a program for later use, you must transfer it to a secondary storage device such as a hard disk or diskette before turning the machine off.

Aside from this physical act of retaining the program on a disk, storage and maintenance consist of two major tasks: (1) upgrading the program in light of experience and (2) ensuring that the program is safely stored. The former is a matter of communicating with users to gather feedback regarding suggested improvements to your design.

The question of safe storage of programs is especially critical when using diskettes. Although these devices offer a handy means for retaining your work and transferring it to others, they can be easily damaged. Table 12.3 outlines a number of tips for effectively maintaining your programs on diskettes. Follow the suggestions on this list carefully in order to avoid losing your valuable programs through carelessness.

TABLE 12.3   *Suggestions and Precautions for Effective Software Maintenance*

- Maintain paper printouts of all finished programs in a file or binder.
- Always maintain at least one backup copy of all diskettes.
- Handle all diskettes gently. Do not bend $5\frac{1}{4}$-in diskettes or touch their exposed surfaces.
- Affix a temporary label describing the contents to all diskettes. When working with $5\frac{1}{4}$-in disks, use only soft felt-tipped pens to write on the label once it is affixed.
- Once a diskette is complete, make it write-protected by covering (for $5\frac{1}{4}$-in diskettes) or moving (for $3\frac{1}{2}$-in diskettes) its write-protect switch.
- Keep $5\frac{1}{4}$-in diskettes in their protective envelope when not in use.
- Store diskettes upright in a diskette box or carrier. Make sure they never get too hot or too cold.
- Keep diskettes away from water and contaminants such as smoke and dust.
- Store diskettes away from magnetic devices such as televisions, tape recorders, and electric motors.
- Remove diskettes from the drive before turning the system off.
- Never remove a diskette when the red light on the drive is lit.
- Be careful not to bend the metal shutter on $3\frac{1}{2}$-in diskettes. It can get caught in the diskette drive and make it impossible to remove without damaging it or obtaining a technician's help.

PROBLEMS

12.1 Describe three advantages of FORTRAN.

12.2 Learn how to access FORTRAN on your computer system. Enter and run the simple addition program from Fig. 12.12. Save it on an external storage device.

12.3 The arithmetic operator for FORTRAN multiplication is the asterisk. Therefore a simple multiplication program can be developed by reformulating the assignment statement as

```
PRODCT = FIRST * SECOND
```

Develop a version of this program patterned after Fig. 12.12. Save the program on your diskette and generate a paper printout or hard copy of the output. Document this program using the format specified in Fig. 12.14. Employ values of FIRST = 20 and SECOND = 16 for your example run.

12.4 Develop, test, and document a program to input the length, width, and height of a box. Have the program compute the volume and the surface area of the box. Also, allow the user to input the units (for example, feet) for the dimensions of the box as string constants and display the correct units with the answers. Document this program using the format specified in Fig. 12.14. Employ values of length, width, and height equal to 5, 1.5, and 1.25 m, respectively, for your example run.

12.5 Develop, test, and document a program to input the external dimensions of a box along with the thickness of its walls. Have the program compute the internal volume, and the internal and external surface areas of the box. Also, allow the user to input the units (for example, feet) for the dimensions of the box as string constants and display the correct units with the answers. Document this program using the format specified in Fig. 12.14. Employ values of length, width, height, and thickness equal to 5, 1.5, 1.25, and 0.1 m, respectively, for your example run.

12.6 Write a FORTRAN program comparable to the one given in Fig. 12.12. The program should prompt the user to input a value for the radius of a circle and should respond with values for the diameter, circumference, and area, displayed in a neat format. Employ a PARAMETER statement to specify a value for π.

12.7 Write a FORTRAN program comparable to the one given in Fig. 12.12. The program should prompt the user for input values for A, B, and C, and should respond with calculated values of Y where

```
Y = A * C + B * A + B * C
```

12.8 Write a FORTRAN program comparable to the one given in Fig. 12.12. The program should prompt the user for input values for the velocity of a car and the duration of a trip and should respond with the calculated distance traveled.

12.9 Repeat Prob. 12.4, except calculate the volume and surface area of a closed cylinder (that is, one with a lid), given its radius and height. Use a radius of 1 m and a height of 10 m to test your program. The FORTRAN formulas to compute the area and volume of a closed cylinder are

```
A = 2. * PI * R * H + 2. * PI * R * R
```

and

```
V = PI * R * R * H
```

where H = height, R = radius, and PI = π.

12.10 Repeat Prob. 12.5, except calculate the internal volume and surface area of a cylinder, given its external radius, height, and wall thickness. Use a radius of 1 m, a height of 10 m, and a thickness of 0.04 m to test your program. See Prob. 12.9 for formulas to compute surface area and volume.

12.11 Repeat Prob. 12.4, except calculate the volume and surface area of a sphere, given its radius. Use a radius of 4 m to test your program. The FORTRAN formulas to compute the area and volume of a sphere are

```
A = 4. * PI * R * R
```

and

```
V = 4. * PI * R * R * R / 3.
```

where R = radius, and PI = π.

12.12 Repeat Prob. 12.5, except calculate the internal volume and surface area of a sphere, given its external radius, and wall thickness. Use a radius of 4 m and a thickness of 0.04 m to test your program. See Prob. 12.11 for formulas to compute surface area and volume.

CHAPTER 13

COMPUTATIONS

As stated on several occasions to this point, one of the primary engineering applications of computers is to manipulate numbers. In the previous chapter, you were introduced to some fundamental concepts that lie at the heart of this endeavor: constants, variables, and assignment. Now we will expand on these concepts and introduce some additional material that will allow you to exploit the further potential of FORTRAN to perform computations.

13.1 ARITHMETIC EXPRESSIONS

Obviously, in the course of engineering problem solving, you will be required to perform more complicated mathematical operations than the simple process of addition introduced in the previous chapter. One key to understanding how these operations are accomplished is to realize that a primary distinction between computer and conventional equations is that the former are constrained to a single line. In other words, a multilevel, or "built-up," formula such as

$$x = \frac{a + b}{y - z^2} + \frac{33.2}{w^3}$$

must be reexpressed so that it can be written as a single-line expression in FORTRAN.

The operators that are used to accomplish this are listed in Table 13.1. Notice how the two operations that are intrinsically multilevel — that is, division and exponentiation — can be expressed in single-line form using these operators. For example, A divided by B, which in algebra is usually written

$$\frac{A}{B}$$

307

TABLE 13.1   Parentheses, Arithmetic Operators, and Their
Associated Priorities in FORTRAN 77

Symbol	Meaning	Priority
( )	Parentheses	1
**	Exponentiation	2
−	Negation	3
*	Multiplication	4
/	Division	4
+	Addition	5
−	Subtraction	5

is expressed in FORTRAN as A/B. Similarly, the cube of A, which in algebra demands that the 3 be raised as a superscript, as in A^3, is expressed in FORTRAN as $A**3$. Note that the power to which the number is raised can be a positive or negative integer or a decimal. However, a negative number can be raised only to an integer power. In general, raising a negative number to a fractional power is not feasible.

13.1.1 Rules for Arithmetic Operators

Although the $**$ and $/$ operators allow us to write algebraic expressions on a single line, they are not sufficient to give the computer all the information that it needs to understand complicated arithmetic expressions. This is due to the fact that a computer can perform only one operation at a time.

To illustrate the difficulties that can occur because of this fact, suppose that you wanted to program the following formula:

$$x = \frac{a + b}{c + d}$$

Using the operators in Table 13.1, we might formulate the following FORTRAN expression:

$X = A + B/C + D$

Now the computer must decide which operation to perform first. Suppose that we decided to make a rule that the computer would evaluate the operations in a left-to-right order. Thus, it would first add A to B. Next it would divide the result by C. Finally, it would add the D and store the final result in X. Unfortunately, this sequence would actually be equivalent to

$$x = \frac{a + b}{c} + d$$

which is quite different from what we intended.

We therefore need a few more rules. First, the computer allows the use of parentheses to group operations and to specify which calculations are to be executed first. In other words, parentheses are given top priority in the sense that the calculations within parentheses are implemented first. For example, the following representation would allow our example to be computed properly:

$$X = (A + B)/(C + D)$$

Thus, the additions are performed first and then the results are divided. Note that if several sets are nested within each other, the operations in the innermost parentheses are performed first.

Parentheses must always come in pairs. A good practice when writing expressions is to check that there are always an equivalent number of left and right parentheses. For example, the expression

$$((X + 2 * (3 - B - (4 * A/B) - 7))$$

is incorrect because there are too many left parentheses. Checking the number of left and right parentheses does not ensure that the expression is correct, but if the number of parentheses does not match, the expression is definitely erroneous.

Note that parentheses cannot be used to imply multiplication as is done in conventional arithmetic. The asterisk must always be included to indicate multiplication. For example, the algebraic multiplication

$$x = c(a + b)$$

could not be expressed as $X = C(A + B)$ in FORTRAN. Rather, an asterisk would be needed, as in $X = C * (A + B)$.

Two arithmetic operators cannot appear consecutively in an expression. For example, although a number can be raised to a negative power, the following would be illegal:

$$X = Y**-2$$

However, parentheses can be employed to rectify the problem, as in

$$X = Y**(-2)$$

Although the parentheses are helpful, additional rules are needed. This is due to the fact that if they were the only criterion for establishing priority, parentheses would clutter up most computer equations. Because this would make expressions extremely difficult to read and interpret, an additional hierarchy of priorities is needed. Such a scheme is embodied in Table 13.1.

After operations within parentheses, the next highest priority goes to exponentiation. If multiple exponentiations are required, as in $X**Y**Z$, they are implemented from right to left.[1] This is followed by negation — that is, the placement of a minus sign before an expression to connote that it is negative. (Negation is formally referred to as the *unary minus operation*.)

Next, all multiplications and divisions are performed from left to right. Finally, the additions and subtractions are performed from left to right. (Subtraction is formally referred to as the *binary minus operation*.) The use of these priorities to construct mathematical expressions is best shown by example.

EXAMPLE 13.1   Arithmetic Expressions

PROBLEM STATEMENT: Determine the proper values of the following arithmetic expressions in FORTRAN:

(*a*) 5. + 6. * 7. + 8.
(*b*) (5. + 6.) * 7. + 8.
(*c*) (5. + 6.) * (7. + 8.)
(*d*) −2. ** 4
(*e*) (−2.) ** 4
(*f*) 5. ** 2 ** 3
(*g*) (5. ** 2) ** 3
(*h*) 8./2. * 4. − 3.
(*i*) (10./(2. + 3.) * 5. ** 2) ** 0.5

SOLUTION:
(*a*) Because multiplication has a higher priority than addition, it is performed first to yield

5. + 42. + 8.

Then the additions can be performed from left to right to yield the correct answer, 55.
(*b*) Although multiplication has a higher priority than addition, the parentheses have even higher priority, and so the parenthetical addition is performed first:

11. * 7. + 8.

followed by multiplication

77. + 8.

and finally addition to give the correct answer, 85.

---

[1]Not all computer languages operate in this fashion. For example, BASIC exponentiation is implemented from left to right.

(c) First evaluate the additions within the parentheses:

11. * 15.

and then multiply to give the correct answer, 165.

(d) Because exponentiation has a higher priority than negation, the 2. is first raised to the power 4 to give 16. Then negation is applied to give the correct answer, −16.

(e) In contrast to the expression in (d), the parentheses here force us to first transform 2 into −2. Then the −2. is raised to the power 4 to give 16.

(f) Because the order of exponentiation is from right to left, 2. is first cubed to give 8. The 5. is then raised to the power 8 to yield the correct answer, 390625.

(g) In contrast to the expression in (f), the normal order of exponentiation is overruled; 5. is first squared to give 25., which in turn is cubed to give 15625.

(h) Because multiplication and division have equal priority, they must be performed from left to right. Therefore, 8. is first divided by 2. to give

4. * 4. − 3

Then 4. is multiplied by 4. to yield

16. − 3.

which can be evaluated for the correct answer, 13. Note that if the left-to-right rule had been disobeyed and the multiplication performed first, an erroneous result of −2. would have been obtained.

(i) Because the parentheses in this expression are nested (that is, one pair is within the other), the innermost is evaluated first:

(10./5. * 5. ** 2) ** 0.5

Next the expression within the parentheses must be evaluated. Because it has the highest priority, the exponentiation is performed first:

(10./5. * 25.) ** 0.5

Then the division and multiplication are performed from left to right to yield

(2. * 25.) ** 0.5

and

50. ** 0.5

which can be evaluated to give the correct answer, 7.071.

When dealing with very complicated equations, it is often good practice to break them down into component parts. Then the components can be combined to come up with the end result. This practice is especially helpful in locating and correcting typing errors that frequently occur when entering a long equation.

EXAMPLE 13.2   Complicated Arithmetic Expressions

PROBLEM STATEMENT: Write a comparable FORTRAN version for the algebraic expression

$$y = \frac{[(x^2 + 2x + 3)/\sqrt{5 + z}] - \sqrt{(15 - 77x^3)/z^{3/2}}}{x^2 - 4xz - 5x^{-4/5}}$$

Treat each major term of the equation as a separate component in order to facilitate programming the equation in FORTRAN.

SOLUTION: The denominator and the two terms in the numerator can be programmed separately as

```
DENOM = X ** 2 - 4. * X * Z - .5 * X ** (-4./5.)
PART1 = (X ** 2 + 2. * X + 3.)/(5. + Z) ** 0.5
PART2 = ((15. - 77. * X ** 3)/Z ** (3./2.)) ** 0.5
```

Then the total equation could be formed from the component parts, as in

```
Y = (PART1 - PART2)/DENOM
```

Whereas the priority rules from Table 13.1 allow you to employ less parentheses, there are cases where they can be used to enhance the clarity of a formula. For example, the following expression

$$\frac{a}{bc}$$

can be represented accurately as

```
a/b/c
```

Because of the left-to-right rule, this relationship will function properly. However, an alternative version can be developed as

```
a/(b * c)
```

It can be argued that the use of parentheses here represents an improvement because this version more faithfully exhibits the structure of the original equation.

While parentheses are useful for increasing the clarity of expressions, their indiscriminate use can be counterproductive. For example, the PART2 expression from Example 13.2 could have been represented as

```
PART2 = ((15. - (77. * (X ** 3)))/(Z ** (3./2.))) ** (.5)
```

Although this version will execute correctly, it is questionable whether the extra parentheses have enhanced the presentation.

Clearly, there is some judgment involved in formulating expressions. The bad news is that it is one of the many instances in engineering involving ambiguities, where no single, clear-cut answer is available. The good news is that such situations allow you the freedom to express yourself and develop your own creativity and "style." In fact, it is usually these instances that will allow you to excel and distinguish yourself from your peers.

On the basis of our own programming experiences, we generally try to follow the rule of thumb: "When building expressions, always make them as simple as possible but not so simple that they are difficult to understand." Additionally, we always try to place ourselves in the position of our user-customer. In other words, we try to write formulas that *we* would be able to grasp easily if we came upon them at a later time. We have found that following these guidelines helps us to develop expressions that are generally clean, lean, and lucid.

13.1.2 Exponentiation in FORTRAN

Earlier in this chapter, we mentioned that negative numbers cannot be raised to noninteger powers. The reason for this lies in the manner in which FORTRAN performs the exponentiation operation.

For cases with an integer power, FORTRAN merely performs the number of multiplications specified by the exponent. Therefore, the cube of a number is evaluated as

$$2 ** 3 = 2 * 2 * 2 = 8$$

Similarly, the cube of a negative number would be evaluated as

$$(-2) ** 3 = (-2) * (-2) * (-2) = -8$$

However, when a noninteger power is used, a logarithm is employed to make the evaluation, as in the following sequence of operations[2]:

$$x = 2^{3.2}$$

$$\log_{10} x = \log_{10} 2^{3.2}$$

$$\log_{10} x = 3.2 \log_{10} 2$$

$$x = 10^{3.2 \log_{10} 2} = 9.189588$$

---

[2]FORTRAN 77 does not actually employ the base 10 or common logarithm to make this evaluation. However, because it is the most familiar of the logarithms, we used it for this example.

Whereas this poses no problem when a positive number is raised to a power, it clearly creates an error when a negative is employed. This is due to the fact that it is impossible to take the logarithm of a negative number.

In summary, whenever the exponent is a whole number, make sure to type it as an integer. That is, type it without a decimal point. This will ensure that the exponentiation will be implemented as a product rather than as a logarithm.

13.1.3 Integer and Mixed-Mode Arithmetic

Integer arithmetic.   All the computations to this point have involved real variables and constants. Expressions with only integers behave similarly. However, integer division presents a special case that demands care. Intermediate results and the final assignment that involve integer division are truncated; that is, any digits following the decimal point are dropped. Unlike most other computer languages, FORTRAN never uses the decimal portion of integer expressions in the intermediate stages of a calculation. This can have a pronounced effect on the final result, as shown in the following example.

EXAMPLE 13.3   Integer Arithmetic

PROBLEM STATEMENT: Determine the values of the following operations involving integers: (*a*) 3/4 * 4, (*b*) 5/3 − 17, (*c*) 7 − 19/5, (*d*) 4 ** (5/10).

SOLUTION:
(*a*) Because precedence is from left to right, the division is performed first to give 0.75. However, because the division involves two integers, the decimal portion of this result is dropped to give 0. Therefore the final result is 0 * 4 = 0.
(*b*) First 5/3 = 1.6667, which is truncated to 1. Then 1 − 17 = −16.
(*c*) The division is implemented first to give 19/5 = 3.8, which is truncated to 3. Then 7 − 3 = 4.
(*d*) The parenthetical division is performed to give 0.5, which is truncated to 0. Then the exponentiation 4 ** 0 gives 1.

Mixed-mode arithmetic.   Expressions involving both real and integer parts are called *mixed-mode expressions.* They follow these rules:

• If an operation involves an integer and a real part, the integer is converted to a real number prior to the operation, and the result is real.

• If an operation involves two integers, the result is an integer. For the case where division of two integers gives an outcome with a decimal portion, the result is truncated to give the final integer value.

• Assignment transforms the result on the right-hand side of the assignment into the type of the variable on the left-hand side.

EXAMPLE 13.4   Mixed-mode Arithmetic

> PROBLEM STATEMENT: Evaluate the following mixed-mode expressions: (*a*) X = 3/4 + 1.0; (*b*) J = 3.0/4 + 1.0; (*c*) X = 3.0/4 * 4; (*d*) X = 3/4 * 4; (*e*) I = 7. − 19/5; (*f*) J = 7 − 19/5.
>
> SOLUTION:
> (*a*) 3/4 = 0.75, but because integers are involved, this result is truncated to 0. Therefore the final result is 1.0.
> (*b*) 3.0/4 = 0.75, and because a real number is involved, this result is not truncated but is added to 1.0 to give 1.75. However, because the result is assigned to the integer J, the final value is truncated to give an integer of 1.
> (*c*) 3.0/4 = 0.75, which is multiplied by 4 to give 3.0.
> (*d*) 3/4 = 0.75, which is truncated to 0 because the operation involves integers exclusively. This result is multiplied by 4 and stored in X to give 0.0.
> (*e*) 19/5 = 3.8, which is truncated to 3 because the operation involves integers. This value is subtracted from 7 to give 4.0. However, the final result is transformed to an integer value of 4 when it is assigned to I.
> (*f*) 19/5. is equal to 3.8. This value is subtracted from 7 to give 3.2. However, the final result is transformed to an integer value of 3 when it is assigned to J.

The easiest way to avoid confusion in arithmetic is to avoid mixed-mode expressions unless you actually desire the results of an expression to be truncated. Most engineering problems require real results, and integer calculations are used almost exclusively for operations such as counting.

13.1.4   Complex Variables

Complex variables can be included in FORTRAN. As mentioned previously in Sec. 12.6.6, a type-specification statement must be employed to declare complex variables. For example,

```
COMPLEX var₁, var₂, . . . . , varₙ
```

tells the computer that the variables var_1 through var_n are complex. Two constants are associated with each complex variable, one for the real part and the other for the imaginary part. These can be assigned to the variable, as in

```
COMPLEX C1, C2
          .
          .
          .
C1 = (1.0,5.0)
C2 = (0.4,2.5)
```

Note that the values must be separated by a comma and enclosed in parentheses as shown. The values can also be entered via input, as in

```
READ *, C1, C2
```

where the input data could be entered as

```
(1.0,5.0),(0.4,2.5)
```

Arithmetic operations can be performed on complex variables resulting in a complex answer. Remember that the variable to which the answer is assigned must also be declared as complex by a type-specification statement. If an operation involves a complex and a real or integer variable, the latter is automatically converted to a complex variable whose imaginary part is zero. Operations between complex and double-precision variables are not allowed. Note that special intrinsic functions are available to manipulate complex variables. Some of these are listed in Appendix C.

13.2 INTRINSIC, OR BUILT-IN, FUNCTIONS

In engineering practice many mathematical functions appear repeatedly. Rather than require the user to develop special programs for each of these functions, manufacturers include them directly as part of the language. These *built-in* (or *intrinsic* or *library*) *functions* are of the general form

name(argument)

where the *name* designates the particular intrinsic function (see Table 13.2) and the *argument* is the value that is to be evaluated. The argument can be a constant, variable, another function, or an algebraic expression. (Note that there are also certain functions with more than one argument.)

There are a great variety of built-in functions available for FORTRAN 77. In the present chapter, we will discuss the more commonly available functions listed in Table 13.2.[3] Additional functions, including those to manipulate complex variables, are compiled in Appendix C.

13.2.1 General Mathematical Functions

The first group of functions deals with some standard mathematical operations that have broad application in engineering.

The square root function (SQRT). To this point we have learned one way to determine the square root by raising a number to the 1/2 power, as in

---

[3]Note that some FORTRAN 77 implementations employ slightly different names for some of the functions in Table 13.2.

TABLE 13.2 Commonly Available Built-in Functions in FORTRAN*

Function	Description
SQRT(X)	Square root of X where X must be positive
ABS(X)	Absolute value of X
ALOG10(X)	Common logarithm of X where X must be positive
ALOG(X)	Natural logarithm of X where X must be positive
EXP(X)	Exponential of X
SIN(X)	Sine of X where X is in radians
COS(X)	Cosine of X where X is in radians
TAN(X)	Tangent of X where X is in radians
ASIN(X)	Inverse sine of X where X is dimensionless and returns a value in radians
ACOS(X)	Inverse cosine of X
ATAN(X)	Inverse tangent of X
SINH(X)	Hyperbolic sine where X is in radians
COSH(X)	Hyperbolic cosine where X is in radians
TANH(X)	Hyperbolic tangent where X is in radians
INT(X)	Truncates X and returns an integer result
NINT(X)	Rounds X and returns an integer result

*The argument of the function X can be a constant, variable, another function, or an algebraic expression.

```
X ** 0.5
```

Because this is such a common operation, the built-in FORTRAN function named *SQRT* is included to perform the same task. An example is

```
SQRT(X)
```

The absolute value function (ABS). There are many situations where you will be interested in the value of an expression or variable but not its sign. The *ABS* function

```
ABS(X)
```

provides the value of X without any sign. If X is greater than or equal to 0, ABS(X) will equal X. If it is less than 0, ABS(X) will equal X times −1.

EXAMPLE 13.5 Using SQRT and ABS Functions to Determine the Complex Roots of a Quadratic Equation

PROBLEM STATEMENT: Given the quadratic equation

$$f(x) = ax^2 + bx + c$$

the roots may be determined by the formula

$$x = \frac{-b \pm \sqrt{b^2 - 4ac}}{2a}$$

If $b^2 - 4ac \geq 0$, this equation can be used directly. However, if $b^2 - 4ac < 0$, then the roots are complex and a revised solution is

$$x = \text{Re} \pm i\,\text{Im}$$

where Re is the real part of the root, which can be calculated as

$$\text{Re} = \frac{-b}{2a}$$

Im is the imaginary part, which is calculated as

$$\text{Im} = \frac{\sqrt{|b^2 - 4ac|}}{2a}$$

and i represents the square root of -1. Use the ABS and SQRT functions to determine the complex roots for the case where $b^2 - 4ac < 0$.

SOLUTION: The code to determine the complex roots of a quadratic equation can be written as

```
REAL XIMAG, XREAL
REAL A, B, C
READ *, A, B, C
XREAL = -B/(2. * A)
XIMAG = SQRT (ABS(B ** 2 - 4. * A * C))/(2. * A)
PRINT *, XREAL, XIMAG
END
```

Notice how the ABS function is located within the argument of the SQRT function. Check the nesting of the parentheses yourself.

The common logarithm function (ALOG10).   Logarithms have numerous applications in engineering. Recall that the general form of the logarithm is

$$y = \log_b x$$

where b is called the *base* of the logarithm. According to this equation, y is the power to which b must be raised to give x:

$$x = b^y$$

The most familiar and easy-to-understand form is the *base 10*, or *common, logarithm*, which corresponds to the power to which 10 must be raised to give x; for example,

$$\log_{10} 10{,}000 = 4$$

because 10^4 is equivalent to 10,000.

Common logarithms define powers of 10. For example, common logs from 1 to 2 correspond to values from 10 to 100. Common logs from 2 to 3 correspond to values from 100 to 1000, and so on. Each of these intervals is called an *order of magnitude*. The common logarithm is computed in FORTRAN as

```
ALOG10(X)
```

The natural logarithm function (ALOG).   Besides the base 10 logarithm, the other important form in engineering is the *base e logarithm*, where $e = 2.71828\ldots$ The base e logarithm, which is also called the *natural,* or *naperian, logarithm,* is extremely important in calculus and other areas of engineering mathematics. In fact, in engineering and most areas of science, the natural logarithm is actually more prevalent than the common logarithm. For this reason, FORTRAN includes the *ALOG* function to compute the natural logarithm, as in

```
ALOG(X)
```

Note that the ALOG or the ALOG10 functions can be employed to determine logarithms with other bases. For example, the following formula can be employed using the natural log:

$$\log_b x = \frac{\ln x}{\ln b}$$

where, as you have already learned in calculus, $\ln x$ is the standard nomenclature for $\log_e x$. Thus, the base e logarithm can be converted to base 10 by

$$\log_{10} x = \frac{\ln x}{\ln 10} = \frac{\ln x}{2.3025}$$

Therefore, if you desire to compute the base 2 logarithm, you can employ a FORTRAN statement like

```
LOG2X = ALOG(X)/ALOG(2.)
```

The exponential function (EXP).   An additional mathematical function that has widespread engineering applications is the exponential function, or e^x, where e is the base of the natural logarithm. Just as 10^x is the inverse of $\log_{10} x$, so also e^x is the inverse of $\ln x$. Therefore

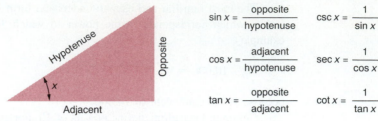

FIGURE 13.1 Definitions of the trigonometric functions.

$$x = e^{\ln x}$$

The exponential formula is often used to characterize processes that grow or decay. For example, the radioactive decay of carbon 14 (^{14}C) can be computed by the formula

$$A = A_0 e^{-0.00012t}$$

where A is the amount of carbon 14 present after t years and A_0 is the initial quantity of carbon 14. In FORTRAN 77, the *EXP* function would be employed to compute this formula as

```
A = A0 * EXP(-0.00012 * T)
```

13.2.2 Trigonometric Functions

Other mathematical functions that have great relevance to engineering are the trigonometric functions. In FORTRAN the sine (SIN), cosine (COS), and tangent (TAN) functions are commonly available. As outlined in Fig. 13.1, other trigonometric functions can be derived by inverting these three. In addition, the inverse trigonometric functions are available on most systems. For example, if the value of the tangent is known, the arctangent (ATAN) can be used to determine the corresponding angle. Finally, functions are available to compute the hyperbolic sine (SINH), cosine (COSH), and tangent (TANH).

For all these functions, the angle must be expressed in radians (Fig. 13.2). A *radian* is the measure of an angle at the center of a circle that corresponds to an arc length that is equal to the circle's radius (Fig. 13.2*b*). The conversion from degrees to radians is (Fig. 13.2*c*)

$$\pi \text{ radians} = 180°$$

Therefore, if X is expressed originally in degrees, the sine of X can be computed in **FORTRAN** as

```
SIN(X * 3.14159/180.)
```

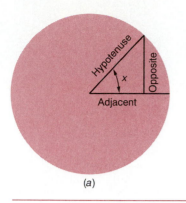

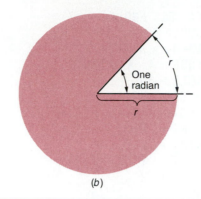

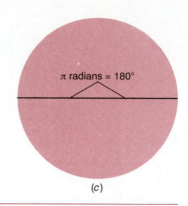

(a)   (b)   (c)

FIGURE 13.2
Relationship of radians to degrees: (a) the angle x of the right triangle placed in a standard position at the center of a circle; (b) the definition of a radian; (c) $\pi = 180°$.

Aside from those listed in Table 13.2, other trigonometric functions are available. For example, there is a function DATAN that returns a double-precision result for the arctangent. In addition, other functions can be evaluated on the basis of the trigonometric intrinsic functions (Appendix A).

EXAMPLE 13.6   How to Determine π Using the ATAN Function

PROBLEM STATEMENT: π is one of the most commonly used constants employed in engineering-oriented programs. Although we all probably remember that it is approximately equal to 3.14, it is more difficult to remember that it is equal to 3.141593 in the 7 significant digits used for single-precision real variables in the IEEE standard. Furthermore, it would be impractical for most of us to try to remember that in double precision it is 3.141592653589793. Develop an assignment statement so that your computer determines π internally.

SOLUTION: Recall that the tangent of 45° is equal to 1. Using radians (45° = $\pi/4$), this identity can be expressed as

$$\tan\left(\frac{\pi}{4}\right) = 1$$

By taking the arctangent of this function we get

$$\frac{\pi}{4} = \tan^{-1}(1)$$

Multiplying both sides by 4 yields

$$\pi = 4\tan^{-1}(1)$$

This relationship can now be added to any computer program to compute π. The following code demonstrates its use:

```
REAL PI, ANGLE, Y, Z
DOUBLE PRECISION PIDBL
PI = 4. * ATAN(1.)
PIDBL = 4. * DATAN(1.)
ANGLE = 30.
Y = SIN(ANGLE)
Z = SIN(ANGLE * PI/180.)
PRINT *, PI
PRINT *, PIDBL
PRINT *, Y
PRINT *, Z
END
```

If this program is run, the following results would be displayed:

```
3.141593
3.141592653589793
-.9880316
.5
```

Observe how the program computes the sine of 30 radians (or 1719°) as −0.9880316. In contrast, when radians are employed, the anticipated result for sin 30° is obtained: 0.5.

13.2.3 Truncating and Rounding Functions

Aside from general mathematical and trigonometric functions, there are other built-in functions that are commonly employed by engineers. A group of these are dedicated to truncating and rounding numbers.

Truncating function (INT). The INT function truncates the argument (that is, it lops off the decimal part and returns an integer result). For example,

```
X = 3.9
Y = INT(X)
```

will give Y equal to 3. For a negative argument,

```
X = -2.7
Y = INT(X)
```

will give Y equal to −2.

EXAMPLE 13.7   Using the INT Function to Truncate and Round Positive Numbers

PROBLEM STATEMENT: It is often necessary to round numbers to a specified number of decimal places. One very practical situation occurs when performing calculations pertaining to money. Whereas an economic formula might yield a result of

$467.9587, in reality we know that such a result must be rounded to the nearest cent. According to rounding rules, because the first discarded digit is greater than 5 (that is, it is 8), the last retained digit would be rounded up, giving a result of $467.96. The INT function provides a way to accomplish this same manipulation with the computer.

SOLUTION: Before developing the rounding formula, we will first derive a formula to truncate, or "chop off," decimal places. First, we can multiply the value by 10 raised to the number of decimal places we desire to retain. For example, in the present situation we desire to retain two decimal places; therefore we multiply the value by 10^2, or 100, as in

$$467.9587 \times 100. = 46795.87 \tag{E13.7.1}$$

Next we can apply the INT function to this result to chop off the decimal portion:

```
INT(46795.87) = 46795
```

Finally we can divide this result by 100 to give

$$46795/100. = 467.95$$

Thus we have retained two decimal places and truncated the rest. The entire process can be represented by the FORTRAN expression

```
INT(X * 10. ** D)/10. ** D                        (E13.7.2)
```

where X is the number to be truncated and D is the number of decimal places to be retained.

Now it is a very simple matter to modify this relationship slightly so that it will round. This can be done by looking more closely at the effect of the multiplication in Eq. (E13.7.2) on the last retained and the first discarded digit in the number being rounded (Fig. 13.3). Multiplying the number by 10^D essentially places the last retained digit in the unit position and the first discarded digit in the first decimal place. For example, as shown in Eq. (E13.7.1), the multiplication moves the decimal point between the last retained digit (5) and the first discarded digit (8). Now it is straightforward to see that if the first discarded digit is above 5, adding 0.5 to the quantity will increase the last retained digit by 1. For example, for Eq. (E13.7.1)

$$46795.87 + 0.5 = 46796.37 \tag{E13.7.3}$$

If the first discarded digit were less than 5, adding 0.5 would have no effect on the last retained digit. For instance, suppose the number were 46795.49. Adding 0.5 yields 46795.99, and therefore the last retained digit (5) is unchanged.

To complete the process, the INT function is applied to the result of Eq. (E13.7.3):

```
INT(46796.37) = 46796
```

FIGURE 13.3
Illustration of the retained and discarded digits for a number rounded to two decimal places. The result after rounding is 467.96.

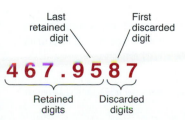

Last retained digit   First discarded digit

4 6 7 . 9 5 8 7

Retained digits   Discarded digits

and the outcome divided by 100 to give the correct answer, 467.96. The entire process is represented by

$$INT(X * 10. ** D + 0.5)/10. ** D \qquad\qquad (E13.7.4)$$

It should be noted that numbers can be rounded to an integer value by setting $D = 0$, as in

$$INT(467.9587 * 10. ** 0 + 0.5)/10. ** 0 = 468.$$

or to the tens, hundreds, or higher places by setting D to negative numbers. For example, to round to the 10's place, $D = -1$ and the result is

$$INT(467.9587 * 10. ** (-1) + 0.5)/10. ** (-1) = 470.$$

Notice that we did not round negative numbers in Example 13.7. Although Eq. (E13.7.4) works properly for positive numbers, it works incorrectly for negative numbers. For example, rounding -467.9587 to two decimal places gives

$$INT (-467.9587 * 10. ** 2 + 0.5)/10. ** 2 = -467.95$$

Thus, instead of rounding to the correct answer of -467.96, it rounds to -467.95.

Because of the way the INT function operates (that is, chopping off the decimal portion), we should be subtracting -0.5 when rounding negative numbers. A slight adjustment of the equation (using the ABS function to determine the sign) can be made to correct this problem:

$$INT(X * 10. ** D + ABS(X)/X * 0.5)/10. ** D \qquad\qquad (13.1)$$

where ABS(X)/X will be -1 if X is negative and $+1$ if X is positive. For the example

$$INT(-467.9587 * 10. ** 1+ABS(-467.9587)/(-467.9587) * 0.5)/10. ** 1$$

when evaluated yields the correct result of -467.96. Thus this equation is superior to Eq. (E13.7.4) in that it adequately rounds both negative and positive numbers. However, it has the flaw that it will not work for the case where $X = 0$.

The previous examples have been expressly designed to illustrate the workings of the INT function. In fact, as described next, there are several intrinsic functions readymade for rounding.

Rounding functions (NINT and ANINT).   The NINT function returns a rounded integer. In contrast, although it also returns a rounded whole number, the ANINT returns a REAL-type result.[4]

---

[4]Note that there is another function, AINT, which behaves the same as INT, except that rather than returning an integer value, it returns a REAL-type truncated whole number.

These functions can be employed to develop rounding functions that perform similarly to Eq. (13.1). For example, the following relationship rounds both positive and negative numbers to D decimal values:

$$\mathrm{ANINT(X * 10. ** D)/10. ** D} \qquad (13.2)$$

Because it is simpler and does not have difficulty with the case where X = 0, this relationship is preferable to Eq. (13.1). For further thoughts on rounding, see Problem 13.24 at the end of the chapter.

13.3 USER-DEFINED STATEMENT (ONE-LINE) FUNCTIONS

Aside from built-in functions, you might be interested in defining functions of your own. The *statement function* is used for this purpose and has the general form

variable(ArgList) = expression

where *variable* is the name of the function (any valid variable name), *ArgList* is the comma-delimited list of arguments for which the function is to be evaluated, and *expression* is the actual arithmetic expression. For example, in the simple addition program (recall Fig. 12.12), we could define a function for the addition of two numbers as

ADD(X, Y) = X + Y

Then the line to perform the addition could be rewritten as

TOTAL = ADD(FIRST, SECOND)

In FORTRAN several arguments may be specified for a statement function. Other variables in the statement have the same value as they had in the program at the point the statement function was invoked (Fig. 13.4). Statement functions must always be placed prior to the first executable statement in a program.

FIGURE 13.4
An example of a statement function.

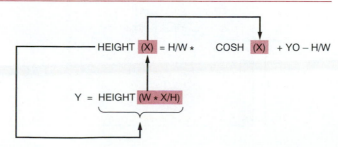

EXAMPLE 13.8 Statement Functions

PROBLEM STATEMENT: The height of a hanging cable is given by the formula (Fig. 13.5)

$$y = \frac{H}{w} \cosh\left(\frac{w}{H}x\right) + y_0 - \frac{H}{w} \qquad (E13.8.1)$$

where H is the horizontal force at $x = a$, y_0 is the height at $x = 0$, and w is the cable's weight per unit length.

Compose a computer program to calculate the height of a cable at equally spaced values of x from $x = 0$ to a maximum x, given w and H. Employ an intrinsic function to compute the hyperbolic cosine and a statement function to calculate the height. Test the program for the following case: $w = 10$, $H = 2000$, and $y_0 = 10$ for $x = 0$ to 40. Display values of y at four equally spaced increments from $x = 0$ to 40.

SOLUTION: Figure 13.6 shows a program to compute the height of a hanging cable. Notice how a line has been set up to calculate the height as a statement function. Several other lines access this function to compute the cable height. Observe how the argument of HEIGHT in these lines is an arithmetic expression, W * X / H, that is passed to the statement function. There its value is ascribed to the dummy variable X which is used to evaluate the height. This result is then passed back and substituted for HEIGHT (W * X / H) in the four lines where the function is invoked. The X is called a *dummy variable* because its use as the argument does not affect its value in the rest of the program. In fact, the value of X in the statement function is actually equal to W * X / H.

FIGURE 13.5
A section of a hanging cable.

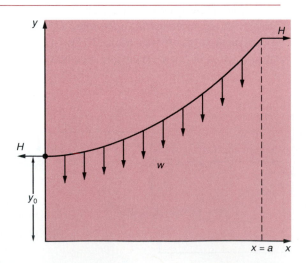

Program:

```
C Program to compute the height of a
C hanging cable
C
C by S.C. Chapra
C June 16, 1992
C
C This program computes the height of a
C hanging cable at four equally-spaced
C points along the X-dimension.
C------------------------------------
C
C Definition of variables
C
C W      = weight per length
C H      = horizontal force
C Y0     = initial height
C XMAX   = cable length along X-dimension
C XINC   = increment of X where Y computed
C HEIGHT = function to compute cable height
C
      REAL W, H, Y0, XMAX, XINC, HEIGHT
C
      HEIGHT(X) = H/W*COSH(X) + Y0 - H/W
C
C identify program
C
      PRINT *, 'Program to compute height'
      PRINT *, 'of hanging cable'
      PRINT *
C
C initialization of variables
C
      X = 0.
C
      READ *, W, H, Y0, XMAX
C
C print headings and initial values
C
      PRINT *, '        DISTANCE           HEIGHT'
      PRINT *
      PRINT *, X, Y0
C
C calculate and print remaining values
C
      XINC = XMAX / 4.
      X = X + XINC
      Y = HEIGHT(W*X/H)
      PRINT *, X, Y
      X = X + XINC
      Y = HEIGHT(W*X/H)
      PRINT *, X, Y
      X = X + XINC
      Y = HEIGHT(W*X/H)
      PRINT *, X, Y
C
      END
```

Input:

```
10.,2000.,10.,40.
```

Output:

DISTANCE	HEIGHT
0.000000E+00	10.000000
10.000000	10.250050
20.000000	11.000830
30.000000	12.254220

FIGURE 13.6
A program to determine the
height of a hanging cable.

Before moving on, we would like to explain a little more clearly how the argument list for the function operates. When a variable appears in a function's argument list, it performs two tasks. Using Fig. 13.4 as an example, the statement function first points back to the line where the function was invoked in order to fetch the value stored in the memory location for the expression $W * X/H$. Second, it indicates how this value is employed on the right side of the function assignment operator. To reinforce the notion that the two X's represent different entities, you should recognize that the statement function could just as easily have been programmed with a Z as the argument,

```
HEIGHT(Z) = H/W * COSH(Z) + YO - H/W
```

We will explore this topic in more detail when we discuss subprograms in Chap. 16.

Although Example 13.8 shows how a user-defined statement function is employed, it does not really illustrate all of its advantages. These advantages become significant when the function is long, and is used numerous times throughout a program. In these instances, the concise representation provided by the statement function is a potent asset. In addition, when the function is extremely long, FORTRAN provides an even more powerful capability called a multiline user-defined function, or *FUNCTION subprogram*. We will also discuss this in Chap. 16.

PROBLEMS

13.1 Which of the following are valid assignment statements in FORTRAN? State the probable errors for those that are incorrect.
(*a*) R1 = -4 ** 1.5
(*b*) A + B = C + D
(*c*) C = C + 22
(*d*) X = Y = 7
(*e*) 4 = X + Y

13.2 If R = 3., S = 5., and T = -2., evaluate the following. Perform the evaluation step by step, following the priority order of the individual operations. Show all your work.
(*a*) -S * R + T
(*b*) S + R/S * T
(*c*) S + T ** T * T ** R
(*d*) R * (R + (R + 1) + 1
(*e*) T + 10 * S - S ** T ** R

13.3 Write FORTRAN statements for the following algebraic formulas.

$$(a)\ x = \frac{a}{c} + \frac{c - d^2}{g}\frac{r^5}{2f^2 - 4}$$

$$(b)\ r = \frac{cb^3}{r^4}(3d + 8x) - \frac{4}{9}x$$

(c) $y = \dfrac{a - b}{x^2 \sin x}$

13.4 In civil engineering the vertical stress under the center of a loaded circular area is given by the following formula. Write this equation in FORTRAN.

$$s = q\left[1 - \frac{1}{(r^2/z^2 + 1)^{3/2}}\right]$$

13.5 In engineering economics the annual payment, A, on a debt that increases linearly with time at a rate G for n years is given by the following equation. Write this equation in FORTRAN.

$$A = G\left[\frac{1}{n} - \frac{n}{(1 + i)^n - 1}\right]$$

13.6 In environmental engineering the following formula is employed to calculate the dissolved oxygen concentration in a river at a location x downstream from a sewage treatment plant effluent. Write this equation in FORTRAN.

$$c = \frac{k_1 D_0}{k_1 - k_2}(e^{-k_2 x/u} - e^{-k_1 x/u})$$

13.7 In chemical engineering van der Waals equation gives pressure of a gas as

$$p = \frac{RT}{V/n - b} - a\left(\frac{n}{V}\right)^2$$

Write this equation in FORTRAN.

13.8 In electrical engineering the electric field at a distance x from the center of a ring-shaped conductor is given by

$$E = \frac{1}{4\pi\varepsilon_0}\frac{Qx}{(x^2 + a^2)^{3/2}}$$

Write this equation as a FORTRAN assignment statement.

13.9 In fluid mechanics the pressure of a gas at the stagnation point is given by

$$p = p_0\left(1 + m^2\frac{k - 1}{2}\right)^{k/(k-1)}$$

Write this equation as a FORTRAN assignment statement.

13.10 The saturation value of dissolved oxygen in fresh water is calculated by the equation

$$c_{sf} = 14.652 - 0.41022T_c + 7.9910 \times 10^{-3}T_c^2 - 7.7774 \times 10^{-5}T_c^3$$

where T_c is temperature in degrees Celsius. Saltwater values can be approximated by multiplying the freshwater result by

$$c_{ss} = [1 - (9 \times 10^{-6})n]c_{sf}$$

where n is salinity in milligrams per liter. Finally, temperature in degrees Celsius is related to temperature in degrees Fahrenheit, T_F, by

$$T_c = \tfrac{5}{9}(T_F - 32)$$

Develop a FORTRAN computer program to calculate c_{ss}, given n and T_F. Test the program for $n = 18,000$ and $T_F = 70$. Note that both c_{sf} and c_{ss} are in milligrams per liter.

13.11 Evaluate the following: (*a*) $X = 6/9 - 4$; (*b*) $J = 5./2 - 4.6$; (*c*) $I = 5/2 - 4.6$; (*d*) $6/9 - 4.0$.

13.12 Evaluate the following expressions. Note which operations involve mixed-mode arithmetic:

(*a*) 3./2 * 2   (*d*) 30/(10 * 1.5)
(*b*) 30/9 * 9   (*e*) 8 ** (2/10)
(*c*) 10 ** 3/2   (*f*) (2 * 4)/10/2

13.13 Describe how the following expressions will be evaluated, and give the numerical result for Y:

(*a*) Y = 10.5 ** 2   (*c*) Y = SQRT(16.) ** 1/2
(*b*) Y = ALOG(17.5) ** 3.   (*d*) Y = 2.5 * (3.5) ** (−2)

13.14 Given X and Y, the coordinates of a point on the circumference, write a statement function that calculates the area of a circle centered on the origin.

13.15 A projectile is fired with a velocity v_0, at an angle of θ above the horizon, off the top of a cliff of height y_0 (Fig. P13.15). The following formulas can be used to compute how far horizontally (x) the projectile will carry prior to impacting on the plane below.

$$(v_y)_0 = v_0 \sin \theta$$

$$(v_x)_0 = v_0 \cos \theta$$

$$t = \frac{-(v_y)_0 - \sqrt{(v_y)_0^2 + 2gy_0}}{g}$$

$$x = (v_x)_0 t$$

where $(v_y)_0$ and $(v_x)_0$ are the vertical and horizontal components of the initial ve-

FIGURE P13.15

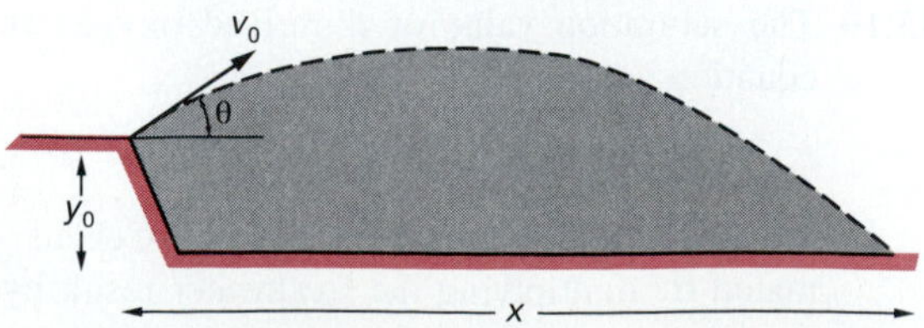

locity, t is the time of flight, and g is the gravitational constant (or -9.81 m/s², where the negative sign designates that we are assigning a negative value to downward acceleration). Prepare a program to compute x, given v_0, θ, and y_0. If $v_0 = 180$ m/s, $\theta = 30°$, and $y_0 = -150$ m, x will be equal to 3100 m. Use this result to test your program. Then employ the program to determine what value of θ results in the greatest value of x. Use trial and error to determine this optimal value of x.

13.16 Figure P13.16 shows a force F acting on a block at an angle θ. The horizontal and vertical components of the force are given by $F \cos \theta$ and $F \sin \theta$, respectively. Write a program similar to the one shown in Fig. 13.6 to compute the horizontal and vertical components of F for given values of F and θ. Use the following values for F and θ:

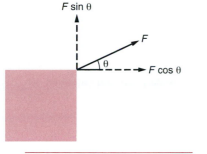

FIGURE P13.16

Force, lb	Angle, degrees
1000	30
50	65
1200	130

Your program should include at least one application of a statement function.

13.17 Write a computer program that requests numerical values to be rounded and the number of places to the right of the decimal point. Employ a statement function to compute the rounded number. Use the following data to test your program:

Number to Be Rounded	Number of Places to the Right of the Decimal Point
-3.14159265	3
$86,400.761$	1
0.6666666	4

13.18 Population growth in an industrial area is given by the following formula:

$$p(t) = P(1 - e^{-at})$$

where $p(t)$ = population at any time ≥ 0, P = maximum population, a = constant [yr], and t = time [yr]. Write a FORTRAN computer program to compute the population at $t = 5$ years with $P = 100,000$ for $a = 0.05$, 0.2, and 0.8. Use a statement function in your program.

13.19 The maximum velocity of a rocket $\bar{v}$ is given by

$$\bar{v} = a \ln\left(\frac{m_0}{m_0 - m_1}\right)$$

where a = relative velocity of exhaust gases, m_0 = total mass of rocket plus fuel at start of flight, and m_1 = mass of fuel. Write a FORTRAN computer pro-

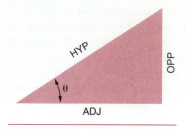

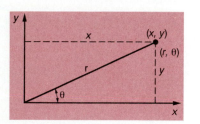

gram to compute $\bar{v}$ for $m_1 = 25{,}000$, $30{,}000$, and $35{,}000$, given $m_0 = 40{,}000$ and $a = 25{,}000$. Use a statement function in your program.

13.20 A triangle is defined by sides ADJ, OPP, and HYP (Fig. P13.20). Write a FORTRAN computer program to calculate HYP and θ using two statement functions. Employ the following data to test your program:

ADJ	OPP
10	10
5	0
2	6

13.21 Sometimes it is useful to make transformations between polar and cartesian coordinates. The following equations are employed for this purpose (Fig. P13.21):

$$x = r \cos \theta$$

and

$$y = r \sin \theta$$

Write a program that reads polar coordinates for a point and displays the corresponding cartesian coordinates. Employ intrinsic and user-defined statement functions to make the conversions.

13.22 Write a FORTRAN program that inputs two complex numbers, `Z1` and `Z2`. Then perform the following operations:

```
Z1 + Z2
Z1 - Z2
Z1 * Z2
Z1 / Z2
```

Display the results. On the basis of the results, explain how each of the arithmetic operators behaves when used in conjunction with complex numbers.

13.23 Write a FORTRAN program that determines the roots of the quadratic equation,

$$y = ax^2 + bx + c$$

Use the quadratic formula to determine the roots, but treat x as a complex variable.

13.24 Section 13.2.3 was not intended as a definitive treatment of the rounding process. Rather, it was intended to illustrate the uses of the `INT`, `NINT`, and `ANINT` intrinsic functions. In fact, it should be noted that the rounding schemes suggested by Eqs. (13.1) and (13.2) are not strictly correct.

The difficulty relates to the special case where the first discarded digit is a 5 and all other discarded digits are 0's. For this situation, rounding up is just as incorrect as rounding down! Either choice is arbitrary. A better approach would be to adopt a rule that would give a 50 percent chance of either rounding up or down. One way to do this is to adopt the following rule:

- When the first discarded digit is exactly 5 and all other discarded digits are 0, round off so as to leave the last retained digit an even number.

If this rule is followed faithfully, errors due to rounding tend to compensate one another.

Perform some numerical experiments to determine how both Eqs. (13.1) and (13.2) round numbers of this sort. For example, you might try rounding values such as 467.05, 467.15, 467.25, and so on, to one decimal place. Determine whether they round up or round down. Also investigate how the negatives of these numbers are rounded. Discuss your results.

SELECTION

To this point we have written programs that perform instructions sequentially. That is, the program statements are executed line by line starting at the top of the program and moving down to the end. However, a strict sequence is only one of the ways in which a program can be executed. The flow of logic can branch (selection) or loop (repetition) depending on certain conditions. The present chapter deals with the statements required to branch, or to reroute the flow of logic, to a statement other than the next in sequence.

Although selection can greatly enhance your power and flexibility, its indiscriminate and undisciplined application can introduce enormous complexity to a program. Because this greatly detracts from program maintenance, computer scientists have developed a number of conventions to impose order on selection. Some of these structured-programming conventions involve the way in which the statements are applied; that is, they are voluntary and involve the "style" with which the programmer chooses to construct the code. Others actually involve specialized FORTRAN statements that provide very clear and coherent ways to perform selection. These are called *control-flow structures.*

Figure 14.1 shows the two principal control-flow structures that are used for selection. Unfortunately, such structures were not available in early versions of FORTRAN. However, because of their great utility, they have been incorporated into all modern dialects. We will stress the use of these structures in this chapter.

14.1   SPAGHETTI CODE

As mentioned above, aside from the control-flow structures, the coherence of a FORTRAN program derives from the way in which the selection statements are applied; that is, their clarity involves the style with which the programmer chooses to construct the code. The lack of style and discipline can lead to *spaghetti code*—or code that is so convoluted that it is extremely difficult to understand. As described next, nowhere is this more manifest than in the use of the GO TO statement.

14.1.1 The GO TO Statement

The GO TO is the simplest of all transfer statements. It allows you to directly override the sequence in which the program is executed. The general form is

GO TO *ln*

where *ln* denotes the line number that is executed next, for example:

```
      READ *, A, B, C
      GO TO 40
   10 D = A * B * C
      PRINT *, A, B, C
   40 READ *, X, Y, Z
```

Under normal circumstances, the program would advance to line 10 after inputting values for A, B, and C. However, because of the GO TO statement, the program jumps to line 40.

By allowing transfer to any other line, the GO TO greatly increases the program's flexibility. However, a word of caution is in order. All the jumping around that results from the indiscriminate use of GO TO statements can lead to programs that are extremely difficult to understand, debug, and maintain. For this reason, structured programming considerations mandate that they be employed only when absolutely necessary.

Negative feelings about GO TO statements are so strong that in some programming languages (Pascal, for one) the GO TO construct is discouraged and is rarely seen. Languages that shun the GO TO statement usually have other constructs that control the flow of program execution in a more coherent and structured manner.

Standard FORTRAN 77 does not presently include a complete set of constructs for

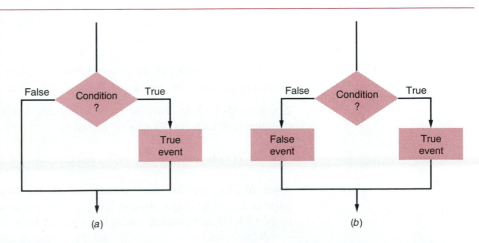

FIGURE 14.1
The two fundamental control constructs used for selection: the single- and double-alternative decision patterns. These are the fundamental building blocks for all types of selection. The combination of these two patterns allows any possible selection process to be implemented.

controlling program flow in this fashion. This is particularly true in reference to the repetition constructs in the following chapter. For this reason, we will use the GO TO to compensate for the lack of certain types of structured control statements in FOR-TRAN. Our use of GO TOs will be limited according to the following guidelines:

- The GO TO statement will be used when it functions as a part of a control structure for which no formal construct is available in FORTRAN 77.
- The GO TO will also be employed for pedagogical purposes to illustrate poor programming habits.

14.1.2   The CONTINUE Statement

The CONTINUE statement performs no specific operation but is sometimes used as the target of a GO TO statement. Once it is reached, the program merely "continues" on to the program's next executable line. Its general form is

ln `CONTINUE`

Because of the deemphasis of the GO TO statement, the CONTINUE statement is employed far less than in the past. However, as described in the next chapter, it is still useful in repetition constructs.

14.1.3   "Pasta" Statements

Aside from the GO TO, there are several other control statements that contributed to the prevalence of spaghetti code in the early days of computing. Among the more notorious are the arithmetic IF and the computed GO TO.

Arithmetic IF.   The arithmetic IF has the general form

`IF` (*ArithmeticExpression*) ln_1, ln_2, ln_3

where *ArithmeticExpression* represents an arithmetic expression, and ln_1, ln_2, and ln_3 are line numbers that the program transfers to if the *ArithmeticExpression* is negative, zero, or positive, respectively.

If the GO TO is bad, the arithmetic IF is three times worse! Not only does it send you off to another part of the program, but it can send you to three different locations depending on the value of the *ArithmeticExpression.*

Computed GO TO.   Now we arrive a the *pièce de résistance* of pasta programming—the computed GO TO. Its general format is

`GO TO` (*ln_1*, *ln_2*, . . . , *ln_n*) , *IntegerExpression*

where *ln_1*, *ln_2*, and *ln_3* are line numbers that the program transfers to depending on whether the *IntegerExpression* is $1, 2, \ldots, n$. Consequently, in one fell swoop, the computed GO TO can catapult you to locations all over your program.

The use of the arithmetic IF and the computed GO TO statements is strongly discouraged. We have merely included them here to warn you against their use. Unfortunately, some Neanderthal programmers still employ them, so you may run across these constructs when reading someone else's code. Hopefully, you can gently instruct these poor souls on the error of their ways and point them toward the promised land of structured programming.

14.2   THE SINGLE-LINE IF STATEMENT

Whereas the statements described in the previous section should be avoided, the single-line logical IF statement has some utility, if employed properly. Its general format is

IF (*LogicalExpression*) *statement*

where *LogicalExpression* represents an expression that is either true or false, and *statement* is a FORTRAN statement that is to be implemented if the expression is true. If the *LogicalExpression* is false, implementation of the *statement* does not occur and control passes automatically to the line following the logical IF statement.

The logical IF statement can control which path a program takes at a branch point. The chosen path could theoretically lead anywhere in the program. An example of a form of the logical IF that branches to another statement is

IF (VAR .EQ. 1.) GO TO 120

For this case, if the value of VAR is equal to 1, control will transfer to line 120. If not, control passes to the next line. For the case where the number following the GO TO is not an existing line number in your program, an error message results.

When used thoughtlessly, such a *conditional transfer* could lead to the same type of confusion that follows from the indiscriminate application of GO TO or arithmetic IF statements. Although such transfers are possible, they are strongly discouraged. It should be noted, however, that because of inadequacies in FORTRAN 77, the logical IF and GO TO statements will come in handy when we develop loop structures in Chap. 15.

A preferable implementation of the single-line logical IF statement involves cases where the *statement* represents an action rather than a transfer.

IF (NUMBER .GT. 0) AVERG = SUM/NUMBER

For this example, if the value of the variable NUMBER is greater than zero (greater than signified by .GT.), then the variable AVERG will be computed as the SUM divided by the NUMBER. After the computation is implemented, control will transfer to the next line. If NUMBER is equal to or less than zero, the statement AVERG = SUM/NUMBER is ignored and control transfers immediately to the next line.

In both examples, the *LogicalExpression* to be tested is located after the IF and before the *statement*. Now we will take a closer look at how these *LogicalExpressions* can be formulated.

14.3 FORMULATING LOGICAL EXPRESSIONS

The *LogicalExpression* is always a quantity that is either true or false. The two simplest cases are a single logical variable or the relation between two expressions.

14.3.1 Logical Variable

In Table 12.2, we introduced the concept of a logical variable. Logical variables may take on only two logical constant values: .TRUE. or .FALSE.. Note that the periods at the beginning and end of each of these words are a part of the constant. An example of using a logical variable as the *LogicalExpression* of an IF statement is

```
      LOGICAL CHECK
      CHECK = .TRUE.
      IF (CHECK) PRINT *, 'Test is true'
   40 CONTINUE
```

The first line is an explicit declaration that is required to tell the computer that the variable CHECK is of the logical type. Then, the second line assigns a value of .TRUE. to CHECK. Consequently, the logical IF in the third line will test true, and therefore, the message will be displayed. If CHECK had been set equal to .FALSE., the *LogicalExpression* would have tested false and the program would proceed to line 40 without displaying the message.

14.3.2 Relation between Expressions

Aside from logical variables, a more common example of a *LogicalExpression* is the comparison of two expressions, which can be represented generally by

$value_1$. *relation* . $value_2$

where the *values* can be variables, constants, or arithmetic expressions. The *relation* is any one of the relational operators in Table 14.1. These include equality, inequality, and testing whether one value is greater or less than another.

The distinction between the "equality" as a relational operator (.EQ.) and its use in assignment (=) now becomes important. This is the reason why the developers of FORTRAN defined different symbols to designate the two processes. For example, the (V .EQ. 0.) portion of the first example in Table 14.1. does not cause a value of zero to be assigned to V. Rather, it merely tests whether V is currently equal to zero.

In contrast, the expression V = 99 causes V to take on the value of 99 if it has previously been equal to zero. If V is not equal to zero when the logical IF statement is reached, it will still maintain its previous value after the statement is executed.

Proceeding with the other relational operators, the inequality tests whether the expression on the right side of the .NE. is different from that on the left. For the example in Table 14.1, the value of the character variable C will be displayed only if C is not equivalent to the character string '9'.

Testing whether one value is less than another is accomplished with the .LT. op-

TABLE 14.1 Relational Operators*

Operator	Relationship	Example
.EQ.	Equal	IF (V.EQ.0.) V = 99
.NE.	Not equal	IF (C.NE.'9') PRINT *,C
.LT.	Less than	IF (A.LT.0.) A =-A
.GT.	Greater than	IF (X.GT.Y) Z=Y/X
.LE.	Less than or equal to	IF (3.9.LE.X/3) GO TO 200
.GE.	Greater than or equal to	IF (VAR.GE.0.) GO TO 160

*Note that the C in the second example is a CHARACTER variable.

erator. Likewise, the .GT. determines whether the value on the left side of the .GT. is greater than that on the right side. The operator .GE. allows the left-hand value to be either exactly equal to or greater than the right-hand value. Similarly, the operator .LE. will be true if the two are equal or if the left-hand value is less than the right-hand value. Remember the rules for working with inequality operators, which require the operator to change if the values on the left and right side of inequality are exchanged. X .GT. Y is the same test as Y .LT. X.

The order of evaluation of the logical IF statement is important to note. First the condition is tested. If the condition is true, then the *statement* is carried out. In the first example in Table 14.1, V is tested to determine whether it is equal to zero prior to performing the assignment. Otherwise V can never equal zero. It would always equal 99.

Arithmetic and relational operators can be combined in a condition being tested. The arithmetic operations are always implemented before testing the relationship. In the fifth example of Table 14.1, X is first divided by 3, and then 3.9 is tested to see if it is equal to or less than the quotient.

Strings and character data can also be tested for their relationship to other string and character data. As in the second example in Table 14.1, character data cannot be directly compared to numeric data. That is why the 9 has to be enclosed in apostrophes. Enclosing the 9 in this manner causes it to be treated as a string character and overrides its numeric value. If the 9 is not set in apostrophes, FORTRAN will stop processing the line and report an error. Just as string variables cannot have numbers assigned to them or vice versa, relational operations require that both sides of the relation have either numeric data or string data, but not both.

Be cautious when testing for the equality or inequality of two real expressions. Remember that real numbers are stored in the computer as series of binary digits and an exponent. If the value to be tested cannot be exactly represented by the significant digits of a single-precision real variable, then the test may fail, even though it is, in fact, "almost" true. For example, the following condition would test as false:

    1/3 .EQ. 0.333

Even though 0.333 is approximately equal to $\frac{1}{3}$, the computer must sense an exact equality for the condition to be true. More subtle differences can occur, as illustrated in the following example.

EXAMPLE 14.1   *Problems with Testing Equalities*

PROBLEM STATEMENT: The following code was written to investigate the testing of equalities between real numbers:

```
      PROGRAM EX141
      REAL A, B
      CHARACTER*25 MESSG1, MESSG2
      A = 1./82.
      B = A * 82.
      MESSG1 = 'equality holds'
      MESSG2 = 'equality does not hold'
      IF (B .EQ. 1.) GO TO 10
      GO TO 20
   10 PRINT *, MESSG1
      GO TO 30
   20 PRINT *, MESSG2
   30 CONTINUE
C output value for B
      PRINT *, 'B =', B
      END
```

Run:

```
equality does not hold
B =        9.999999E-01
```

Although we know that 1/82 multiplied by 82 should be exactly equal to 1, such is not the case for this code. Because of the approximate way that the computer stores real numbers, the equality does not hold.

One way to circumvent this problem is to reformulate it. Rather than testing whether the two numbers are equal, you can test whether their difference is small. For example, the logical IF could be reformulated as

```
IF (ABS(B - 1) .LT. 0.00001) GO TO 10
```

This statement tests whether the absolute value of the difference between B and 1 is less than 0.00001—that is, less than 0.001 percent. Formulated in this way, the logical IF would test true, as desired. This is due to the fact that the discrepancy introduced by the internal binary representation is much smaller than the tolerance used for the comparison. In fact the difference occurs in the seventh significant digit, which corresponds to about a 0.00001 percent difference.

It should also be noted that although it is adequate for illustrative purposes, the code in this example is very poorly structured. In particular, our use of GO TOs is stylistically deficient. Later in this chapter, we will introduce control constructs that are far superior. A problem at the end of the chapter is devoted to structuring this code properly.

Although you should be aware of the problem outlined in the previous example, it is usually limited to equality. Testing whether one value is less than or greater than another can usually be performed without running into this difficulty.

14.4 COMPOUND LOGICAL IF STATEMENTS AND LOGICAL OPERATORS

FORTRAN allows the testing of more than one logical condition in the same logical IF statement. These *compound logical IF statements* employ the logical operators `.AND.`, `.OR.`, and `.NOT.` to simultaneously test conditions (Table 14.2).

14.4.1 Conjunction

The `.AND.` is employed when both conditions must be true, as in the general form

$$IF\ (expression_1\ .AND.\ expression_2)\ statement$$

where the *expressions* are logical. If both *expressions* are true, then the *statement* will be executed. If either or both *expressions* are false, the program will advance to the next line without executing the *statement*.

14.4.2 Disjunction

In contrast, the `.OR.` is employed when either one or the other or both of the *expressions* are true, as in

$$IF\ (expression_1\ OR\ expression_2)\ statement$$

If either or both *expressions* are true, the *statement* will be performed.

TABLE 14.2   Commonly Used Logical Operators*

Operator	Meaning	Example
`.NOT.`	*The logical complement.* This negates a boolean expression. In the example, if `X.LE.0` is false, the expression is true (and vice versa).	`IF (.NOT.(X.LE.0.)) PRINT *, 'positive'`
`.AND.`	*Conjunction.* Both boolean expressions must be true; otherwise, the compound expression is considered false.	`IF ((X.GT.0.).AND.(X.LT.5)) PRINT *, '0 to 5'`
`.OR.`	*Disjunction (inclusive "or").* If either or both of the expressions are true, the compound expression is true.	`IF ((L.EQ.'n').OR.(L.EQ.'N')) PRINT *, 'No'`

*Note that the L in the third example is a CHARACTER variable.

14.4.3 Logical Complement

The .NOT. is used for the case where the *statement* is implemented if an *expression* is false, as in

> IF (.NOT. *expression*) *statement*

If the *expression* is false, the *statement* will be executed.

When a logical expression is false, its logical complement is true, and vice versa. For example, the complement of (X.LE.0.) is (X.GT.0.). Thus, the first example from Table 14.2 could be rewritten as

> IF (X .GT. O.) PRINT *, 'positive'

The complements for the relational operators are summarized in Table 14.3. Note that in most cases, it is clearer and more convenient to use the complement of a relational operator than to employ the .NOT.. On the other hand, there will be other instances where the .NOT. will be desirable.

Compound expressions can contain more than one of the logical operators, for example,

> IF ((A.GT.0.) .OR. (B.LT.0.) .AND. (C.EQ.0.))C = A/B

However, just as with arithmetic expressions, the order in which the operations are evaluated must follow rules. The priorities for the logical operators (along with parentheses and all the other operators established to this point) are specified in Table 14.4. Thus, all other things being equal, the .NOT. would be implemented prior to the .AND. and the .AND. prior to the .OR..

For complicated logical expressions, parentheses and position also dictate the order. Thus operations within parentheses are implemented first. Then the priorities in Table 14.4. are used to determine the next step. Finally, within the same priority class, operations are executed from left to right.

TABLE 14.3 *Complements for the Relational Operators*

Operator	Complement	Identities		
.EQ.	.NE.	A.EQ.B	is equivalent to	.NOT.(A.NE.B)
.LT.	.GE.	A.LT.B	is equivalent to	.NOT.(A.GE.B)
.GT.	.LE.	A.GT.B	is equivalent to	.NOT.(A.LE.B)
.LE.	.GT.	A.LE.B	is equivalent to	.NOT.(A.GT.B)
.GE.	.LT.	A.GE.B	is equivalent to	.NOT.(A.LT.B)
.NE.	.EQ.	A.NE.B	is equivalent to	.NOT.(A.EQ.B)

TABLE 14.4   The Priorities of Operators Used in FORTRAN

Type	Operator	Priority
Arithmetic	Parentheses, ()	Highest
	Exponentiation, **	
	Negation, −	
	Multiplication, *, and Division, /	
	Addition, +, and Subtraction, −	
Relational	.EQ., .LT., .GT., .LE., .GE., .NE.	
Logical	.NOT.	
	.AND.	
	.OR.	Lowest

EXAMPLE 14.2   Priorities in Compound Logical Expressions

PROBLEM STATEMENT: Determine whether the following compound logical expressions are true or false:

(*a*) A * B .GT. 0. .AND. B .EQ. 2. .AND. X .GT. 7. .OR. .NOT. Y .GT. 'd'
(*b*) A * B .GT. 0. .AND. B .EQ. 2. .AND. (X .GT. 7. .OR. .NOT. Y .GT. 'd')

Note that A = −1, B = 2., X = 1., and Y is a character variable that has been assigned a value of 'b'.

SOLUTION: (*a*) Using the priorities from Table 14.4, we can determine that the expression is true by the following sequence of evaluations:

```
 A * B .GT. 0. .AND. B  .EQ. 2. .AND. X  .GT. 7. .OR. .NOT. Y   .GT. 'd'
 -2    .GT. 0. .AND. 2. .EQ. 2. .AND. 1. .GT. 7. .OR. .NOT. 'b' .GT. 'd'
        F      .AND.    T       .AND.    F       .OR. .NOT.      F
                                                                 └───┬───┘
                                                                   True
                                                         └─────┬─────┘
                                                              True
          └────────────┬────────────┘
                    False
        └───────────────────────┬───────────────────────┘
                              False
```

First all the relational operators are evaluated because they have higher priority than the logical operators. Of the logical operators, the .NOT. comparison is implemented first because it has a higher priority than .AND. and .OR.. It negates the last false and converts it to a true. The leftmost .AND. is implemented next because of the left-to-right rule. It is false because both expressions must be true for an .AND. to be true.

Next, the conjunction of two falses leads to a false. Finally, the .OR. comparison is true because one of the conditions being compared is true.

(b) Because of the introduction of the parentheses, we are forced to evaluate the .OR. comparison before the .AND.'s. As a result, the entire operation yields a false result, as in

```
A * B .GT. 0. .AND. B .EQ. 2. .AND. (X  .GT. 7. .OR. .NOT. Y   .GT. 'd')
 -2   .GT. 0. .AND. 2. .EQ. 2. .AND. (1. .GT. 7. .OR. .NOT. 'b' .GT. 'd')
      F     .AND.    T    .AND. (    F     .OR. .NOT.       F    )
                                                              └──────┬──────┘
                                                                    True
                                                   └────────────┬────────────┘
                                                               True
      └──────────────┬──────────────┘
                   False
      └──────────────────────────┬──────────────────────────┘
                               False
```

Modern FORTRAN also employs several other operators (Table 14.5). These are called the *equivalence* and *nonequivalence* operators. Note that the operators in Table 14.5 have lower priority than the .NOT., .AND., and .OR..

In summary, there is a rich array of logical operators in the FORTRAN language. Table 14.6 outlines their behavior in the form of a truth table.

TABLE 14.5 Additional Logical Operators in FORTRAN

Operator	Name	Meaning
EQV	Equivalence	Tests true when either both are true or both are false.
NEQV	Nonequivalence	Tests true if one is true and the other is false. Thus it is the negation of the equivalence statement.

TABLE 14.6 A Truth Table Summarizing the Behavior of the Logical Operators Available in FORTRAN

X	Y	.NOT X.	X .AND. Y	X .OR. Y	X .EQV. Y	X .NEQV. Y
T	T	F	T	T	T	F
T	F	F	F	T	F	T
F	T	T	F	T	F	T
F	F	T	F	F	T	F

14.5   STRUCTURED PROGRAMMING OF SELECTION

The logical IF statement introduced in the previous sections can be orchestrated to introduce all sorts of complicated logic into your computer programs. However, as cautioned at the beginning of this chapter, its indiscriminate application can lead to code that is so complex that it is practically indecipherable (so-called spaghetti code).

14.5.1   "Spaghetti" Selection

There are several ways that the logical IF construct can be employed to program a double-alternative decision pattern. Figure 14.2 shows a flowchart that will serve to illustrate this point. Figure 14.2*a* through *d* presents four sets of FORTRAN code that all implement the algorithm in the flowchart. Figure 14.2*a* and *b* both employ the form of the condition shown in the flowchart; that is, they test whether A is less than zero. They differ in that Fig. 14.2*a* has the false alternative first, whereas Fig. 14.2*b* has the true alternative first. Figure 14.2*c* and *d* are alternatives that result from utilizing the complement of the original condition; that is, they test whether A is greater than or equal to zero. They also differ in that Fig. 14.2*c* has the false alternative first, whereas Fig. 14.2*d* has the true alternative first. Notice how in all cases, GO TO statements are required to transfer control properly.

The fact that four separate versions can be developed[1] to perform the same double-alternative decision is one example of how selection can introduce flexibility and an associated complexity into a program. The result is that the logic of such statements is often very difficult to decipher. In order to reduce this potential confusion, computer scientists have constrained selection by developing the block-structured IF/THEN/ELSE construct.

14.5.2   The Block-Structured IF/THEN/ELSE Construct

The block-structured IF/THEN/ELSE construct can be represented generally as

```
IF LogicalExpression THEN
    TrueStatements
        .
        .
        .
ELSE
    FalseStatements
        .
        .
        .
END IF
```

---

[1]And many more could have been devised!

If the *LogicalExpression* is true, control moves to the *TrueStatements*. After these are executed, control jumps immediately to the line following the END IF statement. If the expression is false, control jumps to the *FalseStatements* following the ELSE. The format is reiterated in Fig. 14.3, which shows a FORTRAN example along with the flowchart and pseudocode representations of the construct. As depicted in Fig. 14.4, a version is also available for the single-decision alternative.

FIGURE 14.2
Four sets of FORTRAN code
that implement the algorithm
specified by the flowchart.

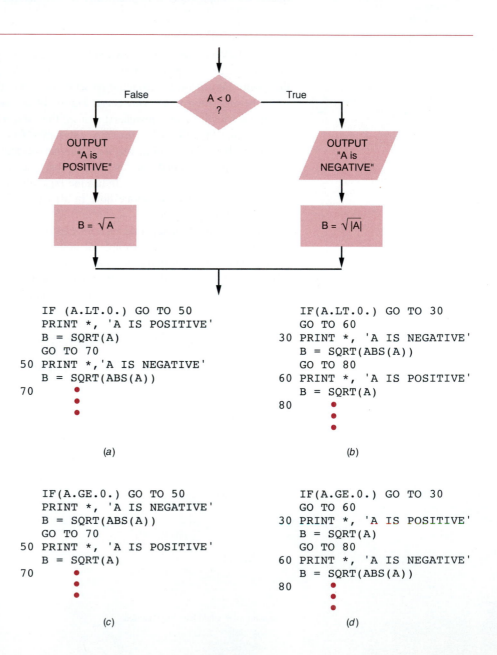

```
    IF (A.LT.0.) GO TO 50
    PRINT *, 'A IS POSITIVE'
    B = SQRT(A)
    GO TO 70
50  PRINT *,'A IS NEGATIVE'
    B = SQRT(ABS(A))
70      •
        •
        •
```

(a)

```
    IF(A.LT.0.) GO TO 30
    GO TO 60
30  PRINT *, 'A IS NEGATIVE'
    B = SQRT(ABS(A))
    GO TO 80
60  PRINT *, 'A IS POSITIVE'
    B = SQRT(A)
80      •
        •
        •
```

(b)

```
    IF(A.GE.0.) GO TO 50
    PRINT *, 'A IS NEGATIVE'
    B = SQRT(ABS(A))
    GO TO 70
50  PRINT *, 'A IS POSITIVE'
    B = SQRT(A)
70      •
        •
        •
```

(c)

```
    IF(A.GE.0.) GO TO 30
    GO TO 60
30  PRINT *, 'A IS POSITIVE'
    B = SQRT(A)
    GO TO 80
60  PRINT *, 'A IS NEGATIVE'
    B = SQRT(ABS(A))
80      •
        •
        •
```

(d)

The formats shown in Figs. 14.3 and 14.4 are referred to as *block-structured IF/THEN/ELSE* constructs. The benefits of using these constructs are great. Each IF/THEN/ELSE construct has only one entry point and one exit. The FORTRAN statements that are a part of the IF/THEN/ELSE construct are immediately apparent, and the true and false alternatives are clearly delineated. This is reinforced by indentation. The primary objective here is to make the resulting code coherent and easily understandable to someone other than the original author of the program. These conventions help standardize the writing of programs so that others may more readily grasp the organization of an unfamiliar program. They also serve to standardize across different languages.

FIGURE 14.3
(*a*) A flowchart for a double-alternative decision pattern; (*b*) the corresponding pseudocode; and (*c*) an example of code in structured FORTRAN.

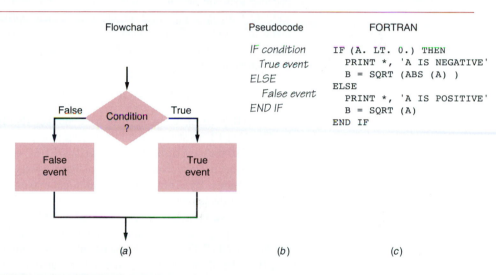

Flowchart

Pseudocode

FORTRAN

```
IF condition
    True event
ELSE
    False event
END IF
```

```
IF (A. LT. 0.) THEN
    PRINT *, 'A IS NEGATIVE'
    B = SQRT (ABS (A) )
ELSE
    PRINT *, 'A IS POSITIVE'
    B = SQRT (A)
END IF
```

(*a*)   (*b*)   (*c*)

FIGURE 14.4
(*a*) A flowchart for a single-alternative decision pattern; (*b*) the corresponding pseudocode; and (*c*) an example of code in structured FORTRAN.

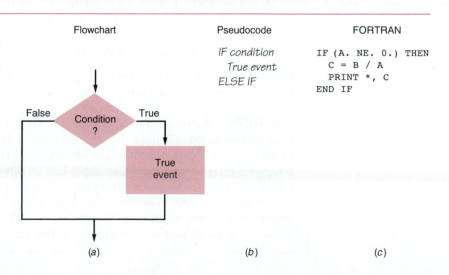

Flowchart

Pseudocode

FORTRAN

```
IF condition
    True event
ELSE IF
```

```
IF (A. NE. 0.) THEN
    C = B / A
    PRINT *, C
END IF
```

(*a*)   (*b*)   (*c*)

FIGURE 14.5
A side-by-side comparison of
Figs. 14.2*a* and 14.3*c*. Both
sets of code accomplish the
same objective, but the
structured version in (*b*) is
much easier to read and
decipher.

```
     IF (A.LT.0.) GO TO 50            IF (A.LT.0.) THEN
     PRINT *, 'A IS POSITIVE'            PRINT *, 'A IS NEGATIVE'
     B = SQRT(A)                         B = SQRT(ABS(A))
     GO TO 70                         ELSE
  50 PRINT *, 'A IS NEGATIVE'            PRINT *, 'A IS POSITIVE'
     B = SQRT(ABS(A))                    B = SQRT(A)
  70 CONTINUE                         END IF

  (a)                                 (b)
```

All forms of selection (with the exception of the single-line logical IF introduced in Sec. 14.2) should be constructed as block structures for easier readability. This is because they are so much easier to understand than versions employing one-line IFs and GO TOs, as in Fig. 14.2. Figure 14.5, which reprints Figs. 14.2*a* and 14.3*c*, should make this evident. Even though this is a very simple example, the superiority of the structured approach is obvious. For more complicated code, the benefits become even more pronounced.

14.5.3   Nesting of Block IF/THEN/ELSE Constructs

An IF/THEN/ELSE construct can have another IF/THEN/ELSE statement as part of its THEN or ELSE statements. The inclusion of a construct within a construct is called *nesting.* For the case of the block IF/THEN/ELSE construct, nesting provides an alternative to having involved conditional expressions with many `.AND.`, `.OR.`, and `.NOT.` clauses. By using simpler conditional expressions it may be easier to follow the logic of the decisions.

The more complex nested IF/THEN/ELSE structures are much easier to understand when proper rules for indentation are followed. The rules for forming and indenting the nested IF/THEN/ELSE statements are essentially the same as those for a single IF/THEN/ELSE structure. Just remember that indentation of the conditional statement ELSE and the END IF is always determined by the IF/THEN statement to which they refer. An example will serve to demonstrate the use of nesting and proper indentation.

EXAMPLE 14.3   Nesting of IF/THEN/ELSE Constructs

PROBLEM STATEMENT: Develop a program to determine the real and complex roots of the quadratic equation.

SOLUTION: A structured flowchart for this problem is shown in Fig. 14.6. Note that there are two nested selections. The first selection is designed to avoid division by zero in the quadratic equation for the case where $a = 0$. The second selection branches to compute the real or complex roots, depending on whether the term within the square root (called the "discriminant") is positive or negative.

A structured code to implement the flowchart is listed in Fig. 14.7. Indentation is used to show clearly the true and false options of each IF/THEN/ELSE. Problem 14.10 at the end of the chapter deals with refining this program.

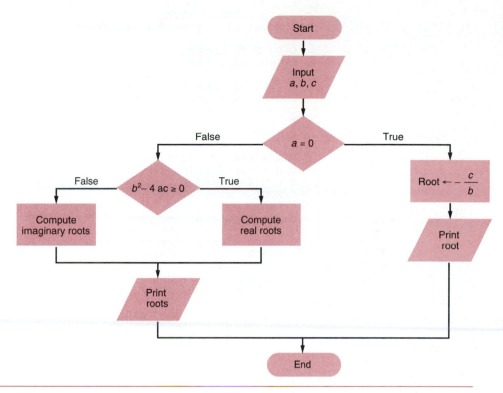

FIGURE 14.6
Structured flowchart to determine the real and complex roots of a quadratic equation. Note that the first selection prevents division by zero in the equation if $a = 0$.

At this point you may wonder why all FORTRAN programs have not been written in a structured form. What are the disadvantages of using this type of program organization? For one thing, early versions of FORTRAN did not have the proper statements to program in a structured fashion. Beyond this, the amount of typing is increased in the structured code. In addition, the speed of execution of the structured code may be slightly slower than that of a nonstructured version. Thus there is a trade-off here. The structured program is easier to understand and change later, but more time may be consumed developing it initially. However, in the overall scheme of program development and use, the time and effort are almost always a worthwhile investment.

Additionally, as you employ structured programming constructs on a regular basis, they will become second nature to you. As your experience grows, you will realize that the structured constructs are actually simpler to program initially and unquestionably easier to modify in the future.

14.5.4 The CASE Construct

The CASE construct is a special type of nested IF/THEN/ELSE block. As depicted in the flowchart and pseudocode in Fig. 14.8, the CASE construct selects one path from a group of alternative paths based on the values of a single variable. This is a very use-

```
          PROGRAM ROOTS
C Roots of a quadratic equation
C
C by S.C. Chapra
C June 16, 1992
C
C This program determines the real or
C complex roots of a quadratic.
C-----------------------------------
C
C Definition of variables
C
C A = second-order term
C B = first-order term
C C = zero-order term
C ROOT = single root
C REAL1 = real part of first root
C IMAG1 = imaginary part of first root
C REAL2 = real part of second root
C IMAG2 = imaginary part of second root
C DISCR = discriminant

      REAL A, B, C
      REAL ROOT, REAL1, IMAG1, IMAG2, REAL2

C     input data and print headings
      PRINT *, 'a, b, c ='
      READ *, A, B, C
      PRINT *
      PRINT *, 'Results:'
      PRINT *

      IF (A. EQ. 0.) THEN
        ROOT = -C / B
        PRINT *, 'single root = ', ROOT
      ELSE
        DISCR = B**2 - 4. * A * C
        IF (DISCR .GT. 0.) THEN
C         compute real roots
          REAL1 = (-B + SQRT(DISCR)) / (2.*A)
          REAL2 = (-B - SQRT(DISCR)) / (2.*A)
        ELSE
C         compute complex roots
          REAL1 = -B/(2.*A)
          REAL2 = REAL1
          IMAG1 = SQRT( ABS(DISCR) ) / (2.*A)
          IMAG2 = - IMAG1
        END IF
C         output results
          PRINT *, 'root1 = ', REAL1, IMAG1, 'i'
          PRINT *, 'root2 = ', REAL2, IMAG2, 'i'
      END IF

      END
```

FIGURE 14.7
Structured FORTRAN program to implement the algorithm from Fig. 14.6. Notice how indentation clearly sets off the program's structure.

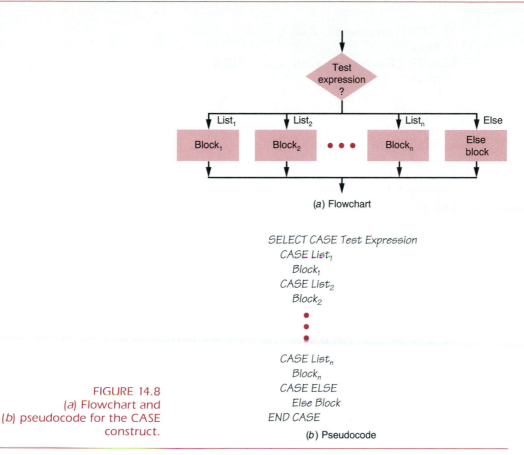

(a) Flowchart

SELECT CASE Test Expression
   CASE List₁
     Block₁
   CASE List₂
     Block₂

      •
      •
      •

   CASE Listₙ
     Blockₙ
   CASE ELSE
     Else Block
END CASE

(b) Pseudocode

FIGURE 14.8
(a) Flowchart and
(b) pseudocode for the CASE
construct.

ful construction in situations where the program must branch to perform several independent tasks and then come together again.

As seen in Fig. 14.8, the CASE statement tests to find which of the *Lists*, 1 through *n*, matches the *TestExpression*. Only the statements associated with the matched *List* are then executed. The statements following ELSE would be carried out if none of the *Lists* matched the contents of the *TestExpression*.

Although it is a part of many advanced languages such as Pascal and QuickBASIC, the CASE statement is not formally a part of standard FORTRAN 77. However, it can be simulated with a series of IF/THEN/ELSE statements. In addition, FORTRAN 77 has a feature that facilitates building a CASE structure—the ELSE IF. Instead of each IF statement having a paired END IF, only the last IF in a group has an END IF and all intermediate IFs end with ELSE IFs.

To make this construction more obvious to the reader, slightly different indenting and branching in the IF/THEN/ELSEIF structure are helpful. The approach is depicted in Fig. 14.9. Again statements following the condition are executed when the *TestExpression* equals the specific *List*.

```
IF (TestExpression.EQ.List₁) THEN
    Block₁
ELSEIF (TestExpression.EQ.List₂) THEN
    Block₂
        .
        .
        .
ELSEIF (TestExpression.EQ.Listₙ) THEN
    Blockₙ
ELSE
    ElseBlock
END IF
```

FIGURE 14.9
Generalized format for the
CASE construct using
IF/THEN/ELSEIF statements.

EXAMPLE 14.4   The CASE Construct

PROBLEM STATEMENT: Suppose that each class at a university is to go to a different room to register for courses. The classroom assignments are

Class	Identification Code	Classroom
Freshmen	1	CR2–3
Sophomores	2	CE1–4
Juniors	3	ME1–6
Seniors	4	CR2–6
All others		CH2–8

The identification code is a number that signifies the student's class. Design a program that uses the CASE construct to display the proper classroom number in response to the student's identification code.

SOLUTION: The FORTRAN program to accomplish the objective is listed in Fig. 14.10a. Notice that the CASE construct is used to display the correct message on the basis of the student's code.

14.6   FORTRAN 90 ENHANCEMENTS

FORTRAN 90 (as well as some dialects such as Microsoft FORTRAN version 5.1) explicitly includes the CASE construct. The general format of the CASE construct in FORTRAN 90 is

```
C Program fragment to illustrate        C Program fragment to illustrate
C CASE decisions (FORTRAN 77)            C CASE decisions (FORTRAN 90)
C-----------------------------------     C-----------------------------------
C                                        C
      INTEGER CLASS                            INTEGER CLASS
C                                        C
      READ *, CLASS                            READ *, CLASS
C                                        C
C     CASE construct                     C     CASE construct
C                                        C
      IF (CLASS.EQ.1) THEN                     SELECT CASE (CLASS)
        PRINT *, 'Proceed to Room CR2-3'          CASE (1)
      ELSE IF (CLASS.EQ.2) THEN                     PRINT *, 'Proceed to Room CR2-3'
        PRINT *, 'Proceed to Room CE1-4'        CASE (2)
      ELSE IF (CLASS.EQ.3) THEN                     PRINT *, 'Proceed to Room CE1-4'
        PRINT *, 'Proceed to Room ME1-6'        CASE (3)
      ELSE IF (CLASS.EQ.4) THEN                     PRINT *, 'Proceed to Room ME1-6'
        PRINT *, 'Proceed to Room CR2-6'        CASE (4)
      ELSE                                          PRINT *, 'Proceed to Room CR2-6'
        PRINT *, 'Proceed to Room CH2-8'        CASE DEFAULT
      END IF                                        PRINT *, 'Proceed to Room CH2-8'
C                                              END SELECT
      END                                 C
                                                END
```

(a)   *(b)*

FIGURE 14.10
A CASE construct
programmed with *(a)*
IF/THEN/ELSEIF and *(b)* the
FORTRAN 90 SELECT CASE.

```
SELECT CASE (TestExpression)
    CASE (List₁)
        Block₁
    CASE (List₂)
        Block₂
            .
            .
            .
    CASE (Listₙ)
        Blockₙ
    CASE DEFAULT
        ElseBlock
END Select
```

where *TestExpression* = any numeric or string expression, *Block$_i$* = the group of statements corresponding to the *i*th case, and *List$_i$* = the value(s) or range of the *TestExpression* corresponding to the *i*th case. The *ElseBlock* statements following the CASE DEFAULT would be carried out if none of the *Lists* matched the contents of the *TestExpression*.

Note that individual values are delimited by commas,

```
CASE (5, 8, -1)
```

and ranges of values by a colon,

```
CASE (15:20)
```

Also, combinations of ranges and individual values can be specified,

```
CASE (2, 8:15, 20)
```

As shown in Fig. 14.10*b*, the CASE construct looks very similar to the version using IF/THEN/ELSEIF statements in Fig. 14.10*a*. As programmed in Fig. 14.10, the two versions would execute in precisely the same manner. The CASE would be preferred because it is more descriptive of the underlying structure.

Although the CASE is preferred for situations where the branching depends on a single variable, the IF/THEN/ELSEIF must be employed when the branching depends on several variables. In addition, the CASE cannot use relational operators (such as .GT., .LT., and so on). For such situations, the IF/THEN/ELSEIF version is required.

PROBLEMS

14.1 The following code is designed to display a message to a student according to his or her final exam grade (FINAL) and total points accured in a course (POINTS).

```
IF (FINAL.LE.60.) PRINT *, 'Failing Grade'
IF (FINAL.GT.60.) PRINT *, 'Passing Grade'
IF (FINAL.GT.60. .AND. POINTS.LT.200.) PRINT *, 'Poor'
IF (FINAL.GT.60. .AND. POINTS.GE.200. .AND. POINTS.LT.400.) PRINT 'Good
IF (FINAL.GT.60. .AND. POINTS.GE.400.) PRINT *, 'Excellent'
```

Reformulate this code using block-structured IF/THEN/ELSE constructs. Test the program for the following cases:

```
FINAL = 50, POINTS = 250
FINAL = 70, POINTS = 321
FINAL = 90, POINTS = 561
```

14.2 As an industrial engineer you must develop a program to accept or reject a rod-shaped machine part according to the following criteria: its length must be not less than 7.75 cm or greater than 7.85 cm; its diameter must be not less than 0.335 cm or greater than 0.346 cm. In addition, under no circumstances may its mass exceed 5.6 g. Note that the mass is equal to the volume (cross-sectional area times length) multiplied by the rod's density (7.8 g/cm³). Write a structured computer program to input the length and diameter of a rod and display whether it is accepted or rejected. Test it for the following cases. If it is rejected, display the reason (or reasons) for rejection.

  (*a*) Length = 7.71, diameter = 0.338
  (*b*) Length = 7.81, diameter = 0.341

    (c) Length = 7.83, diameter = 0.343

    (d) Length = 7.86, diameter = 0.344

    (e) Length = 7.63, diameter = 0.351

14.3 An air conditioner both cools and removes humidity from a factory. The system is designed to turn on (1) between 7 A.M. and 6 P.M. if the temperature exceeds 75°F and the humidity exceeds 40 percent, or if the temperature exceeds 70°F and the humidity exceeds 80 percent; or (2) between 6 P.M. and 7 A.M. if the temperature exceeds 80°F and the humidity exceeds 80 percent, or if the temperature exceeds 85°F, regardless of the humidity.

    Write a structured computer program that inputs temperature, humidity, and time of day and displays a message specifying whether the air conditioner is on or off. Test the program with the following data:

    (a) Time = 7:40 P.M., temperature = 81°F, humidity = 68 percent.

    (b) Time = 1:30 P.M., temperature = 72°F, humidity = 79 percent.

    (c) Time = 8:30 A.M., temperature = 77°F, humidity = 50 percent.

    (d) Time = 2:45 A.M., temperature = 88°F, humidity = 28 percent.

14.4 Develop a structured program that uses the IF/THEN/ELSE construct to determine whether a number is positive, negative, or zero.

14.5 Develop a structured program that uses the CASE construct to determine whether a number is positive, negative, or zero.

14.6 The American Association of State Highway and Transportation Officials provides the following criteria for classifying soils in accordance with their suitability for use in highway subgrades and embankment construction:

Grain Size, mm	Classification
>75	Boulder
2–75	Gravel
0.05–2	Sand
0.002–0.05	Silt
<0.002	Clay

Develop a structured program that uses the CASE construct to classify a soil on the basis of grain size. Test the program with the following data:

Soil Sample	Grain Size, mm
1	2×10^{-4}
2	10
3	0.6
4	120
5	0.01

14.7   Develop a structured program that uses the CASE construct to assign a letter grade to students on the basis of the following scheme: A (90–100), B (80–90), C (70–80), D (60–70), and F (<60). For the situation where the student's grade falls on the boundary (for example, 80), have the program give the student the higher letter grade (for example, a B).

14.8   Write a structured FORTRAN program that reads values of X and outputs absolute values of X without using the ABS function.

14.9   Determine whether the following are true or false. Show all the steps in the evaluation of the condition in a manner similar to Example 14.2. (*Note:* $X = 10$, $Y = 20$, and $Z = 30$.)
   (*a*) (X .LE. 10 .OR. Y .EQ. 20)
   (*b*) (X .LE. 20 .OR. .NOT. Y .GE. 50)
   (*c*) (Y .EQ. 20 .AND. Z .EQ. 30) .AND. (X .GT. 10 .OR. Y .GT. 30)
   (*d*) (Y .EQ. 20 .OR. Z .EQ. 50) .AND. X .GE. 10 .AND. Y .LE. 20
   (*e*) (Y .EQ. 50 .AND. X .EQ. 10) .OR. (Z .LT. 40 .OR. Y .GT. 30)

14.10 Develop a program to determine the real and complex roots of the quadratic equation. Base your code on Fig. 14.7, but:
   (*a*) Modify the code so that it covers every possible numerical input. (*Note:* The version in Fig. 14.7 does not cover every possible case.)
   (*b*) Improve the output. For example, as programmed in Fig. 14.7, the output looks like

```
root1 = -.5857865 0 i
root2 = -3.414214 0 i
```

Reprogram the code so that the imaginary part is omitted when it is equal to zero. Similarly, for the case where there are complex roots, display the sign between the real and the imaginary components of the root. Use selection constructs to accomplish these enhancements.

14.11 As previously reviewed in Prob. 13.21, it is relatively straightforward to compute cartesian coordinates (x, y) on the basis of polar coordinates (r, θ). The reverse process is not so simple. The radius can be computed with the following formula:

$$r = \sqrt{x^2 + y^2}$$

If the coordinates lie within the first or fourth quadrants (that is, $x > 0$), then a simple formula can be used:

$$\theta = \tan^{-1}\left(\frac{y}{x}\right)$$

The difficulty arises for all other situations. The following table summarizes the possible cases that can occur:

x	y	θ
<0	>0	$\tan^{-1}\left(\dfrac{y}{x}\right) + \pi$
<0	<0	$\tan^{-1}\left(\dfrac{y}{x}\right) - \pi$
<0	=0	π
=0	>0	$\dfrac{\pi}{2}$
=0	<0	$-\dfrac{\pi}{2}$
=0	=0	0

Write a well-structured program to compute (r, θ) given (x, y). Express the final result for θ in degrees. Test the program for the following combinations of x and y:

x	y
0	1
−1	0
−1	1
0	−1
−1	−1
0	0

14.12 Develop a well-structured program where the user can enter a number and one of the following units of time: second, minute, hour, and day. Design the program so that it will then display the time in the remaining units.

14.13 Develop a well-structured program where the user can input the number of seconds past midnight (0 through 86,400 s). Design the program so that it will then display the time in hours, minutes, and seconds. Use the 12-h format for displaying the result (that is, midnight is 12:00:00 P.M. and noon is 12:00:00 A.M.). If the user inputs a negative time or a time that is too large (>86,400 s), display an appropriate error message.

14.14 Develop a well-structured program that converts Julian day (1, 2, . . . , 365) into calendar day (Jan. 1, Jan. 2, . . . , Dec. 31) for a nonleap year. If the user inputs a negative day or a day that is too large (>365), display an appropriate error message.

14.15 Write the code from Example 14.1 in a well-structured format.

14.16 The following CASE statement can be written in FORTRAN 90:

```
SELECT CASE (I)
  CASE (1)
    PRINT *, 'The first =', I
  CASE (2, 3, 9)
    PRINT *, 'The second, third and ninth =', I
  CASE (4 : 8)
    PRINT *, 'The fourth through eighth =', I
  CASE DEFAULT
    PRINT *, 'All others', I
END SELECT
```

Rewrite this code using IF/THEN/ELSEIF statements.

14.17 Develop a structured program based on the flowchart shown in Fig. P14.17.

FIGURE P14.17

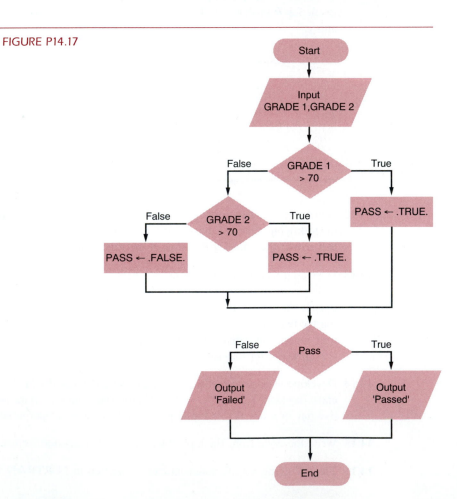

Note that the variable PASS is of the LOGICAL type. Test the program for the following cases:

GRADE1	GRADE2
100	100
100	50
50	100
50	50

14.18 The nested IF/THEN/ELSE structures at the top of Fig. P14.17 can be simplified. Do this by reprogramming them as a single IF/THEN/ELSE structure. Redraw the flowchart to represent your new algorithm.

REPETITION

Computers excel at performing tasks that are repetitive, boring, and time-consuming. Even very simple programs can accomplish huge tasks by executing the same chore over and over again. One of the earliest mechanical calculations was to tally and categorize the U.S. population for the 1890 census, a very repetitive task. Even today, one of the primary applications of computers is to maintain, count, and organize many pieces of information into smaller, more meaningful collections.

Nowhere is the repetitive power of the computer more valuable than in engineering. A large variety of our applications hinge on the ability of computers to implement portions of programs repetitively. This is true in many areas of our discipline but is especially relevant to computationally intensive or *"number-crunching"* calculations. Before discussing the actual statements and constructs used for repetition, we will present a simple example to demonstrate the major concept underlying repetition—the loop.

15.1   GO TO, OR INFINITE, LOOPS

Suppose that you want to execute a statement or a group of statements many times. One simple way of accomplishing this is to write the set of statements over and over again. For example, a repetitive version of the simple addition program is

```
READ *, A, B
PRINT *, A + B
READ *, A, B
PRINT *, A + B
READ *, A, B
PRINT *, A + B
        .
        .
        .
```

Such a sequential approach is obviously inefficient. A much more concise alternative can be developed using a GO TO statement, as in

```
10 READ *, A, B
   PRINT *, A + B
   GO TO 10
```

As depicted in Fig. 15.1, the use of the GO TO statement directs the program to automatically circle or "loop" back and repeat the READ and PRINT statements. Such repetitive execution of a statement or a set of statements is called a *loop.* Because one of the primary strengths of computers is their ability to perform large numbers of repetitive calculations, loops are among the most important operations in programming.

A flaw of Fig. 15.1, called a *GO TO loop,* is that it is a *closed,* or *infinite, loop.* Once it starts, it never stops. Consequently, provisions must be made so that the loop is exited after the repetitive computation is performed satisfactorily.

Two different approaches are employed to terminate loops. *Decision loops* are terminated on the basis of the state of a logical expression (in other words, an expression that is either true or false). In this sense they are related to the selection constructs discussed in the previous chapter. Because they are based on a decision, they may repeat a different number of times on every occasion that they are executed. In contrast, *counter loops* can be preset to repeat a fixed number of times.

15.2 DECISION LOOPS

There are two fundamental types of decision loop constructs—the DO WHILE and the DO UNTIL structures. As depicted in Fig. 15.2*a,* the DO WHILE is used to create a

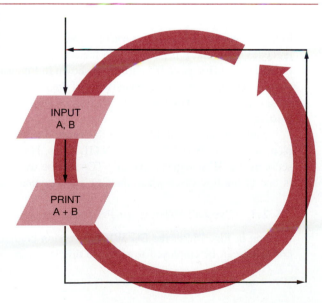

FIGURE 15.1
A GO TO loop. Once it starts, it never stops.

INPUT
A, B

PRINT
A + B

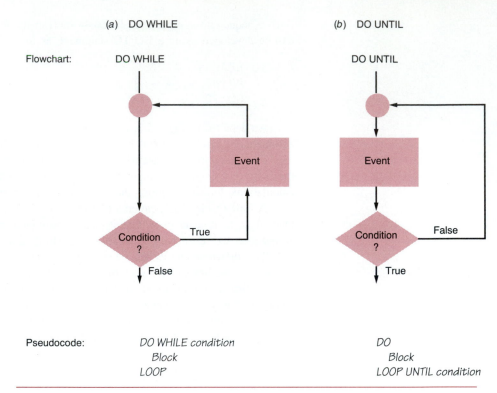

FIGURE 15.2
Comparison of the DO WHILE and the DO UNTIL loop structures.

loop that repeats as long as, or *while,* a logical condition is true. Because the condition precedes the event, it is possible that the action will not be implemented if the condition is false on the first pass. In contrast, as shown in Fig. 15.2*b,* the DO UNTIL is designed to have the event execute repeatedly *until* the condition is true. Then control is passed outside the loop. A key feature of this type of loop is that the event is always executed at least once.

Unfortunately, FORTRAN 77 has no structures that are expressly designed to implement the DO WHILE or DO UNTIL loops. However, as described in the following sections, the IF statements and GO TOs can be used to simulate these structures. This is one of the few cases where the GO TO statement can be applied positively.

15.2.1   The DO WHILE (or PreTest) Loop

As seen in Fig. 15.2*a,* the DO WHILE version begins with a decision. The IF/THEN construct can be employed to simulate this structure in FORTRAN 77:

```
In CONTINUE
      IF LogicalExpression THEN

         StatementBlock

      GO TO In
END IF
```

The loop will execute only if the *LogicalExpression* is true. When it becomes false, control transfers to the next statement following the END IF. Because the test is at the beginning of the loop, this construct is also called a *PreTest loop*.

In addition, a GO TO is used to loop back to the head of the construct. Note that the body of the loop is indented to make the structure more apparent. This indentation scheme is intentionally different than the one used for the IF/THEN/ENDIF selection structure (contrast with Fig. 14.4). We chose this style to make it apparent that a loop is being represented.

As with all decision loops, something must happen within the loop to make the *LogicalExpression* false. Otherwise, the loop would never terminate. This important point is illustrated in the following example.

EXAMPLE 15.1   Using a DO WHILE Loop to Input, Count, and Total Values

PROBLEM STATEMENT: A common application of repetition is to enter data from the keyboard. There are occasions when the user will have prior knowledge of the number of values that are to be input to a program. However, there are just as many instances when the user will either not know or not desire to count the data beforehand. This is typically the case when large quantities of information are to be entered. In these situations, it is preferable to have the program count the data.

Employ the DO WHILE loop to input, count, and total a set of values. Use the results to compute the average.

SOLUTION: The important issue that needs to be resolved in order to develop this program is how to signal when the last item is entered. A common way to do this is to have the user enter an otherwise impossible value to signal that data entry is complete. Such impossible data values are sometimes called *sentinel,* or *signal, values.* An IF/THEN statement can then be used to test whether the sentinel value has been entered and whether to transfer control out of the loop. The sentinel value can be a negative number if all data values are known to be positive. A large number, say 999, might be employed if all valid data were small in magnitude. In the following code the number −999 is used as the sentinel value:

```
PROGRAM EX151
REAL TOTAL, AVERAG, VALUE
INTEGER COUNT

COUNT = 0
TOTAL = 0.

PRINT *, 'Value = (-999 to terminate)'
READ *, VALUE

10 CONTINUE
   IF (VALUE .NE. -999.) THEN
   COUNT = COUNT + 1
   TOTAL = TOTAL + VALUE
   PRINT *, 'Value = (-999 to terminate)'
   READ *, VALUE
   GO TO 10
END IF

IF (COUNT .GT. 0) THEN
   AVERAG = TOTAL/COUNT
   PRINT *
   PRINT *, 'number of values = ', COUNT
   PRINT *, 'average = ', AVERAG
ELSE
   PRINT *
   PRINT *, 'no data entered; average not computed'
END IF
END
```

Output:

```
Value = (-999 to terminate)?
20
Value = (-999 to terminate)?
100
Value = (-999 to terminate)?
60
Value = (-999 to terminate)?
40
Value = (-999 to terminate)?
80
Value = (-999 to terminate)?
-999

number of values = 5
average = 50
```

Before the loop is implemented, `COUNT` and `TOTAL` are set to zero. This is called *initialization*. On each pass through the loop, the values are summed by accumulating them in the variable `TOTAL`. On the initial pass, the first value is added to the `TOTAL` on the right side of the assignment operator (=) and the result assigned to the `TOTAL` on the left. Therefore, after the first pass `TOTAL` will contain the first value. On the next pass the second value is added, and `TOTAL` will then contain the sum of the first and second values. As the iterations continue, additional values are input and added to `TOTAL`. Thus after the loop is terminated, `TOTAL` will contain the sum of all the values. The variable `COUNT` operates in a similar fashion but accumulates an additional value of 1 on each pass. Finally, both `TOTAL` and `COUNT` are used to compute the average in the IF/THEN/ELSE block.

The first READ statement, which is called a *priming input,* is required so that the user has the option of entering no values at all. If the first entry were −999, then the body of the loop would not be implemented and `COUNT` would remain at zero. In the same spirit, the computation of the average value is conditional in order to avoid division by zero for this case.

Also observe how the input of data is the last statement in the loop. This ordering prevents the sentinel value from being added to `TOTAL` or tallied in `COUNT`.

Note that FORTRAN 90 (as well as other modern FORTRAN implementations such as Microsoft FORTRAN) has an alternative form to implement this control structure without using GO TOs. The general structure is

```
DO WHILE LogicalExpression

    StatementBlock

END DO
```

This construct operates in a fashion identical with the IF/THEN version and is interchangeable with it. For example, the loop from Example 15.1 could be reprogrammed as

```
DO WHILE (VALUE .NE. -999.)
   COUNT = COUNT + 1
   TOTAL = TOTAL + VALUE
   PRINT *, 'Value = (-999 to terminate)'
   READ *, VALUE
END DO
```

Figure 15.3, which presents a side-by-side comparison of the two alternatives, suggests that the FORTRAN 90 version is superior. This is because (*a*) it is more concise and (*b*) its first line clearly identifies the structure as a DO WHILE loop. However, the FORTRAN 77 version is certainly adequate.

Although the DO WHILE is superior to the IF/THEN, the former should be used

FIGURE 15.3
Side-by-side comparison of
the (a) FORTRAN 77 and
(b) FORTRAN 90 versions of
the DO WHILE loop
structures.

```
10 CONTINUE                          DO WHILE (VALUE.NE.-999.)
   IF (VALUE.NE.-999.) THEN            COUNT = COUNT + 1
   COUNT = COUNT + 1                   TOTAL = TOTAL + VALUE
   TOTAL = TOTAL + VALUE               PRINT *, 'Value = '
   PRINT *, 'Value = '                 READ *, VALUE
   READ *, VALUE                    END DO
   GO TO 10
   END IF
```

(a) *Fortran 77*   (b) *Fortran 90*

with some discretion: because it is not part of the FORTRAN 77 standard, including the DO WHILE might make your programs less portable. In other words, programs containing it will not be compatible with FORTRAN compilers that do not support this feature.

15.2.2 The DO UNTIL (or PostTest) Loop

As seen in Fig. 15.2*b,* the DO UNTIL loop ends with a decision. A one-line logical IF statement can be employed to simulate this structure in FORTRAN 77,

ln CONTINUE

   StatementBlock

   IF (.NOT. *LogicalExpression*) GO TO *ln*

where the *LogicalExpression* is the condition that must be true to terminate the loop.

In contrast to the DO WHILE construct, this loop will cycle at least one time. This is a consequence of the test residing at the end of the loop. Hence, this construct is sometimes called the *PostTest loop.*[1]

Observe that the .NOT. is included so that the structure behaves in the same fashion as the flowchart and pseudocode representations from Fig. 15.2*b.* Consequently, it will continue to execute only if the *LogicalExpression* is false. When it becomes true, control transfers to the next statement following the loop. Thus, the reason for using the word "UNTIL" to describe the loop should be clear.

EXAMPLE 15.2   *Employing a DO UNTIL Loop to Check the Validity of Data*

PROBLEM STATEMENT: One situation where a DO UNTIL loop would be of use is to ensure that a valid entry is made from the keyboard. For example, suppose that you developed a program that required nonzero data. Such would be the case where the data might serve as the divisor in a formula, and hence division by zero had to be avoided. It would be advisable to have the program check for zero values upon input. Employ the DO UNTIL loop to accomplish this objective.

---

[1] Note that it is called the *repeat until* structure in Pascal.

SOLUTION: The following program fragment inputs a value and checks to ensure that it is greater than zero:

```
10 CONTINUE
    PRINT *, 'Value = '
    READ *, VALUE
  IF (.NOT. VALUE .GT. 0.) GO TO 10
```

Clearly, the user would not be able to proceed beyond this set of statements unless a positive VALUE was entered.

It should be pointed out that a DO WHILE statement could have been employed for the same purpose by using a priming input to make it execute at least once and by changing the *LogicalExpression,* as in

```
    VALUE = -1.
10  IF (VALUE .LE. 0.) THEN
      PRINT *, 'Value = '
      READ *, VALUE
      GO TO 10
    END IF
```

Notice that more lines are required for this version because of the need for the priming assignment statement and the END IF. From the perspectives of structured programming usage and economy of effort, the DO UNTIL would be preferable.

Also, recognize that the *LogicalExpression* in this version is actually equivalent to the one used previously (recall the presentation of logical complements in Table 14.3). It was formulated in this way to make it more consistent with the fact that the DO WHILE terminates on a false, whereas the DO UNTIL terminates on a true.

In contrast to the DO WHILE, there is no special construct for the DO UNTIL in modern FORTRANs (including FORTRAN 90). However, the FORTRAN 90 DO WHILE statement can be employed to simulate the DO UNTIL structure. For the case described in Example 15.2, this could be done, as in

```
VALUE = -1
DO WHILE (VALUE .LE. 0.)
  PRINT *, 'Value = '
  READ *, VALUE
END DO
```

Notice how a priming input must be used to force the loop to cycle at least once. In addition, the sense of the *LogicalExpression* has been modified so that a false result terminates the loop.

Although the DO WHILE version certainly gets the job done in a "GO TO–less"

FIGURE 15.4		

FIGURE 15.4
Side-by-side comparison of
the (a) FORTRAN 77 and
(b) FORTRAN 90 versions of
the DO UNTIL loop
structures.

```
10 CONTINUE
     PRINT *, 'Value = '
     READ *, VALUE
   IF (.NOT.VALUE .GT. 0.) GO TO 10
```

(a) *Fortran 77*

```
VALUE = -1.
DO WHILE (VALUE .LE. 0.)
   PRINT *, 'Value = '
   READ *, VALUE
END DO
```

(b) *Fortran 90*

fashion, it has a serious deficiency. This can be seen clearly from the side-by-side comparison in Fig. 15.4. Besides being more verbose, the DO WHILE version is deficient in that its exit point is at the loop's top rather than at its bottom. Because this contradicts the underlying structure of the DO UNTIL (recall Fig. 15.2b), the IF/THEN version is superior. In addition, the IF/THEN representation has the added advantage that it is compatible with a larger proportion of presently existing compilers.

Before moving on, we should point out that there is another way to formulate the DO UNTIL structure in FORTRAN 90 that is superior to both methods in Fig. 15.4. This design employs the break loop, which is described next.

15.2.3 The Break (or MidTest) Loop

Aside from ending a loop at its beginning or end, we can also terminate in the middle (Fig. 15.5). As with all the other decision loops, standard FORTRAN 77 does not include statements to directly represent this structure. However, as with the other examples, IF statements and GO TOs can be employed to simulate it, as in

FIGURE 15.5
The EXIT DO or break loop.

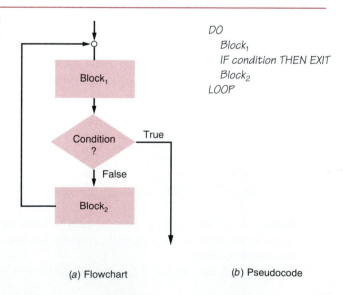

```
DO
    Block₁
    IF condition THEN EXIT
    Block₂
LOOP
```

(a) Flowchart

(b) Pseudocode

In CONTINUE

StatementBlock$_1$

IF (.NOT. *LogicalExpression*) THEN

StatementBlock$_2$

GO TO *In*

END IF

Thus, this construct repeats until a *LogicalExpression* located in the body of the loop tests true. At this point, control transfers to the statement following the loop. For this reason, it is sometimes referred to as a *break loop* because it involves "breaking out" in the middle of a loop. It is also called a *MidTest loop* because of the position of the logical test in the body of the loop rather than at either end.

EXAMPLE 15.3   Using a Break Loop to Check the Validity of Data

PROBLEM STATEMENT: One problem with the program fragment in Example 15.2 is that no guidance was given to the user regarding what she or he had done wrong. This could be remedied by including additional information in the body of the loop, as in

```
10 CONTINUE
      PRINT *, 'Value = (must be positive)'
      READ *, VALUE
   IF (.NOT. VALUE .GT. 0.) GO TO 10
```

One problem with this approach is that the guidance is embedded in the query. In some cases, you might want to take a different approach and offer feedback only when the user errs. This mechanism is referred to as an *error trap* or *diagnostic*. Employ a break loop to design such a program fragment.

SOLUTION: The break loop can be employed to separate the input action from the error message, as in

```
10 CONTINUE
      PRINT *, 'Value = '
      READ *, VALUE
      IF (.NOT. VALUE .GT. 0.) THEN
      PRINT *, 'Value must be positive, try again'
      GO TO 10
      END IF
```

The .NOT. is again employed so that the exit occurs on an outcome (in this case, true) that conforms to the underlying structure (Fig. 15.5).

```
10 CONTINUE                               DO
   PRINT *, 'Value = '                       PRINT *, 'Value = '
   READ *, VALUE    .                        READ *, VALUE
   IF (.NOT.VALUE .GT. 0.) THEN              IF (VALUE .GT. 0.) EXIT
   PRINT *, 'Value must be pos.'             PRINT *, 'Value must be pos.'
   GO TO 10                               END DO
   END IF
```

(*a*) *Fortran 77*   (*b*) *Fortran 90*

Note that FORTRAN 90 offers an alternative form to implement this construct without using GO TOs.[2] The general structure is

```
DO
    StatementBlock₁
    IF (LogicalExpression) EXIT
    StatementBlock₂
END DO
```

This code is perfectly consistent with Fig. 15.5. Using it, the loop from Example 15.3 could be reprogrammed as

```
DO
    PRINT *, 'Value = '
    READ *, VALUE
    IF (VALUE .GT. 0.) EXIT
    PRINT *, 'Value must be positive, try again'
END DO
```

For this case, the FORTRAN 90 version is vastly superior to the FORTRAN 77 alternative. This can be seen clearly from the side-by-side comparison in Fig. 15.6. Besides being slightly more verbose, the IF/THEN version (Fig. 15.6*a*) is deficient in that its exit point is at the loop's bottom rather than in its body. Because this contradicts the underlying structure of the break loop (Fig. 15.5), the FORTRAN 90 alternative is preferable.

It should be noted that, aside from being unavailable in standard FORTRAN 77, the break loop is not an established control construct in strongly structured high-level languages such as Pascal. Additionally, an alternative version of the break loop can be programmed with the single-line logical IF statement previously discussed in Section 14.2. Unfortunately, careless use of this statement can allow you to jump out of a loop to a distant line in your program. This results in spaghetti code.

Despite these caveats, when utilized properly, the break loop is actually well structured. As implemented in Fig. 15.5, it exits at a single point and transfers to the

---

[2]At the time of this book's publication, Microsoft FORTRAN 5.1 did not have this capability.

statement immediately following the loop. Therefore, although it is not one of the fundamental control structures, its design is consistent with the spirit of structured programming. As a consequence, the midtest loop is now an accepted construct in modern FORTRAN. In fact, as we shall discuss later, it is actually the only type of logical loop structure that is recommended by designers of the FORTRAN 90 standard.

Formulating the DO UNTIL structure with a break loop. Before moving on, we should illustrate how the break loop can be used to formulate the DO UNTIL structure. For example, the code in Fig. 15.4 could be reprogrammed as

```
DO
  PRINT *, 'Value = '
  READ *, VALUE
  IF (VALUE .GT. 0.) EXIT
END DO
```

As with Fig. 15.4*a,* this version has its exit point at the bottom of the structure. In addition, its exit test is consistent with the sense of the underlying structure (Fig. 15.2*b*). Consequently, the awkward use of the .NOT. is unnecessary. Thus, even though it is one statement longer, we feel that it is an advance over Fig. 15.4*a*. However, as with all the other FORTRAN 90 extensions, it should not be employed if portability is a major concern. For those cases, the IF/THEN version provides an acceptable alternative.

15.2.4 Logical Loops and "What If?" Computations

Most early mainframe computers (prior to the 1970s) operated as *batch-processing systems*. This mode allowed little direct interaction between the user and the machine. Often there were significant delays between the time a batch job was submitted and the time the output was received. Consequently, these batch programs were usually written in computer languages that were compatible with this type of environment.

Today, most computing applications take place in an *interactive environment*. The user typically sits in front of a display monitor and communicates directly with the computer via the keyboard or a mouse. Because response from the computer is direct, this interaction amounts to a dialogue between the user and the machine. Contemporary programs and computer languages must adapt to this environment.

As illustrated in Example 15.3, one facet of the user-machine interaction is providing the user with feedback regarding mistakes. Another example involves the case where a user wants to repeatedly implement a computation and view the result. These are commonly referred to as "*what if?*" or interactive computations. The following example is designed to illustrate some features of decision loops and other aspects of FORTRAN that can expedite such computations.

EXAMPLE 15.4   *Designing an Interactive Version of the Simple Addition Program*

PROBLEM STATEMENT: In Chap. 12, we presented the simple addition program as an example of a very elementary FORTRAN program. Develop a more sophisticated version employing decision loops and other facets of FORTRAN 77, such as selection constructs. Design the program so that it is interactive. In particular, provide the user with feedback and permit him or her to repeat the program as many times as desired.

SOLUTION: Before designing the program, we will first explore how loops can be employed to allow the user to perform multiple runs. Because most programs would usually perform the action once before questioning the user, this is clearly a case where the DO UNTIL loop would be used. One way to set up the algorithm would be

```
     CHARACTER Q
 10 CONTINUE
          .
          .
          .
      action
          .
          .
          .
     PRINT *, 'Do you want to repeat the action (''y'' or ''n'')'
     READ *, Q
   IF (.NOT. (Q .EQ. 'N')) GO TO 10
```

Notice how we instruct the user to enter the character response enclosed in apostrophes. For example,

```
PRINT *, 'Do you want to repeat the action (''Y'' or ''N'')'
```

We use double apostrophes to designate the apostrophes that are a part of the query message. This is the way that FORTRAN distinguishes between apostrophes within a string and those used to enclose the string. Upon output, only single apostrophes are displayed, as in

```
Do you want to repeat the action ('Y' or 'N')
```

The loop will cycle until the user enters an 'N' to signify a negative response to the query. However, a problem with this scheme is that the user might enter a lowercase 'n'. If this is done, the loop test will not work as expected because to the computer 'N' $\neq$ 'n'. Therefore, the user might want to terminate the action but be forced to perform it again because of the computer's distinction between uppercase and lowercase.
   An alternative involves reformulating the UNTIL statement as

```
IF (.NOT. ( Q.EQ.'N' .OR. Q.EQ.'n')) GO TO 10
```

For this version, if the user enters either 'n' or 'N', the test is satisfied.

However, a problem still exists. What happens if the user wants to type an 'N' but mistakenly strikes one of the adjacent keys (for example an 'M' or a 'B')? Close inspection of the loop indicates that because the struck key was not an 'N', the loop will cycle again as though the user had typed a 'Y'! Clearly, this is unacceptable.

A way to avoid this is to ensure that the user enters only an 'N' or a 'Y'. This can be accomplished with a break loop, as previously described in Example 15.3. One way to do this is shown here:

```
      CHARACTER Q
   10 CONTINUE
          .
          .
          .
          action
          .
          .
          .
   20     CONTINUE
            PRINT *, 'Do you want to repeat the action (''y'' or ''n'')'
            READ *, Q
            IF (.NOT.(Q.EQ.'N'.OR.Q.EQ.'n'.OR.Q.EQ.'Y'.OR.Q.EQ.'y')) THEN
            PRINT *, 'You did not answer a ''y'' OR ''n''. Try again.'
            GO TO 20
          END IF
      IF (.NOT. (Q.EQ.'N' .OR. Q.EQ.'n')) GO TO 10
```

Now we can use this structure to refine the simple addition program. Code to solve this problem is listed in Fig. 15.7, and a sample session is depicted in Fig. 15.8. Observe how an IF/THEN/ELSE structure is employed to give the user feedback regarding the correct answer for the summation. In particular, if the user enters the incorrect answer, we provide a tip to help him or her on the next try.

Also, observe how nested logical loops are employed to query the user to determine whether another run is desired. As described above, the inner break loop ensures that the user has provided either a 'Y' or 'N' signifying yes and no, respectively, in answer to the query. Then, if the response is 'Y', the outer DO UNTIL loop repeats. If it is 'N', the outer loop terminates.

In addition, see that we have used a LOGICAL variable, YESNO, to hold the outcome of the evaluation of whether a 'Y' or an 'N' is entered. We did this because of the length of the logical expression. Although we could have employed a character in column 6 to continue the statement on a new line, we felt that using the LOGICAL variable makes the action easier to follow.

```
                    PROGRAM EX154
         C          Simple Addition Program

         C          by S.C. Chapra
         C          August 17, 1992

         C          This program queries the user for two
         C          numbers, asks for the summation, and
         C          then provides feedback on the validity
         C          of the user's response

         C          Definition of variables

         C          FIRST, SECOND = variables to be added
         C          TOTAL = sum of the variables
         C          ANSWER = user's response for the sum
         C          YESNO = test variable for yes/no response
         C          Q = query variable

                    REAL FIRST, SECOND, TOTAL, ANSWER
                    CHARACTER Q
                    LOGICAL YESNO

         10 CONTINUE
                    PRINT *, 'First number = '
                    READ *, FIRST
                    PRINT *, 'Second number = '
                    READ *, SECOND
                    TOTAL = FIRST + SECOND
                    PRINT *
                    PRINT *, 'Please type in the answer = '
                    READ *, ANSWER
                    PRINT *

                    IF (ANSWER .EQ. TOTAL) THEN
                      PRINT *, 'Good job!'
                    ELSE
                      PRINT *, 'Sorry, the correct answer is ', TOTAL
                      PRINT *
                      PRINT *, 'Remember, if you get stuck'
                      PRINT *, 'use those old fingers and toes'
                    END IF

                    PRINT *

         20    CONTINUE
                    PRINT *, 'another run (''Y'' or ''N'')?'
                    READ *, Q
                    YESNO=Q.EQ.'N'.OR.Q.EQ.'n'.OR.Q.EQ.'Y'.OR.Q.EQ.'y'
                    IF (.NOT. YESNO) THEN
                    PRINT *
                    PRINT *, 'You did not enter a ''Y'' or a ''N''. Try again'
                    PRINT *
                    GO TO 20
                    END IF

                    PRINT *
              IF (.NOT. (Q.EQ.'N' .OR. Q.EQ.'n') ) GO TO 10

              END
```

FIGURE 15.7
An interactive version of the simple addition program. This version illustrates how selection and repetition constructs are employed to provide feedback and allow the user to perform multiple runs. See Example 15.4 for further explanation of the program.

```
First number =
1
Second number =
2

Please type in the answer =
4

Sorry, the correct answer is =        3.000000

Remember, if you get stuck
use those old fingers and toes

Another run ('Y' or 'N')
'y'

First number =
2
Second number =
3

Please type in the answer =
5

Good job!

Another run ('Y' or 'N')
'h'

You did not enter a 'Y' or an 'N'. Try again.'

Another run ('Y' or 'N')
'N'
```

FIGURE 15.8
Input/output from a session
using the program from
Fig. 15.7. Note: The
user-supplied values are
highlighted.

15.3   COUNTER (DO) LOOPS

Suppose that you wanted to perform a specified number of repetitions, or *iterations*, of a loop. One way, employing a DO WHILE construct, is depicted in the flowchart and program fragment in Fig. 15.9. The loop is designed to repeat 10 times. The variable I is a counter that keeps track of the number of iterations. If I is less than or equal to 10, an iteration is performed. On each pass, the value of I is displayed and I is incremented by 1. After the tenth iteration, I becomes 11; therefore, the *LogicalExpression* will fail and control is transferred out of the loop.

Although the DO WHILE loop is certainly a feasible option for performing a specified number of iterations, the operation is so common that a special set of statements is available in FORTRAN for accomplishing the same objective in a more efficient manner. Called the *DO* loop, its general format is (Fig. 15.10)

DO *ln index = start, finish, increment*

StatementBlock

ln   CONTINUE

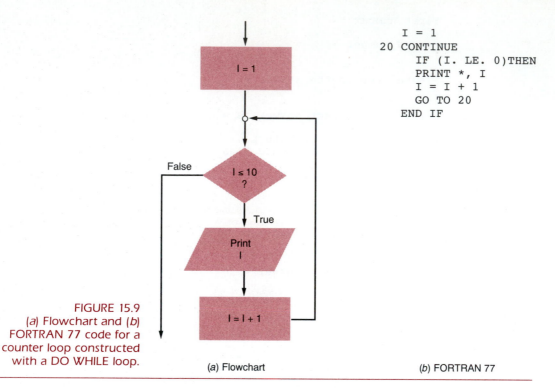

```
         I = 1
      20 CONTINUE
           IF (I. LE. 0)THEN
           PRINT *, I
           I = I + 1
           GO TO 20
         END IF
```

FIGURE 15.9
(a) Flowchart and (b)
FORTRAN 77 code for a
counter loop constructed
with a DO WHILE loop.

(a) Flowchart

(b) FORTRAN 77

The DO loop works as follows. The *index* is a variable that is set at an initial value (*start*). The program then executes the body of the loop, moves to the line number (*ln*), and then cycles back to the DO statement. Every time the *ln* statement is encountered, the *index* is automatically increased by the *increment*. Thus the *index* acts as a counter. Then, when the *index* is greater than the final value (*finish*), the computer automatically transfers control out of the loop to the line following the *ln* statement.[3]

An example of the DO loop is provided in Fig. 15.10c. Notice how this version is much more concise than the version in Fig. 15.9b. Also notice that the *increment* is omitted from the DO statement. If this term is not included, the computer automatically assumes a default value of 1 for the *increment*.

As shown in Fig. 15.10c, all FORTRAN statements within a DO loop should be indented two spaces from that of the DO statement. The CONTINUE statement should be indented the same number of spaces as its associated DO statement. Any number of FORTRAN statements can be included within the loop.

[3]In reality, it should be noted that the number of iterations is actually precomputed before the loop is implemented. The explanation presented above is useful pedagogically in communicating accurately how the FORTRAN DO loop behaves.

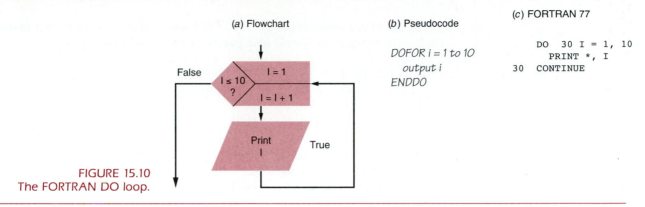

(a) Flowchart   (b) Pseudocode   (c) FORTRAN 77

FIGURE 15.10
The FORTRAN DO loop.

EXAMPLE 15.5   Using a DO Loop to Input and Total Values

PROBLEM STATEMENT: As in Example 15.1, a common application of repetition is to enter data from the keyboard. The DO loop is particularly well suited for cases where the user has foreknowledge of the number of values that are to be entered. Employ the DO loop to input a set number of values to a program and determine their sum.

SOLUTION: A program fragment to perform the above task is

```
      PROGRAM EX155
      REAL TOTAL, VALUE
      INTEGER I, N
C
      TOTAL = 0.
C
      PRINT *, 'Number of values = '
      READ *, N
C
      DO 10 I = 1, N
         PRINT *, 'Value = '
         READ *, VALUE
         TOTAL = TOTAL + VALUE
   10 CONTINUE
```

The number of data values is input and stored in the variable N. Thereafter, N serves as the *finish* parameter for the loop. Because the *increment* is not specified, a step size of 1 is assumed. Therefore the loop repeats N times while inputting one value per iteration.

It should be mentioned that the CONTINUE statement is not required. For example, the loop in Example 15.5 could have just as well been programmed as

```
    DO 10 I = 1, N
        PRINT *, 'Value = '
        READ *, VALUE
10 TOTAL = TOTAL + VALUE
```

However, for reasons of style, we prefer to use the CONTINUE statement as a nice means to clearly define the end of the loop.

The size of the *increment* can be changed from the default of 1 to any other value. The *increment* does not have to be an integer, nor does it even have to be positive. If a negative value is used, then the logic for stopping the loop is reversed. With a negative *increment,* the loop terminates when the value of the *index* is less than that of the *finish.* Step sizes such as −5, 10, −1, or 60 would all be possible.

EXAMPLE 15.6   **Examples of Using DO Loops**

PROBLEM STATEMENT: Illustrate the use of DO loops with (*a*) an increment greater than 1 and (*b*) a negative increment.

SOLUTION: (*a*) The following program cycles from 2 to 15 with an increment of 4:

```
    DO 30 J = 2, 15, 4
        PRINT *, J
30 CONTINUE
    PRINT *, 'After exiting loop, J = ', J
```

Output:

```
    2
    6
   10
   14
After exiting loop, J =    18
```

Notice that upon exit from the loop, J has been incremented by 4. If you do not recognize this, you might run into problems later on in your code. For example, you might misuse J in a computation.

(*b*) The following program cycles from 20 down to −3, using an increment of −6:

```
    DO 4 K = 20, -3, -6
        PRINT *, K
4 CONTINUE
    PRINT *, 'After exiting loop, K = ', K
```

Output:

```
        20
        14
         8
         2
 After exiting loop, K =     -4
```

Note that FORTRAN 90 (as well as other modern FORTRAN implementations such as Microsoft FORTRAN) has an alternative form to implement this control structure without using line numbers. The general structure is

DO *index = start, finish, increment*

    StatementBlock

END DO

This construct operates in a fashion identical with the line number version and is interchangeable with it. For example, the loop from Example 15.5 could be reprogrammed as

```
DO I = 1, N
  PRINT *, 'Value = '
  READ *, VALUE
  TOTAL = TOTAL + VALUE
END DO
```

FIGURE 15.11
Side-by-side comparison of the (a) FORTRAN 77 and (b) FORTRAN 90 versions of DO-loop structures. Note that by "hanging off" on the left side, the line number, 10, interferes slightly with the way indentation sets off the loop. In contrast, the FORTRAN 90 version is cleaner.

Because it does not involve line numbers, the FORTRAN 90 version is slightly superior. This is because the existence of the line numbers in columns 1 through 5 interferes visually with the loop indentation scheme. This is evident from the side-by-side comparison in Fig. 15.11. However, it is not a serious drawback, and the FORTRAN 77 version is certainly adequate.

Although the DO/END DO is superior to the FORTRAN 77 construct, the former should be used with some discretion: because it is not part of the FORTRAN 77 standard, including the DO/END DO might make your program less portable. In other words, programs containing it will not be compatible with FORTRAN compilers that do not support this feature.

```
DO 10 I = 1, N
   PRINT *, 'Value = '
   READ *, VALUE
   TOTAL = TOTAL + VALUE
10 CONTINUE
```
(a) Fortran 77

```
DO I = 1, N
   PRINT *, 'Value = '
   READ *, VALUE
   TOTAL = TOTAL + VALUE
END DO
```
(b) Fortran 90

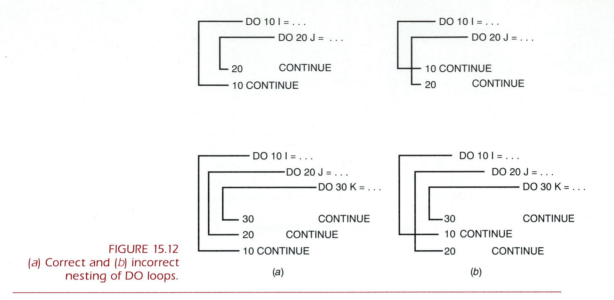

FIGURE 15.12
(a) Correct and (b) incorrect
nesting of DO loops.

15.4   NESTING CONTROL STRUCTURES

Repetition and selection structures may be enclosed completely within other repetition and selection structures. This arrangement, called *nesting,* allows algorithms of great power and complexity to be designed. In this section, we will present several examples of how this can be done.

15.4.1   Nesting Counter Loops

DO loops may be enclosed completely within other DO loops. This arrangement is valid only if it follows the rule:

- If a DO loop contains either the DO or the CONTINUE parts of another loop, it must contain both of them.

Figure 15.12 presents some correct and incorrect versions. Notice how indentation is used to clearly delineate the extent of the loops.

EXAMPLE 15.7   Example of Nested DO Loops

PROBLEM STATEMENT: Illustrate the nesting of loops with several nested DO loops.

SOLUTION: The following program illustrates both the nesting of DO loops and the associated indentation:

```
PROGRAM EX157
DO 10 I = 1, 2
   DO 20 J = 1, 3
      PRINT *, I * J
20    CONTINUE
      PRINT *
10 CONTINUE
   PRINT *, I, J
   END
```

Output:

```
1
2
3

2
4
6

3      4
```

For the first pass of the outer loop, I = 1 and the inner loop cycles though J = 1, 2, 3. Therefore, the product I * J is displayed as 1 2 3. On the second pass, I = 2 and the inner loop again cycles though J = 1, 2, 3. Therefore, the product is displayed as 2 4 6. Finally, upon exit for the loop, I and J are both incremented by 1 beyond their *finish* values to yield 3 and 4, respectively.

15.4.2 Nesting Decision and Counter Loops

It should be mentioned that besides counter loops, logical loops can also be nested. In fact, the two types of loops can be combined. Such a structure will be illustrated in the following example.

EXAMPLE 15.8   *Nested Decision and Counter Loops*

PROBLEM STATEMENT: Loops are commonly used to increment and display values in a computation. For example, suppose that you wanted to display a value for the beginning of a calculation (time = 0) and then display values for the following time increments. A DO loop can be employed to accomplish this, as in

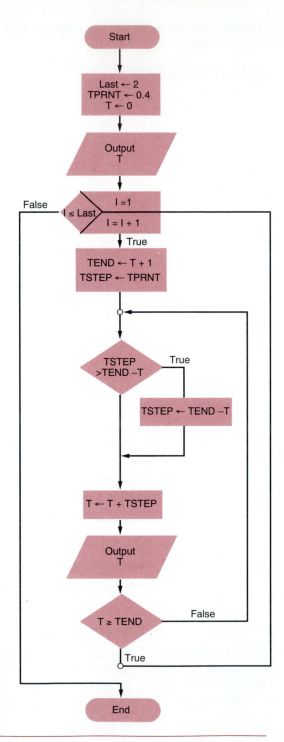

FIGURE 15.13
Nested DO and LOOP UNTIL
constructs to solve
Example 15.8.

```
              LAST = 2
              T = 0.
              PRINT *, 'time = ', T
              DO 30 I = 1, LAST
                 T = T + 1.
                 PRINT *, 'time = ', T
           30 CONTINUE
```

Output:

```
    time = 0.000000E+00
    time = 1.000000
    time = 2.000000
```

Now suppose that you want to display intermediate times within a unit increment. If you did this using increments that are evenly divisible into 1 (for example, 0.1, 0.2, 0.25), it would be relatively simple to modify the DO loop to accomplish this action.

However, suppose that you wanted to use a fractional increment that was not an even factor of 1, say 0.4. Design code to accomplish this objective.

SOLUTION: One way to do this involves nesting a decision loop within the DO loop as delineated in the flowchart in Fig. 15.13. The corresponding FORTRAN 77 code is

```
              LAST = 2
              TPRNT = 0.4
              T = 0.
              PRINT *, 'time = ', T
              DO 30 I = 1, LAST
                 TEND = T + 1.
                 TSTEP = TPRNT
           10    CONTINUE
                    IF (TSTEP .GT. TEND - T) TSTEP = TEND - T
                    T = T + TSTEP
                    PRINT *, 'time = ', T
                 IF (.NOT. T .GE. TEND - T) GO TO 10
           30 CONTINUE
              END
```

Output:

```
    time = 0.000000E+00
    time = 4.000000E-01
    time = 8.000000E-01
    time =    1.000000
    time =    1.400000
    time =    1.800000
    time =    2.000000
```

Notice how the statement

```
IF (TSTEP .GT. TEND - T) TSTEP = TEND - T
```

is used to "shorten up" the increment TSTEP when using TSTEP = TPRNT would overshoot TEND.

15.4.3 Nesting Repetition and Selection Structures

Repetition and selection constructs can also be nested. As with loops, each internal structure is indented. As a rule of thumb, we usually indent two spaces. The following example provides a simple illustration of how this is done.

EXAMPLE 15.9 Design of an Interactive Menu

PROBLEM STATEMENT: Everyone is familiar with ordering from a restaurant menu. Similarly, we might want the user to choose among a variety of options in a computer program. Use selection and repetition structures to design a menu for a simple addition/subtraction program.

SOLUTION: One way to do this is outlined in Figs. 15.14 and 15.15. After entering two numbers, the program moves into a group of nested structures.

The user is first presented with a menu consisting of three choices which are selected by entering a number: (1) add, (2) subtract, or (3) terminate. The menu is enclosed within a DO UNTIL loop that is designed to ensure that the user makes one of these choices before proceeding.

After making one of the three choices, the flowchart moves on to a CASE structure that either adds or subtracts and displays the result. The outer DO UNTIL loop then tests whether the user had made menu selection 3 (terminate). If so, the flow drops out of the loops and the program ends. Otherwise, the loop cycles back and reprints the menu so that the user can make another selection.

As illustrated by the previous example, nested repetition and selection constructs can be employed to develop a user-friendly interactive program. Although the present example was simple, such nesting provides a powerful capability that can become logically complex. Close adherence to structured programming conventions such as indentation, as manifested by Figs. 15.14 and 15.15, is essential to keeping such algorithms lucid and maintainable.

15.5 REPETITION IN FORTRAN 90

In most of the other chapters in this part of the book, we have included a short section (such as the one you are presently reading) to summarize the new developments in FOR-

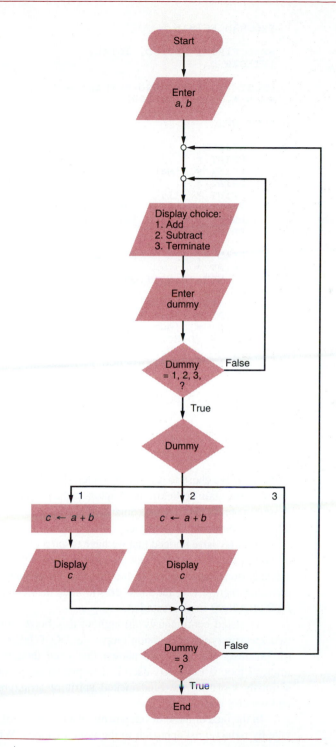

FIGURE 15.14
A flowchart for an interactive addition/subtraction program. This code includes a menu that provides the user with a number of options.

```
         PROGRAM FG1515

         REAL FIRST, SECOND, TOTAL
         INTEGER D

         PRINT *, 'Enter two numbers = '
         READ *, FIRST, SECOND

   10 CONTINUE

   20    CONTINUE
            PRINT *
            PRINT *, 'Make a menu selection'
            PRINT *
            PRINT *, '1.    Add'
            PRINT *, '2.    Subtract'
            PRINT *, '3.    Terminate'
            PRINT *
            PRINT *, 'Enter a number'
            READ *, D
            PRINT *
         IF (.NOT.(D.EQ.1.OR.D.EQ.2.OR.D.EQ.3)) GO TO 20

         IF (D.EQ.1) THEN
            TOTAL = FIRST + SECOND
            PRINT *, 'sum =', TOTAL
         ELSE IF (D.EQ.2) THEN
            TOTAL = FIRST - SECOND
            PRINT *, 'difference =', TOTAL
         END IF

         IF (.NOT. (D.EQ.3) ) GO TO 10

         END
```

FIGURE 15.15
Computer code
corresponding to the
algorithm depicted in
Fig. 15.14.

TRAN 90. We chose to present the material in this fashion because of our philosophy of devoting the main bodies of the chapters to the most fundamental form of FORTRAN 77. As stated previously, this was done so that your programs are as compatible with as many different compilers as possible.

You may have noticed that we have taken a different approach in the present chapter by contrasting the old and new options as each construct was presented. We have consciously done this because of the fact that loops (and in particular decision loops) are presently manifesting the most drastic and, in a certain sense, radical changes in the FORTRAN 90 implementation.

As stated continuously throughout this book, computer scientists have identified two specific types of decision loops: the DO WHILE (or pretest) and the DO UNTIL (or posttest). We have also shown that even though it is not orthodox, a break (or midtest) loop is extremely useful; if implemented properly, it can be employed without seriously violating the fundamental spirit of structured programming—that is, clear and concise code.

In the face of these developments, the designers of FORTRAN 90 have taken a decidedly reductionist approach to the issue of loops. When they devised the break loop, they recognized that it could mimic the DO WHILE and DO UNTIL structures by merely positioning the test at the very beginning or the very end of the loop. Therefore, although FORTRAN 90 does include a formal DO WHILE structure, some of its designers (for example, Brainard, Goldberg, and Adams, 1990) already deem the DO WHILE obsolete.

For example, instead of the DO WHILE code from Section 15.2.1

```
DO WHILE (VALUE.NE.-999.)
   COUNT = COUNT + 1
   TOTAL = TOTAL + VALUE
   PRINT *, 'Value = '
   READ *, VALUE
END DO
```

the FORTRAN 90 EXIT statement could be used to formulate

```
DO
   IF (VALUE.EQ.-999.) EXIT
   COUNT = COUNT + 1
   TOTAL = TOTAL + VALUE
   PRINT *, 'Value = '
   READ *, VALUE
END DO
```

Similarly, rather than the DO UNTIL example from Sec. 15.2.2

```
10 CONTINUE
     PRINT *, 'Value = '
     READ *, VALUE
   IF (.NOT. VALUE .GT. 0.) GO TO 10
```

the FORTRAN 90 EXIT statement could be employed to develop

```
DO
   PRINT *, 'Value = '
   READ *, VALUE
   IF (VALUE .GT. 0.) EXIT
END DO
```

The designers have justified their stance on the basis that employing the DO/END DO and the EXIT actually simplifies matters. They contend that there is only one fundamental type of loop, the DO/END DO. Then customized versions are possible either by prespecifying the number of repetitions (as in the counter formulation of the DO state-

QuickBASIC   Turbo Pascal

```
DO WHILE value <> -999       while (value <> -999) do
   count = count + 1         begin
   total = total + value       count := count + 1;
   PRINT "Value = "            total := total + value;
   INPUT value                 Write('Value = ');
LOOP                           Readln(value);
                             end;
```

(a) Do While

```
DO                           repeat
   PRINT "Value = "            Write('Value = ');
   INPUT value                 Readln(value);
LOOP UNTIL value > 0         until (value > 0);
```

(b) Do Until

FIGURE 15.16
Examples of the fundamental
repetition control structures
in QuickBASIC and
Turbo Pascal.

ment) or by using a logical EXIT statement to terminate from some position within the body of the loop.

We have ambivalent feelings regarding this point of view. On the one hand, we agree that the modification *does* simplify matters. For example, we no longer need to worry about nomenclature such as WHILE and UNTIL.

On the other hand, we believe that some clarity may have been discarded in the process. This can be seen most clearly by contrasting FORTRAN 90 loops with other widely used languages such as Pascal and modern BASIC. Both of these languages include specific constructs for the DO WHILE and DO UNTIL structures. For example, Fig. 15.16 shows versions programmed in QuickBASIC and Turbo Pascal.

Inspection of Fig. 15.16 leads us to conclude that although the FORTRAN 90 approach is certainly acceptable, a case can be made that the explicit inclusion of WHILE and UNTIL structures is slightly better. The argument can be made that the three types of decision loops—WHILE, UNTIL, AND EXIT—*do* represent three different types of actions. If this is true, distinguishing among them by nomenclature makes the resulting code more transparent.

Although the foregoing argument can be made, it does not represent a major reservation. Furthermore, FORTRAN 90 is now a fact of life. Because its loop structures lead to programs that are better structured than FORTRAN 77, it is a major step forward in the quest for more elegant and clear FORTRAN codes.

PROBLEMS

15.1 What will be displayed when the following loops are implemented?

(a)
```
      A = 2
   10 CONTINUE
      A = A * 3.
      PRINT *, 2. * A
      IF (.NOT. A .GT. 18.) GO TO 10
```

(b)
```
      X = 5.
      DO 10 X = 2, 4, 0.8
      PRINT *, X ** 2
   10 CONTINUE
      PRINT *, X
```

(c)
```
      M = 4
      N = -10
      DO 10 I = 1, M, 2
         DO 20 K = 9, N, -6
            PRINT *, I * K
   20    CONTINUE
         PRINT *
   10 CONTINUE
```

(d)
```
      DO 10 I = 1, 10, 4
         J = I
   20    CONTINUE
         J = J - 3
         PRINT *, I, J
         IF (.NOT. J .LE. 0) GO TO 20
         PRINT *
   10 CONTINUE
```

15.2 The exponential function can be represented by the following infinite series:

$$e^x = 1 + x + \frac{x^2}{2!} + \frac{x^3}{3!} + \frac{x^4}{4!} + \ldots$$

Write a program to implement this formula so that it computes the value of e^x as each term in the series is added. In other words, compute and display in sequence

$$e^x \cong 1$$

$$e^x \cong 1 + x$$

$$e^x \cong 1 + x + \frac{x^2}{2}$$

.

.

.

up to the order term of your choosing. For each of the above, compute the percent relative error as

$$\% \text{ error} = \frac{\text{true} - \text{series approximation}}{\text{true}} \times 100\%$$

Use the library function for e^x in your computer to determine the "true value."

Have the program display the series approximations and the error at each step along the way. Employ loops to perform the analysis. As a test case, use the program to evaluate $e^{1.5}$ up to the term $x^{10}/10!$.

15.3 Repeat Prob. 15.2, but make the order of the series expansion dependent on a user-supplied stopping criterion. In other words, have the user specify that the series should be expanded until the percent relative error falls below a prescribed level. Estimate the percent relative error with the following formula:

$$\% \text{ error} = \left| \frac{\text{present} - \text{previous}}{\text{present}} \right| \times 100\%$$

where the present estimate is the latest value of the series approximation and the previous estimate is the value calculated with one fewer term. Test your program for the same case as Prob. 15.2, using a stopping criterion of 0.1 percent. Have the program display the series approximations and the error at each step along the way.

15.4 Repeat Prob. 15.2 for the hyperbolic cosine which can be approximated by

$$\cosh x = 1 + \frac{x^2}{2!} + \frac{x^4}{4!} + \ldots$$

As a test case, employ the program to evaluate $\cosh 78°$ up to the term $x^{10}/10!$. Remember that both the cosh and the above formula use radians. Also, recall that the cosh can be computed with an intrinsic function, COSH. You can employ this function to check your program.

15.5 Repeat Prob. 15.4, but use the stopping criterion as specified in Prob. 15.3.

15.6 Repeat Prob. 15.2 for the sine, which can be approximated by

$$\sin x = x - \frac{x^3}{3!} + \frac{x^5}{5!} - \frac{x^7}{7!} + \ldots$$

As a test case, employ the program to evaluate $\sin 30°$. Remember, both the sine function and the above formula use radians.

15.7 Write a computer program to evaluate the following series:

$$sum = \sum_{i=0}^{n} \frac{1}{2^i} \qquad \text{for } n = 20$$

Display *sum* and *n* during each iteration. Does the series converge to a constant value or diverge to infinity?

15.8 Repeat Prob. 15.7, except with the following formula ($n = 20$):

$$sum = \sum_{i=1}^{n} \frac{1}{i^2}$$

15.9 Repeat Prob. 15.7, given ($n = 20$):

$$\text{sum} = \sum_{i=0}^{n} \left(\frac{2}{5}\right)^i$$

15.10 Consider the following formula:

$$\frac{1}{1 - x} = 1 + x + x^2 + \cdots$$

Write a computer program to test the validity of this formula for values of 0.6, -0.3, 2, and $-3/2$.

15.11 The series

$$f(x) = \frac{4}{\pi}\left(\sin x + \frac{\sin 3x}{3} + \frac{\sin 5x}{5} + \cdots\right)$$

is an approximation of $f(x)$ as shown in Fig. P15.11. Write a computer program to evaluate the accuracy of the series as a function of the number of terms at $x = \pi/2$. The program should display the error in the approximation and the number of terms in the series.

FIGURE P15.11

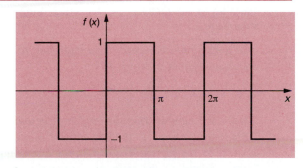

15.12 Economic formulas are employed extensively in engineering projects. For example, the following formula allows you to calculate the annual payment, A, needed for a loan:

$$A = \frac{i(1 + i)^n}{(1 + i)^n - 1}P$$

where P is the loan value, i is the annual interest rate, and n is the number of years for the loan.

Every year part of your payment is used to cover the interest. This amount is equal to the interest rate multiplied by the principal (that is, the amount left to pay at any particular time). The rest of your annual payment goes to reduce the principal.

Suppose that you are involved in the construction of a $450,000 storage tank for a gas processing facility. You can obtain a 10-year loan ($n = 10$) at an annual interest rate of 12 percent ($i = 0.12$). Develop a program to display a table showing the cumulative payment, interest, and principal (that is, the amount left to pay) for each year of the loan.

15.13 Develop a well-structured FORTRAN program to display a table of temperatures in degrees Celsius and Fahrenheit. Query the user for the lower bound, upper bound, and increment for the Celsius temperatures. Use error traps to ensure that (1) the lower bound is less than the upper bound and (2) the increment is less than the range from the lower to the upper bound. Test the program for the case where the lower and upper bounds are 0 and 100, respectively, and the increment is 15°C. Also set up tests to ensure that your error traps are working properly.

FIGURE P15.14
The golden ratio can be used to generate the rectangle shown below.

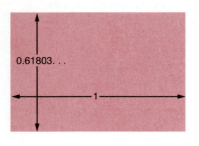

15.14 In many cultures, certain numbers are ascribed qualities. For example, we are all familiar with "lucky 7" and "Friday the 13th." Ancient Greeks called the following number the "golden ratio":

$$\frac{\sqrt{5} - 1}{2} = 0.61803\ldots$$

This ratio was employed for a number of purposes, including the development of the rectangle in Fig. P15.14. These proportions were considered aesthetically pleasing to the Greeks. Among other things, many of their temples followed this shape.

The golden ratio is related to an important mathematical sequence known as the Fibonacci numbers, which are

0, 1, 1, 2, 3, 5, 8, 13, 21, 34, . . .

Thus, each number after the first two represents the sum of the preceding two. This sequence pops up in many diverse areas of science and engineering. In the context of the present discussion, an interesting property of the Fibonacci sequence relates to the ratio of consecutive numbers in the sequence; that is, $0/1 = 0$, $1/1 = 1$, $1/2 = 0.5$, $2/3 = 0.667$, $3/5 = 0.6$, $5/8 = 0.625$, and so on. If the series is carrier out far enough, the ratio approaches the golden ratio! Develop a well-structured FORTRAN program to compute and display the Fibonacci sequence and the corresponding ratios of its terms. Allow the user to enter the number of terms in the sequence. Employ error traps along with a diagnostic message to prevent the user from entering numbers less than 1. Also, in the event that the user requests more than 10 terms, limit the display to 10 terms at a time and display subsequent pages at the user's discretion. Finally, design the program so that the entire computation can be repeated as many times as the user desires.

15.15 The following equation is a simple model of the velocity of a falling object as a function of time:

$$v = \frac{gm}{c}[1 - e^{-(c/m)t}] \tag{P15.15.1}$$

where v = velocity [m/s], c = a drag coefficient [kg/s], m = mass [kg], g = gravitational acceleration = 9.8 m/s², and t = time [s]. Suppose that you had to determine the drag coefficient required in order that a 68.1-kg object would reach a v = 44.87 m/s in 10 s. A problem with making such an estimate is that it is mathematically impossible to solve Eq. (P15.15.1) for c explicitly.

An alternative is to solve the problem interactively. In other words, the equation can be programmed to compute v repeatedly as the user makes guesses until the correct result for c is obtained. Design the program to solve the equation as a "what if?" type of computation. Also, design it so that the key input parameters are within the following bounds:

$0 < m < 100$ kg

$0 < t < 50$ s

$0 < c < 20$ kg/s

15.16 Develop an interactive program that allows the user to perform addition, subtraction, multiplication, and division of two numbers. Employ ideas from Examples 15.4 and 15.9 when designing your code. In particular, employ a menu to allow the user to select from among addition, subtraction, multiplication, and division. Try to be as creative as possible in providing feedback and help to the user.

15.17 Write structured FORTRAN code for the flowchart shown in Fig. P15.17.

FIGURE P15.17

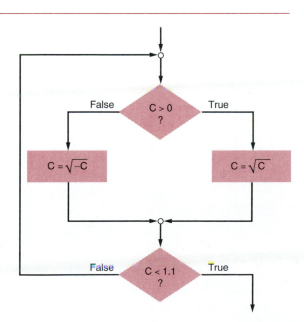

15.18 Write structured FORTRAN code to input, count, and sum a collection of positive numbers. Use an appropriate sentinel value to signal that all the numbers have been entered. Also, keep track of the maximum and the minimum values that are entered. After performing this task, print out the number of values, the average, and the minimum and maximum.

MODULAR PROGRAMMING

It is often desirable to execute a particular set of statements several times or duplicate them at different points in a program. It would be inconvenient and awkward to rewrite this set separately every time it is required. Fortunately, it is possible in FORTRAN to write such sections once and then have them available at different locations within the program to use as many times as necessary. These "miniprograms" are referred to as *subprograms*. In the present chapter, we will introduce the two types that are available in FORTRAN: subroutine and function subprograms.[1]

16.1 SUBROUTINE SUBPROGRAMS

The *subroutine* is a program module that is designed to perform a specific task. It is invoked by a CALL statement in the main program. The general form of the CALL statement is

CALL *name* (*arg*$_1$, *arg*$_2$, . . . , *arg*$_n$)

where *name* is a legal FORTRAN name that identifies the subroutine and *arg*$_i$ is the ith argument passed to the subroutine. These arguments can be either constants, variables, or expressions.

The subroutine itself has the following general organization:

---

[1]Note that in some other languages, terminology for the subroutine and function subprogram differs from FORTRAN. For example, in QuickBASIC, the subroutine subprogram is called a SUB procedure and the function subprogram is called a FUNCTION procedure. In Turbo Pascal, they are called procedures and functions, respectively.

SUBROUTINE *name(arg₁ , arg₂ , . . . , argₙ)*
Declaration Statements
.

.

.

Body of Subroutine (Executable Statements)

.

.

RETURN
END

where *name* is the same *name* as was used in the CALL statement and each of the arguments corresponds to the arguments listed in the CALL. In the SUBROUTINE statement all the arguments must be variable names. These arguments are not required to have the same names used in the CALL statement. Only their position determines which values get passed to which variables. Although the names may differ, the variable types must match.

The subroutine ends with a RETURN and an END statement. The CALL statement transfers control to the first line of the subroutine. This operation of transferring control is referred to as *calling* the subroutine. Within the subroutine, execution proceeds as normal until the RETURN is encountered, whereupon control is transferred back to the line immediately following the CALL statement. The process is depicted graphically in Fig. 16.1.

The effect of calling a subroutine is like having a GO TO that branches from the

FIGURE 16.1
Typical sequence of a subroutine subprogram call and return. The subprogram determines values from the calling program via its parameter list. After the execution of the subroutine, the value of the sum will be contained in the variable Z in the calling program. The parameter list is like a "window" through which information is selectively passed between the calling program and the subroutine.

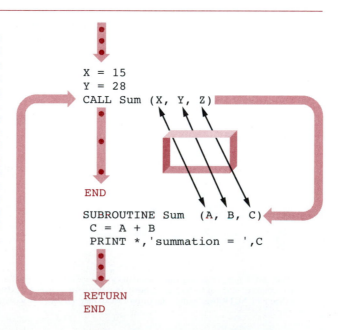

main program to another area of the program and a GO TO that then branches back to the statement immediately after the calling GO TO. The advantage of using CALL and SUBROUTINE statements instead of two GO TO statements is that the former conveniently allows the subroutine to be called from several locations. The RETURN ensures that control is automatically transferred back to the correct line in the main program, whereas the GO TO would have to be modified to return correctly for each particular case.

Strictly speaking, subroutines may have several RETURN statements for routing back to the main program. However, because of structured programming considerations, this practice is discouraged. Rather, each subroutine should have a single entrance at its beginning and a single exit at its end. In this way, the logical flow is kept simple and is easy to understand.

Subroutines should always be considered major structures in the program and, consequently, should be delineated clearly using blank spaces and lines. Structures should be identified by indentations and comment statements as needed. In addition, the statements in the body of the subroutine should be organized in a fashion similar to that of the main program. For example, declarations such as type statements must be placed before the executable statements.

All subroutines must be located together after the END statement of the main program. This organization reflects the top-down nature of program design, with the broad ideas in the main program followed by the detailed operations in the subroutines.

EXAMPLE 16.1   **Subroutine to Compute Volume and Surface Area of a Rectangular Prism**

PROBLEM STATEMENT: Figure 16.2a summarizes formulas for the volume and surface area of a rectangular prism. Develop a subroutine to compute these values.

SOLUTION: Figure 16.3 shows FORTRAN code to solve this problem. The main program is organized into three subroutine CALLS to a preprocessor (inputs data), a processor (performs computation), and a postprocessor (displays results). Note that all the data needed by each subroutine is included as arguments.

In the subroutines themselves, notice that abbreviated variable names are used for the arguments. Although the same names as were used in the main program could have been chosen, different names are employed to highlight the modular nature of the subroutine.

Finally, we have used comments to label each module and define variables. Although this is not really necessary for such a simple program, it is absolutely essential for large and complex programs. Even though it takes extra effort, it is recommended that you document all your programs in this manner.

16.2   PASSING PARAMETERS TO SUBPROGRAMS

Note that the subprogram's arguments are collectively referred to as its *parameter list* and each argument is called a *parameter*. A *formal parameter* is the name by which the argument is known within the subprogram. The *actual parameter* is the specific variable, expression, or constant passed to the subprogram when it is called.

The parameter list provides a means to control the flow of information between the calling program and the subroutine. This quality is illustrated by the following example.

FIGURE 16.2
Formulas to compute the
volume and the surface areas
of a number of geometric
solids: (a) rectangular prism,
(b) cone, (c) cylinder, and
(d) sphere.

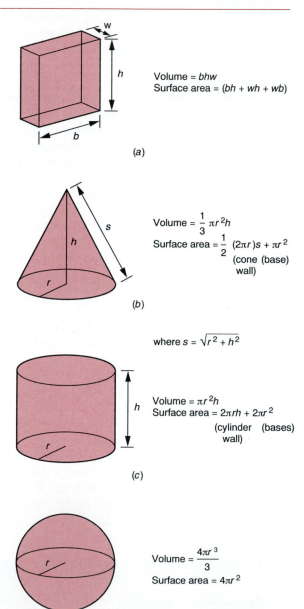

(a)

Volume = bhw
Surface area = $(bh + wh + wb)$

(b)

Volume = $\frac{1}{3}\pi r^2 h$
Surface area = $\frac{1}{2}(2\pi r)s + \pi r^2$
                      (cone (base)
                       wall)

where $s = \sqrt{r^2 + h^2}$

(c)

Volume = $\pi r^2 h$
Surface area = $2\pi rh + 2\pi r^2$
                 (cylinder   (bases)
                  wall)

(d)

Volume = $\dfrac{4\pi r^3}{3}$
Surface area = $4\pi r^2$

```
        PROGRAM PRISM
C Rectangular Prism Program
C .
C by S.C. Chapra
C June 18, 1992
C
C This program calculates the volume and
C surface area of a rectangular prism.
C-----------------------------------
C
C Definition of variables:
C PLENGT = prism base length
C PWIDTH = prism base width
C PHEIT = prism height
C PVOL = prism volume
C PAREA = prism area
C
      REAL PLENGT, PWIDTH, PHEIT, PVOL, PAREA
C
      CALL PREPRC(PLENGT, PWIDTH, PHEIT)
      CALL PROC(PLENGT, PWIDTH, PHEIT, PVOL, PAREA)
      CALL PSTPRC(PVOL, PAREA)
C
      END


      SUBROUTINE PREPRC(L, W, H)

C inputs dimensions
C-----------------------------------
C
C Definition of variables:
C L = prism base length
C W = prism base width
C H = prism height
C
      REAL L, W, H
C
      PRINT *, 'Enter dimensions:'
      PRINT *
      PRINT *, 'base length = '
      READ *, L
      PRINT *, 'base width = '
      READ *, W
      PRINT *, 'height = '
      READ *, H
      RETURN
      END
```

```
      SUBROUTINE PROC(L, W, H, V, A)

C computes volume and area
C-----------------------------------
C
C Definition of variables:
C L = prism base length
C W = prism base width
C H = prism height
C V = prism volume
C A = prism area
C
      REAL L, W, H, V, A
C
      V = L * W * H
      A = 2. * (L * H + W * H + W * L)
      RETURN
      END


      SUBROUTINE PSTPRC(V, A)

C displays results
C-----------------------------------
C
C Definition of variables:
C V = prism volume
C A = prism area
C
      REAL V, A
C
      PRINT *
      PRINT *, 'Results:'
      PRINT *
      PRINT *, 'volume = ', V
      PRINT *, 'area = ', A
      RETURN
      END
```

```
Enter dimensions:

base length =
1
base width =
2
height =
3

Results:

volume =          6.000000
area =           22.000000
```

FIGURE 16.3
A program using subroutine subprograms to compute the volume and surface area of a rectangular prism.

EXAMPLE 16.2   The Behavior of the Parameter List

PROBLEM STATEMENT: Investigate how the parameter list controls the flow of information by examining the output of the following program:

```
PROGRAM PARAM
INTEGER A, B, C, D
A = 1
B = 2
C = 3
PRINT *, A, B, C, D
CALL TEST(A, B, C)
PRINT *, A, B, C, D
END
```
Main program

```
SUBROUTINE TEST(B, A, D)
INTEGER A, B, C, D
PRINT *, A, B, C, D
D = 4
C = B
RETURN
END
```
Subroutine

SOLUTION: The first PRINT statement yields

1   2   3   0

Thus, A, B, and C have their original assigned values of 1, 2, and 3, respectively. The variable D is equal to 0 at this point because it has yet to be assigned a value.

Next, the values A, B, and C are passed to the subroutine, where they are temporarily assigned to the variables B, A, and D. Consequently, when the second PRINT statement is implemented, the result is

2   1   0   3

Notice the "local" nature of variable C in the subroutine. The labels that are passed as parameters (B, A, and D) initially have values related to the variables in the main program (A, B, and C). Consequently, a label such as C has a value of zero when it is first printed in the subroutine. It is then assigned a value of 1 by the statement, C = B. However, because C is not "passed" back to the main program, this assignment holds only in the SUB and has no effect on the value of C in the main program.

After the two assignment statements in the subroutine are implemented, control transfers back to the main program, where the final PRINT yields

```
 1      2     4     0
```

Observe how the value of 4 is assigned to the variable C because of its position in the parameter list. Similarly, the subroutine assignment statement, D = 4, has no effect on the variable D in the main program. This is because D is not included in the CALL parameter list. Therefore, the two D's are distinct.

Although the foregoing example is not very realistic, it serves to illustrate how the parameter list operates. In particular, it demonstrates how actions and variables in a subroutine can be either isolated from or connected to other parts of a program. This quality is essential for implementation of modular programming. Before exploring how this is done, we will first discuss how the *ParameterList* actually works within the computer.

As described earlier, when a variable appears in the main program, a location is defined in the computer's memory. When an action is taken to assign a value to the variable, this is where the value resides. Now, when a variable name is passed to a subprogram via the *ParameterList,* a new memory location is *not* created. Rather, the computer knows to go back to the main program to the original memory location in order to obtain the value. It "knows" this because of the presence of the variable in the parameter list. This is referred to as passing the parameter *by reference.*

In contrast, suppose that the argument in the parameter list is a constant. For example, if you wanted to pass a value of 2 as the first argument in the subroutine from Example 16.2, you could merely include the statement

```
CALL TEST(2., B, C)
```

For this case, the 2. is copied to a temporary memory location; then the address of this location is passed to the subprogram. This is referred to as passing the parameter *by value.*

This operation has special significance because variables can also be passed by value. This is done by enclosing the variable in parentheses. For example, suppose that we wanted to pass the variable A by value. This could be accomplished by

```
CALL TEST((A), B, C)
```

The following example provides a further illustration of this feature and its behavior.

EXAMPLE 16.3   *Passing by Value and by Reference*

PROBLEM STATEMENT: Develop a short program that employs passing by value and by reference in the same subprogram.

SOLUTION: The following program can be written:

```
REAL X, Y
X = 1.
Y = 0.
PRINT *, 'before sub: x = ', X, ' y = ', Y
CALL ADD((X), Y)
PRINT *, 'after sub: x = ', X,' y = ', Y
END

SUB ADD (X, Y)
Y = X + 1.
X = 0.
PRINT *, 'in sub: x = ', X, ' y = ', Y
RETURN
END
```

The output for this program would be

```
before sub: x =           1.000000 y =       0.000000E + 00
in sub: x =    0.000000E + 00 y =          2.000000
after sub:  x =           1.000000 y =          2.000000
```

Notice that because X is passed by value, even though it is modified in the subprogram, its original value is maintained in the main program. The same is not true for Y, which is passed by reference.

Another example of passing by value is when an expression is passed to a subprogram. For example,

```
CALL TEST(A + 5, B, C)
```

Just as with the constant, the computer first evaluates the expression and places the result in a temporary memory location. Then the address of this location is passed to the subprogram. Now it should be clear why the parentheses are used to pass a variable by value. In essence, the parentheses "trick" the computer into thinking that you are passing an expression!

Before proceeding, we should point out why you might want to pass a variable by value. There will be occasions where you may want to perform operations on a variable in a subprogram that change its value. At the same time, you might want to retain the original value for some other purpose. On such occasions, passing by value provides a handy means to perform the desired operation while maintaining the integrity of the original variable.

16.3 FUNCTION SUBPROGRAMS

In Sec. 13.3, we introduced the statement function to convenienetly program an expression that was used repeatedly. Although these one-line functions are useful, there are often occasions when several lines are needed to compute the desired quantity. The function subprogram provides a convenient way to do this.

In contrast to the statement function, the *function subprogram* is a miniprogram whose general format is

> *type* FUNCTION *name (ParameterList)*
> *Declaration Statements*
> .
> .
> .
>
> *name = expression*
> .
> .
> .
>
> RETURN
> END

where *type* declares the data type of the result returned (this can be omitted if the *type* is REAL or INTEGER and the naming conventions are observed), *name* is the name of the function, *ParameterList* is the comma-delimited list of arguments passed to the function, and *expression* is the value that is assigned to the function. Note that, as was the case with the subroutine subprogram, the *ParameterList* in the FUNCTION statement must consist of variables. However, when the function is invoked, it may consist of constants, variables, expressions, or a combination of these.

Notice that somewhere in the body of the function, the *expression* must be assigned to the function's *name*. As illustrated by the following example, this is the way that the result is returned to the place where the function was invoked.

EXAMPLE 16.4 | A Simple Function Subprogram to Calculate a Factorial

PROBLEM STATEMENT: Employ a function subprogram to determine the factorial. Recall that the factorial is computed as

$$n! = 1 * 2 * 3 * \ldots * (n - 2) * (n - 1) * n$$

For example, 4 factorial would be calculated as

$$4! = 1 * 2 * 3 * 4 = 24$$

Note that by definition, $0! = 1$.

SOLUTION: The following program illustrates how the factorial can be computed as a function subprogram:

```
PROGRAM EXAMPL
INTEGER I, FACTOR
DO 10 I = 0.5
   PRINT *, I, '! = ', FACTOR(I)
10 CONTINUE
END

INTEGER FUNCTION FACTOR(N)
INTEGER X, N, I
X = 1
DO 10 I = 1, N
   X = X * I
10 CONTINUE
FACTOR = X
RETURN
END
```

Output:

```
0! =          1
1! =          1
2! =          2
3! =          6
4! =         24
5! =        120
```

Notice how the function is invoked by name in the PRINT statement. If you wanted to save the result in a memory location, you could add an assignment statement, such as

```
L3 = FACTOR (3)
```

In addition, you can use the function as part of an expression, as in

```
SINAPP = X - X ** 3/FACTOR(3) + X ** 5/FACTOR(5)
```

Recognize that because the name of the integer function starts with the letter F, we must declare the function name, FACTOR, as an INTEGER. Otherwise, an error message would result because the proper type of memory location would not be set aside for the function's result.

Finally, see that the number 10 is employed as a line number in both the main program and the function. This underscores the notion that the subprogram is a miniprogram. Within either the main program or a subprogram, it would be illegal to use the same line number more than once.

16.3.1 Comparison between Functions and Subroutines

Note that function and subroutine subprograms are similar in many respects. For example, information can be passed to a function by either reference or value with a parameter list.

Although they behave similarly in many respects, subroutine and function subprograms differ in some significant ways. Figure 16.4 delineates the differences among subroutine subprograms, function subprograms, and the other types of functions that are available in FORTRAN.

The major difference between a subroutine and function subprogram is that the subroutine can pass many results back to the calling program via its parameter list. In contrast, the function subprogram returns a single result that can be used directly, as though it were a variable.

FIGURE 16.4
Summary and comparison of the subroutine and the various types of functions available in FORTRAN.

INTRINSIC OR LIBRARY FUNCTIONS: Built-in functions that perform commonly employed mathematical or trigonometric operations; they generate a single value that can be used as part of a mathematical expression.

```
Definition:    Defined by system
Application:   Y = SIN(X)
```

STATEMENT FUNCTIONS: User-defined functions that can be employed as part of a mathematical expression; the function is defined by a single line, which must be placed prior to the first executable statement.

```
Definition:    ROUND(X) = INT(X + 0.5)
Application:   Y = ROUND(X)
```

FUNCTION SUBPROGRAMS: User-defined miniprograms that consist of several lines, which must be placed after the main program; they generate a single value that can be used as part of a mathematical expression.

```
Application:   Y = AREA(R)
Definition:    FUNCTION AREA(R)
               AREA = 3.14159 * R ** 2
               RETURN
               END
```

SUBROUTINE SUBPROGRAMS: User-defined miniprograms that consist of several lines, which must be placed after the main program. Several results may be generated and passed back to the main program as variables. The results are passed between the subroutine and the main program via the subroutine's arguments or via COMMON statements.

```
Application:   CALL SUM(X,Y,Z)
Definition:    SUBROUTINE SUM(A,B,C)
               C = A + B
               PRINT *, C
               RETURN
               END
```

To this point, we have covered two other types of statements that behave in a similar fashion to the function subprogram. These are the intrinsic functions and statement functions discussed in Chap. 13. Recall how an intrinsic function such as `SIN (X)` can be used as part of, say, a mathematical expression. The FUNCTION subprogram can be employed in a similar fashion. However, they differ from the intrinsic and the one-line user-defined statement functions in some fundamental ways. In contrast to the intrinsic function, which is supplied by the system, the FUNCTION subprogram is concocted by the user. In contrast to the user-defined statement function, which is limited to a single line, the FUNCTION subprogram can consist of many statements. Thus, as seen in Fig. 16.4, the FUNCTION subprogram lies midway between statement functions and subroutines.

16.4 LOCAL AND GLOBAL VARIABLES

As described in the previous sections, two-way communication is required between the subprogram and the main program. The subprogram needs data to act on and must return its results for utilization in the main program.

In FORTRAN, the parameter list provides a controlled means for transferring information. In other words, the information that is passed in and out of the subprogram must be specified explicitly by the programmer. Because both the main program and the subprogram have direct access to them, the variables in the list can be thought of as being global to the calling program and that subprogram specifically. In contrast, variables that reside exclusively in the subprogram are called *local variables*.

It should be mentioned at this point that when a local variable appears in a subprogram, a location is defined in the computer's memory. When an action is taken to assign a value to the variable, this is where the value for this variable resides. This is in contrast to parameter-list variables, which (as noted previously) do not have independent memory locations.

The ability to have local variables in a subprogram can be a great asset. In essence, they serve to make each subprogram a truly autonomous entity. For example, changing a local variable will have no effect on variables outside the subprogram. In fact, true local variables can even have the same name as variables in other parts of the program but still be distinct entities. One advantage of this is that it decreases the likelihood of "side effects" that occur when the value of a global variable is changed in a subprogram and is then used erroneously elsewhere in another part of the program. In addition, the existence of local variables also faciltates the debugging of a program. The subprograms can be debugged individually without fear that a working subprogram will cease functioning because of an error in another subprogram.

16.4.1 The COMMON Statement

Although local variables are a great benefit, there will be occasions when you will want your subprograms to have direct access to certain variables in the calling program.

Any type of variable may be made available for use by a subprogram by employing the COMMON statement, which allows variables to be defined as global variables. The COMMON statement directs the compiler to set aside a certain area in the computer's memory which the main program and subprograms may access to obtain values for the variable. COMMON statements can be unnamed blank COMMONs, or they can be given explicit names.

For a blank COMMON, the general form is

    COMMON var_1, . . . , var_n

whereas for an explicitly named COMMON, the general form is

    COMMON /$name$/var_1, . . . , var_n

Both the main program and the subprogram must contain the COMMON statement in order for both parts to have access to a variable. Also, note that there can be only one blank COMMON statement in the main program.

Within a subprogram the number of variables in a COMMON statement can be fewer than in the main program, and the variables do not have to have the same name as in the main program. The correspondence between a variable in the main program and in a subprogram is by position in the COMMON statement. If the main program has a named COMMON statement

    COMMON /SUBSET/ ARR, X, VEL, NAME

the subprogram could have a COMMON statement such as

    COMMON /SUBSET/XMATR, Y

or

    COMMON /SUBSET/ARR, Y, VAL, LNAME

The values of ARR in the main program would be found in XMATR in the first subprogram COMMON statement and ARR in the second. In the first subprogram COMMON statement, VEL and NAME would not be available to the subprogram. In fact, these same names could be used by the subprogram for local variable names with entirely different values from those in the main program. In the subprogram, Y would refer to the same data as X in the main program.

The COMMON statement should follow type-declaration statements, but it should be placed prior to the executable statements of the program or subprograms.

For example, in the program from Fig. 16.3, the subroutine PSTPRC could have been represented as (note that comments and blank PRINTs have been omitted for conciseness here)

```
SUBROUTINE PSTPRC
REAL V, A
COMMON V, A
PRINT *, 'volume = ", V
PRINT *, 'area = ', A
RETURN
END
```

The main program would be modified to

```
PROGRAM PRISM
REAL PLENGT, PWIDTH, PHEIT, PVOL, PAREA
COMMON PVOL, PAREA
CALL PREPRC(PLENGT, PWIDTH, PHEIT)
CALL PROC(PLENGT, PWIDTH, PHEIT, PVOL, PAREA)
CALL PSTPRC
END
```

EXAMPLE 16.5   Use of COMMON in a Subprogram

PROBLEM STATEMENT: Show how COMMON and the parameter list invoke different behavior in subprogram variables.

SOLUTION: The following program fragment can be written:

```
    PROGRAM EX165
    INTEGER I, J, K, L
    COMMON K
    DO 10 I = 1, 2
      CALL ADDER(J)
 10 CONTINUE
    PRINT *, 'in main program: j=', J, ", k=', K, ', l=', L
    END
    SUBROUTINE ADDER(J)
    INTEGER J, K, L
    COMMON K
    J = J + 1
    K = K + 2
    L = L + 3
    PRINT *, 'in subprogram: j=', J, ", k=', K, ', l=', L
    RETURN
    END
```

The output for this program would be

```
in subprogram: j=      1, k=      2, l=      3
in subprogram: j=      2, k=      4, l=      6
in main program: j=      2, k=      4, l=      0
```

Observe how J and K have the same values in the main program and the subprogram. This is because J is included in the parameter list and K is included in the COMMON statement. In contrast, even though L is incremented in the subroutine, its value never changes in the main program. Thus, the L in the main program is clearly shown to be a distinct variable from the local variable L in the subprogram.

At this point, you might be wondering why there are seemingly identical means for sharing information between the main program and subprograms: the *Parameter-List* and the COMMON statement. Although they may seem identical, there are some subtle differences that might make one preferable to another in a particular problem context. One example relates to the fact that the *ParameterList* allows you to pass an argument to a subprogram "by value." The COMMON statement cannot be employed in this fashion.

In addition, named COMMON is often preferred for cases where several program modules require the same information. For these cases, named COMMON provides a neat modular way to make this data available to each of the modules.

16.5 MODULAR PROGRAMMING

Just as chapters are the building blocks of a novel, subprograms are the building blocks of a computer program. Each subprogram can have all the elements of a whole program; however, in general, it is smaller and more specialized. Some computer experts suggest that a reasonable size limit for a subprogram should be no more than about 50 to 100 lines. Because this size will fit on a few monitor screens, it allows quick viewing of the whole subprogram. It is a reasonable guideline because it facilitates visualizing the purpose and the composition of the subprogram.

In addition to allowing you to view the subprogram easily, the 50- to 100-line limit will also encourage you to restrict the subprogram to one or two well-defined tasks. It is this characteristic of subprograms that makes them the ideal vehicle for expressing a modular design in program form.

Subprograms should always be laid out clearly using blank spaces and lines. Every subprogram should begin with a short description of the purpose of the subprogram, and definitions of local variables (that is, those used exclusively in the subprogram) should be included after the title and before the body of the subprogram.

The statements in the body of the subprogram should be organized in a fashion similar to that of the main program. Structures should be identified by indentations, and remarks should be included as needed. Such style is illustrated in the following example.

EXAMPLE 16.6   *Menu-Driven Program to Access Subroutines*

PROBLEM STATEMENT: Modify the program from Fig. 16.3 so that it provides the user with a choice of determining the area and volume of a cone or a cylinder.

SOLUTION: A well-structured modular program to solve this problem is displayed in Fig. 16.5. A sample run is shown in Fig. 16.6.

Several important features of the program bear mentioning:

1. A PARAMETER statement is used to define π in the subroutines where it is needed. Specifically, it is employed in the subprograms to calculate the volumes and surface areas according to the formulas from Fig. 16.2.

2. DO UNTIL loop and CASE structures are used in tandem to present and implement a menu in the main program. The interior DO loop cycles until the user makes an appropriate menu selection (1, 2, or 3). Then the CASE employs the entered value to CALL the appropriate subprogram. When the user enters a value of 3, the outer DO loop is exited and the program terminates.

3. Separate subprograms are employed for the cone and the cylinder. Notice that neither of these subprograms has a parameter list or a COMMON. These were omitted because there is no need to pass any information between these subprograms and the main program.

4. Each of these subprograms, CONE and CYLIND, performs input and calculations. However, because their output is very similar, a separate subprogram, DISPLY, is employed to display results. Notice that this subprogram does have a parameter list in order to receive the computed area and volume for each case. Also, see that we have passed a string, SOLID, so that DISPLY can output a customized label for each case.

As in Fig. 16.6, the program's output consists of a series of queries that allow the user to interactively compute areas and volumes for the two types of solids. Of course the program could be refined in many ways (see Prob. 16.7). However, it serves to illustrate how subprograms can be employed to create a modular program.

Subprograms facilitate the process of program development for two major reasons. First, they are discrete, task-oriented modules that are, therefore, easier to understand, write, and debug. Second, they can be used over and over both within a program and among programs. Testing a small subprogram to prove it works properly is far easier than testing an entire program. Once a subprogram is thoroughly debugged and tested, it can be incorporated into other programs. Then, if errors occur, the subprogram is unlikely to be the culprit, and the debugging effort can focus on other parts of the program.

Main program:

```
      PROGRAM SOLIDS
C Volume/Area Program
C
C by S.C. Chapra
C June 18, 1992
C
C This program calculates the volume and
C surface area of a cone and a cylinder.
C------------------------------------
C
C Definition of variables:
C D = dummy variable
C
      INTEGER D
C
   10 CONTINUE
   20    CONTINUE
         PRINT *
         PRINT *, 'Select one of the following options'
         PRINT *, 'to determine the volume and area for:'
         PRINT *
         PRINT *, '1. Cone'
         PRINT *, '2. Cylinder'
         PRINT *, '3. Terminate the program'
         PRINT *
         PRINT *, 'Enter a number'
         READ *, D
      IF (.NOT.(D.EQ.1.OR.D.EQ.2.OR.D.EQ.3)) GO TO 20
      IF (D.EQ.1) THEN
         CALL CONE
      ELSE IF (D.EQ.2) THEN
         CALL CYLIND
      END IF
      IF (.NOT. (D.EQ.3)) GO TO 10
      END
```

Subroutines:

```
      SUBROUTINE DISPLY (SOLID, AREA, VOLUME)

C displays results
C------------------------------------
C
C Definition of variables:
C SOLID = type of solid
C AREA = solid's area
C VOLUME = solid's volume
C D = dummy variable
C
      CHARACTER*8 SOLID
      REAL AREA, VOLUME, D
C
      PRINT *
      PRINT *, 'Results:'
      PRINT *
      PRINT *, 'For the ', SOLID
      PRINT *, 'area = ', AREA
      PRINT *, 'volume = ', VOLUME
      PRINT *
      PRINT *, 'Enter any number to continue'
      READ *, D
      RETURN
      END
```

FIGURE 16.5

A program using subroutines to compute the volume and surface area of a cone and a sphere.

```
      SUBROUTINE CONE

C Inputs dimensions and computes volume
C and area for a cone
C------------------------------------
C
C Definition of variables:
C R = cone radius
C H = cone height
C S = cone slope
C V = cone volume
C A = cone area
C
      REAL R, H, S, V, A, PI
      PARAMETER (PI = 3.14159)
C
C input dimensions
C
      PRINT *
      PRINT *, 'Enter dimensions for a cone:'
      PRINT *, 'radius = '
      READ *, R
      PRINT *, 'height = '
      READ *, H
C
C perform computations
C
      S = SQRT(R ** 2 + H ** 2)
      V = PI * R ** 2 * H / 3
      A = 2. * PI * R * S / 2 + PI * R ** 2
C
C display results
C
      CALL DISPLY('cone    ', A, V)
C
      RETURN
      END
```

```
      SUBROUTINE CYLIND

C Inputs dimensions and computes volume
C and area for a cylinder
C------------------------------------
C
C Definition of variables:
C R = cylinder radius
C H = cylinder height
C V = cylinder volume
C A = cylinder area
C
      REAL R, H, V, A, PI
      PARAMETER (PI = 3.14159)
C
C input dimensions
C
      PRINT *
      PRINT *, 'Enter dimensions for a cylinder:'
      PRINT *, 'radius = '
      READ *, R
      PRINT *, 'height = '
      READ *, H
C
C perform computations
C
      V = PI * R ** 2 * H
      A = 2. * PI * R * H + 2. * PI * R ** 2
C
C display results
C
      CALL DISPLY('cylinder', A, V)
C
      RETURN
      END
```

```
Select one of the following options
to determine the volume and area for:

1. Cone
2. Cylinder
3. Terminate the program

Enter a number
1

Enter dimensions for a cone
radius =
1
height =
2

Results:

For the cone
area =        10.166400
volume =        2.094393

Enter any number to continue
1

Select one of the following options
to determine the volume and area for:

1. Cone
2. Cylinder
3. Terminate the program

Enter a number
2

Enter dimensions for a cylinder
radius =
1
height =
2

Results:

For the cylinder
area =        18.849540
volume =        6.283180

Enter any number to continue
1

Select one of the following options
to determine the volume and area for:

1. Cone
2. Cylinder
3. Terminate the program

Enter a number
3
```

FIGURE 16.6
Output generated by the
program from Fig. 16.5.
Notice that values input by
the user are highlighted.

Many professional programmers accumulate a collection, or library, of useful subprograms that they access as needed. Thus instead of programming many lines of code, they combine and manipulate larger, more powerful pieces of code—that is, subprograms. Programming with these custom modules saves overall effort and, more important, improves program reliability. The modules in a library can be thoroughly debugged under a number of different conditions so as to ensure that they work properly. Such thorough testing can rarely be afforded for sections of code that are used only once.

It should be mentioned that FORTRAN is an ideal language for this approach. Because it has been around so long, a vast number of programs have been developed in the language. In particular, several "libraries" of algorithms are available to implement a variety of numerical and statistical methods. Two of the most useful are the International Mathematical and Statistical Library (IMSL) and Numerical Algorithms Group (NAG). Although the use of these libraries is beyond the scope of the present book, you should be aware of this powerful capability.

In the remainder of the book we will be developing a number of programs that have general applicability in engineering. You should view these programs as the beginning of your own subprogram library, with an eye to their utility in your future academic and professional efforts.

16.6 FORTRAN 90

Recursion is a powerful problem-solving tool that is available in certain computer languages such as Pascal and modern dialects of BASIC. A *recursive process* is one that calls itself (see Fig. 16.7). Although it is not available in FORTRAN 77, recursion is part of the FORTRAN 90 standard. In addition, it is so important in modern computer science that we will introduce it at this point.

The factorial calculation provides a good example for the use of recursion. This can be seen by realizing that it can be represented by the sequence

$$5! = 5 \cdot 4!$$

$$4! = 4 \cdot 3!$$

$$3! = 3 \cdot 2!$$

$$2! = 2 \cdot 1!$$

$$1! = 1 \cdot 0!$$

The following example illustrates how this pattern can be concisely expressed by exploiting the recursive ability of FORTRAN 90.

EXAMPLE 16.7   A Recursive Function Subprogram to Calculate a Factorial

PROBLEM STATEMENT: Employ recursion to determine the factorial.

FIGURE 16.7
A visual example of recursion.

SOLUTION: The following FORTRAN 90 program illustrates how the factorial can be computed using a recursive FUNCTION subprogram:

```
      PROGRAM RECURS
      INTEGER :: I, FACTOR
      DO 10 I = 0, 5
        PRINT *, I, '! = ', FACTOR(I)
   10 CONTINUE
      END

      RECURSIVE FUNCTION FACTOR (N) RESULT (FACRES)
      INTEGER :: FACRES, X, N
      IF (N.GT.O) THEN
        FACRES = N * FACTOR (N - 1)
      ELSE
        FACRES = 1
      END IF
      RETURN
      END
```

The output for this program would be identical with that from Example 16.4.

Notice how the function must be declared as being RECURSIVE. Also, see that the result of the computation, FACRES, must be distinguished from the subprogram's name, FACTOR. Otherwise, in FORTRAN 90, there would be no way to distinguish the use of FACTOR as a result variable from its use as a function.

Although Example 16.7 employed a function, subroutines can also call themselves recursively in FORTRAN 90. Note that a recursive process must always have a terminating condition. Otherwise, it would go on forever. In the case of the factorial, the condition arises naturally from the fact that $0! = 1$. Consequently, when $0!$ is called, the IF statement shifts to an assignment statement, FACRES = 1. At this point, because the function is no longer calling itself, the process terminates.

PROBLEMS

16.1 Write a program to determine the roots of a quadratic equation. Pattern the computation after Fig. 14.7. However, use separate subprograms to (*a*) input the coefficients, (*b*) compute the roots, and (*c*) display the results.

16.2 Reprogram Fig. 13.6 so that it uses subprograms.

16.3 Write a function subprogram that determines the sign of a number. Design it so that it returns a value of 1 if the number is positive, a -1 if it is negative, and a 0 if it is zero.

16.4 Write a subprogram to input and convert temperatures from degrees Fahrenheit to degrees Celsius and kelvins by the formulas

$$T_C = \tfrac{5}{9}(T_F - 32)$$

and

$$T_K = T_C + 273.15$$

Test it with the following data: $T_F = 20$, 203, and 98.

16.5 Write a subprogram to input and convert temperatures from degrees Celsius to degrees Fahrenheit and Rankine by the formulas

$$T_F = \tfrac{9}{5}T_C + 32$$

and

$$T_R = T_F + 460$$

Test it with the following data: $T_C = -10$, 27, and 90.

16.6 Develop the subprograms in Probs. 16.4 and 16.5. Then merge them into a menu-driven program to perform general temperature conversions. The main program menu should include such choices as (*a*) to convert from Fahrenheit to Celsius and kelvins, (*b*) to convert from Celsius to Fahrenheit and Rankine, and

(c) to terminate the program. Test the program with the following data: 10°C, 25°F, 120°C, and 800°F.

16.7 Using Fig. 16.5 as your starting point, develop a well-structured program to compute the volumes and surface areas for all the geometric solids from Fig. 16.2. Enhance your program in any way you see fit. For example, you might include error traps to prevent the user from entering negative values or zeros for the dimensions. Another suggestion might be to allow the user to specify units which could then be displayed with the results. You should also try to develop ideas of your own to make the program as user-friendly and reliable as possible.

16.8 Write a function subprogram to interpolate linearly between two points: x_1, y_1 and x_2, y_2. Place the interpolated value as the value of the function. Specifically, the user-defined function will have five values in its argument list: x_1, y_1, x_2, y_2, and the value of x at which the interpolation is to be performed. Note also that your function should check whether the value of x is within the interval from x_1 to x_2 and should display an error message if not. (*Hint:* Note that the function name should always be assigned a value in the subprogram. Guard against the possibility that it might not.)

16.9 Write a function subprogram that checks whether the first value in the argument list is equal to one of the four remaining values. If the first value is identical with one of the arguments, assign a value of .TRUE. to the function's name; otherwise, assign it a value of .FALSE. *Note:* This requires that the function name itself be declared a logical variable in the main program. For example, if the function is named CHECK, then the main program must declare CHECK to be logical with the statement

```
LOGICAL CHECK
```

Additionally, at the start of the subprogram, the function name must be declared as a logical function. This can be done with the statement

```
LOGICAL FUNCTION CHECK(TEST, X1, X2, X3, X4)
```

Test your main program and function by assigning different values to the arguments; after the function is executed, have the program write a message stating whether a match was found. Note that on some compilers, the above subprogram may yield an error message. If this occurs on your system, determine how you would circumvent the error.

16.10 Even though FORTRAN has the capability to handle complex numbers, write individual FORTRAN subroutines to perform complex addition, subtraction, multiplication, and division. Use only REAL type variables in these subprograms. Integrate the subroutines into a modular program that allows a user to enter two complex numbers. Then have the user select the desired operation from a menu and display the result. Allow the user to repeat the process as many times as he or she desires. Note that if the two complex numbers are $z_1 = x_1 + iy_1$ and $z_2 = x_2 + iy_2$, the operations are defined as

$$z_1 + z_2 = (x_1 + x_2) + i(y_1 + y_2)$$

$$z_1 - z_2 = (x_1 - x_2) + i(y_1 - y_2)$$

$$z_1 z_2 = x_1 x_2 - y_1 y_2 + i(x_1 y_2 + x_2 y_1)$$

$$\frac{z_1}{z_2} = \frac{x_1 x_2 + y_1 y_2 + i(x_2 y_1 - x_1 y_2)}{x_2^2 + y_2^2}$$

16.11 Write a program to determine the range of a projectile fired from the top of a cliff as described in Prob. 13.15. Employ a modular approach, and use three subprograms to (*a*) input the data, (*b*) perform the computation, and (*c*) output the results. Design both the main program and the input subprogram so that they are menu-driven. The main program menu should include such choices as to input or modify data, to perform the computation and output results, and to terminate the program. Figure 16.5 can serve as the model for how this can be done. The input subprogram menu should include such choices as to input or modify (1) the initial velocity, (2) the angle, and (3) the cliff height and to (4) return to the main menu.

16.12 (*FORTRAN 90*) Although it is already available by using an operator (**), integer exponentiation is an excellent example of a recursive function. Exponentiation can be expressed recursively as

$$x^0 = 1$$

$$x^1 = x \cdot x^0$$

$$x^2 = x \cdot x^1$$

.

.

.

$$x^n = x \cdot x^{n-1}$$

Develop a recursive FUNCTION subprogram to implement this algorithm. Call the subprogram XPOWER. Make sure to specify the power as an integer variable.

16.13 (*FORTRAN 90*) As described previously in Prob. 15.14, the golden ratio is related to an important mathematical sequence known as the Fibonacci numbers, which are

0, 1, 1, 2, 3, 5, 8, 13, 21, 34, . . .

Thus, each number after the first two represents the sum of the preceding two. An interesting property of the Fibonacci sequence relates to the ratio of consecutive numbers in the sequence; that is, $0/1 = 0$, $1/1 = 1$, $1/2 = 0.5$,

2/3 = 0.667, 3/5 = 0.6, 5/8 = 0.625, and so on. If the series is carried out far enough, the ratio approaches the golden ratio:

$$\frac{\sqrt{5} - 1}{2} = 0.61803\ldots$$

The Fibonacci sequence can be expressed recursively, as in

$f(0) = 0$

$f(1) = 1$

$f(2) = f(1) + f(0)$

$f(3) = f(2) + f(1)$

.

.

.

$f(n) = f(n - 1) + f(n - 2)$

Develop a recursive INTEGER FUNCTION subprogram to implement this algorithm. Call the subprogram FIBON. Specify the argument as an integer. Ensure that the user cannot enter a negative value or a noninteger value for the argument. Test it by computing the golden ratio.

16.14 *(FORTRAN 90)* Compound interest is computed by the following sequence:

$A_1 = (1 + i)A_0$

$A_2 = (1 + i)A_1$

.

.

.

$A_n = (1 + i)A_{n-1}$

Close inspection shows that the process can be expressed concisely by the general formula

$$A_n = (1 + i)^n A_0 \qquad\qquad\qquad\qquad (P16.14.1)$$

where A_0 is the original amount that was borrowed, A_n is the amount owed after n periods, and i is the fractional rate of interest. Use a recursive function to program the process. Employ Eq. (P16.14.1) to verify that your subprogram yields correct results.

16.15 *(FORTRAN 90)* Write a recursive function to compute the sum of the integers

1, 2, 3, . . . , n

Specify the argument as an integer. Ensure that the user cannot enter a negative value or a noninteger value for the argument.

DATA STRUCTURE

In earlier chapters, we have seen how algorithm structure makes computer code much easier to use and to modify. In a similar fashion, data can also be organized to make its use more efficient. Such organization is referred to as *data structuring*.

The present chapter focuses on two features that lead to more efficient and succinct organization of information in FORTRAN programs. First, DATA statements allow constants to be assigned to variables in a very concise manner. Second, arrays permit the use of subscripted variables. These allow us to access a group of values with a single variable name. The subscript provides a handy way to distinguish among the individual values. Along with their other advantages, these features can be employed to input information into a program in a neat and effective manner.

Finally, although a record is available only in FORTRAN 90 implementations, we will also introduce the notion here. This programming feature allows several types of information to be accessed using a single name. Hence, the record represents a powerful vehicle for data structuring.

17.1 THE DATA STATEMENT

Data can be entered into a FORTRAN program in a variety of ways. For example, the READ statement allows the user to enter different values for a variable on each run without modifying the program itself. Although this capability has its advantages, there will be times when it will be inconvenient to enter all or a part of the data from the keyboard. This is particularly true when entering large amounts of data.

In addition to large data sets, there will also be times when you will need the same information to be entered every time a program is run. For example, a table of physical properties for a group of materials might be required in programs used for engineering design. Again, it is inefficient to make the user input such information one value at a time with individual READ statements.

```
DATA N /17 /
DATA I, GRADE /7, 89.3/
DATA LETTER, PI, E /'Y', 0.31416E1, 2.71828/
DATA X, Y, Z /3*0.0/
DATA I, J, A, B, C, STUNAM /2*1, 3*0.0, 'Mary J'/
DATA R, J, S, K, T, L /3*(0.0, 1) /
```

FIGURE 17.1 Examples of acceptable DATA statements. These examples assume that variable type follows the FORTRAN naming convention, where variables starting with the letters I through N are integers and all others are real. The exceptions are the variables LETTER and STUNAM, which should be explicitly declared as character variables.

One remedy might be to assign values to the variables within the body of the program using assignment statements. Although this circumvents the problem of retyping the data for each run, it can add quite a few lines to your program. DATA statements provide a more concise and efficient alternative for incorporating moderately large amounts of information into the program code. They allow the data to be kept together, apart from the body of the program so that it is uncluttered. Having data in a separate place may also facilitate program maintenance, especially if the data is to be changed. The general form of the DATA statement is

DATA *VariableList/ConstantList/*

where *ConstantList* are the numeric or string constants that are being entered and *VariableList* are the variable names to which the constants are assigned.

Functionally, the DATA statement

DATA NUMBER/7/

is equivalent to the assignment statement NUMBER = 7.

Examples of some acceptable DATA statements are shown in Fig. 17.1. Notice how the number of variables must match the number of constants. Also, when the same value is to be assigned to several variables, the * can be employed to implement the assignment succinctly.

EXAMPLE 17.1   Using DATA Statements

PROBLEM STATEMENT: Employ DATA statements to enter the first four values of flow from Table 17.1. Then determine the average flow. Finally, display the average along with the years and flows in tabular form.

TABLE 17.1
Thirty Years of
Water-Flow Data for
a River

Year	Flow*
1954	125
1955	102
1956	147
1957	76
1958	95
1959	119
1960	62
1961	41
1962	104
1963	81
1964	128
1965	110
1966	83
1967	97
1968	79
1969	63
1970	115
1971	71
1972	106
1973	53
1974	99
1975	153
1976	112
1977	82
1978	132
1979	96
1980	102
1981	90
1982	122
1983	138

*In billions of gallons
per year.

SOLUTION:

```
      PROGRAM EX171
      REAL F1, F2, F3, F4, AVERAG, SUM
      CHARACTER*4 Y1, Y2, Y3, Y4
C
      DATA Y1, F1, Y2, F2/'1954', 125., '1955', 102./
      DATA Y3, F3, Y4, F4/'1956', 147., '1957', 76./
C
      SUM = F1 + F2 + F3 + F4
      AVERAG = SUM/4.
      PRINT *, 'year    flow'
      PRINT *, Y1, F1
      PRINT *, Y2, F2
      PRINT *, Y3, F3
      PRINT *, Y4, F4
      PRINT *, 'average = ', AVERAG
C
      END
```

Output:

```
  year     flow
  1954     125.000000
  1955     102.000000
  1956     147.000000
  1957      76.000000
  average =      112.500000
```

As in the previous example, the DATA statement must be placed just after the type-declaration statements at the beginning of the program. Later in this chapter, we will have more to say on this issue.

In summary, DATA statements provide an excellent vehicle for neatly organizing and assigning information within a program. We now move on to another means for organizing information—the array.

17.2 ARRAYS

To this point, we have learned that a variable name conforms to a storage location in primary memory. Whenever the variable name is encountered in a program, the computer can access the information stored in the location. This is an adequate arrangement for many small programs. However, when dealing with large amounts of information, it often proves too limiting.

For example, suppose that in the course of an engineering project you were using stream-flow measurements to determine how much water you might be able to store in a projected reservoir. Because some years are wet and some are dry, natural stream flow varies from year to year in a seemingly random fashion.[1] Thus you would require about 25 to 50 years of measurements to estimate the long-term average flow accurately. Data of this type is listed in Table 17.1.

A computer would come in handy in such an analysis. However, it would obviously be inconvenient to come up with a different variable name for each year's flow. For example, recall that in Example 17.1, we used the variable names

F1   F2   F3   F4

to refer to the four flows used in the example. For greater quantity of data, such a scheme would obviously be onerous.

Similar problems involving large amounts of data arise continually in engineering and other computer applications. For this reason, subscripted variables, or arrays, have been developed.

An *array* is an ordered collection of data. That is, it consists of a first item, second item, third item, and so on. For the stream-flow example, an array would consist of the first year's flow, the second year's flow, the third year's flow, and the like.

All the individual items, or *elements,* of the array are referenced by the same variable name. In order to distinguish between the elements, each is given a subscript. This is a standard practice in mathematics. For example, if we assign flow the variable name *F,* an index or subscript can be attached to distinguish each year's flow, as in

$$F_1 \qquad F_2 \qquad F_3 \qquad F_4 \qquad F_5$$

where F_i refers to the flow in year i. These are referred to as *subscripted variables.* In FORTRAN a similar arrangement is employed, but because subscript symbols are not allowed, the index is placed in parentheses, as in

F(1)   F(2)   F(3)   F(4)   F(5)

and the parentheses notation is still called a "subscript."

Note that the elements F(1) and F(2) are different from F1, F2, and so on. In fact F(1) and F1 are considered to be different variables in FORTRAN and can be used in the same program. Each would be assigned a different storage location, as depicted in Fig. 17.2. However, if you use a particular name for an array, you cannot use that same name for another variable. For example, the use of the unsubscripted variable F in Fig. 17.2 would produce an error message.

Assignment statements can be employed to assign values to the elements of an array in a fashion similar to assigning them to nonarray variables. For example, using the data in Table 17.1,

---

[1] Natural phenomena like stream flows are actually not as random as they might first appear. The field of *time-series analysis* is used to detect the underlying structure or trends exhibited by such data.

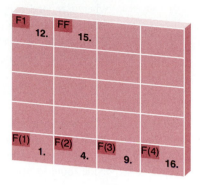

```
    F1 = 12.
    FF = 15.
    DO 2 I = 1,4
       F(I) = I ** 2
  2 CONTINUE
```

FIGURE 17.2 Visual depiction of memory locations for unsubscripted variables and subscripted variables. The program on the left assigns values to F1, FF, and F(1) through F(4), with the resulting constants being stored in the memory locations depicted on the right.

```
F(1) = 125.
F(2) = 102.
F(3) = 147.
       .
       .
       .
```

READ statements can also be employed, as in

```
READ *, F(1)
READ *, F(2)
READ *, F(3)
       .
       .
       .
```

Data statements can also be employed to assign values to arrays. Before showing how this is done we must introduce the notion of dimensioning.

17.2.1 Dimensioning Arrays

The DIMENSION statement is a nonexecutable statement that is used to declare the name and size of arrays. Its general form is

DIMENSION $var_1(n_1)$, $var_2(n_2)$, . . .

where var_1 and var_2 are the variable names of different arrays and n_1 and n_2 are integer values defining the upper limits on the number of elements in each array. For example, if we wanted to perform an analysis on all the data in Table 17.1, we would have to include the following DIMENSION statement in our program:

```
DIMENSION YEAR(30), FLOW(30)
```

This would set aside 30 memory locations (subscripts 1 through 30), which would be adequate to hold the 30 years and flows in Table 17.1.

Although there are only 30 values in Table 17.1, we could just as well have set aside more locations, as in

```
DIMENSION YEAR(100), FLOW(100)
```

This might be done if we anticipated that at a later time additional data would be obtained (say, from the years after 1983). Thus we do not have to specify the exact size of an array in the DIMENSION statement; we merely have to make sure that there are enough elements available to store our data. Otherwise an error will occur.

Note that a lower limit other than 1 can be specified for the dimension. For example, the following DIMENSION statement could be used to make the subscript of FLOW equal to the corresponding range of years from Table 17.1:

```
DIMENSION FLOW(1954:1983)
```

In a similar fashion, negative subscripts could be established as in

```
DIMENSION DISPLC(-20:20)
```

The limit on the number of elements for subscripted variables depends on the amount of memory available on your computer. Therefore, for large programs it is advisable to estimate the size of your array realistically so that you do not exceed your computer's memory capacity.

It should be noted that an array can be dimensioned as part of an explicit type declaration. For example, the following statements would set aside 30 locations for each variable and define their type at the same time:

```
CHARACTER*4 YEAR(30)
REAL FLOW(30)
```

If this is done, a DIMENSION statement would not be used.

EXAMPLE 17.2   *Employing Loops, Arrays, and DATA Statements*

PROBLEM STATEMENT: Solve the same problem as in Example 17.1 but utilize arrays, loops, and DATA statements to make the code more concise.

SOLUTION:

```
        PROGRAM EX172
        REAL FLOW(4), AVERAG, SUM
        CHARACTER*4 YEAR(4)
        INTEGER I
C
        DATA YEAR/'1954', '1955', '1956', '1957'/
        DATA FLOW/125., 102., 147., 76./
C
        DO 10 I = 1, 4
           SUM = SUM + FLOW(I)
    10  CONTINUE
        AVERAG = SUM/4.
        PRINT *, 'year     flow'
        DO 20 I = 1, 4
           PRINT *, YEAR(I), FLOW(I)
    20  CONTINUE
        PRINT *, 'average = '; AVERAG
C
        END
```

The output for this program would be identical with that of Example 17.1. Notice how we have used more descriptive names, YEAR and FLOW, for the arrays. In addition, notice how the index (I) of the loop is used as a subscript for the arrays. Thus, on the first pass through the loop, I is 1, and the program will sum the value FLOW(1). Then, on subsequent passes, I will be incremented in order to sum FLOW(2), FLOW(3), and so on.

Notice how the DATA statement in the previous example does not include the subscript. This is valid when you want to use the DATA statement to specify every member of the array. In the event that you wanted to assign only some of the values, you have to include the subscript. For example, the DATA statement for FLOW could have been expressed as

```
DATA FLOW(1), FLOW(2)/125.,102./
DATA FLOW(3), FLOW(4)/147.,76./
```

17.2.2 Specification Statements

To this point, we have now introduced several types of statements that we have said must be placed at the beginning of a program. These are the PROGRAM, PARAMETER, DATA statements, type declarations (including DIMENSION and COMMON), and user-defined statement functions.

Before proceeding further, we must note that some of these specification statements need to be placed in a particular order. We would suggest that you adopt the following scheme:

1. PROGRAM statement
2. PARAMETER statements
3. IMPLICIT type statements
4. EXPLICIT type statements
5. DIMENSION statements
6. COMMON statements
7. DATA statements
8. Statement functions
9. The body of the program (executable statements)
10. END statement

A similar scheme would be followed in subprograms, with the exception that a SUB-ROUTINE or FUNCTION statement would take the place of the PROGRAM statement and a RETURN would be required before the END statement.

It should be noted that some of the above ordering is not mandatory. For example, PARAMETER statements can be intermingled with the type specifications. Similarly, DATA statements can be interspersed with statement functions and executable statements. Although there are times when it will make sense to mix the statements in these ways, we have found that as a rule this can result in programs that are sometimes difficult to understand. Consequently, we believe that generally following the above guidelines will lead to FORTRAN codes that are clearer and easier to use.

17.2.3 Implied DO Loops

Aside from DATA statements, arrays can be input to a program using READ statements. An implied DO loop provides a means to make this action more concise. In addition, the implied DO loop can be employed for output.

The implied DO loop has the general form

(*i/oList,index* = *start,finish,increment*)

where the *i/oList* is the input-output list of variables. The values of the *index* are established from *start, finish,* and *increment* exactly as for the DO loop.

An illustration of the implied DO loop can be developed by recalling that in Example 17.2, flows were entered with the DATA statement,

```
DATA FLOW/125., 102., 147., 76./
```

If you would rather have the data entered by the user, a READ statement could be employed:

```
READ *, FLOW(1), FLOW(2), FLOW(3), FLOW(4)
```

The user would then enter the values when the program was run as

```
125., 102., 147., 76.
```

An equivalent READ utilizing the implied DO loop could be written as

```
READ *, (FLOW(I), I = 1, 4)
```

The implied DO loop is especially useful when a large number of values must be input or output. For example, the 30 values from Table 17.1 could be input with

```
N = 30
READ *, (FLOW(I), I = 1, N)
```

Therefore, rather than using a DO loop, this single line would accomplish the same action.

As mentioned above, the implied DO loop can be combined with other values that are to be input or output. This attribute is illustrated in the following example.

EXAMPLE 17.3   Employing an Implied DO Loop to Enter Values

PROBLEM STATEMENT: Solve the same problem as in Examples 17.1 and 17.2 but utilize implied DO loops to enter the years and flows.

SOLUTION:

```
      PROGRAM EX173
      REAL FLOW(4), AVERAG, SUM
      CHARACTER*4 YEAR(4)
      INTEGER I, J, N
C
      PRINT *, 'Enter number of years of flow'
      PRINT *, 'followed by the years entered as strings.'
      PRINT *, 'Enter as many values per line as desired'
      READ *, N, (YEAR(J), J = 1, N)
      PRINT *
      PRINT *, 'Enter the flows.'
      PRINT *, 'Enter as many values per line as desired'
      READ *, (FLOW(J), J = 1, N)
C
      DO 10 I = 1, N
         SUM = SUM + FLOW(I)
   10 CONTINUE
      AVERAG = SUM/N
      PRINT *, 'year    flow'
      DO 20 I = 1, N
         PRINT *, YEAR(I), FLOW(I)
   20 CONTINUE
      PRINT *, 'average = '; AVERAG
C
      END
```

When the program was run, the data could be entered, as in

```
Enter the number of years of flow
followed by the years entered as strings.
Enter as many values per line as desired
4 '1954' '1955' '1956'
'1957'

Enter the flows.
Enter as many values per line as desired
125. 102. 147. 76.
```

The output for this program would be identical with that of Examples 17.1 and 17.2.

Aside from READ and PRINT statements, the implied DO loop can also be used in conjunction with the DATA statement. This is usually done when only part of an array is to be initialized. For example, suppose that you wanted to initialize the first half of a 50-element array, X, to the same value. An example illustrating how this could be done is

```
DATA (X(I), I = 1, 25)/25*1.0/
```

This statement would set the first 25 elements of X to a value of 1.0.

Finally, it should be mentioned that implied DO loops can be nested. However, to truly appreciate where this is useful, we will first introduce multidimensional arrays. Then, in Example 17.4, we will illustrate their use.

17.2.4 Two-Dimensional Arrays

There are many engineering problems where data is arranged in tabular or rectangular form. For example, temperature measurements on a heated plate (Fig. 17.3) are often taken at equally spaced horizontal and vertical intervals. Such a two-dimensional arrangement is formally referred to as a *matrix*. It should be noted that in mathematics the one-dimensional array is often called a *vector*.

As shown in Figure 17.4, the horizontal sets of elements are called *rows,* whereas the vertical sets are called *columns.*[2] The first subscript designates the number of the row in which the element lies. The second subscript designates the column. For example, element A(2, 3) is in row 2 and column 3.

Matrices are so common in mathematically-oriented fields like engineering that FORTRAN has the capability of storing data in *two-dimensional arrays*. As in the following example, these are extremely useful when dealing with information such as the temperatures from Fig. 17.3.

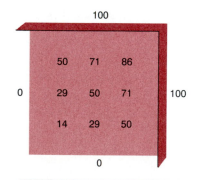

FIGURE 17.3
Temperature measurements
on a heated plate.

---

[2]A handy way to keep the notions of a row and a column straight is to remember that the rows of a theater are horizontal and the columns of a temple are vertical.

FIGURE 17.4
A two-dimensional array, or matrix, used to store the temperatures from Fig. 17.3. Note that the first and second subscripts designate the row and the column of the elements, respectively.

EXAMPLE 17.4   Two-Dimensional Arrays

PROBLEM STATEMENT: Write a computer program to input the temperature data from Fig. 17.3. Then convert it to degrees Fahrenheit by the formula

$$T_F = \tfrac{9}{5} T_c + 32$$

Employ a user-defined function to make this conversion. Display the results as rounded integers.

SOLUTION: Figure 17.5 shows the program. Observe how a user-defined function is employed to make the temperature conversion (TCTOTF) and an intrinsic function is used to round the results (NINT). Also note that the rounded results are stored in a special variable, TFINT. This was done so that the values stored in the array TEMPF are unaffected by the rounding process.

Finally, see that an implied loop is employed to expedite data entry. For this case, nested loops are used to capitalize on the two-dimensional nature of the array. For output, only a single implied loop is utilized because we wanted each displayed line to consist of a single row of temperatures. This result is due to the fact that each time a PRINT statement executes, it starts at the beginning of a new line. Nested inplied loops would have resulted in an output with a different appearance; that is, each new line would begin at the end of the previous line.

Note that arrays with more than two dimensions can be employed. These are also useful in engineering. For example, if a third dimension of information were available for the plate in Fig. 17.3 (suppose temperatures were available at several levels into the plate), a three-dimensional array would come in handy. If the plate's temperatures were measured at several different times, a fourth dimension could be added to specify each

Program:

```
              PROGRAM PLATE
C
C Heated plate program
C
C by S.C. Chapra
C July 1, 1992
C
C This program enters temperatures in
C in degree C for a two-dimensional
C heated plate.  It converts these
C values to degree F and outputs the
C results as rounded integers
C
C Definition of variables
C TEMPC() = temperatures (C)
C TEMPF() = temperatures (F)
C TFINT() = rounded integer temps (F)
C NR = number of rows of temperatures
C NC = number of columns of temperatures
C I, J = loop counters
C
      REAL TEMPC(20, 20), TEMPF(20, 20)
      INTEGER NR, NC, I, J, TFINT(20, 20)
C
      TCTOTF(T) = (9. / 5.) * T + 32.
C
C input temperatures
      READ *, NR, NC
      READ *, ((TEMPC(I, J), J = 1, NC), I = 1, NR)
C
C perform conversion to rounded Fahrenheit
      DO 30 I = 1, NR
        DO 20 J = 1, NC
          TEMPF(I, J) = TCTOTF(TEMPC(I, J))
          TFINT(I, J) = NINT(TEMPF(I, J))
  20    CONTINUE
  30 CONTINUE
C
C output results
      PRINT *
      PRINT *, 'Results:'
      PRINT *
      DO 40 I = 1, NR
        PRINT *, (TFINT(I, J), J = 1, NC)
  40 CONTINUE

      END
```

Data:

```
5 5
100. 100. 100. 100. 100.
0. 50. 71. 86. 100.
0. 29. 50. 71. 100.
0. 14. 29. 50. 100.
```

Output:

```
Results:
```

212	212	212	212	212
32	122	160	187	212
32	84	122	160	212
32	57	84	122	212
32	32	32	32	212

FIGURE 17.5
Program to input,
convert, and output a
two-dimensional array for the
temperatures from Fig. 17.3.

case. Other problem contexts might require additional dimensions. FORTRAN allows the use of such multidimensional arrays.

17.2.5 How Arrays Are Stored in Memory

Although it is valid for illustrating the two-dimensional structure of a matrix, Fig. 17.4 is somewhat misleading regarding the way in which the computer actually stores such information in its memory. In the present section, we would like to elaborate on this topic. We do this because there are instances where an accurate understanding of this process can be useful. In particular, it has relevance to the actual manner in which the computer passes arrays to subprograms.

If you think about it, there are two simple ways to keep track of the values of an array in a computer's memory. One way would be to assign labels to individual memory locations as in Fig. 17.4. Although this would certainly work, storing all this information would itself take up memory.

An alternative, which requires less memory, would be to design the language so that the elements are stored sequentially. Then, the location (or address) of the first element would be all that would be required to find the array. An individual element could then be easily located by determining how far down it was located from the first element. This downward distance is dubbed the element's *displacement.*

A hypothetical example might help clarify this alternative. Suppose that the first five flow values from Table 17.1 are stored with the following statements:

```
REAL FLOW(5)
DATA FLOW/125., 102., 147., 76., 95./
```

The way that these values are stored in the computer can be represented by a *storage table* of the form

Name	Value	Address[3]
flow(1)	125	0305
flow(2)	102	0306
flow(3)	147	0307
flow(4)	76	0308
flow(5)	95	0309

This table shows how the array would be stored in a series of sequential memory locations. The locations are represented by their *address,* which is the code employed by the computer to label a specific memory location. When a computer compiles code, it replaces the variable names with these assigned memory addresses.

The first address is especially important since it represents the means whereby the

---

[3]Note that the actual codes would be represented in binary machine language. We have employed the more familiar base 10 numbers for illustrative purposes.

computer locates the array in memory. Consequently, it is called the array's *base address*. Also, observe that this initial element is considered to have a displacement of 0. If we wanted to find the address of the fourth element, we could merely calculate it by the simple formula

$$a_i = a_1 + d_i$$

where a_1 is the base address, a_i is the address of the ith element, and d_i is the displacement of the ith element. For this simple case, the displacement is simply the subscript minus 1, or

$$d_i = i - 1$$

Consequently, the address of the element, FLOW(4), could be calculated as

$$a_4 = 0305 + 4 - 1 = 0308$$

Such a system becomes a little more complicated when storing a two-dimensional array. Because the computer's memory addresses are accessed sequentially, the information would still be stored in sequence. This means that a decision has to be made regarding how this is done: by columns or by rows? As illustrated in Fig. 17.6, storage

FIGURE 17.6
Column storage of a
two-dimensional array in a
computer.

(a) Program

```
PROGRAM ARRAY
PARAMETER (N = 3, M = 2)
REAL A (N,M)
INTEGER I,J
DO 10 I = 1,N
    DO 20 J = 1,M
        READ *, A(I,J)
20    CONTINUE
10 CONTINUE
END
```

(b) Matrix

$$\begin{array}{c c c} & j = 1 & j = 2 \\ i = 1 & \begin{bmatrix} 4 & -1 \\ i = 2 & 7 & 3 \\ i = 3 & 2 & 5 \end{bmatrix} \end{array}$$

(c) Storage table

element	value	displacement	address
a(1,1)	4	0	5063
a(2,1)	7	1	5064
a(3,1)	2	2	5065
a(1,2)	-1	3	5066
a(2,2)	3	4	5067
a(3,2)	5	5	5068

is customarily done by columns. As we will see later, this convention is extremely significant.

Next, a formula for determining displacement must be devised. Suppose that we are dealing with an array with n rows and m columns. Our intent is to come up with a displacement to substitute into the following formula:

$$a_{i,j} = a_{1,1} + d_{i,j} \tag{17.1}$$

For column one (second subscript, $j = 1$), the displacement is simply the first subscript minus 1, or $i - 1$. For column two ($j = 2$), the displacement would amount to the number of elements in column one (n) plus the first subscript minus 1, $i - 1$. If there were a third column ($j = 3$), the displacement would be equal to the number of elements in the first and second columns $[(j - 1) \cdot n]$ plus the first subscript, $i - 1$. Generalizing this scheme, a simple formula can be developed as

$$d_{i,j} = (j - 1) \cdot n + (i - 1) \tag{17.2}$$

The application of this formula will be illustrated in the following example.

EXAMPLE 17.5   Displacement for Two-Dimensional Arrays

PROBLEM STATEMENT: Test Eqs. (17.1) and (17.2) for the matrix shown in Fig. 17.6 by verifying that element A(3, 2) has an address of 5068.

SOLUTION: For this example, the base address is $a_{1,1} = 5063$; and the parameters are $n = 3$, $m = 2$, $i = 3$, and $j = 2$. Substituting these values into the displacement equation (17.2) yields

$$d_{3,1} = (2 - 1) \cdot 3 + (3 - 1) = 5$$

This result can be substituted into the address equation (17.1) to give

$$a_{3,1} = 5063 + 5 = 5068$$

Substitution of other values from the table will confirm that the formula is generally valid.

We have reviewed the foregoing material to acquaint you with how arrays are handled within the computer. As described in Section 17.2.7, such information has relevance to the manner in which arrays are passed to subprograms. However, before learning how this is done, we will briefly show how arrays are used in conjunction with DATA statements.

17.2.6 Two-Dimensional Arrays and DATA Statements

Aside from how it affects passing values to subprograms, the way arrays are stored has relevance to the way in which they are initialized with DATA statements. All that needs to be understood is that the values in the DATA statement's *ConstantList* must be ordered in the same way that the values are stored. For example, an equivalent version of the program from Fig. 17.6*a* would be

```
PROGRAM ARRAY
PARAMETER (N = 3, M = 2)
REAL A(N, M)
DATA A/4., 7., 2., -1., 3., 5./
END
```

Thus, the data is assigned according to the manner in which it is stored; that is, by columns.

17.2.7 Arrays in Subprograms

Arrays can be passed to subprograms via argument or by using COMMON statements. Furthermore, arrays can also exist in subprograms as local variables. In this section, we will briefly review these features.

Arguments.   Arrays are passed to subprograms in a fashion similar to simple variables. The only difference is that they must be dimensioned in a special way in the subprogram.

Suppose that we were transferring a one-dimensional array TEMP of length N = 20 to a subroutine named PROC. The relevant parts of the calling program might look like

```
DIMENSION TEMP(20)
CALL PROC(TEMP, 20)
```

The beginning of the corresponding subroutine could be written as

```
SUBROUTINE PROC(T, N)
DIMENSION T(N)
```

Notice that just as with unsubscripted variables, different variable names can be used for arrays in the main program and the subprogram.

When an array is transferred to a subprogram as an argument, it is *passed by reference* (recall Sec. 16.2). That is, a new memory location is not set up to store the value. Rather, the subprogram uses the array's base address and displacement to locate the proper values in the memory locations for the variable *in the main program*. The presence of the variable name in the subprogram's argument list provides a path that the computer follows to determine the original memory location. The DIMENSION statement informs the compiler that it is dealing with an array. It, therefore, knows that it has to manage base addresses and displacements to locate the values it needs.

In addition, observe that the array must be dimensioned in the subprogram. However, see that we have done this in a somewhat peculiar fashion. Rather than use a constant (20), as in the main program, we employ the variable (N). The use of a variable suggests that we could pass a part of an array to a subprogram. For example, the CALL statement might be represented as

```
CALL PROC(TEMP, 10)
```

Before exploring the implications of such a capability, we must clearly delineate the somewhat different role of the DIMENSION statement in the main program and the subprogram. In the main program, the DIMENSION statement has three functions: (1) to name the array, (2) to identify how many subscripts the array has, and (3) to specify the ranges of the subscripts. The final function itself has two aspects. First, specifying the ranges provides the necessary information that allows the computer to determine the memory location of each element (as discussed in Sec. 17.2.5). Second, it instructs the computer regarding how many memory locations must be set aside to store the entire array.

The DIMENSION in the subroutine has the first two functions. That is, it provides a name and the number of subscripts for the array. Furthermore, it also provides the necessary information to locate the individual elements. However, it has the important distinction that it does not actually set aside any new memory locations. Because the arrays in both DIMENSION statements refer to the same information, this would obviously be a wasteful expenditure of memory. Rather, the DIMENSION in the calling program solely performs this job. The significance of this distinction can be best illustrated by example.

EXAMPLE 17.6   Argument and Nonargument Array Declaration

PROBLEM STATEMENT: Problems involving passing arrays to subprograms sometimes involve passing part of a multidimensional array. Write a program to enter and display the matrix of numbers

$$\begin{bmatrix} 1 & 2 & 3 & 4 \\ 5 & 6 & 7 & 8 \\ 9 & 10 & 11 & 12 \end{bmatrix}$$

Then, develop a subroutine to display the upper left section of the matrix

$$\begin{bmatrix} 1 & 2 & 3 \\ 5 & 6 & 7 \end{bmatrix}$$

SOLUTION: The following program could be developed by someone who did not understand the nuances of passing parts of arrays to subprograms:

```
PROGRAM EX176
INTEGER I, J, K, L
PARAMETER (K = 3, L = 4)
INTEGER ICALL(K, L)
DATA ICALL/1,5,9,2,6,10,3,7,11,4,8,12/
PRINT *, 'total array in calling program:'
DO I = 1, K
   PRINT *, (ICALL(I, J), J = 1, L)
10 CONTINUE
CALL DISPLY(ICALL, 2, 3)
END

SUB DISPLY(ISUBR, N, M)
INTEGER I, J, M, N, ISUBR(N, M)
PRINT *, 'portion of array in subprogram:'
DO I = 1, N
   PRINT *, (ISUBR(I, J), J = 1, M)
10 CONTINUE
RETURN
END
```

When this program is run, the following output results:

```
total array in calling program:
1       2       3       4
5       6       7       8
9       10      11      12
portion of array in subprogram:
1       9       6
5       2       10
```

Before the subroutine is implemented, the array appears in the original, nicely ordered way in which it was entered. However, because the dimensions (2 × 3) that are passed are smaller than the original array, only the first 6 elements are transferred to the subroutine. Because of the column-storage convention, the first two columns of the original array are actually passed. Then, because of the way in which we set up the PRINT statement, the resulting output is laid out as shown.

The foregoing program was meant to illustrate that you should be very careful when passing multidimensional arrays to subprograms. One safe way to avoid problems is to always pass the entire array to the subprogram. For instance, the CALL statement in the above program could have been rewritten as

```
CALL DISPLY(ICALL, K, L, 2, 3)
```

and the first few lines of the subroutine changed to

```
SUBROUTINE DISPLY(ISUBR, K, L, N, M)
INTEGER I, J, K, L, M, N, ISUBR(K, L)
```

After these modifications, the relevant program output would give the desired result

```
portion of array in subprogram:
1    2    3
5    6    7
```

There are three options for passing the entire array to the subroutine. As in the previous example, the array dimensions can be passed as variables. A second alternative would be to merely set the dimensions in the subprogram to the same constant values as in the main program. For example, the relevant statements could have been coded as

```
CALL DISPLY(ICALL, 2, 3)
        .
        .
        .
SUBROUTINE DISPLY(ISUBR, N, M)
INTEGER I, J, K, L, M, N, ISUBR(3, 4)
```

It should be noted that FORTRAN provides a third option to do this that is a little more concise. This involves using an asterisk in the subprogram declaration. For one-dimensional arrays, this option is straightforward. For example, the relevant parts of the calling program might look like

```
DIMENSION TEMP(20)
CALL PROC(TEMP, 20)
```

The beginning of the corresponding subroutine could be written as

```
SUBROUTINE PROC(T, N)
DIMENSION T(*)
```

The compiler will know that the array T has the same dimension as the variable TEMP in the calling program.

The use of an asterisk for multidimensional arrays is a little more tricky. For example,

```
CALL DISPLY(ICALL, 2, 3)
        .
        .
        .
SUBROUTINE DISPLY(ISUBR, N, M)
INTEGER I, J, M, N, ISUBR(3, *)
```

For this case, the value of the first dimension must be specified. The asterisk then automatically tells the compiler that you desire the array in the subprogram to have the same second dimension as in the main program. In the case of higher dimensional arrays, an asterisk can be used only for the final dimension.

It should be noted that individual elements of arrays can be passed to subprograms. If this is done, the individual element is listed in the CALL and a nonsubscripted variable is used in the corresponding position in the subprogram parameter list. For example,

```
CALL EXMSUB(X(3))
```

and

```
SUBROUTINE EXMSUB(A)
```

Notice that, when individual elements are passed, the subroutine must use nonsubscripted variables to represent the values. In this example, the SUB would go to the memory locations for X(3) in order to determine the value of A used in the subprogram. If the value of A is modified in the subroutine, this will result in X(3) being changed in the calling program.

Finally, it should be mentioned that entire arrays cannot be passed by value (recall Sec. 16.2). This is important if you desire to retain the original values for the case where the subprogram modifies array elements. In these situations, an option is to assign the original values to another array before you invoke the subprogram.

COMMON.   Arrays can also be passed to subprograms using COMMON statements. For instance, returning to the example of transferring a one-dimensional array TEMP of length 20 to a subroutine named PROC, the relevant parts of the calling program might look like

```
DIMENSION TEMP(20)
COMMON TEMP
CALL PROC
```

The following COMMON statement could be placed in the subprogram in order to give it access to the temperatures:

```
SUBROUTINE PROC
COMMON T(20)
```

It should be noted that the COMMON statement can be used to specify the size of the array. For example,

```
COMMON TEMP(20)
CALL PROC
       .

       .

       .
SUBROUTINE PROC
COMMON T(20)
```

Thus, for this case, a DIMENSION statement is not required in the main program. However, it should be stressed that you cannot specify an array's size in both a COMMON and a type statement.

Arrays as local variables in subprograms.   Aside from being passed as arguments or shared in COMMON, arrays can also exist as local variables in subprograms. For these cases, the array is not dimensioned in the calling program, but it must be dimensioned at the beginning of the subprogram. This and other aspects of arrays and subprograms are explored in the following example.

EXAMPLE 17.7   *Modular Temperature Conversion Program*

PROBLEM STATEMENT: Write a computer program similar to the one previously developed in Example 17.4 to input the temperature data from Fig. 17.3. Then convert it to degrees Fahrenheit by the formula

$$T_F = \tfrac{9}{5}T_C + 32$$

Employ a user-defined function to make this conversion. Take the results and convert them to degrees Rankine by the formula

$$T_R = T_F + 460$$

Display the results as rounded integers. Develop the program in a modular fashion by employing subroutines.

SOLUTION: Figure 17.7 shows the program, and Fig. 17.8 displays the resulting output for an 80-column printer. Observe that two subroutines are employed to process (PROC) and postprocess (PSTPRC) the information. The following features of the program are noteworthy:

1. The DATA statement employs nested implied DO loops to enter the temperatures in a concise and efficient fashion.
2. A local array, TF, is employed to hold the Fahrenheit temperatures prior to their conversion into degrees Rankine. Notice how this array is dimensioned in the subroutine.
3. The outputs of the initial and the converted values are decidedly different. This is because the original data is input into real variables, whereas the final rounded values

```
      PROGRAM PLATE2                              REAL TC(20, *)
C                                                 INTEGER NR, NC, TR(20, *)
C Heated plate program                            REAL TF(20, 20)
C                                                 INTEGER I, J, TCTOTF
C by S.C. Chapra                      C
C July 1, 1992                                    TCTOTF(T) = (9. / 5.) * T + 32.
C                                     C
C This program enters temperatures in
C in degree C for a two-dimensional               DO 30 I = 1, NR
C heated plate.  It converts these                   DO 20 J = 1, NC
C values to degree R and outputs the                    TF(I, J) = TCTOTF(TC(I, J))
C results as rounded integers                           TR(I, J) = NINT(TF(I, J) + 460.)
C                                          20      CONTINUE
C Definition of variables              30  CONTINUE
C TEMPC() = temperatures (C)               RETURN
C TEMPR() = rounded integer temps (R)      END
C NR = number of rows of temperatures
C NC = number of columns of temperatures   SUBROUTINE PSTPRC(TC, TR, NR, NC)
C                                           REAL TC(20, 20)
      REAL TEMPC(20, 20)                    INTEGER NR, NC, I, J, TR(20, *)
      INTEGER NR, NC, TEMPR(20, 20)         PRINT *
      DATA NR, NC / 5,  5/                  PRINT *, 'Input data (C):'
      DATA ((TEMPC(I, J), J = 1, 5), I =1, 5)  PRINT *
     &   / 100., 100., 100., 100., 100.,   DO 10 I = 1, NR
     &     0., 50., 71., 86., 100.,           PRINT *, (TC(I, J), J = 1, NC)
     &     0., 29., 50., 71., 100.,       10  CONTINUE
     &     0., 14., 29., 50., 100.,          PRINT *
     &     0., 0., 0., 0., 100. /           PRINT *, 'Results (R):'
C                                           PRINT *
      CALL PROC(TEMPC, NR, NC, TEMPR)       DO 20 I = 1, NR
      CALL PSTPRC(TEMPC, TEMPR, NR, NC)       PRINT *, (TR(I, J), J = 1, NC)
      END                                 20  CONTINUE
C                                           RETURN
      SUBROUTINE PROC(TC, NR, NC, TR)       END
```

FIGURE 17.7 Revised program to input, convert, and output a
two-dimensional array for the temperatures from Fig. 17.3.

FIGURE 17.8
Output from the program in
Fig. 17.7.

Input data (C):

100.000000	100.000000	100.000000	100.000000
100.000000			
0.000000E+00	50.000000	71.000000	86.000000
100.000000			
0.000000E+00	29.000000	50.000000	71.000000
100.000000			
0.000000E+00	14.000000	29.000000	50.000000
100.000000			
0.000000E+00	0.000000E+00	0.000000E+00	0.000000E+00
100.000000			

Results (R):

671	671	671	671	671
492	581	619	646	671
492	544	581	619	671
492	517	544	581	671
492	492	492	492	671

are stored as integers. Hence, when the temperatures are output in free format, the results for the real values do not align nicely into columns. In the next chapter, we will introduce methods to control the format of such output.

17.3 FORTRAN 90: RECORDS

FORTRAN 90 has much greater capabilities for data structuring than FORTRAN 77. In the present section, we will introduce the most fundamental of these capabilities: the record.[4]

The arrays discussed to this point represent an example of data structuring. They have allowed us to access a group of numbers using a single name. Furthermore, the two-dimensional organization of information in a matrix embodies information over and above the actual values of the numbers themselves. For example, the way the array was used to store the temperatures in the program from Fig. 17.5 corresponds to the spatial distribution shown in Fig. 17.3.

As described to this point, an array has the limitation that all of its information must be the same type. Information does not always come in this form. Often it consists of different types of information. For example, think of the type of information that is associated with you as an individual. Aside from your name, you also have several identification numbers such as your social security and driver's license number. Furthermore, you could be described by your sex, height, weight, and eye and hair color. Such information could be compiled on a database called a "file."

As seen in Fig. 17.9, each individual piece of information on a file is called an *item*. Examples might be a name, a social security number, a flow rate, or a price. A *record* is a group of items that relate to the same object or individual. For example, in your university's central computer there is undoubtedly a record that refers to your academic standing. It probably consists of a number of items, including your name, social security or identification number, class, department, grades in individual courses, credit hours, and the like. This record is in turn a part of a larger entity called a *file,* which contains the academic records of all the students at your school. Finally, this file is one of many files that are maintained in the *storage* or *file system* of the university's mainframe computer.

In the following chapter, we will show how FORTRAN will allow you to generate a file of information. However, beyond formally developing a file, FORTRAN also allows us to make an array into a file of sorts. That is, it allows us to view the individual rows of a matrix as a record.

To do this, two types of statements are used: the STRUCTURE and the RECORD statements. The STRUCTURE statement has the general form

---

[4]The material in this section corresponds to the Microsoft FORTRAN 5.1 implementation.

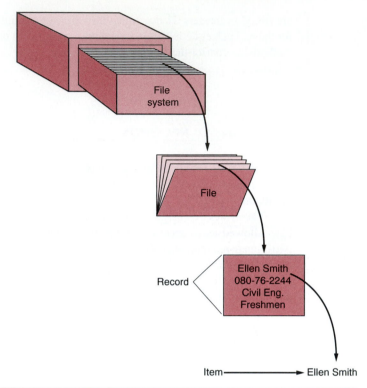

FIGURE 17.9
The various components
associated with data files. The
figure uses the analogy
between an office file system
and a file system employed
on a computer.

```
STRUCTURE /UserType/
   type   ElementName
              .
              .
              .
END STRUCTURE
```

These statements allow programmers to set up their own customized type. The *UserType* represents the name given to this type. Each element of the record is then given a *type* and an *ElementName*.

After establishing the user type, the RECORD statement is employed to assign this new data type to a variable name, as in

```
RECORD /UserType/ VariableName(n), . . .
```

where *VariableName* is the variable name of an array that is being defined as *UserType* and *n* is the integer value defining the upper limit on the number of elements in this array.

Finally, the *j*th member of a particular element can be invoked in the program through *VariableName(j).ElementName*. The following example shows how these statements are employed in a program.

EXAMPLE 17.8   Using a Record in Conjunction with an Equation of State

PROBLEM STATEMENT: The ideal gas law models the behavior of a single mole of gas as

$$P = \frac{RT}{V}$$

where P = pressure [atm], V = volume [L], R = the universal gas constant [= 0.08205 (atm · L)/(mole · K)], and T = temperature [K].
A modification of this formula, which can handle nonideal conditions, is the Redlich and Kwong equation of state:

$$P = \frac{RT}{V - b} - \frac{a}{\sqrt{T}V(V + b)}$$

where

$$a = \frac{0.4278R^2T_c^{2.5}}{P_c}$$

$$b = \frac{0.0867RT_c}{P_c}$$

where P_c = the critical pressure of the substance [atm] and T_c = the critical temperature [K]. These parameters have been compiled for numerous gases. For example:

Compound	T_c	P_c
Methane	190.6	45.4
Ethylene	282.4	49.7
Nitrogen	126.2	33.5
Water	647.1	217.6

Develop a program to compute pressure using both formulas for each of these gases for a particular temperature and pressure. Test it for $T = 400°K$ and $V = 100$ L. Employ DATA statements to input the critical data into record arrays.

SOLUTION: The program is shown in Fig. 17.10. The output from the program would be

```
Temperature (K) = 400.000000
Volume (L) = 100.000000

Compound                P(ideal)        P(R&K)
Methane                 0.328200        0.328139
Ethylene                0.328200        0.3279446
Nitrogen                0.328200        0.3282111
Water                   0.328200        0.3275647
```

FIGURE 17.10
Program to illustrate the use
of records in FORTRAN 90.
See Example 17.8 for an
explanation of the program.

```
      PROGRAM EX178
      PARAMETER (R = 0.08205, NVAL = 4)
      STRUCTURE /ChemData/
         CHARACTER*15 Compound
         REAL         Tc
         REAL         Pc
      END STRUCTURE
      RECORD /ChemData/ Chem(NVAL)
      REAL T, V, Pideal, Prk, a, b
      INTEGER I
C
C initialization of record
C
      Chem(1).Compound = 'Methane'
      Chem(1).Tc = 190.6
      Chem(1).Pc = 45.4
      Chem(2).Compound = 'Ethylene'
      Chem(2).Tc = 282.4
      Chem(2).Pc = 49.7
      Chem(3).Compound = 'Nitrogen'
      Chem(3).Tc = 126.2
      Chem(3).Pc = 33.5
      Chem(4).Compound = 'Water'
      Chem(4).Tc = 647.1
      Chem(4).Pc = 217.6
      PRINT *, 'Temperature (K) ='
      READ *, T
      PRINT *, 'Volume (L) ='
      READ *, V
      PRINT *
      PRINT *, 'Compound            P(ideal)          P(R&K)'
      DO 10 I = 1, NVAL
         Pideal = R * T / V
         a = 0.4278 * R ** 2 * Chem(I).Tc ** 2.5 / Chem(I).Pc
         b = 0.0867 * R * Chem(I).Tc / Chem(I).Pc
         Prk = R * T /(V - b) - a / (SQRT(T) * V * (V + b))
         PRINT *, Chem(I).Compound, Pideal, Prk
   10 CONTINUE
      END
```

PROBLEMS

17.1 Given the following code, what is the output?

```
      DO 10 I = 1, 3
        DO 20 J = 1, 3
          A(I, J) = I * J
20      CONTINUE
10 CONTINUE
      DO 30 I = 1, 3
        DO 40 J = 1, 3
          PRINT *, A(J, I)
40      CONTINUE
30 CONTINUE
```

17.2 Write a FORTRAN DO loop to fill each element of a one-dimensional array of 100 elements with the element's subscript.

17.3 (a) What is the largest integer array (that is, the largest dimension) you can use on your computer (to the nearest kilobyte)?
(b) How big is the largest array of real numbers you can use on your computer (to the nearest kilobyte)?
(c) How big is the largest N×N×N array of real numbers you can use on your computer (to the nearest kilobyte)?
(d) Contrast the results of (a) and (b), and draw conclusions related to the use of integer and real arrays in computer programs.

17.4 Write FORTRAN statements to display the contents of an array, A. Note that the array begins with the element A(1) and ends with the element A(150). Write the statements so that the array is displayed in reverse order, (a) placing each element on a line by itself and (b) placing two elements on each line.

17.5 Write a well-structured modular program to input a series of numbers and compute their average and standard deviation. Whereas the average (or *arithmetic mean*) provides an estimate of the "location" or center of a group of data, the *standard deviation* is a statistic that quantifies its "spread" around the mean. A formula for the standard deviation is

$$s_x = \sqrt{\frac{\sum x_i^2 - \left[\left(\sum x_i \right)^2 / n \right]}{n - 1}}$$

Where all summations are from $i = 1, n$. Employ the following algorithm as the basis for the program:

- Step 1 Use a loop to input, count, and sum the numbers. Example 15.1 might serve as a model for inputting and counting. Employ an array to store the values as subscripted variables.
- Step 2 Compute the average value as the sum divided by the total number of values. Also calculate the standard deviation using the formula shown above.
- Step 3 Display the individual values.
- Step 4 Display the average and the standard deviation. Test the program by determining the average flow for the data in Table 17.1.

17.6 Repeat Prob. 17.5, but rather than inputting the data via the keyboard, use DATA statements to incorporate the data into the program. Employ subscripted variables to store the values for the flows.

17.7 Repeat Prob. 17.6, but use an array that has subscripts equal to the years at which the flows were measured.

17.8 Suppose that you have an array A of dimension N which contains values that are increasing. An example of such data, for the case where n = 12, is

A(1) = -6   A(5) = 9.3   A(9) = 22
A(2) = 0   A(6) = 12   A(10) = 33
A(3) = 2   A(7) = 15   A(11) = 49
A(4) = 6.5   A(8) = 21   A(12) = 100

Develop a modular program that employs a subprogram to enter this array using READ statements. Then, set up a query in the main program that prompts the user for a number, X. Pass this number to another subprogram that determines a value K such that X is between A(K) and A(K+1). Return values of 0 or N if X is outside the range. Display this number for the user. Employ a DO UNTIL structure to allow the user to repeat the process as many times as desired.

17.9 Repeat Prob. 17.8, but design it to work for arrays containing values that are monotonic (that is, either increasing *or* decreasing).

17.10 Suppose that FORTRAN employed default arrays that started with a 0 instead of a 1 subscript. Rederive Eq. (17.2) for this case.

17.11 Employ arrays and DATA statements to develop a modular computer program to determine your grade for this course.

17.12 Employ arrays and DATA statements to create a modular program to determine the statistics for a sports team.

17.13 When dealing with numbers, it is often necessary to sort them in ascending order. There are many methods for accomplishing this objective. A simple approach, called a *selection sort,* is probably very close to the commonsense approach you might use to solve the problem. First you would scan the set of numbers and pick the smallest. Next you would bring this value to the head of the list. Then you would repeat the subprogram on the remaining numbers and bring the next smallest value to the second place on the list. This subprogram would be continued until the numbers were ordered. Develop a well-structured modular program to implement the selection sort. Test it on the data from Table 17.1.

17.14 Another simple sorting algorithm, called the *bubble sort,* consists of starting at the beginning of the array and comparing adjacent values. If the first value is smaller than the second, nothing is done. However, if the first is larger than the second, the two values are switched. Next the second and third values are compared, and if they are not in the correct order (that is, smallest first), they are also switched. This process is repeated until the end of the array is reached. This entire sequence is called a *pass.*

After the first pass the process returns to the top of the list and another pass is implemented. However, on this pass, you only have to go down to the next-to-the-last element. The subprogram is repeated until a pass with no switches occurs. The name "bubble sort" comes from the fact that the smallest value tends to rise to the top of the list like a bubble rising through a liquid. Develop a well-structured modular program to implement the bubble sort. Test it on the data from Table 17.1.

17.15 Aside from sorting, another important operation performed on data is searching—that is, locating an individual item in a group of data. One common method for accomplishing this objective is the *binary search.* Suppose that you are given an array A of length N and must determine whether the value V is included in this array. After sorting the array in ascending order (see Probs. 17.13 and 17.14), you employ the following algorithm to conduct a binary search:

Compare V with the middle term A(M). Three results can occur: (*a*) If V = A(M), the search is successful; (*b*) if V < A(M), V is in the first half; and (*c*) if V > A(M), V is in the second half.

In the case of (*b*) or (*c*), the search is repeated in the half in which V is known to lie. Then V is again compared with A(M), where M is now redefined as the middle value of the smaller list. The process is repeated until V is located or the array is exhausted.

Develop a subroutine subprogram to implement a binary search. Use structured programming techniques and a modular design so that the subroutine can be easily integrated into other programs.

17.16 Write nested FORTRAN DO loops to fill each element of a three-dimensional array with the product of its subscripts. Test your code for a $4 \times 5 \times 3$ matrix.

17.17 (*FORTRAN 90*) Select 10 of your favorite issues from the New York Stock Exchange. Write, debug, and document a FORTRAN program that reads the name of the stock, its selling price today, and its 52-week low. Use a record to hold this information. The program should then display the name of the stock, its selling price, and the percentage gain over the 52-week low.

17.18 *(FORTRAN 90)* Employ records to enter the following Student Aptitude Test (SAT) data into a computer program:

ID Number	Name	Math	Verbal
551556681	Reckhow, Sarah	736	710
337653878	Chen, David	650	680
263748294	Morales, Juan	690	550
876770183	Hornacek, Heather	620	640
627839874	Kelly, Carlos	720	480
863454637	Jones, Isaac	780	715

Include a subprogram in your program which determines and displays the average and standard deviation (see Prob. 17.5) for the math and verbal scores. Include another subprogram that displays the name, ID number, and scores for the student with the highest total score (that is, math plus verbal).

17.19 *(FORTRAN 90)* Employ records to enter the following information into a computer program:

Name	Team	AB (At Bats)	H (Hits)
Kruk	Phil	355	121
Van Slyke	Pitt	378	126
Sheffield	SD	390	127
Grace	Chi	380	121
Butler	LA	376	119
DeShields	Mon	408	128

Include a subprogram in your program which determines the batting average ($= H/AB$) for each player. Include another subprogram that displays the name, team, and batting average for all the players on the list. Note that batting averages are typically expressed to three decimal places (for example, 0.305).

17.20 *(FORTRAN 90)* Members of a sales force receive commissions of 5 percent for the first $1000, 7 percent for the next $1000, and 10 percent for anything over $2000. Compute and display the commissions and social security numbers for the following salespeople: John Smith, who has $5325.76 in sales and a social security number of 080-55-6642; Bubba Holottashakingoinon, who has $8050.63 in sales and whose number is 306-77-5520; and Bubbette Schumacker, who has $6643.96 in sales and a social security number of 707-00-1122. Use records to input the information.

17.21 *(FORTRAN 90)* Write a program to input your friend's and family's names and birthdays into an array of records.

17.22 *(FORTRAN 90)* Modify the program from Prob. 17.21 so that when the user enters today's date, the record containing the next birthday is displayed.

ADVANCED INPUT-OUTPUT

Because it represents the interface between the user and the program, input-output (or *I/O*) is a critical aspect of software design. In the early days of computing, input was via punched cards and output was on paper produced by large line printers. As a result, the first computer languages had limited capabilities for input-output. Today, we enter data into computer programs via a keyboard, with a mouse, or directly from measuring instruments. We display our results on a monitor, laser printer, or pen plotter. Consequently, most modern languages have well-developed capabilities for implementing I/O. Today, FORTRAN provides the ability to design a machine-user interface that is expressive, efficient, and friendly.

The present chapter is devoted to the most important aspect of FORTRAN related to refined I/O: formatting. In addition, we also explore how data can be stored on and retrieved from external files.

18.1 LIST-DIRECTED I/O

There are many cases in engineering where you would want to control the format of a program's input and output. This is particularly important when the program is to be used for report generation. To this point, you have been introduced to list-directed or free-format I/O. Before learning how we can gain more control, we will first take a closer look at how the list-directed PRINT operates.

Using the free-format or list-directed PRINT statement, the output device, such as the monitor screen or printer, can then be thought of as consisting of vertical columns, or display fields, that are an equal number of characters wide (Fig. 18.1). The first numeric value in a PRINT statement is placed at the far right of the first display field, and subsequent values are placed sequentially at the far right of the subsequent fields.[1]

---

[1] Note that the display fields may differ in width depending on the compiler. In addition, REAL and INTEGER values usually have different field widths. However, all numbers are right-justified.

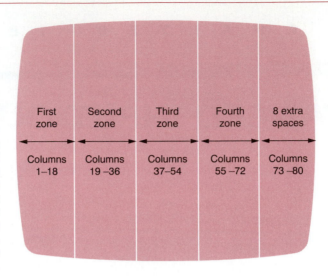

FIGURE 18.1
When values are output
in a list-directed PRINT
statement, the output device
can be thought of as
consisting of equal-width
display fields. For the compiler
used in the above case, the
fields are 18 characters wide.
Other systems may use
somewhat different widths.

When numbers are displayed in this way, they are said to be *right-justified*. String values are handled differently. They are merely displayed end to end, starting at the left side of the screen or printed page.

EXAMPLE 18.1   Outputting a Table with the List-Directed PRINT

PROBLEM STATEMENT: The students in a class have obtained the following grades:

Student Name	Average Quiz	Average Homework	Final Exam
Smith, Jane	82.00	83.90	100
Johnson, Fred	96.55	92.00	77
Horvath, Sarah	97.30	89.67	87

Their final grade can be computed with the following weighted average:

$$FG = 0.5AQ + 0.2AH + 0.3FE$$

where FG = final grade, AQ = average quiz, AH = average homework, and FE = final exam grade. Develop a computer program to compute the final grade for each student. Display the students' grades in a tabular format as shown above, with the final grade added as the last column.

SOLUTION: The computer code to determine the final grade and to output the table is shown in Fig. 18.2. The results illustrate some of the idiosyncrasies of list-directed output. First, the character strings used for the title of the table are all jammed together.

Program listing:

```
PROGRAM FIG182
PARAMETER (NS = 3)
CHARACTER*18 NAME(NS)
REAL AQ(NS), AH(NS), FE(NS), FG(NS)
INTEGER I
DATA (NAME(I), AQ(I), AH(I), FE(I), I = 1, NS)
  &      /'Smith, Jane', 82., 83.9, 100.,
  &       'Jones, Fred', 96.55, 92., 77.,
  &       'Horvath, Sarah', 97.3, 89.67, 87./
  PRINT *, 'student', 'avg', 'avg', 'final', 'final'
  PRINT *, 'name', 'quiz', 'hmwk', 'exam', 'grade'
  PRINT *
  DO 10 I = 1, NS
     FG(I) = 0.5 * AQ(I) + 0.2 * AH(I) + 0.3 * FE(I)
     PRINT *, NAME(I), AQ(I), AH(I), FE(I), FG(I)
10 CONTINUE
  END
```

FIGURE 18.2
The computer code to determine the final grade and output the results in tabular form for Example 18.1. Because these results were output on an 80-column display, each line of numbers extends, or wraps around, to the next line.

Output:

```
studentavgavgfinalfinal
namequizhmwkexamgrade

Smith, Jane               82.000000      83.900000      100.000000
        87.780000
Jones, Fred               96.550000      92.000000       77.000000
        89.775000
Horvath, Sarah            97.300000      89.670000       87.000000
        92.684010
```

Second, although the numbers are output in nice equal-width fields, the fact that each line of numbers exceeds the total width of the output device (in this case, an 80-column screen) means that the last numbers extend to the next line. This tendency to "wrap around" means that for this case we are limited to displaying an 18-character name and three fields of numerical information per line.

If the output device were a 132-column printer, we would be able to output a greater number of fields. For this case, the four numbers in the present example would fit nicely. An attempt could be made to correct the table heading by assigning the string constants used for this purpose to fixed-length string variables using a CHARACTER type-specification statement. Unfortunately, outputting these assignment statements in a free format would result in *left-justified* headings; that is, they would be placed at the far left of the field. Therefore, although the headings would be spaced apart, they would not align with the numbers.

List-directed I/O is often quite satisfactory. List-directed input is especially useful for programs that use the keyboard for data entry. Free format, with entries separated by commas or spaces, is probably the manner that would be employed by most individuals in the absence of any other guidelines. In addition, free formatting can also be used when simple tables consisting of a few columns are to be output. However, as shown in the previous example, additional control of the output is often required. As described next, FORTRAN provides such control through format specification.

18.2   FORMATTED INPUT AND OUTPUT

Format-directed I/O allows the programmer to exert control over the manner in which data is input and output. The format of numeric real data is specified as to the number of digits, the number of decimal places, the location of data on a line, and whether scientific notation is to be used. Integer and character data is handled in a somewhat simpler fashion because no decimal or scientific notation is possible. Therefore only the number and position of digits or characters need be specified. All this information is included in the format specification. Each READ or PRINT statement can reference the format specification to keep the input and output under rigid control.

The reasons for formatting output should be obvious; formatted output allows us to control the way the information is displayed or printed. However, the formatting of input may not seem justified. Formatting requirements for input made sense when data was entered on punch cards that had clearly marked columns. In addition, formatted entry of data was employed to put as much information on a punched card as possible. Free format wastes a certain amount of space for the delimiters. By specifying exactly where one number begins and another ends, a FORMAT statement can permit numbers to be placed end to end with no comma or space in between.

In fact, when data is entered via a keyboard, free format is superior to formatted input. As stated at the beginning of this section, list-directed or free-formatted entry with comma or space delimiters is probably the way most people would naturally enter data at a keyboard. In contrast, a FORMAT statement would require that numbers begin in specific columns on a line. Trying to keep track of such spacing on a monitor screen is very inconvenient. Thus inputting formatted data on a keyboard is usually inefficient. Consequently, although formatted input is treated in the following sections, our emphasis will be on output. (An exception will be presented in Example 18.3.)

18.2.1   Format Specification

The general form of the READ statement is

> READ *FormatSpec,VariableList*

whereas for the PRINT, the general form is

> PRINT *FormatSpec,ItemList*

where *FormatSpec* is the format specifier. To this point we have dealt only with the list-directed case where *FormatSpec* is an asterisk. An alternative that provides more control of I/O is available when a string or a format number is used for the *FormatSpec*.

Direct specification with a string. The simplest way to specify the format is to embed the *FormatSpec* directly in the PRINT or READ statement. This is done by representing the *FormatSpec* as a string. For example,

```
PRINT '(1X, I8)', NUMBER
```

For this case, the string '(1X, I8)' provides a guide to how the value of NUMBER is to be printed. The 1X specifies that one space is to be skipped. Then the I8 designates that the variable NUMBER will be displayed as an integer in the following eight spaces.

The use of a string to specify the format is ideal for cases where the particular *FormatSpec* is (1) short and (2) not used repeatedly for many different I/O statements. For a long or widely used specification, the FORMAT statement is preferable.

FORMAT statements and numbers. The *FormatNumber* refers to a line number of a FORMAT statement which can be employed to set the layout of the I/O. The FORMAT statement has the general form

FormatNumber FORMAT (*FormatCode₁*, . . . , *FormatCodeₙ*)

where the *FormatNumber* is a positive whole number that is employed to connect the FORMAT statement to a particular READ and PRINT statement. For example, as seen in Fig. 18.3, the FORMAT statement provides explicit instructions as to the manner in which each variable is to be output.

The *FormatNumber* must be located in the first five columns of a line. It serves to link each READ or PRINT with its corresponding FORMAT statement. Note that several READ or PRINT statements may access the same FORMAT statement.

It should be noted that the *FormatNumber* must be unique. In other words, the same number cannot be used for more than one FORMAT statement. However, the same number *can* be employed as a line number. Although the compiler will allow you

FIGURE 18.3
An example of how a FORMAT statement provides the information to guide the output of data by a PRINT statement. Each variable has a corresponding format code that specifies how the variable is to be output.

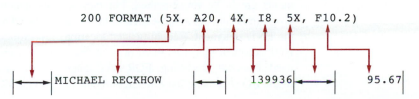

```
INTEGER ID
REAL GRADE
CHARACTER*20 NAME
DATA NAME/'MICHAEL RECKHOW'/
ID = 139936
GRADE = 574.0/6.
PRINT 200, NAME, ID, GRADE


200 FORMAT (5X, A20, 4X, I8, 5X, F10.2)
```

to do this, it is recommended that you use different numbers for line and format numbers to avoid confusion. In the present chapter, we employ line numbers in the range 1 to 99. The range above 100 is dedicated to format numbers.

The *FormatCode*s supply a symbolic description of the form of the data. As seen in Fig. 18.3, they always follow the word "FORMAT" and are enclosed in parentheses. Each format code is separated from the next by a comma. The codes themselves have several forms. Some examples are

```
101 FORMAT (10X, 2I6, T40, E20.3)
102 FORMAT (5X, F5.2)
103 FORMAT (F7.0, F8.3)
104 FORMAT (A30)
105 FORMAT (E16.8)
```

In the above, the X and T are used to skip over column spaces in the line, while the rest of the letters specify how the numbers or characters are to be displayed. For example, the letter 5X skips five spaces on the output device, whereas I6 tells the computer that the data is an integer (I) that can be up to six spaces long. Several of the codes have numbers in front of them. This merely tells the computer how many times to repeat the code. For example, 2I6 tells the computer that there will be two integers that are each six spaces long. Thus

```
308 FORMAT (I6, I6)
```

and

```
308 FORMAT (2I6)
```

are equivalent. If a FORMAT statement has more format codes than are needed by a READ or PRINT statement, the extra codes are ignored. For example,

```
    PRINT 100, NUM1, NUM2
100 FORMAT (3I8)
```

Because there are only two variables to be written, the third I8 is unnecessary and is ignored. The statement 100 FORMAT (2I8) would work in an identical fashion for this case.

FORMAT statements are not executable FORTRAN statements. That is, the compiler makes use of them to obtain information regarding how READ and PRINT statements are to be implemented, but they are not executed by the computer when the program is run. As a consequence, they can be placed anywhere in the program prior to the END statement. Recommended locations are immediately after the input-output (or I/O) statement that uses them or at the beginning or end of the module where they are used. Generally speaking, FORMAT statements that are used once can be placed next to the I/O statement that references them. In contrast, FORMAT statements that are employed by more than one I/O statement should be clustered in one location. In the present chapter, we place them at the end of the major module where they occur, but preceding the END statement.

Now that you have assimilated some preliminary information on FORMAT specifications, we can look in detail at the format codes that are at the heart of their operation. The codes can be divided into different categories: (1) codes to control spacing, (2) codes for character data, (3) codes for integers, and (4) codes for real numbers.

18.2.2 Codes to Control Spacing

Lines can be spaced in two dimensions on a display or printer: vertically and horizontally. As described next, the vertical spacing between lines can be dependent on the way your compiler treats the output devices for your computer.

Spacing between lines.   As mentioned at the beginning of this chapter, most output in the early years of computing was directed to line printers. As a consequence, spacing between lines was designed to accomodate these devices. In particular, vertical spacing of output was specified by placing a special control code in the first column on a line. This first character, called a *carriage-control character,* is not actually displayed but instead controls the vertical line spacing.

The various codes are listed in Table 18.1. Note that if the first character is not one of the five given in Table 18.1, the results will be unpredictable; that is, they are compiler-dependent. Consequently, a carriage-control character should start each format specification that relates to displayed output.

For example, if we desire to skip a line and display the message `'Enter more data'`, the following PRINT and FORMAT statements would be used:

```
      PRINT 100
      PRINT 200
  100 FORMAT ('')
  200 FORMAT ('Enter more data')
```

which cause a line to be skipped and the message displayed. A single PRINT and FORMAT could also be used to perform the same action:

```
      PRINT 100
  100 FORMAT ('0Enter more data')
```

Because the zero is in column 1, it is not displayed.

It should be noted that even today when most output is directed to display monitors, many FORTRAN compilers treat the display as if it were a printer. However, some FORTRAN implementations either ignore carriage control, allow you to choose to ignore it, or provide other means to control vertical spacing of lines. Because of its prevalence, the remainder of this chapter is written for a computer where carriage control is in effect. However, you should investigate to learn whether or not carriage control applies for your system.

TABLE 18.1   First Column "Carriage-Control" Codes to Control Spaces between Lines

First Character	Effect	Spacing
' '	Advances one line and then displays	Single
'+'	No line feed before displaying	Overstrike
'0'	Skips to second line and then displays	Double
'—'	Skips to third line and then displays	Triple
'1'	Skips to top of next page and then displays	New page

In addition to carriage-control characters, the slash (/) can also be used to control vertical line spacing. When a slash is encountered, it directs the computer to advance to the next line. For example,

```
    PRINT 800
800 FORMAT (' Play', /, ' ball')
```

results in the output

```
Play
ball
```

In contrast to other FORMAT codes, the commas before and after the slash are optional. Thus the FORMAT statement above could also be written as

```
800 FORMAT (' Play' / ' ball')
```

Consecutive slashes can be employed to skip several lines, but their effect is dependent on their position in the line. If n consecutive slashes appear at the beginning or end of a FORMAT statement, n lines will be skipped. However, if they are in the interior of the statement, $n - 1$ lines will be skipped.

Spacing within a line.   Spacing between characters is determined by special codes, as outlined in Table 18.2. Remember that the first column is not displayed, and so the location specified by the first format code in the FORMAT statements in Table 18.2 should be one space greater than the actual desired column for output.

The T and the X codes are especially useful for creating tabular output. The T code is analogous to a typewriter tab setting in that it specifies the starting position of the following value as measured from the left margin. In contrast, the X code is analogous to a typewriter spaced bar in that it inserts spaces between values.

EXAMPLE 18.2   Table Headings with the T and X Codes

PROBLEM STATEMENT: Use both T and the X codes to create headings for the table previously developed in Example 18.1.

TABLE 18.2   Format Codes to Control Spaces between Characters

Format Code	Effect	Example
nX	Skip n spaces	10 FORMAT (10X, 'Name', 25X, 'Grade')
Tn	Tab to column n	20 FORMAT (T11, 'Name', T30, 'Grade')
/	Begin new line	30 FORMAT (1X, 'NAME', /, ' Grade')

Output from FORMAT statements 10, 20, and 30:

```
    12345678901234567890123456789012345678901234567890
        Name                            Grade
        Name                Grade
    NAME
    Grade
```

SOLUTION: A program to create the headings with the T code is shown in Fig. 18.4. The comparable FORMAT statements using the X code are

```
100 FORMAT (' Student', 16X, 'Average', 4X, 'Average', 4X, 'Final', 6X, 'Final')
200 FORMAT (' Name', 19X, 'Quiz', 7X, 'Homework', 3X, 'Exam', 7X, 'Grade')
```

Statement 100 for both cases is displayed in Fig. 18.5 with the output. This diagram clearly demonstrates the difference between the T and the X codes.

18.2.3   Codes for Character Data

The input or output of character data is accomplished with the alphanumeric, or A, format code. The A code can be preceded by an integer number to indicate the repetitions of the A format code (see Table 18.3), for example,

```
500 FORMAT (4A30, A40)
```

The 4A30 would allow for four 30-character-long strings and the A40 for one 40 characters long. If a string contains more characters than the FORMAT statement permits, then only those characters described in the FORMAT statement will be displayed. Figure 18.6 shows an example.

EXAMPLE 18.3   Using A Codes to Facilitate Character Input

PROBLEM STATEMENT: Demonstrate how the A code and format specification can be employed to enter character data without apostrophes.

SOLUTION: Recall from Example 15.4 that we queried the user to respond by typing a 'y' or an 'n' to signify a yes or no response, respectively, to a query. Similar code that performs the same action is

Program listing:

```
        PROGRAM FIG184
        PARAMETER (NS = 3)
        CHARACTER*18 NAME(NS)
        REAL AQ(NS), AH(NS), FE(NS), FG(NS)
        INTEGER I
        DATA (NAME(I), AQ(I), AH(I), FE(I), I = 1, NS)
     &      /'Smith, Jane', 82., 83.9, 100.,
     &       'Jones, Fred', 96.55, 92., 77.,
     &       'Horvath, Sarah', 97.3, 89.67, 87./
        PRINT 100
        PRINT 200
        PRINT *
        DO 10 I = 1, NS
          FG(I) = 0.5 * AQ(I) + 0.2 * AH(I) + 0.3 * FE(I)
          PRINT *, NAME(I), AQ(I), AH(I), FE(I), FG(I)
     10 CONTINUE

C     format statements:
    100 FORMAT(' student', T25, 'average', T36, 'average', T47, 'final',
     &          T58, 'final')
    200 FORMAT(' name', T25, 'quiz', T36, 'homework', T47, 'exam', T58,
     &          'grade')

        END
```

Output:

```
student                  average  average   final    final
name                     quiz     homework  exam     grade

Smith, Jane              82.000000  83.900000  100.000000
    87.780000
Jones, Fred              96.550000  92.000000   77.000000
    89.775000
Horvath, Sarah           97.300000  89.670000   87.000000
    92.684010
```

FIGURE 18.4
A FORTRAN program to create headings using the T code to set the spacing.

FIGURE 18.5
Comparison of FORMAT statements using the T and the X codes to implement identical output.

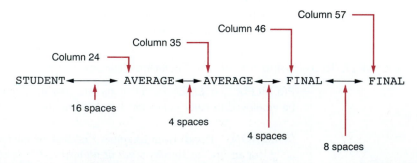

TABLE 18.3   General Forms of the Format Codes to Input and Output Letters and Numbers

Format Code	Name	Interpretation
nAw	Alphanumeric	This defines n alphanumeric fields, each w columns wide, each holding left-justified character data.
nIw	Integer	This defines n integer fields, each w columns wide, each holding a right-justified integer value.
nFw.d	Floating point	This defines n real fields, each w columns wide, each containing a right-justified value with a decimal point and with d decimal places.
nEw.d	Exponential	This defines n real fields, each w columns wide, each containing a right-justified value in scientific notation. The exponent is placed at the far right of the field preceded by the letter E. The decimal point of the mantissa is d columns to the left of the E.

Program fragment:

```
    CHARACTER*80 STRING
    STRING = 'Mary Jo'
    PRINT 600, STRING
    PRINT 601, STRING
600 FORMAT(' First Initial = ', A1)
601 FORMAT(' Name = ', A7)
    END
```

FIGURE 18.6
A program fragment
illustrating the use of the A
code for outputting
character data.

Output:

```
First Initial = M
Name = Mary Jo
```

```
CHARACTER Q
PRINT *, 'Enter y or n'
READ *, Q
```

When these statements are executed, the response must be enclosed in apostrophes, as in

```
Enter y or n
'y'                              user response
```

Because we employ free format, the apostrophes are required to let the compiler know that character-type information is being entered.

By using a format specification and the A code, the apostrophes can be omitted, as in the following code:

```
CHARACTER Q
PRINT *, 'Enter y or n'
READ '(A1)', Q
```

The query and response would now be

```
Enter y or n
y  ←————————————————————— user response
```

Note that the READ could have also been represented as

```
READ '(A)', Q
```

because only one character is to be entered. In both cases, because the format specification tells the compiler that a single character string is to be entered, the apostrophes can be omitted.

18.2.4 Codes for Integers

Integer numbers are formatted with the I format code (Table 18.3). The I is always followed by an integer number indicating the allocated width.

An example of the use of the I code is provided in Fig. 18.7. The integer values are displayed right-justified in the field. For the I4, only four columns are allocated; whereas for 2I10, two areas of ten columns each are set aside for the numbers. If a number does not fill the entire field, the empty spaces are merely left blank. On the other hand, if the number exceeds the field, asterisks are displayed to signify that the space allocated was insufficient. Such is the case for 32,000 in the third PRINT statement in Fig. 18.7. When the I format codes are used for READ statements, the entries have to be right-justified in the field when they are typed.

FIGURE 18.7
A program fragment
illustrating the use of the I
code for outputting
integer data.

Program fragment:
```
      PRINT 100, 20, 199
      PRINT 100, 4300, 7, 39
      PRINT 100, 32000, -2786, -32000
100   FORMAT(' ', I4, 2I10)
      END
```

Output:
```
1234567890123456789012345678890 <------- columns
  20       199
4300         7        39
****     -2786    -32000
```

(*a*) F Format codes

Program fragment:

```
     PRINT 200, 387.945, 387.945, 387.945
     PRINT 200, -4.876, -592.333, 0.269E7
     PRINT 200, -0.521E-2, 0.000722, 0.099
200  FORMAT(1X, F12.4, ';', F8.2, ';', F6.0)
     END
```

Output:

```
12345678901234567890123456789012345678901234567890 <------- columns
     387.9450;  387.95;  388.
      -4.8760; -592.33;******
       -.0052;     .00    0.
```

(*b*) E Format codes

Program fragment:

```
     PRINT 300, 387.945, 387.945, 387.945
     PRINT 300, -4.876, -592.333, 0.269E7
     PRINT 300, -0.521E-2, 0.000722,  0.099
300  FORMAT(1X, E12.4, ';', E9.3, ';', E14.1)
     END
```

FIGURE 18.8
Program fragment illustrating
the use of the (*a*) F and (*b*) E
codes to output real numbers
in decimal and exponential
form.

Output:

```
12345678901234567890123456789012345678901234567890 <------- columns
   .3879E+03;  .388E+03;          .4E+03
  -.4876E+01;-.592E+03;          .3E+07
  -.5210E-02;  .722E-03;          .1E+00
```

18.2.5 Codes for Real Numbers

Real numbers may be displayed or entered either in scientific or exponential notation (E format code) or as decimal numbers (F, or floating-point, format code). For these cases, both the width of the field *and* the number of decimal places must be specified. The general forms are shown in Table 18.3. The number must not exceed the width allocated. If the number to be displayed is too large, then the space will be filled with asterisks to indicate the program's inability to display the number.

Figure 18.8 presents examples of both the F and E codes. These examples illustrate why these codes are especially well suited for devising tables. Because the decimal of a number is always placed in a specific location, when the same FORMAT statement is used repeatedly (as in Fig. 18.8), the decimals automatically line up.

In Fig. 18.8 the semicolons are inserted between numbers to highlight where one number ends and another begins. If these semicolons are left out, the numbers will run together. For example, the last PRINT (without semicolons) looks like

```
-.5210E-02 .722E-03      .1E+00
```

which is not very easy to read. Another way (besides using semicolons) to make such numbers clearer is to leave spaces between the numbers, using the X code.

The row of asterisks occurs when the allocated field cannot accommodate a number. The value 0.269E7 exceeds the width of the F6.0 format code. The value −592.333 is displayed with the E9.3 format code, even though after the four places for the exponent and the five places for the number and the decimal are set aside, there is no room for the minus sign. This occurred because the FORTRAN compiler used to prepare this ouput automatically dropped the zero before the decimal in order to make room for the minus sign. Another compiler might not have done this, and a row of asterisks would have been displayed. You should experiment with your compiler to learn how it handles such situations. In any event, always leave adequate space for the fields in your output.

EXAMPLE 18.4   Using FORMAT Statements to Output a Table

PROBLEM STATEMENT: Employ FORMAT statements to output the table from Example 18.1.

FIGURE 18.9
A computer program to
display a table of student

Program fragment:

```
      PROGRAM FIG189
      PARAMETER (NS = 3)
      CHARACTER*18 NAME(NS)
      REAL AQ(NS), AH(NS), FE(NS), FG(NS)
      INTEGER I
      DATA (NAME(I), AQ(I), AH(I), FE(I), I = 1, NS)
     &     /'Smith, Jane', 82., 83.9, 100.,
     &      'Jones, Fred', 96.55, 92., 77.,
     &      'Horvath, Sarah', 97.3, 89.67, 87./
      PRINT 100
      PRINT 200
      PRINT *
      DO 10 I = 1, NS
        FG(I) = 0.5 * AQ(I) + 0.2 * AH(I) + 0.3 * FE(I)
        PRINT 300, NAME(I), AQ(I), AH(I), FE(I), FG(I)
   10 CONTINUE

C     format statements:
  100 FORMAT(' student', T25, 'average', T36, 'average', T47, 'final',
     &       T58, 'final')
  200 FORMAT(' name', T25, 'quiz', T36, 'homework', T47, 'exam', T58,
     &       'grade')
  300 FORMAT(1X, A15, 6X, 4(F7.1,4X))

      END
```

Output:

student name	average quiz	average homework	final exam	final grade
Smith, Jane	82.0	83.9	100.0	87.8
Jones, Fred	96.6	92.0	77.0	89.8
Horvath, Sarah	97.3	89.7	87.0	92.7

SOLUTION: Figure 18.9 shows the program and the output for the table. Notice how for the FORMAT statement labeled 300, a space is used to avoid having a character output in column 1. Also, compare this program with that in Figs. 18.2 and 18.4 to demonstrate to yourself why the version in Fig. 18.9 is superior.

18.3   FILES

To this point, we have covered three methods for entering data into a program (see Table 18.4). Notice in the table that all of them have disadvantages related to large quantities of data and data for which frequent change might be required. In addition, all the methods deal exclusively with the information requirements of a single program. If the same data is to be used for another program, it must be retyped.

Another option is to store information on an external memory device such as a magnetic disk. This is conceptually similar to the way we have been saving our programs on disks so that they are not destroyed when the computer is shut off. Along with our programs, we can store data in external memory files just as office information is stored in file cabinets for later use.

A number of advantages result from storing data on files. First, data entry and modification become easier. In addition, files make it possible to run several programs using the data from the same file and, alternatively, to run the same program from several different files. Before proceeding to the specifics of manipulating files, we will introduce some new terminology.

18.3.1   File Types

In its most general sense, a file can be defined as a section of external storage that has a name. There are two types of files used on computers: sequential and direct-access files. *Sequential files* are those in which items are entered and accessed in sequence. That is, there is a first item, a second item, and so on. The key concept is that when we want to access one of these items, we have to start at the beginning of the file and make our way through each item until we come across the one we want. Using an analogy from musical recording devices, sequential files are very much like cassette tapes. If you want to listen to a tune at the end of a tape, you have to fast-forward through all the other music you want to skip.

In contrast, for *direct-access,* or *random-access, files,* we can access a record directly without having to make our way through all previous records. Continuing the musical analogy, random-access files are like a compact disk player. For CDs, you punch in the selection you want and the player directly moves to that selection. Random access works in a similar fashion by allowing direct movement to a specific location in the file.

Direct-access files have a number of advantages, not the least of which is the fact that they retrieve data more quickly and more efficiently than sequential files. They are especially attractive for large-scale databases such as a bank's billing or an airline's ticketing program.

TABLE 18.4   Advantages and Disadvantages of Four Ways in Which Information Is Entered into FORTRAN Programs

Method	Advantages	Disadvantages	Primary Applications
Assignment and and PARAMETER statements	Are simple and direct. Values are "wired" into the program and do not have to be entered on each run.	Take up space in the program. If a new value is to be assigned, the user must edit the assignment statement.	Appropriate for some simple programs. Good for assigning values to true constants (such as π or the acceleration of gravity) that are used repeatedly throughout a program and never change from run to run.
INPUT statement	Permits different values to be assigned to variables on each run. Allows interaction between user and the program.	Values must be entered by user on each run. This can become tedious.	Appropriate for those variables that take on different values on each run.
DATA statements	Provide a concise way to assign values.	If a new value is to be assigned, the user must edit the DATA statement.	Same as for the assignment statement but for cases where many (but not hundreds of) constants are to be assigned.
Files	Provide a concise way to handle large amounts of input data. Allow data to be transferred between programs.	Are somewhat more complicated than the other input options. Involve other devices in the computer system.	Appropriate for large quantitites of data and where data must be shared between programs.

Although sequential files are inefficient, they have their place in data processing. They are especially well suited for data that is to be used sequentially anyway, such as mailing labels. In engineering, data is often in sequential form; for example, data is frequently listed in temporal order. Consequently, many engineering applications are adequately handled by sequential files. Finally, and most important in the present context, they are easier to explain than random-access files. Because of the scope of this text, we have chosen to devote this chapter to sequential files. You can consult your computer's user's manual in the event that you desire or need to employ direct-access files.

18.3.2 FORTRAN Sequential Files

Before describing how sequential files are used, it must be noted that they are not as standardized as other elements of FORTRAN. Fortunately, most of the differences relate to nomenclature—that is, the specific statements that are used to designate the fundamental file manipulations. From a functional perspective, the actual operations themselves are fairly similar from system to system. Therefore in the following discussion we have tried to keep the description of the procedures as general as possible. Where we do use specific nomenclature, we have opted to use statements compatible with the Microsoft FORTRAN 5.1 implemented in an MS-DOS environment. It is possible that some of the specific names that we employ may differ somewhat from those used on your system. You may therefore have to obtain details from your dialect's user's manual in order to implement the material effectively.

To this point, you have learned how to input and output data to your computer's default devices using the READ and PRINT statements. At the simplest level, creating and using a sequential file in FORTRAN is very similar. The only major difference is that you must employ somewhat different statements.

For input, a modified READ statement is used. It has the general form

READ *(UnitSpec, FormatSpec) VariableList*

where the *UnitSpec* is a positive whole number that is employed to specify the particular file that you want to access. When this statement is executed, it acts just like any other READ statement, except that rather than inputting values from the keyboard, it directs the compiler to read values from a file.

For output, a WRITE statement is used. It has the general form

WRITE *(UnitSpec, FormatSpec) ItemList*

This statement acts just like a PRINT statement, except that rather than directing output to a monitor or a printer, it transfers output to a file.

When reading from or writing to a file, the *FormatSpec* can be set to an asterisk, in which case the compiler will determine the format of the information. Alternatively, you can use it to designate a FORMAT statement which you can devise to control the format.

Before actually using these statements, the *UnitSpec* must be defined. This is done with an OPEN statement. For the simplest applications, it has the general form

```
OPEN (UNIT = UnitSpec, FILE = FileSpec)
```

where the file specification or *FileSpec* is a character string used to identify the file in auxiliary storage. Most generally, any unique, legal FORTRAN name can be employed for the *FileSpec*. It should be noted that, on personal computers, the *FileSpec* can include the specific storage device (for example, a particular disk drive) where the file resides. If the device specification is omitted (as will be the case for all the following material), FORTRAN assumes that the device is the default storage device for your computer.

It is important to realize that the *UnitSpec* identifies the file *within the program,* whereas the *FileSpec* identifies it *on the external storage device.* Therefore, if a file is used in two different programs, it could have different file numbers, but the same *File-Spec* would be required in both instances. For example, for the first program the file might be opened for output of data from the program with

```
OPEN (UNIT = 8, FILE = 'FLOW')
```

whereas for the second program it could be opened for input of data to the program with

```
OPEN (UNIT = 7, FILE = 'FLOW')
```

Finally, all files are automatically closed when the program terminates normally. There may be occasions when you would like to close a file before this point. You might desire to do this because you would like to release the *UnitSpec* and the file for other purposes. The CLOSE statement can be used for this purpose. For the simplest applications, it has the general form

```
CLOSE (UNIT = UnitSpec)
```

18.3.3 Standard File Manipulations

Now that we have introduced some of the FORTRAN file commands, we can show how they are used to perform some of the standard file manipulations that are needed for routine engineering applications. These are

- Creating a file
- Retrieving and displaying a file

Creating a file.   Figure 18.10 shows a simple program to transfer the data from Table 17.1 into a file. Notice that the file name is specified as a string constant, 'FLOW'. This name would have to be changed if we were to use this program to store

```
            PROGRAM FIG1810
            CHARACTER*4 YEAR
            REAL FLOW
            OPEN (UNIT=1,FILE='FLOW',STATUS='NEW')
        10  CONTINUE
            PRINT *, 'year = (enter -999 to terminate)'
            READ 100, YEAR
            IF (.NOT. YEAR .EQ. '-999') THEN
            PRINT *, 'flow = '
            READ *, FLOW
            WRITE(1, 110) YEAR, FLOW
            GO TO 10
            END IF
            CLOSE (1)
        100 FORMAT (A4)
        110 FORMAT (A4,F10.2)
            END
```

FIGURE 18.10
A program to output data
from a program to a file.

other information consisting of pairs of string [YEAR(I)] and numeric [FLOW(I)] constants. For example, if we had up to 50 years of flow data from another stream we could use this program to store it in a file. However, to accomplish this correctly, we would have to give the file a new name other than 'FLOW'. Otherwise the new data would be transferred onto the file and supplant (and hence destroy) the old data.

Also, observe that we have added an additional specification to the OPEN statement:

$$STATUS = 'NEW'$$

This specification tells the compiler that we are generating a new file. After the file is established on your disk, this specification would be changed to STATUS = 'OLD'. This option is sometimes handy in preventing you from destroying an already existing file by creating a new file with the same name.

Finally, note how a mid-test loop has been employed to enter the values. A sentinel value of -999 is used to designate when the last value has been entered.

Displaying the contents of a file. Figure 18.11 shows a program to transfer the flow data back from the file to the program. This version also displays the data so that we can check that the data transfer has been performed as expected. Notice that in this example, we used the same file number of 1 as in the previous example (Fig. 18.10). Also notice that both examples employ the same name ('FLOW'), which is how the file is referenced on the external storage device.

As the data is being input, we must develop a way to detect when there is no more data to be read from the file. Otherwise, if we try to read data that is not present, we will obtain an error and the program will stop.

There are several ways to detect that we are at the end of a file. One way is to define an end-of-file variable. When a file is created, an end-of-file mark is placed after the last item of data. The end-of-file variable is designed to detect this mark. The IOSTAT clause can be employed to do this. The IOSTAT clause has the general form,

FIGURE 18.11
The various components
associated with data files. The
figure uses the analogy
between an office file system
and a file system employed
on a computer.

```
PROGRAM FIG1811
CHARACTER*4 YEAR
INTEGER EOF
REAL FLOW
OPEN (UNIT=1,FILE='FLOW',STATUS='OLD')
PRINT *, ' YEAR        FLOW'
10 CONTINUE
   READ(1, 110, IOSTAT=EOF) YEAR, FLOW
   IF (EOF.EQ.0) THEN
   PRINT 120, YEAR, FLOW
   GO TO 10
END IF
CLOSE (1)
110 FORMAT (A4, F10.2)
120 FORMAT (1X, A5, F10.2)
END
```

FIGURE 18.11
The various components associated with data files. The figure uses the analogy between an office file system and a file system employed on a computer.

$$IOSTAT = IntegerVariable$$

where the *IntegerVariable* is positive if an input error occurs, negative if the end-of-data is encountered, and zero if neither an input error nor an end-of-data occurs.

This clause can then be employed to reformulate the READ statement as

```
READ(1, 110, IOSTAT = EOF) YEAR, FLOW
```

Therefore, until the end of the file is reached, the variable EOF will equal zero. When the end is encountered, it will be set to a negative value.

As seen in Fig. 18.11, as long as we have not read all the data from the file, EOF will be zero and control will transfer to the next line. However, when the last item is read, the EOF becomes negative and control transfers out of the loop. Thus, if we are looping to read the data, this statement allows us to transfer control out of the loop before we attempt to read a nonexistent record.

PROBLEMS

18.1 What is the difference between printing numeric and character data in correctly specified fields?

18.2 Suppose that the following specifications and assignments have been made:

```
REAL A, B
INTEGER I, J, L, N
A = 545.6789
B = -0.05678
N = 6058
L = 3
```

What output would be produced by the following statements, assuming that control characters are in effect?

```
(a)        PRINT 100, N, N*2
     100 FORMAT (3(2X, I5))
(b)        PRINT 110, B, B, A
     110 FORMAT (1X, F7.2, T20, 2E10.2)
(c)        DO 1 I = 1, L
               DO 2 J = 1, 2*L, 2
                   PRINT '(3(1X, I4))', I - J, I + J
         2     CONTINUE
               PRINT *
         1 CONTINUE
```

18.3   What output would be produced if the following statements were executed?

```
(a)  X = 3.5
     R = 3.
     PRINT 301, X, R
301 FORMAT (' The value of X is', F10.3/'0The value of R is',F10.0)
(b)  SUM = 35.5
     NUM = 35
     PRINT 302, NUM, SUM
302 FORMAT (' The sum of the', I3, ' numbers = ', F12.5)
(c)  X = 10E02
     Y = 12.6E 2
     PRINT 203, X, Y
203 FORMAT (1X, F10.2/1X, '0', T20, E12.1)
```

18.4   Develop a program to add a pair of 2-digit integers. Format the output so that it appears as:

```
   15
 + 28
   43
```

18.5   Write a short program to produce the following output. Include the top-line column index as a part of the output.

```
12345678901234567890123456789012345678 90
    NUMBER       X's        X - average
      1          2.0           -3.5
      2          3.0           -2.5
      3          4.0           -1.5
      4          7.0            1.5
      5          8.0            2.5
      6          9.0            3.5
      7          7.0            1.5
      8          6.0            0.5
      9          5.0           -0.5
     10          4.0           -1.5

           Average = 5.5
```

18.6 Write a short program that employs DATA statements to input the following character constants into seven-character variables:

```
'angle', 'degrees', 'radians', 'x', 'y', 'z'
```

Then output the information so that each character constant is a heading in a six-field table. Each field should be 10 spaces wide, with the character constant printed right-justified in the 10-space field.

18.7 The saturation value of dissolved oxygen, c_s, in fresh water is calculated by the equation

$$c_s = 14.652 - 4.1022 \times 10^{-1}T_c + 7.9910 \times 10^{-3}T_c^2 - 7.7774 \times 10^{-5}T_c^3$$

where T_c is temperature in degrees Celsius. Saltwater values can be approximated by multiplying the freshwater result by $1 - (9 \times 10^{-6})n$, where n is salinity in milligrams per liter. Finally, temperature in degrees Celsius is related to temperature in degrees Fahrenheit, T_f, by

$$T_c = \tfrac{5}{9}(T_f - 32)$$

Develop a computer program to compute c_s, given n and T_f. Use these formulas and the capabilities you have gathered in the present chapter to compute and print out the following table of saturation concentrations:

	$n = 0$	$n = 5000$
$T = 40F$		
$T = 50F$		

18.8 Members of a sales force receive commissions of 5 percent for the first $1000, 7 percent for the next $1000, and 10 percent for anything over $2000. Compute and display the commissions and social security numbers for the following salespeople: John Smith, who has $5325.76 in sales and a social security number of 080-55-6642; Priscilla Olkowski, who has $8050.63 in sales and whose number is 306-77-5520; and Bubba Schumacker, who has $6643.96 in sales and a social security number of 707-00-1112. Output the results in a neat tabular format.

18.9 Write, debug, and document a computer program to determine statistics for your favorite sport. Pick anything from softball to bowling to basketball. Design the program so that it is user-friendly and provides valuable and interesting information to anyone (for example, a coach or player) who might use it to evaluate athletic performance. In particular, display the results in a neat tabular format and retain all data on a sequential file.

18.10 Expand the program in Example 18.4 so that it also includes a class participation grade (*PG*). The final grade would be recalculated as

$$FG = .5AQ + .2AH + .25FE + .05PG$$

Give Sarah and Jane *PG*s equal to 95, and give Fred a 75. Include this new information on the table along with their identification numbers (Jane: 073-99-3336; Fred: 544-01-9969; and Sarah: 139-93-4446).

18.11 Select several of your favorite issues from the New York Stock Exchange. Write, debug, and document a FORTRAN program that reads the name of the stock, its selling price today, and its 52-week low. Use arrays to incorporate the data into the program. The program should then display the name of the stock, its selling price, and the percentage gain over the 52-week low. Output the results in a neat tabular format.

18.12 Repeat Prob. 18.11, but employ a sequential file to store the information on an external storage device. Employ code similar to the program fragments from Figs. 18.10 and 18.11. Utilize subscripted variables to store the data.

18.13 Using Figs. 18.10 and 18.11 as models, develop a program fragment to delete a record from a file.

18.14 Develop a modular computer program to determine your grade for this course. Store all relevant data in a sequential file.

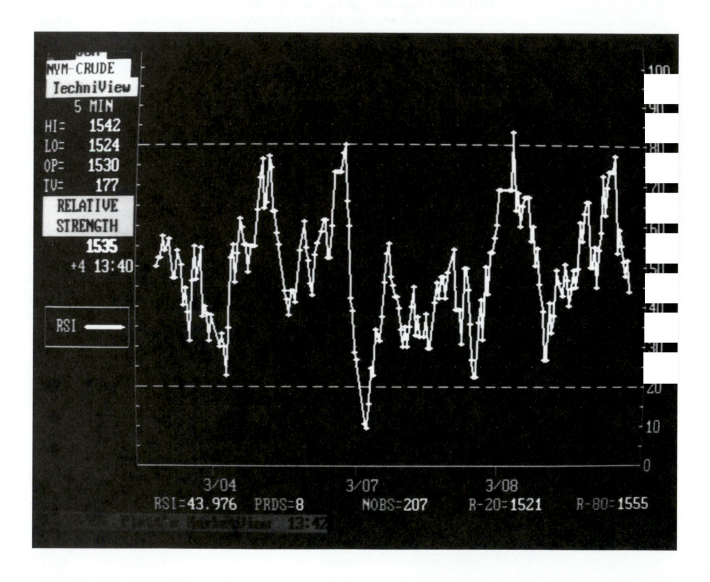

DATA ANALYSIS

Now that you have mastered the rudiments of programming, we turn to the ways in which this capability can be applied in engineering practice. Most of the remainder of this book is devoted to skills and tools that contribute to the problem-solving process. In all cases we emphasize the ways in which computers are used to facilitate the process. The first topic we cover is data analysis.

Problem solving with the computer is a futile exercise if it is based on inadequate information. In computer jargon, this is sometimes referred to as GIGO, or "garbage in, garbage out." Consequently, data analysis is one of the most important skills of the practicing engineer. Although many projects rely on existing data and formulas, there are a great many more situations where information is not available in these convenient forms. Special cases and new problem contexts often require that you measure your own data and develop your own predictive relationships.

The present part of the book introduces a number of specific tools and techniques that can prove useful in these tasks. *Chapter 19* is devoted to graphical methods that provide a visual expression of data and relationships. *Chapter 20* deals with algorithms for sorting and searching techniques. Chapters 21 and 22 deal with quantifying a measurement's reliability. *Chapter 21* reviews significant digits and error analysis, and *Chap. 22* explores the area of statistics.

C H A P T E R   1 9

GRAPHICS

It is often remarked that "one picture is worth a thousand words." This saying holds true whether we are assimilating information or conveying it to others. Graphical depictions are invaluable communication devices for conveying data and concepts to an audience (Fig. 19.1). In addition, it is well known that visualization is a key prerequisite for our individual efforts to understand otherwise abstruse concepts and phenomena. Because engineers must continually deal with abstract information and must communicate findings to clients, we make extensive use of graphics in our profession.

The microelectronic revolution adds an exciting new dimension to these endeavors. Before computers, hand drafting was a time-consuming and, hence, a costly process. Even after computers were invented, the cost and logistical problems connected with utilizing mainframe computers discouraged the widespread application of computer graphics. Because of their speed, memory, low cost, and accessibility, personal computers offer a significant increase in our capability to produce and modify drawings. As such, they will not only make us more efficient but also broaden and modify the role of graphics in our work.

The present chapter is devoted to ways in which engineers present information visually and how computers can facilitate the process. The first sections deal with precomputer methods for developing graphs. This is necessary because many characteristics of well-designed computer graphs are related to the characteristics of graphs that are concocted by hand. This review is followed by an introduction to computer graphics.

19.1 LINE GRAPHS

A routine task in engineering practice is to establish the relationship between two variables. Line graphs are the most common visual displays employed for this purpose. As seen in Fig. 19.2, line graphs display the relationship on a two-dimensional space. They are useful because the pattern on the graphs illustrates qualitatively how one variable changes with respect to another—that is, whether it is increasing, decreasing, or oscillating.

475

FIGURE 19.1
An engineer using personal computer graphics to present technical information to clients.

FIGURE 19.2
Three examples of line graphics showing (*a*) increasing, (*b*) decreasing, and (*c*) oscillating relationships between the dependent variable *y* and the independent variable *x*.

Figure 19.3 illustrates the two primary coordinate systems used for line drawings in engineering: cartesian and polar. *Cartesian coordinate systems* define position in terms of distance along a vertical and a horizontal axis (Fig. 19.3*a*). In contrast, *polar coordinate systems* define position in terms of an angle and a distance from the center of the space (Fig. 19.3*b*).

Cartesian coordinate systems are further distinguished by the manner in which distance along each axis is delineated. The type shown in Fig. 19.3*a*, called a *rectilinear*

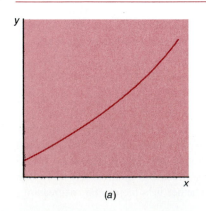

(*a*)

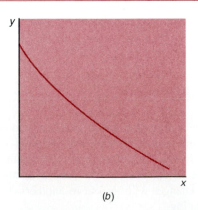

(*b*)

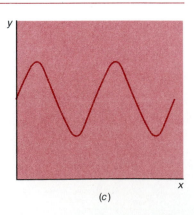

(*c*)

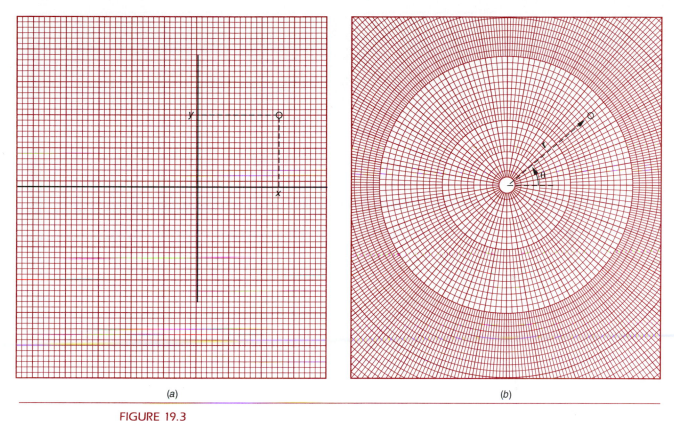

(a)   (b)

FIGURE 19.3
The difference between
(a) cartesian and (b) polar
coordinate systems. In both
cases, two values are required
to specify location on the
two-dimensional space.

line graph, divides each axis into intervals of equal length. In contrast, the *semiloga-rithmic* and *logarithmic* line graphs in Fig. 19.4a and b use logarithmic distances for one or both of the axes.

19.2 RECTILINEAR GRAPHS

For cartesian coordinate systems, it is conventional to plot the *dependent variable* on the vertical axis, or *ordinate,* and the *independent variable* on the horizontal axis, or *abscissa* (Fig. 19.5). As the names imply, these variables usually characterize a cause-effect relationship, where the dependent variable (the effect) "depends" on the value of the independent, or "given," variable (the cause). Thus the dependent variable is often said to be plotted "versus," or "as a function of," the independent variable.

As seen in Fig. 19.6, several curves can be plotted. In Fig. 19.6a, several depen-dent variables are plotted versus a single independent variable. However, because this necessitates including additional ordinates and labels, the number of dependent vari-ables plotted in this fashion is usually limited to two or three. As shown in Fig. 19.6b, several curves can also be plotted to relate a single dependent variable to a single inde-

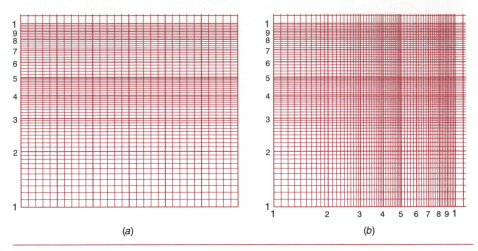

FIGURE 19.4 (a) Semilog and (b) log-log graph paper.

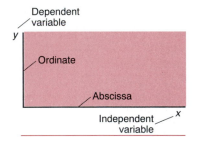

FIGURE 19.5
Conventions regarding the placement of the dependent and independent variables for cartesian coordinate systems.

pendent variable. Each curve corresponds to the relationship between the variables for a fixed parameter value. In the present context, a *parameter* is a constant or variable that characterizes or identifies a particular curve on a line graph. For example, each curve in Fig. 19.6b corresponds to a unique value of the parameter, temperature (T).

The procedure for constructing a graph can be reduced to the following steps:

- Tabulating the data
- Selecting the correct type of graph paper
- Positioning the axes
- Graduating and calibrating the axes
- Plotting the points and drawing the lines
- Adding labels and titles

FIGURE 19.6
Two cases where more than one curve is plotted on a graph. (a) Three dependent variables—y_1, y_2, and y_3—versus an independent variable x. Note how the extra ordinates may be added to the left or on the opposite side from the primary ordinate. (b) Graphs of a dependent variable y versus an independent variable x for three values of the parameter T.

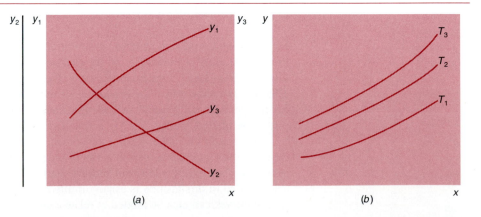

19.2.1 Tabulating Data

It is good practice to tabulate the data before plotting it on a graph. Two situations normally arise. When the data originates from an experiment or a test, it is usually in the form of a set of readings that can be directly assembled in tabular form. By convention, the independent variable is usually listed in the first column and dependent variables in the adjacent columns. For cases where there are several dependent variables conforming to different parameter values, the parameters should be clearly specified in the table's heading.

A second source of data for plotting is an equation. Because the equation allows the determination of the value of the dependent variable on the basis of any value of the independent variable, an unlimited number of points can be generated in order to plot the continuous curve represented by the equation. Thus the equation is used to generate a table of values that can be employed to plot the curve. The actual data values in the generated table are not plotted as points that are visible on the final graph but are used as the basis for sketching a smooth, continuous curve. The number of data points needed for this purpose depends on the equation. A smooth equation would require few points, whereas a complicated equation with oscillations would require many.

EXAMPLE 19.1 Tabulating Data from Experiments and Equations

PROBLEM STATEMENT: The fall velocity of a 68.1-kg parachutist is measured as a function of time; the results are summarized in Table 19.1. Notice that two experiments were performed using jumpsuits with different air resistance. This effect is designated by a parameter called the drag coefficient c, which has units of kilograms per second. In addition to these measurements, a theoretical equation has been developed to predict fall velocity as a function of time:

$$v = \frac{gm}{c}[1 - e^{-(c/m)t}]$$

where v = velocity, m/s; g = acceleration of gravity, 9.8 m/s^2; m = mass, kg; and t = time, s.

Use this equation to generate a table of values to plot and compare with the experimental data.

TABLE 19.1 The Measured Fall Velocity of a Parachutist versus Time for Two Different Drag Coefficients

	Fall Velocity, m/s	
Time, s	c = 12 kg/s	c = 18 kg/s
3	22	19
6	34	27
9	43	33
12	47	34

SOLUTION: The equation can be used to compute a velocity for each of the times and each of the drag coefficients in Table 19.1. For example, for $t = 6$ s and $c = 12$ kg/s,

$$v = \frac{9.8(68.1)}{12}[1 - e^{-(12/68.1)6}] = 36.3 \text{ m/s}$$

This process is repeated for the other values in the table. However, in order to more effectively characterize the curve, additional points must be generated. As seen in Table 19.2, we compute values from $t = 0$ to $t = 14$ using increments of 2 s. Notice that this extends several seconds beyond either side of the measured data range. This amount of detail is required to effectively characterize this curve. If the curve oscillates, even more detail is required.

TABLE 19.2 The Computed Fall Velocity of a Parachutist versus Time for Two Different Drag Coefficients

Time, s	Fall Velocity, m/s	
	$c = 12$ kg/s	$c = 18$ kg/s
0	0.0	0.0
2	16.5	15.2
4	28.1	24.2
6	36.3	29.5
8	42.0	32.6
10	46.1	34.4
12	48.9	35.5
14	50.9	36.2

19.2.2   Choosing Graph Paper

Printed graph paper is available in a variety of colors, paper types, and grid sizes. The choice of the proper kind of paper is predicated on a number of factors, including the purpose of the graph and the range and the precision of the data. For example, semilog and log-log paper is often required when one or more of the variables range over many orders of magnitude. Additionally, such paper is often used for curve fitting. Obviously, logarithmic scales cannot be employed for variables with both positive and negative values. We will return to these types of graphs and their applications in later sections of this chapter.

For rectilinear graphs, a number of different grid spacings are available. Figure 19.7 shows three examples, ranging from a very fine to a very coarse grid. The very fine grids are usually avoided in cases where the graph is to be reproduced or photographed. However, fine grids are required when the data must be plotted to a high degree of precision. Because the data in Example 19.1 is not very precise, we would use a fairly coarse grid, one with perhaps 10 divisions per inch, for a plot of the fall velocity versus time.

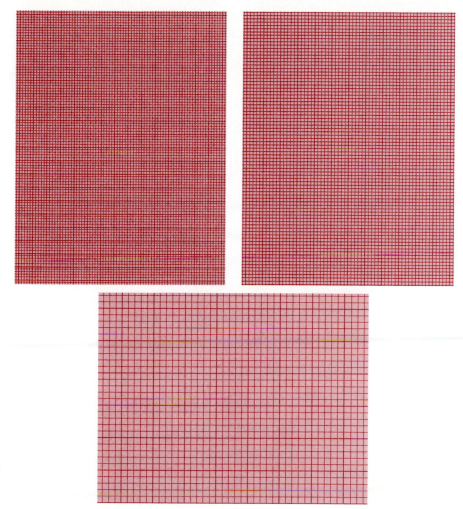

FIGURE 19.7
Three examples of grid
spacing for rectilinear
graph paper.

19.2.3 Positioning the Axes

Because data often contains both positive and negative values for the variables, rectilinear graphs may fall within any or all of the four quadrants depicted in Fig. 19.8. The positioning of the axis is determined by the range of the data and the objective of the graph. For situations where all the values are positive, we can limit the plot to the first quadrant of Fig. 19.8. In such cases it is usually desirable to place the origin (that is, the values of zero for the variables) at the lower left corner of the graph. An exception is for situations where such positioning results in the data being compressed drastically (Fig. 19.9a). If the object of the graph is to depict the pattern in which the points fall with respect to each other, such compression will make it difficult to perceive the con-

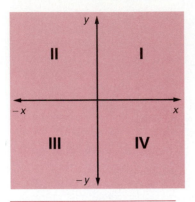

FIGURE 19.8
The four quadrants of a
rectilinear graph.

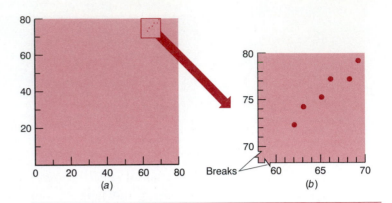

FIGURE 19.9 (a) An example of how placement of the origin at
the lower left corner can lead to a compressed data set. (b) A
remedy for this situation can be developed by using axis
breaks so that the points fill the space of the graph.

figuration of the points. For such cases, the minimum and maximum values of the axes can be chosen so that the points fill the space of the graph (Fig. 19.9b). However, if the object of the graph is, in fact, to show that the data is grouped in a cluster far from the origin, the original version (Fig. 19.9a) is perfectly valid. For situations where both positive and negative values are to be plotted, the origin should be positioned so that all values can be shown.

19.2.4 Graduating and Calibrating the Axes

After the axes are properly positioned, they must be *scaled*. That is, a series of marks, called *graduations*, must be used in order to specify distances along each axis. It is convenient when the graduations are chosen so that the smallest divisions are a power 10 times 1, 2, or 5. This is done to facilitate *interpolation*—that is, estimation of values falling between graduations (Fig. 19.10).

FIGURE 19.10
Examples of graduations:
(a) Follows the rule of making
the graduations of a power of
10 times 1, 2, or 5; hence
interpolation is facilitated. In
contrast, (b) violates the rule
and interpolation is difficult.

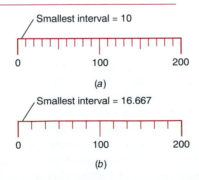

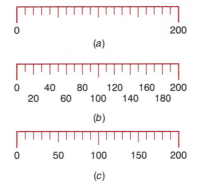

FIGURE 19.11
Examples of scale calibrations: (*a*) too sparse, (*b*) too crowded, and (*c*) the ideal case, neither too crowded nor too sparse, and thus easy to read and understand.

After the graduations are added, numerical values are included to *calibrate* the scale. As seen in Fig. 19.11, calibrations should be neither too crowded nor too spread out. The object is to make the graph as easy to understand as possible.

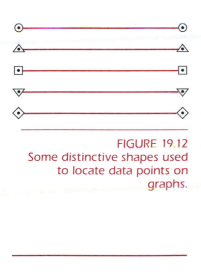

FIGURE 19.12
Some distinctive shapes used to locate data points on graphs.

FIGURE 19.13
Some styles used to represent different lines on the same graph.

19.2.5 Plotting Points and Drawing Lines

Data to be plotted on rectilinear graphs may consist of points or lines, or both. Points are usually employed to represent experimental data, whereas lines or curves are typically used to represent theoretical or empirical equations. Points for a particular set of data are identified by using one of the distinctive shapes shown in Fig. 19.12. The exact location of each point is specified by the dot in the center of each symbol. The dot may be omitted, provided the symbol is centered on the data point. Note that for graphs involving several sets of data, a different symbol can be used to designate each set.

Lines are employed to represent continuous trends, such as those specified by an equation. As with points, several lines may be included on the same graph. When these lines overlap, it is often convenient to use different types, as shown in Fig. 19.13, in order to visually distinguish between the various curves.

There are several instances where lines and points appear on the same curve. One example is the case where a line is used to trace out, or "fit," the trend suggested by a set of points. Notice in Fig. 19.14 that the lines do not penetrate the points. Another example occurs when the lines and points represent independent measures of the same quantity. This was the case for the theoretical and empirical data recorded for the velocity of the falling parachutist in Tables 19.1 and 19.2. As shown in Fig. 19.15, plotting both points and lines on the same graph provides a visual comparison of the theoretical and empirical results. Different symbols are used to distinguish between the two cases. The same solid line is employed for the equations because it is clear which line is associated with which data. However, if the lines overlapped, it would be preferable to use a dashed line for one of them. Notice that Fig. 19.15 indicates that the theoretical relationships (the lines) consistently overestimate the empirical data (the points). Also no-

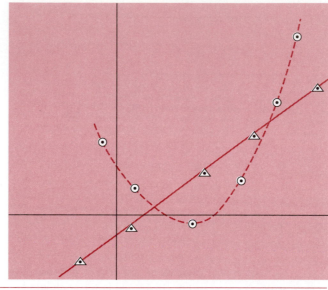

FIGURE 19.14
Example of lines and points
on the same graph. In this
example, the lines are
intended to trace out the
trend suggested by the
points. Notice that the lines
are not allowed to penetrate
the points.

FIGURE 19.15
Plot of data from Tables 19.1
and 19.2 showing
graduations, calibrations,
points, and lines.

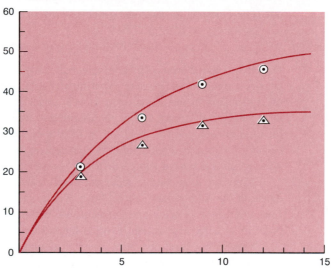

tice that the points from Table 19.2 are not shown explicitly on the figure but, rather, are used as the basis for sketching the curves.

19.2.6   Adding Labels, an Identification Key, and a Title

The final phase in producing a graph is to add words to identify the plot and its elements. Your general objective should be to make the graph as easy as possible to understand. The process involves adding labels, an identification key, and a title for the plot.

The axes are labeled with the variable names and the units of measurement. In cer-

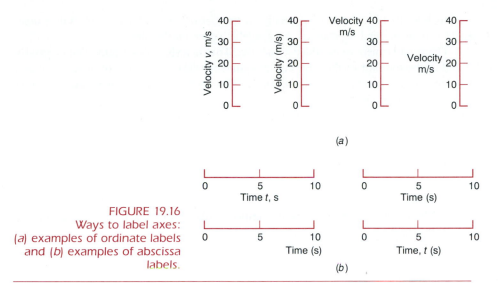

FIGURE 19.16
Ways to label axes:
(a) examples of ordinate labels
and (b) examples of abscissa
labels.

tain cases, the mathematical symbols [such as t for time or $f(x)$ for a function of x] may be used in place of the variable names. However, because your intent should be to communicate easily, this practice is usually not recommended. Figure 19.16 provides examples of some acceptable ways to label axes.

Where several curves and types of points are included, labels can be added to designate their identities. This can be done in two ways. Where the curves are sufficiently separate, the identifier (such as a run number, parameter value, date, and the like) may be placed adjacent to the curve (Fig. 19.17a). An alternative is to include a key. As

FIGURE 19.17
Two ways to add titles and
identification keys.

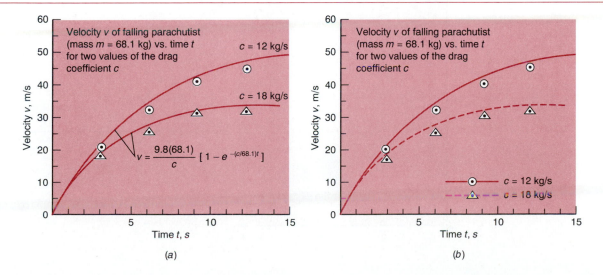

seen in Fig. 19.17*b,* the key is usually in the bottom half of the graph. Of course, where it is positioned depends on the available space on the plot.

The final element to be added is the title. This is a short description that is usually placed in the top half of the plot. For experimental data, it is common practice to add the name of the experimenter and the data and location of the experiment. For plots that are reproduced in reports, papers, or books, the title is typically expressed as a caption placed below or alongside the plot. Figures in this book use this approach.

19.3   OTHER TYPES OF GRAPHS

The rectilinear graphs described in the previous section are but one, albeit the most common, of the graphs you will use as an engineer. Other important types are logarithmic and polar graphs, nomographs, and charts.

19.3.1   Logarithmic Graphs

Two types of logarithmic graphs are commonly used in engineering: semilog and log-log plots. These are sometimes employed when one or more of the variables range over many orders of magnitude. For such cases, using logarithms permits a wide range of values to be graphed without compressing the smaller values (Fig. 19.18). Another reason, as described next, is to develop mathematical relationships between the variables.

FIGURE 19.18
(*a*) Untransformed data ranges over several orders of magnitude. For this case, the smaller values tend to be bunched together. (*b*) When such data is transformed logarithmically, the points are spread out. Sometimes, as in this example, a predictable relationship such as a straight-line, or linear, trend emerges.

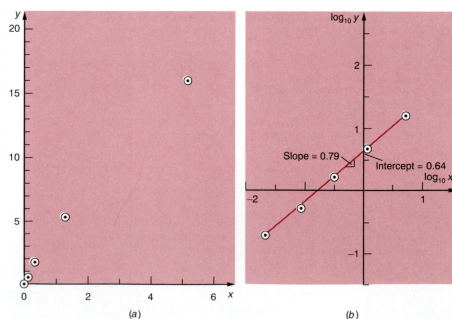

(a)   (b)

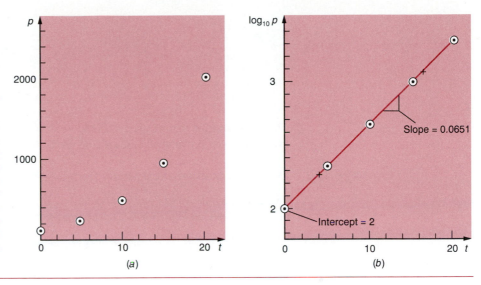

FIGURE 19.19
(a) A curving plot of population versus time. (b) When this data is graphed on a semilog plot, it is transformed into a straight line. This line in (b) can be used to determine a mathematical equation or model to predict values of the curved function in (a). The additional points designated by x's in (b) are employed to determine the slope.

Semilogarithmic plots. A graph in which the ordinate is plotted with a logarithmic scale and the abscissa with a linear scale is referred to as a semilogarithmic, or *semilog*, plot. Oftentimes data that is curved when plotted on rectilinear graph paper (Fig. 19.19a) will plot as a straight line on a semilog plot (Fig. 19.19b). The intercept and the slope of the straight line can be used to formulate the following equation:

$$\log_{10} y = \log_{10} y_0 + (\text{slope})x \tag{19.1}$$

where y_0 is the value of the dependent variable at the intercept (that is, the y value at $x = 0$) and the slope is calculated as

$$\text{Slope} = \frac{\log_{10} y_2 - \log_{10} y_1}{x_2 - x_1} \tag{19.2}$$

where $(x_1, \log_{10} y_1)$ and $(x_2, \log_{10} y_2)$ are two points from the straight line.

In order to transform this equation back to a form that is appropriate for nonlogarithmic scales, we must first subtract $\log_{10} y_0$ from both sides of Eq. (19.1) to give

$$\log_{10} y - \log_{10} y_0 = (\text{slope})x$$

or

$$\log_{10} \left(\frac{y}{y_0} \right) = (\text{slope})x$$

The antilogarithm of this equation is

$$\frac{y}{y_0} = 10^{(\text{slope})x}$$

or

$$y = y_0 10^{(\text{slope})x} \tag{19.3}$$

Thus this equation can be used to compute values for Fig. 19.19a. As seen in the next example, it can prove useful in making predictions.

EXAMPLE 19.2   **Deriving a Mathematical Model from a Semilog Plot**

PROBLEM STATEMENT: The population (p) of a small community on the outskirts of a city grows rapidly over a 20-year period:

t	0	5	10	15	20
p	100	212	448	949	2009

This data is plotted in Fig. 19.19a. As an engineer working for a utility company, you must forecast the population 5 years into the future in order to anticipate the demand for power. Employ a semilog plot to make this prediction.

SOLUTION: A semilog plot of the data can be developed by taking the logarithm of the population:

t	0	5	10	15	20
$\log_{10} p$	2	2.326	2.651	2.977	3.303

These values can be graphed (Fig. 19.19b) to verify that the data follows a straight-line relationship. Then the graph can be used to determine an intercept of 2 and a slope of

$$\text{Slope} = \frac{3.041 - 2.260}{16 - 4} = 0.0651$$

Therefore, according to Eq. (19.3), the model for Fig. 19.19a is

$$p = 100(10)^{0.0651t}$$

Substituting $t = 25$ into this equation allows us to predict

$$p = 100(10)^{0.0651(25)} = 4241 \tag{E19.2.1}$$

Therefore, in 5 years the population will more than double if present trends continue.

The utility of semilogarithmic paper is that it can be used to produce plots such as Fig. 19.19*b* without actually taking the logarithms of the data. Figure 19.20 shows how the data from Example 19.2 can be directly plotted on such paper. If, as is the case in Fig. 19.20, a straight line results, then the analysis from Example 19.2 can be performed to derive the predictive equation. The only need for logarithms is in determining the slope with Eq. (19.2).

Note that the foregoing analysis could have also been accomplished with natural logarithms. If so, the equation for the transformed line [comparable to Eq. (19.1)] is

$$\ln y = \ln y_0 + (slope')x \tag{19.4}$$

The slope would be [comparable to Eq. (19.2)]

$$Slope' = \frac{\ln y_2 - \ln y_1}{x_2 - x_1} \tag{19.5}$$

and the final predictive equation is [comparable to Eq. (19.3)]

$$y = y_0 e^{(slope')x} \tag{19.6}$$

Either Eqs. (19.1) through (19.3) or Eqs. (19.4) through (19.6) can be fit to the same

FIGURE 19.20
Plot of the data from Example 19.2 on semilog graph paper.

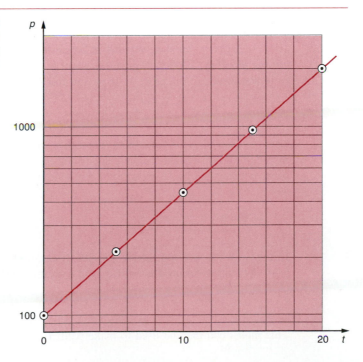

data. For example, applying the natural log version to the data in Example 19.2 yields the following predictive equation:

$$p = 100e^{0.1499t}$$

This equation produces results identical with those given by Eq. (E19.2.1). For example,

$$p = 100e^{0.1499(25)} = 4241$$

The slope' is marked with a prime in Eqs. (19.5) and (19.6) to distinguish it from the slope in Eqs. (19.2) and (19.3). The two slopes are related by the formula

$$\text{Slope} = \frac{\text{slope}'}{2.3025}$$

which derives from the fact that

$$\log_{10} x = \frac{\ln x}{2.3025} \tag{19.7}$$

Note that the use of semilog paper does not hinge on whether you employ common (base 10) or natural (base e) logarithms. This is because the two types of logarithms are linearly related by Eq. (19.7). However, it is very important to determine the slope consistently. If the base 10 model [Eq. (19.3)] is used, the slope must be detemined with Eq. (19.2), whereas the base e model [Eq. (19.6)] requires use of Eq. (19.5).

Log-log plots.   Another way that an equation can be fit to curved data is to employ a log-log plot (Fig. 19.18b). Again, the intent is to see whether the data "straightens out," or is linearized, in the process. If so, the following equation holds:

$$\log_{10} y = \log_{10} y_0 + (\text{slope}) \log_{10} x \tag{19.8}$$

where y_0 is equal to the value of the dependent variable at the intercept. For a log-log plot the intercept corresponds to the value of $\log_{10} y$ at $\log_{10} x = 0$ (that is, at $x = 1$). The slope is computed by [compare with Eq. (19.2)]

$$\text{Slope} = \frac{\log_{10} y_2 - \log_{10} y_1}{\log_{10} x_2 - \log_{10} x_1} \tag{19.9}$$

Just as with the semilog model, Eq. (19.8) can be reexpressed by subtracting $\log_{10} y_0$ from both sides, taking the antilog, and rearranging the result to yield

$$y = y_0 x^{\text{slope}} \tag{19.10}$$

This model can then be used to make predictions of y as a function of x.

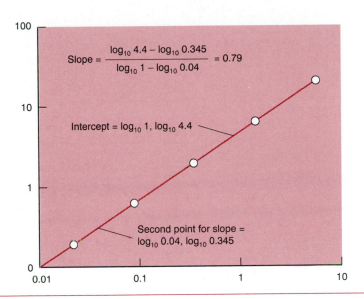

FIGURE 19.21
A plot of the data from
Fig. 19.18 on log-log graph
paper.

For this case, special log-log graph paper is available (recall Fig. 19.4b) to facilitate making the initial plot without having to actually take the logarithm of the data. Then if you ascertain that the data traces out a straight line, the slope and the intercept can be determined. Both the plot and the evaluation of the slope and intercept are shown in Fig. 19.21. These values can be substituted into Eq. (19.10) to give the model equation,

$$y = 4.4x^{0.79}$$

This equation can then be used to make predictions. Note that again it does not matter whether common or natural logs are employed as long as you are consistent. Equation (19.9) yields the same slope, regardless of the type of logarithm employed.

19.3.2 Polar Graphs, Nomographs, and Charts

Polar graphs. Polar coordinates are employed when the magnitude of a variable is to be displayed with regard to angular position. One application would be to display the intensity of light or sound at distances around a source such as a lamp or speaker (Fig. 19.22). Other applications range from the study of rotating objects to the analysis of mathematical functions. Sometimes it is useful to make transformations between

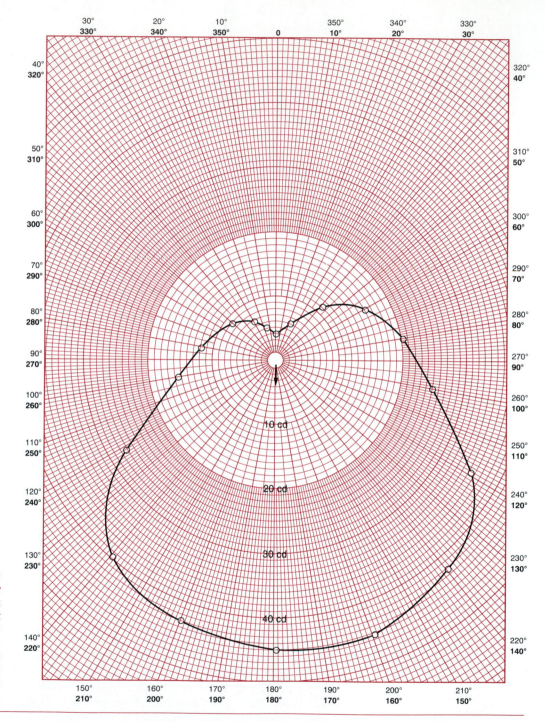

FIGURE 19.22
Data for the intensity
of light at distances
around a light
source, as displayed
on polar coordinates.
The arrow indicates
the direction in
which the source is
oriented.

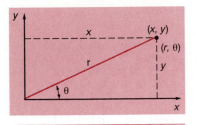

FIGURE 19.23
The relationship of cartesian
and polar coordinates.

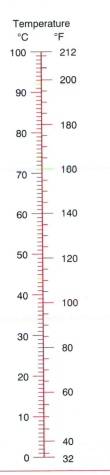

FIGURE 19.24
A simple nomograph relating
the Celsius and Fahrenheit
scales for temperature.

cartesian and polar coordinates. The following equations are used for this purpose (see Fig. 19.23). To convert from polar to cartesian,

$$x = r \sin \theta \qquad (19.11)$$

and

$$y = r \cos \theta \qquad (19.12)$$

Although it is relatively straightforward to compute cartestian coordinates (x, y) on the basis of polar coordinates (r, θ), the reverse process is not so simple. The radius can be computed with the following formula:

$$r = \sqrt{x^2 + y^2} \qquad (19.13)$$

If the coordinates lie within the first or fourth quadrants (that is, $x > 0$), then a simple formula can be used to compute θ:

$$\theta = \arctan\left(\frac{y}{x}\right) \qquad (19.14)$$

The difficulty arises for all other situations. The following table summarizes the possible cases that can occur:

x	y	θ
<0	>0	$\tan^{-1}\left(\dfrac{y}{x}\right) + \pi$
<0	<0	$\tan^{-1}\left(\dfrac{y}{x}\right) - \pi$
<0	=0	π
=0	>0	$\dfrac{\pi}{2}$
=0	<0	$-\dfrac{\pi}{2}$
=0	=0	0

Nomographs.   A nomograph is a graphical display of an established relationship between variables. Figure 19.24 shows a very simple example for the conversion of temperature between the Fahrenheit and the Celsius scales. This nomographs is made by simply placing the scales side by side vertically so that the corresponding values line up. Thus, if you know the Celsius value, you merely have to read the corresponding Fahrenheit value from the adjacent scale. More complex nomographs include several vertical scales and intricate schemes for predicting values. However, the fundamental idea exemplified by Fig. 19.24—that is, the visual alignment of the relationship—holds.

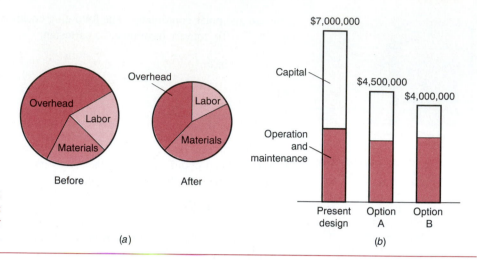

FIGURE 19.25
Some engineering examples
of (*a*) pie charts and (*b*) bar
diagrams.

Bar and pie charts. Although more commonly applied in the economic and social sciences, bar and pie charts can prove useful in engineering. Figure 19.25 shows some examples. We will not discuss rules or guidelines for their construction here. In addition, computer software packages are available to generate such charts accurately and efficiently. These displays are particularly useful when communicating technical information to clients and the general public.

19.4   COMPUTER GRAPHICS

As with other aspects of engineering problem solving, the construction of graphs can be facilitated by the computer. Computer graphics are based on a simple idea that relates back to the very definition of plotting itself. As should be clear to you by now, plotting consists of locating points on a two-dimensional space. Computer output media, such as printer paper or monitor screens, also represent two-dimensional spaces. Figure 19.26 shows the screen of a typical computer monitor. Such a screen is said to be in the *text mode*. As an example, this figure shows a text mode that consists of 25 rows and 80 columns, defining a total of 2000 rectangles, or cells.

Computer graphics in text mode are built up from the fundamental operation of placing a character into one of these cells. For example, as seen in Fig. 19.27*a,* a character such as an asterisk may be placed in one of the cells. If several asterisks are placed in a row of cells, a line can be formed (Fig. 19.27*b*). Similarly, one vertical column and one horizontal row may be positioned to form a set of axes (Fig. 19.27*c*). In a similar fashion, all manner of images may be constructed on the screen.

Beyond the text mode, higher-resolution graphic modes that employ finer grids are also available to attain more refined patterns. In addition, canned software has been developed to facilitate graphics. Before illustrating such canned software, we will briefly review the ways in which computer monitors operate.

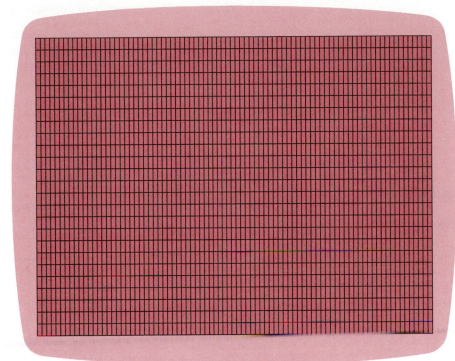

FIGURE 19.26
The screen of a typical computer monitor. When in text mode, the screen consists of 25 rows and 80 columns. It should be recognized that some computers have different numbers of rows and columns. However, the basic idea of the monitor screen as a two-dimensional space holds for all computers.

FIGURE 19.27
Examples of how images are formed by placing characters on a screen: (a) a single character, (b) a line formed by a series of characters, and (c) a set of axes formed by two lines of characters.

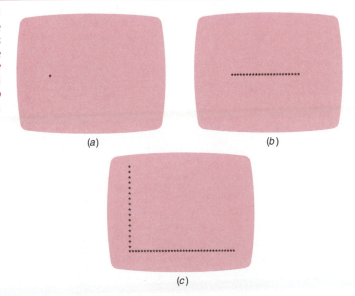

(a)

(b)

(c)

19.4.1 Graphics Equipment and Terminology

Most personal-computer monitors are based on the technology of the *cathode-ray tube,* or *CRT.* A beam of electrons is directed at a coat of phosphors on the interior surface of the screen. These phosphors glow momentarily wherever the beam strikes the surface. It is the composite of all the individual glowing phosphors that produces the image that we view on the other side of the screen.

Because the glow fades when the beam moves elsewhere on the screen, the image must be redrawn or refreshed continuously. There are two basic ways that the beam moves about the screen to refresh the images. The first, called a *vector display,* produces images by moving the beam between two points on the screen to create a straight line or vector. The final image consists of a composite of these individual vectors. Figure 19.28*a* indicates how a set of axes is created on a vector display.

FIGURE 19.28
The fundamental difference between the ways that images are created for (*a*) vector and (*b*) raster displays.

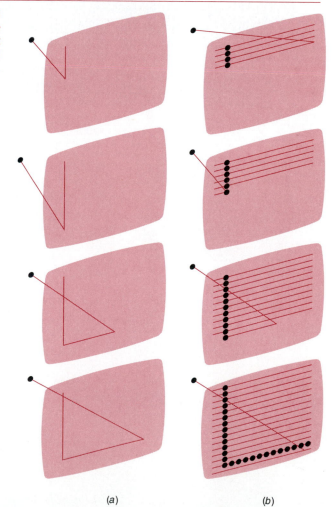

(a)   (b)

The second mode of operation, called the *raster display,* produces images by moving the beam over a set of horizontal lines. The beam's intensity is heightened to leave a glow wherever an image is desired. The common television set uses this technology. Figure 19.28*b* shows how a set of axes is created by a raster display.

Vector displays were originally the most common mode used for graphics. They have a number of advantages, including the fact that they require little memory, produce very crisp images, and are well-suited for dynamic or moving graphics. However, they are also relatively costly. Because they make use of television technology, the raster displays are much less expensive. As a consequence, they are the type of display used in most personal computers.

As shown in Fig. 19.28*b,* the image on a raster display is composed of the individual energized points. These are sometimes referred to as *picture elements,* or *pixels.* One deficiency of raster displays is that images composed of discrete pixels can sometimes appear jagged. As seen in Fig. 19.29, this effect, which is also called *staircasing,* is dependent on the density or number of pixels that make up the screen. The degree of density is referred to as the screen's *resolution.* A larger number of pixels—that is, a higher resolution—yields smoother and crisper images. However, the better the resolution is, the more expensive the monitor will be. Also, as the number of pixels increases, the required memory increases and the speed of operation decreases. Thus there are trade-offs that must be considered when choosing a monitor. If the primary use of the system is for word processing or simple computations, a low-resolution monitor might be adequate. However, if the main purpose is to obtain graphics displays, a high-resolution screen will be required.

A screen's resolution is specified quantitatively by its dimensions—that is, by the number of columns and rows of pixels it contains. For example, the text mode used for the plots in Fig. 19.27 had dimensions of 80 columns by 25 rows, or 80×25 cells.

As its name implies, the text mode is not intended for intricate graphic displays but is designed for dealing with textual material, as in word processing or simple computer programming. Higher-resolution graphics modes typically range from 280×192 to upwards of 1000×1000 pixels.

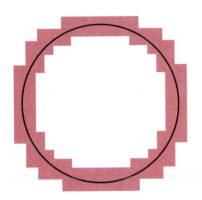

FIGURE 19.29
An exaggerated example of how the use of discrete pixels to create an image can lead to jagged or "staircased" surfaces. In this case, because of low resolution, the circle appears very jagged.

19.4.2   The Electronic TOOLKIT

Because of the differences between graphics commands for computers, it is inappropriate to present a comprehensive exposition on the topic here. We have, therefore, chosen to focus the present discussion on the graphics software that accompanies this text.

The Electronic TOOLKIT software that accompanies this text includes a data-and-function plotter program. As shown in the next example, it represents a potent tool for gaining visual insight into engineering-related problems.

EXAMPLE 19.3   Locating Roots with Computer Graphics

PROBLEM STATEMENT: The following type of function frequently occurs in the analysis of vibrating systems such as harmonic oscillators:

$$f(x) = \sin 10x + \cos 3x$$

Use computer graphics to gain insight into the behavior of this function.

SOLUTION: Figure 19.30a shows the menu for the plotting program from the Electronic TOOLKIT. Notice that you have four options: (1) edit functions, (2) enter data, (3) plot data and functions, and (4) return to the main menu of the Electronic TOOLKIT. Select option 1 to enter the function. Type in the function, as shown in Fig. 19.30b, and strike [End] to enter it into the computer and exit. Note that X is the only variable permitted. Standard mathematical operators (such as $+$, $-$, $*$, $/$, and $^\wedge$ for exponentiation) and intrinsic functions (such as EXP, SIN, COS, and so on) are allowed. You can use the [F1] key to toggle on and off a help screen that describes all the mathematical functions and special operators available on the TOOLKIT. The function can contain up to 256 characters. The TOOLKIT can plot two different functions on the same graph. The [PgUp] and [PgDn] keys are used to toggle between the two functions.

The program checks for syntax errors when you enter the function. If syntax errors are detected, a diagnostic message is displayed at the bottom of the screen. In these situ-

FIGURE 19.30
The menus and screens of the GRAPHICS program of the Electronic TOOLKIT used to generate a line graph: (a) the main menu; (b) an equation being entered; and (c) the selection of type of graph, axes dimensions, and labels.

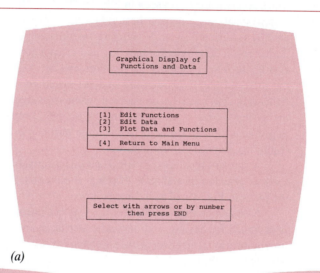

(a)

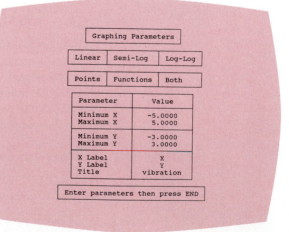

(b)   (c)

ations, the program may also display the error-free part of your function on the screen. This gives you a clue to the character that prompted the error message. Upon identifying the error, you can make corrections by inputting characters with appropriate modifications. You can also use the [Backspace] or [Del] keys to erase a character or the [F8] key to erase the entire function. This process of entering and editing the function is a convenient feature of the Electronic TOOLKIT software. It is an example of the type of time-saving features that are characteristic of high-quality canned software.

When the function has been successfully entered into the program, you are automatically transferred back to the plotting menu. Now you may choose option 3, where you can select a linear, semilog, or log-log form of the plot. In addition, you can opt to plot functions alone, data alone, or both. Before the plot can be constructed, it is also necessary to supply minimum and maximum values for both the independent variable X and the dependent variables (called Y in the program). The program prompts you for the dimensions with the wide horizontal bar. Figure 19.30c shows the screen after the limits are entered. Any or all of these values can be changed by moving the cursor to them and entering new numbers. If you enter an erroneous limit—for example, if minimum X > maximum X—the program will alert you with an error message at the bottom of the screen. This is another example of how high-quality software facilitates your efforts.

When you are satisfied that you have entered the desired limits, the program gives you the opportunity to label the X and Y axes and provide a title for the plot. The input session is terminated using the [End] key. When you have struck this key, a plot is automatically constructed, as shown in Fig. 19.31a. Notice how this plot has many of the features of a good line graph. The axes are numbered and scaled, and the plot is drawn in high-resolution color mode. Notice also how secondary dashed axes are included to designate the zero values of the variables. The axes are labeled, and the plot has a title.

The utility of the software can be demonstrated by using it to gain insight into the behavior of the function. Close inspection of Fig. 19.31a indicates that the function crosses the x axis at various points, connoting several roots in the interval from $f(x) = -5$ to 5. In addition, the function appears to be tangential to the axis at about $x = 4.2$. Such behavior is referred to as a *double root*.

In order to examine this possibility in more detail, the plot can be "blown up" by entering a new set of axis dimensions. The result, corresponding to a narrowed range of the X axis from $x = 3$ to $x = 5$, is shown in Fig. 19.31b. This version still suggests a multiple root at about $x = 4.2$.

Finally, as seen in Fig. 19.31c, the vertical scale is narrowed further to $f(x) = -0.15$ to $f(x) = 0.15$, and the horizontal scale to $x = 4.2$ to $x = 4.3$. The plot shows clearly that a double root does not exist in this region and that in fact there are two distinct roots at about $x = 4.23$ and $x = 4.26$.

The time and effort required to manually construct the graphs in Fig. 19.31 would be enormous and might even discourage you from performing the analyses. In contrast, the Electronic TOOLKIT GRAPHICS program accomplishes the same task easily. The effort to develop your own homemade software of comparable quality would be considerable. Thus the canned program provides a significant resource that might warrant its inclusion in your arsenal of problem-solving tools.

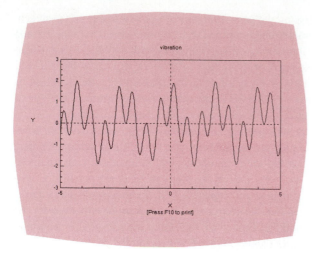

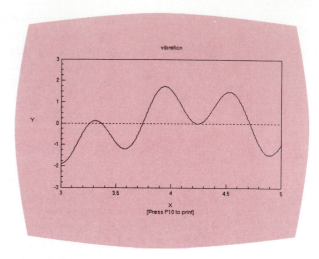

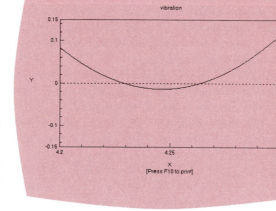

FIGURE 19.31
The progressive enlargement
of $f(x) = \sin 10x + \cos 3x$ by
the GRAPHICS program of the
Electronic TOOLKIT. Such
interactive graphics permit
the analyst to determine that
two roots exist between
$x = 4.2$ and $x = 4.3$.

The Electronic TOOLKIT's GRAPHICS program is but one example of the types of computer graphics software that are available or will be available for engineering problem solving. Other examples range from three-dimensional plotters to computer-aided drafting systems. Additionally, sophisticated graphics are often an important component of other types of canned software. For example, most spreadsheets have plotting options. As described in the next section, these capabilities represent an immense increase in our ability to produce visual expressions of our work.

19.4.3 Computer Graphics and Engineering

In the present chapter, we have stressed how graphs can be employed to easily generate visual depictions of data and mathematical functions. Because visualization is often a critical step in gaining insight into otherwise abstract information, we have chosen to emphasize this quality of computer graphics. Beyond line graphs, however, there are many other areas and applications where computer-generated graphics can benefit engineering.

Examples are found in the areas of computer-aided design and computer-aided manufacture (CAD/CAM). Graphics figure prominently in several facets of the processes but are particularly relevant to CAD. Computer-aided drafting facilitates the preparation of diagrams, plans, and other drawings that are needed to bring engineered works and products to fruition.

Aside from two-dimensional graphs, an interesting capability offered by computers is the rendering of drawings in three dimensions. Among other things, the coordinates of an object can be entered and the computer employed to display it from a number of different perspectives.

Finally, computer graphics can be employed for animation—that is, to add the dimension of time to the spatial dimensions. "Moving pictures" can be of great value in understanding the dynamics of objects and systems. Because computers provide the vehicle for producing such images, graphics will figure prominently in the development and advancement of engineering in the coming years.

PROBLEMS

19.1 For a graph in rectangular cartesian coordinates, on which axes are the dependent and the independent variables plotted?

19.2 Determine the polar coordinates for a point with the following rectangular coordinates: $(5, 2)$, $(8, -1)$, $(-5, -6)$.

19.3 Name the six steps for constructing a graph.

19.4 Why is it desirable to scale a graph so that the graduations are a power of 10 times 1, 2, or 5?

19.5 Give some reasons for employing semilog and log-log plots.

19.6 Will the equation $y = 5e^{-x}$ plot as a straight line on log-log paper? If not, is there a type of paper on which it will plot straight?

19.7 The following data was gathered in an experiment to determine how the flow rate through a valve (Q) varied with the valve opening (V):

Q, ft^3/s	V (% valve opened)	Q, ft^3/s	V (% valve opened)
0.000	0	0.770	50
0.021	5	0.980	60
0.096	10	1.100	70
0.200	20	1.200	80
0.350	30	1.300	90
0.560	40	1.400	100

Plot the data on a graph with appropriate labels and titles, and determine visually the best-fit curve through the points.

19.8   Fill in the steps omitted in the text to derive Eq. (19.10). Why is the intercept of a line on a log-log plot taken to be at the value of $x = 1$?

19.9   The following data was taken during an experiment to determine the strength of a new concrete mixture as a function of the time after it has been poured:

Time, d	3	4	5	6	7	10	14	21
Strength, ksi	0.7	1.9	3.2	4.1	5.4	7.6	8.3	9.1

Plot this data on rectilinear and semilog graph paper, and extrapolate from these plots the compression strength after 28 days. Interpret your results.

19.10 A variable y is said to be inversely proportional to a variable x if the relationship between the two can be expressed as

$$y = \frac{1}{x}$$

On what graph paper would this equation plot as a straight line?

19.11 On what graph paper would

$$y = \frac{1}{x^3}$$

plot as a straight line?

19.12 Determine all the types of graph paper on which the following functions would plot as straight lines:

(a) $y = -5 + x$ (d) $y = 23e^{-x}$

(b) $y = \dfrac{30}{x}$ (e) $y = 23e^{x^2}$

(c) $y = \dfrac{10^x}{x}$ (f) $y = \dfrac{x}{33 + x}$

Employ transformations if necessary.

19.13 (*Computer-oriented problem*) Develop your own printer-plot program based on the material in this chapter. Express the program in subroutine form so that in later chapters you can use it to add a plotting capability to other programs. Have the following data serve as input to the subroutine: one-dimensional arrays, X and Y, holding the x, y coordinates of the points; N, the total number of points; and the minimum and maximum limits of the X and Y axes—XMIN, XMAX, YMIN, YMAX. Use a structured approach for the program, and develop both external and internal documentation. Try to incorporate as many features of a well-designed hand-drawn graph as possible.

19.14 Test the program developed in Prob. 19.13 by using it to plot the points from Probs. 19.7 and 19.9.

19.15 Develop a program as described in Prob. 19.13, but design it to plot a function.

19.16 Test the program developed in Prob. 19.15 by using it to plot the functions in Probs. 19.10 and 19.11 from $x = 0.5$ to 5 and from $x = -3$ to 3.

19.17 Develop a program as described in Prob. 19.13, but design it to plot both a function and discrete data points on the same graph.

19.18 Employ the most advanced graphics capabilities available on your computer to develop a graphics program similar to the one on the Electronic TOOLKIT. That is, develop a program that can plot a rectilinear line graph of discrete data points, a continuous function, or both. If possible, include the option to plot semilog and log-log graphs. This program can be employed in place of or along with the Electronic TOOLKIT to solve many of the following problems.

19.19 Use the Electronic TOOLKIT or your own program to obtain linear plots of the following functions:

(a) $y = \sin x + \cos x$ $x = -10$ to 10
(b) $y = 10e^{-x}\cos 5x$ $x = 0$ to 3
(c) $y = 10x/(5 + x)$ $x = 0$ to 50
(d) $y = 10e^{-x} + 15(1 - e^{-0.5x})$ $x = 0$ to 8
(e) $y = 5 + 4x$ $x = -5$ to 5
(f) $y = x^2 + 2x - 8$ $x = -10$ to 10

In each case try different scales for the vertical and horizontal axes. Change the labels on the x and y axes. Use your own title for the plots. Use a printer, if available, to obtain a hard copy of any of the above.

19.20 Use the Electronic TOOLKIT or your own program to obtain both linear and semilog plots of the following functions:

(a) $y = e^{-3x}$ $x = 0$ to 1
(b) $y = 3e^{-x/2}$ $x = 0$ to 10
(c) $y = 10^{x/10}$ $x = 0$ to 10
(d) $y = 2^x$ $x = 0$ to 10

Discuss the relationship between the linear and semilog plots.

19.21 Use the Electronic TOOLKIT or your own program to obtain both linear and log-log plots of the following functions:

(a) $y = 2x^2$ $x = 0.1$ to 10
(b) $y = x^{-0.5}$ $x = 0.1$ to 10
(c) $y = 4.1x^{1.3}$ $x = 0.1$ to 10

Discuss the relationship between the linear and the log-log plots.

19.22 The following is experimentally derived data that describes the relationship between the growth rate of a fermentation yeast (r) and the concentration of an organic compound (c):

Concentration, mg/L	0	1	2	3	4	8	12	16
Growth rate, day^{-1}	0	0.05	0.15	0.15	0.12	0.10	0.07	0.05

The following model has been proposed to represent the data:

$$r = \frac{c}{c^2 + 2c + 5}$$

Use the Electronic TOOLKIT or your own program to plot both the data and the model. How well do the data and the plot compare on the basis of the plot?

19.23 The population of a city p_c is decreasing according to the formula

$$p_c = 100{,}000e^{-t/30} + 40{,}000$$

where t is in years and $t = 0$ corresponds to 1985. The population of a suburb p_s is increasing according to the formula

$$p_s = 50{,}000 + 70{,}000(1 - e^{-t/15})$$

for $t = 0$ to 100. Employ the Electronic TOOLKIT or your own program to graphically estimate the year in which the two populations will be equal.

19.24 The following function characterizes the oscillations of an electric relay:

$$y = \sin 1.1t + \sin 1.4t$$

Use the Electronic TOOLKIT or your own program to determine how many times the relay opens or closes as t varies from 0 to 30 s. The relay is closed when y is positive and open when y is negative. Thus the number of times the relay opens or closes is equal to the number of times the function crosses the t axis.

19.25 Use the Electronic TOOLKIT or your own program to determine the number of real roots of the following functions in the range of $x = -10$ to $x = 10$:

(a) $y = 4x^3 - 3x^2 + x - 5$
(b) $y = x^4 - 3x^3 + 2x^2 - x - 8$
(c) $y = x^5 + 3x^4 - 5x^3 - 15x^2 + 4x + 10$
(d) $y = x^4 - 16x^3 + 60x^2 - 50x + 4$
(e) $y = x^5 - 8x^4 - 40x^3 + 300x^2 + 100x - 500$

19.26 Graphically approximate the smallest positive root of each of the functions from Prob. 19.25. Employ the Electronic TOOLKIT or your own program to estimate the root to within two significant figures.

SORTING AND SEARCHING

Data will not be much use to you if you are not able to manipulate it. Two of the most common such manipulations are the processes of sorting and searching. Sorting involves rearranging data so that it is in ascending or descending order. Searching involves determining whether a particular value is contained within a collection of data.

20.1 SORTING

```
DO FOR  i ← 1 to n–1
    small ← x_i
    k ← i
    DO FOR j ← i + 1 to n
        IF x_j < small
            small ← x_j
            k ← j
        END IF
    END DO
    x_k ← x_i
    x_i ← small
END DO
```

FIGURE 20.1
Algorithm to perform a selection sort of an array of numbers.

A fundamental programming operation that has a variety of engineering applications is the sorting of a group of items in some specified order. Aside from its relevance to engineering problem solving, sorting also serves to demonstrate the utility of arrays.

20.1.1 Selection Sort

When dealing with numbers, it is often necessary to sort them in ascending order. There are many methods for accomplishing this objective. A simple approach, called a *selection sort,* is probably very close to the commonsense approach you might use to solve the problem. First you would scan the set of numbers and pick the smallest. Next you would bring this value to the head of the list. Then you would repeat the procedure on the remaining numbers and bring the next smallest value to the second place on the list. This procedure would be continued until the numbers were ordered. A simple algorithm to perform the selection sort is displayed in Fig. 20.1. The following example illustrates how it works.

EXAMPLE 20.1   Selection Sort

PROBLEM STATEMENT: Use a selection sort to arrange the following five numbers in ascending order. The flows are

| 125 | 102 | 147 | 76 | 95 |

SOLUTION: First we must find the lowest number, which we designate as $small$. We start by setting $small$ equal to the first number on the list, x_1, and noting that its subscript, which we designate as k, is equal to 1. Then we compare $small$ with the remaining numbers on the list. If we find a smaller number, we set $small$ equal to this number and store the subscript as k. When this search is completed, $small = 76$ and $k = 4$. Then we switch $x_1 = 125$ with $x_4 = 76$ in order to bring the smallest number to the head of the list. This switch is also required to retain the number (125) that was originally first. After this is completed, the list looks like

76	102	147	125	95

Now we repeat the process, but with $small = x_2 = 102$. On this pass, we find that x_5 is the smallest remaining number, and so we switch x_5 and x_2 to give

76	95	147	125	102

Next we start at $small = x_3 = 147$ and switch x_3 with x_5 to yield

76	95	102	125	147

At this point, the numbers are ordered. However, the computer cannot perceive this, and so it must take the final step of $small = x_4 = 125$, even though this does not result in a switch.

20.1.2 Bubble Sort

Although the selection sort works well enough, there are other algorithms that perform more efficiently (Knuth, 1973). One such method, the Shell sort, is relatively simple to understand and executes approximately twice as fast as the selection sort for arrays of about 200 members. For larger arrays, its relative advantage becomes even more pronounced.

Before describing this technique, we will first briefly present another method that is very slow but will help us to understand the Shell sort. Called the *bubble sort,* this method consists of starting at the beginning of the array and comparing adjacent values. If the first value is smaller than the second, nothing is done. However, if as seen in Fig. 20.2a, the first is larger than the second, the two values are switched. Next the second and third values are compared, and if they are not in the correct order (that is, smallest first), they are also switched. This process is repeated until the end of the array is reached. This entire sequence is called a *pass*.

Figure 20.2a through d shows the switches that occur on the first pass through an array of five numbers. After the first pass the process returns to the top of the list, and another pass is implemented. The procedure is repeated until a pass with no switches occurs (Fig. 20.2g). At this point, the sort is complete. The name "bubble sort" comes from the fact that, as depicted in Fig. 20.2, the smallest value tends to rise to the top of the list like a bubble rising through a liquid.

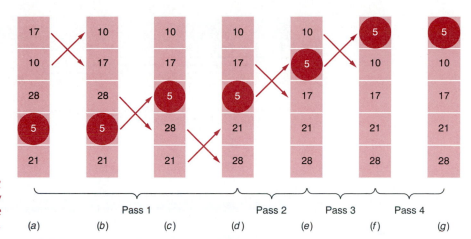

FIGURE 20.2
The bubble sort. Notice how the smallest number rises like a bubble as the sort proceeds.

Pass 1 Pass 2 Pass 3 Pass 4

(a) (b) (c) (d) (e) (f) (g)

```
range ← n
DO
    switch ← 0
    DO FOR i ← 1 to range − 1
        IF x_i > x_{i+1}
            dummy ← x_i
            x_i ← x_{i+1}
            x_{i+1} ← dummy
            switch ← i
        END IF
        range ← i
    END DO
LOOP UNTIL switch = 0
```

FIGURE 20.3
Algorithm to perform a bubble sort of an array of numbers.

Note that after each pass, the numbers below the last switch are in the correct order. Therefore, on the subsequent pass, it is unnecessary to make comparisons beyond this point. This behavior is characteristic of the bubble sort and is exploited in the computer algorithm outlined in Fig. 20.3. Note that we have defined a variable `range` to keep track of where the last switch occurs on a particular pass. This variable is then employed as the upper limit on the counter loop used to define the length of the succeeding pass.

20.1.3 Shell Sort

The *Shell sort* (named for its originator, Donald Shell) is similar to the bubble sort in that values in an array are compared and switched. However, in contrast to the bubble sort, nonadjacent elements are swapped. On the first pass the interval over which the comparison is made is one-half the entire length of the array. As seen in Fig. 20.4a, in the first pass through a 16-element array, the first and ninth items are compared, then the second and tenth, and so on. On the second pass, the comparison interval is halved. As seen in Fig. 20.4b, the first is compared with the fifth, the second with the sixth, and so forth. For each pass these comparisons are repeated until no switches occur. (Note that, if switches occur during a pass, the pass is repeated.) Then the next pass with a halved comparison interval is implemented. Thus, as shown in Fig. 20.4d, the last pass amounts to a bubble sort.

Although the Shell sort is slightly more complicated than either the bubble or the selection sort, its associated computer algorithm is still quite simple (Fig. 20.5). Its increased efficiency more than justifies the slight increase in complexity.

20.1.4 Tabular Data and Sorting in Place

In all the aforementioned algorithms, we sorted an array by switching its elements. This was done using a dummy variable in a manner similar to the following pseudocode:

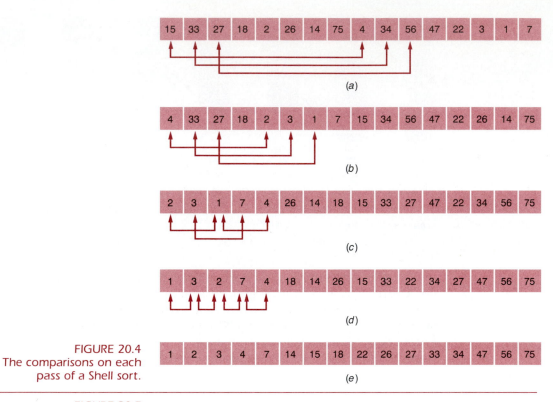

FIGURE 20.4
The comparisons on each
pass of a Shell sort.

FIGURE 20.5
Algorithm for the Shell sort.

```
gap ← INT (n / 2)
DO WHILE gap > 0
    DO
        switch ← 0
        FOR j ← 1 TO n − gap
            IF x_j > x_{j + gap}
                dummy ← x_j
                x_j ← x_{j + gap}
                x_{j + gap} dummy
                switch ← j
            END IF
        END DO
    LOOP UNTIL switch = 0
    gap ← INT (gap / 2)
LOOP
```

$$IF\ x_i > x_{i+1}$$
$$dummy\ \leftarrow x_i$$
$$x_i \leftarrow x_{i+1}$$
$$x_{i+1} \leftarrow dummy$$
$$END\ IF$$

It should be noted that certain languages have intrinsic functions to perform this operation in a single statement. For example, the QuickBASIC statement

```
SWAP (x(i), x(i + 1))
```

performs the same operation as the pseudocode.

Although these operations will accomplish our objective, they become highly inefficient when sorting tabular data of the type shown in Table 20.1. Suppose that we wanted to sort these student records by grade. Every time that a switch of grades was made, we would have to concurrently switch the names and identification numbers. For large arrays, this added switching can be extremely time-consuming.

An efficient alternative is to "sort in place." To do this, an *index array* (also called an *order array*) is set up. This is an array that is the same length as the arrays that are to be sorted, but it initially contains the ordered integers $1, 2, \ldots, n$. These integers are sometimes referred to as *pointers*. When a switch is to be made, only these values are switched. As in the following example, this means that at the end of the sort, the index array will be in the proper order. Its elements can then be employed as subscripts for the arrays that were sorted.

TABLE 20.1 Names, Identification Numbers, and Grades for Six Students

Index	Name	ID Number	Grade
1	Osborne, Oswald	050–56–7728	54
2	Chen, Cynthia	454–33–8866	92
3	Garza, Maria	550–72–8602	95
4	Hansen, James	112–55–8395	100
5	Callahan, Erin	063–92–4680	85
6	Helmsley, Leona	335–82–3957	68

EXAMPLE 20.2 Sorting in Place

PROBLEM STATEMENT: Table 20.1 contains the names, ID numbers, and grades for six students. Show how the bubble sort would be used to sort this table by grade.

SOLUTION: Before sorting, the index array would be

```
order(1) = 1
order(2) = 2
order(3) = 3
order(4) = 4
order(5) = 5
order(6) = 6
```

The pseudocode for the switch segment of the bubble sort would be written as

$$IF\ grade_{order(j)} < grade_{order(j+1)}$$
$$dummy \leftarrow order(j)$$
$$order(j) \leftarrow order(j+1)$$
$$order(j+1) \leftarrow dummy$$
$$END\ IF$$

Notice how the decision to sort is based on the magnitude of the grade. A "less than" comparison is employed because, for this case, the grades are to be sorted in descending order. Also notice how the order array acts as a subscript for the comparison. However, during the switching phase, it is the elements of the order array that are actually swapped. After sorting in place, the order array would be

```
order(1) = 4
order(2) = 3
order(3) = 2
order(4) = 5
order(5) = 6
order(6) = 1
```

Consequently, if you wanted to output the student with the highest grade, you would write pseudocode of the form

$$DISPLAY\ student.name_{order(1)}$$

which is equivalent to

$$DISPLAY\ student.name_4$$

which, for Table 20.1, results in the output

```
Hansen, James
```

20.1.5 Comparisons and Advanced Algorithms

The efficiency of sorting algorithms (as measured by the number of operations) varies widely. As described in the following example, the choice of a sorting algorithm can have great significance.

EXAMPLE 20.3 Comparison of Sorting Algorithms

PROBLEM STATEMENT: Establish the dependency of the bubble, selection, and Shell sorts on the number of data.

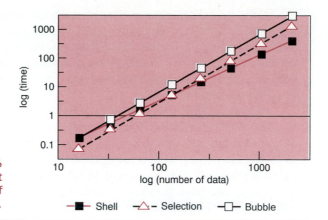

FIGURE 20.6
Performance of three sort
algorithms as a function of
the size of the array.

SOLUTION: A computer program was developed to fill an array with random numbers. This array was then sorted with the resulting performance:

	Sort Time (s)		
n	Bubble	Selection	Shell
16	0.168	0.109	0.168
32	0.711	0.328	0.430
64	2.848	1.371	1.320
128	12.031	5.160	4.539
256	46.129	19.918	12.410
512	180.492	78.320	37.969
1024	717.930	310.332	103.801

If these values are plotted on a log-log plot (Fig. 20.6), a distinctive pattern emerges. The efficiencies of all the algorithms are proportional to n to a power:

Method	Power
bubble	2.01
selection	1.86
Shell	1.52

At small values of n, the three methods have comparable efficiencies. However, as n grows, the Shell sort rapidly becomes superior. It is, therefore, greatly preferred. This is especially true because it is not much more complicated than the selection or bubble sorts.

Aside from the number of elements, it should be noted that the efficiency of sorting methods depends in part on the state of the original array; that is, whether or not it is highly disordered. For example, if the original array is already well ordered, the bubble sort performs efficiently. Some techniques actually perform poorer when the original array is highly ordered! A technique that performs well regardless of the state of the data is said to be *robust*.

Beyond the Shell sort, there are a variety of other algorithms for putting information in order. Although it is somewhat more complicated than the algorithms we have described to this point, the *heapsort* is especially good. Not only is it very efficient for large amounts of data, but it is very robust. Another popular algorithm is the *quicksort*. On most computers, it is on average the most efficient algorithm for large arrays. This is especially true when the array is totally disordered. Interestingly, if the array is originally in order, the standard quicksort actually exhibits its poorest performance!

Press and associates (1986) present a nice review of some of the methods, along with code to implement the more advanced ones. Knuth (1973) offers a comprehensive treatment of the subject.

20.2   SEARCHING

```
i ← 1
found ← 0
DO
   IF  value = x_i
       found ← 1
   ELSE
       i ← i + 1
   END IF
LOOP UNTIL i > n OR found = 1
```

FIGURE 20.7
Algorithm for the sequential search.

Aside from sorting, another important operation performed on data is searching—that is, locating an individual item in a group of data.

20.2.1   Sequential Search

The simplest way to determine whether a value occurs in an array is the *sequential search*. As the name implies, this method involves comparing the value with each element of the array in sequence. A pseudocode for this purpose is displayed in Fig. 20.7.

Although the sequential search is simple to understand and program, it can be extremely inefficient for large arrays. In particular, it is inadequate for situations where the array is ordered. Consequently, alternative approaches such as the binary search are preferred.

20.2.2   The Binary Search

In contrast to the sequential search, the binary search capitalizes on the fact that an array is ordered. Suppose that you are given the task of determining whether a value is contained in an array A. Assume that the array has been sorted in ascending order and has dimensions from 1 to n. The value can be initially compared with the middle term of the array, A(m), and the following assessments made: (*a*) If value = A(m), the search is terminated; (*b*) if value < A(m), the value is in the first half; and (*c*) if value > A(m), the value is in the second half.

In the case of (*b*) or (*c*), the search is repeated in the half in which value is known to lie. Then value is again compared with A(m) where m is now redefined as the middle value of the smaller list. The process is repeated until value is located or the array is exhausted. An algorithm for the binary search is outlined in Fig. 20.8.

$$first \leftarrow 1$$
$$last \leftarrow n$$
$$found \leftarrow 0$$
$$DO$$
$$\quad middle \leftarrow INT \left(\frac{first + last}{2} \right)$$
$$\quad CASE\ value$$
$$\quad\quad CASE = x_{middle}$$
$$\quad\quad\quad found \leftarrow 1$$
$$\quad\quad CASE < x_{middle}$$
$$\quad\quad\quad last \leftarrow middle - 1$$
$$\quad\quad CASE\ ELSE$$
$$\quad\quad\quad first \leftarrow middle + 1$$
$$\quad END\ CASE$$
$$LOOP\ UNTIL\ first > last\ OR\ found = 1$$

FIGURE 20.8
Algorithm for the binary
search.

20.2.3 Comparisons

For cases where the array is small, the sequential search is actually on the average more efficient than the binary search. However, for large ordered arrays, the binary search is superior. In fact, if searches are to be made repeatedly on a single large list, it is worthwhile to expend the effort to presort the array so as to be able to use the binary search.

PROBLEMS

20.1 In the course of a flood analysis, it is often necessary to analyze past rainfall records to determine the maximum rainfall intensities. As part of this analysis, it is your assignment to develop a program to sort such data and then print it out. The data is given as:

Amount	Duration	Date
2.4	1.50	4–12–50
3.5	1.75	5–27–52
7.6	5.20	1–01–55
5.2	2.40	6–26–62
3.5	3.10	5–21–65
10.2	26.20	6–20–68
8.4	2.10	3–29–73
3.9	2.70	4–29–76
2.5	0.40	7–05–78
7.5	3.10	6–28–82

Use this data to compute storm intensity (amount/duration) for each reading. Then sort the data in ascending order according to storm intensity. Print out the intensities in ascending order.

20.2   Repeat Prob. 20.1, but print out the date, amount, and duration, along with the intensity, in a tabular format. (*Note:* The easiest way to handle the dates is with a character-variable array.)

20.3   Develop a computer program to sort a group of names in alphabetical order. Employ the Shell sort as the algorithm for ordering the names.

20.4   In engineering it is sometimes necessary to acquire data from many sources and then consolidate and sort it. For example, suppose that a chemical manufacturing firm wants to access the efficiency of each of its five manufacturing plants. Efficiency tests are conducted at each of the factories on 5 successive days. Because of unforeseen circumstances, plants 2 and 4 are able to successfully conduct only 3 and 4 of the tests, respectively. Therefore the data is returned as:

Day	Plant 1	Plant 2	Plant 3	Plant 4	Plant 5
1	70	—	53	72	44
2	55	—	48	70	48
3	62	52	52	65	52
4	73	37	61	75	51
5	57	50	56	—	47

Enter this data into your program, and print it out as shown. Determine the average efficiency for each plant, sort these averages, and then print out the plant number, the number of successful tests, and the average efficiency in order of performance. Use a Shell sort as the algorithm to order the data.

20.5   Repeat Prob. 20.4, but replace the Shell with a selection sort. Note that if good modular programming technique has been employed, the modification will only involve changing the sort subroutine.

20.6   Develop a subroutine to implement the Shell sort for a one-dimensional array. Use structured programming techniques and modular design so that the subroutine can be integrated into other programs throughout the remainder of the book. Document the subroutine both internally and externally.

20.7   Repeat Prob. 20.6, but for the bubble sort.

20.8   Repeat Prob. 20.6, but for the selection sort.

20.9   Modify Fig. 20.1 so that it employs sorting in place.

20.10   Modify Fig. 20.3 so that is employs sorting in place.

20.11   Develop a subroutine to implement the Shell sort for tabulated data using sorting in place. Use structured programming techniques and modular design so that the subroutine can be integrated into other programs throughout the remainder of the book. Document the subroutine both internally and externally. Test your code by solving Prob. 20.1.

20.12 Repeat Prob. 20.11, but for the bubble sort.

20.13 Repeat Prob. 20.11, but for the selection sort.

20.14 Develop a sequential search routine for searching a one-dimensional array. Use structured programming techniques and modular design so that the subroutine can be integrated into other programs throughout the remainder of the book. Document the subroutine both internally and externally.

20.15 Develop a binary search routine for searching a one-dimensional array. Include a sorting option in the code. Use structured programming techniques and modular design so that the subroutine can be integrated into other programs throughout the remainder of the book. Document the subroutine both internally and externally.

CHAPTER 21

SIGNIFICANT DIGITS AND ERROR ANALYSIS

Engineers must almost always deal with data that contains errors. This statement at first might seem contrary to what one normally conceives of as sound engineering. Students and practicing engineers constantly strive to limit errors in their work. When taking examinations or doing homework problems, you are penalized, not rewarded, for your errors. In professional practice, errors can be costly and sometimes catastrophic. If a structure or device fails, lives can be lost.

Although perfection is a laudable goal, it is rarely, if ever, attained. For example, it is unlikely that an engineering survey crew would be able to measure the exact length of an automobile test track. There would always be some discrepancy or error between the measurement and the true value. If the discrepancy is large, the measurement is unacceptable. However, if it is sufficiently small, the error is considered negligible and the measurement may be deemed adequate. In every case, the key question is, "How much error is tolerable?"

In the next two chapters, we will deal with this question by showing how errors can be quantified. In the present chapter we will discuss some general issues related to working with numbers. We will also focus on the error of a single measurement. Then, in the next chapter, we will cover some of the statistical techniques used when repeated measurements are performed to obtain an estimate.

21.1 RELIABILITY OF NUMERICAL QUANTITIES

Engineers make many claims based on numerical quantities. Often these are data collected in experiments, but frequently they are numbers calculated from basic data which may come from numerical tables in handbooks. An elementary example would be to calculate the volume of a spheroid, also known as a round ball. One could measure the diameter of the ball in various ways and then calculate the volume using the formula

$$V = \frac{4\pi r^3}{3}$$

where r is the radius.

In this case, the number reported would be the volume, but it would depend on the measured diameter, divided by 2 to get the radius, and the value used for the constant π, which could be obtained from a table in a handbook. Even for this simple example, when a value for the volume of the ball is written down, a claim is made implicitly regarding the reliability of the numerical quantity. For instance, the two results for the volume

4.3   and   4.3294

imply a much higher degree of reliability in the second number. Note that it would be misleading to express the result using a large number of digits if, for instance, there was a large error associated with measurement of the ball's radius. The following sections will help you express numerical results in a manner consistent with the reliability of the input values.

Failure to consider the reliability of numerical quantities can lead to poor communication, technical blunders, and erroneous decisions. As an engineer you may be asked to provide specifications regarding the reliability of numerical quantities. Your decisions should consider two principles:

• The degree of reliability demanded in any numerical work should depend on the projected use of the results. The reliability of the data and calculations must be consistent with this use.
• There may be a large penalty in cost paid to obtain small increases in reliability.

The second point may present itself as a constraint on the first, and there may be a trade-off required, that is, giving up some reliability to save costs.

21.1.1 Rounding

Your calculator probably gives 8 digits when expressing the results of numerical calculations. However, the previous discussion has shown that it may not be appropriate to use all 8 digits when reporting the results. A procedure called *rounding* is used to reduce the digits to a meaningful number when expressing the results of numerical computations.

Rounding to n digits is discarding all digits to the right of the nth position. The rules for rounding are:

• Increase the digit in the nth place by 1 if the number in the succeeding place (that is, the next one to the right) is greater than 5 or is 5 followed by at least one nonzero digit.
• Leave the digit in the nth place unaltered if the number in the succeeding place is less than 5. Discard or replace the numbers in the succeeding places with zeros.

- When the numbers in the succeeding places are all zeros, leave the *n*th digit unchanged. Discard or leave the zeros in the succeeding places intact.
- When the number in the succeeding place is exactly equal to 5 (that is, 5 followed by zeros), round so as to leave the *n*th digit an even number. Discard or replace the numbers in the succeeding places with zeros.

In all these cases, the numbers in the succeeding places should either be discarded (when they are to the right of the decimal) or replaced with zeros (when they are to the left of the decimal).

The reason for the fourth rule can be illustrated by examining the number

486.85000...

Suppose that we were to round this quantity to 4 digits. Because 486.85 lies exactly midway between 486.8 and 486.9, we could just as correctly round up to 486.9 as round down to 486.8. Either choice would be arbitrary. Thus, by always rounding so that the last retained digit is even, we have a 50-50 chance of either rounding up or down. Consequently, the errors due to rounding would tend to compensate one another in the long run.

EXAMPLE 21.1   Rounding

PROBLEM STATEMENT: The number of cars passing a toll booth in a year is 865,250. Round this value to 1, 2, 3, 4, and 5 digits.

SOLUTION: The following table summarizes the results:

Digits	Rounded Value	Comment
1	900,000	discarded > 5
2	870,000	discarded = 5 followed by nonzeros
3	865,000	discarded < 5
4	865,200	make number even
5	865,250	discarded exactly zero

21.1.2   Significant Figures

Almost all of engineering deals with the manipulation of numbers. Consequently, before discussing errors, it is useful to further examine basic concepts related to the approximate representation of numbers themselves.

Whenever we employ a number in a computation, we must have assurance that it can be used with confidence. For example, Fig. 21.1 depicts a speedometer and an odometer from an automobile. Visual inspection of the speedometer indicates that the car is traveling between 48 and 49 km/h. We have confidence in this result because two or more reasonable individuals with normal eyesight reading this gauge would ar-

FIGURE 21.1
An automobile speedometer and odometer.

rive at the same conclusion. However, let's say that we insist that the speed be estimated to one decimal place. For this case, one person might read 48.6, whereas another might read 48.7 km/h. Therefore, because of the limits of this instrument, only the first 2 digits can be considered certain. Estimates of the third digit (or higher) must be viewed as suspect. It would be ludicrous to claim, on the basis of the speedometer, that the automobile is traveling at 48.6642138 km/h.

In contrast, the odometer provides 6 certain digits. From Fig. 21.1, we can conclude that the car has traveled slightly more than 87,324.4 km during its lifetime. In this case, if we estimated a value of 87,324.43, the seventh digit (and higher) is uncertain. Note that because of the way the odometer scale is configured, it is very difficult to estimate the seventh digit. In cases such as this, it is conventional to simply use a value midway between the smallest measurement unit. Consequently, for our example, a 7-digit estimate of the distance becomes 87,324.45 km.

The concept of a significant figure, or digit, has been developed to formally designate the reliability of a numerical value. The *significant digits* of a measurement are those that can be used with confidence. They correspond to the certain digits plus 1 estimated digit. For example, the speedometer reading in Fig. 21.1 could be reported as 3 significant digits, 48.7, where the 4 and 8 are certain digits and the 7 is an estimated digit. Similarly, the odometer would be reported as the 7 significant digits: 87,324.45.

Significant figures must be considered in all engineering work involving numbers. Whether we are reading an instrument or performing calculations, we should report only the significant digits, 48.7, where the 4 and 8 are certain digits and the 7 is an estimated digit. Similarly, the odometer would be reported as the 7 significant digits: 87,324.45.

Significant figures must be considered in all engineering work involving numbers. Whether we are reading an instrument or performing calculations, we should report only the significant figures plus at most 1 estimated digit of a result so that anyone who uses the numbers can do so with confidence.

Although it is usually a straightforward procedure to ascertain the significant figures of a number, some cases can lead to confusion. For example, zeros are not always significant figures because they may be necessary just to locate a decimal point. The numbers 0.00001845, 0.0001845, and 0.001845 all have 4 significant figures.

Similarly, when trailing zeros are used in large numbers, it is not clear how many, if any, of the zeros are significant. For example, at face value, the number 45,300 may have 3, 4, or 5 significant digits, depending on whether the zeros are known with confidence. Such confusion can be resolved by using scientific notation, where 4.53×10^4, 4.530×10^4, and 4.5300×10^4 designate that the number is known to 3, 4, and 5 significant figures, respectively.

Sometimes it is possible to report a measured quantity with an explicit estimate of the error

$$x \pm \Delta x$$

where Δx is the estimated error or uncertainty. In general numerical work, it is sufficient to retain only 1 significant digit and 1 estimated digit in expressing the value of the error itself (that is, Δx).

Consider the measurement

$$7500 \pm 18$$

The error, 18, consists of 2 significant digits. It would make no sense to express the uncertainty as something like 18.234.

What is communicated when no error is presented? For instance, suppose that you measure temperature with a thermometer whose smallest division is 1°C, and you report the following result:

$$\text{Temperature} = 29.1°\text{C}$$

First, a claim is being made that the temperature is known to a precision of 3 significant digits. Thus, we are sure that the true temperature is greater than 29 and less than 30. How accurate is the estimated digit? It depends on the skill and judgment of the observer. Certainly the error is less than ±1°C, and perhaps much less. But without more information, it is difficult to characterize the reliability of the error further.

EXAMPLE 21.2   *Significant Figures*

PROBLEM STATEMENT: Suppose that you are given the population of a city as 258,526 and that the counting error could be as much as ±2 percent. How would you express the result to reflect this level of uncertainty?

SOLUTION: A 2 percent error represents a $\Delta x = 5170.52$, or approximately 5171. Therefore, the maximum population could be as high as 263,697 or as low as 253,355.

Given the 2 percent error, it is seen that the second digit is uncertain. Therefore, it would be most appropriate to express the population using only 1 certain digit and 1 estimated digit rounded to a value of 260,000. Note how the appropriate number of significant figures used to represent the population is directly related to the reliability of the count.

21.1.3   Significant Digits and Calculations

Addition and subtraction.   When several quantities are added, the sum should contain only as many significant digits as the quantity with the greatest uncertainty. The sum of 123, 32.3, 0.276, and 0.0324 is 155.6084, as carried out on a calculator; but the result should be reported only as 156.

There can occur a great loss in accuracy in subtraction. For example,

$$9654 \pm 40 - 9435 \pm 50 = 219 \pm 90$$

The relative uncertainty has increased from about 0.5 percent in the initial values to 40 percent in the result.

Multiplication and division.   The result should be given the same number of significant digits as the value entering in the calculation with the least number of significant digits. For example,

$$16.024 * 4.12/3.9 = 16.92792$$

on a given calculator, but should be reported only as 17, given the limiting 3.9. Note that the same result, 17, would have been achieved by rounding the 16.024 to 16.0 and the 4.12 to 4.1, before the calculation was done.

It is acceptable, and most common, to retain more significant digits for intermediate, unseen results in a calculation. In fact, a computer or calculator will do this automatically. Care must be taken, however, in expressing any visible results.

Nonlinear functions.   Errors may be amplified greatly or attenuated by nonlinear functions. Although calculus provides a systematic means to analyze such functions, numerical investigations may also be used.

For example, the nonlinear function

$$\exp(-10000/(T + 273.2))$$

for a value of temperature 29.1 evaluates to

$$4.30 \times 10^{-15}$$

when rounding to 3 digits.

But how should the result be expressed? Suppose that the uncertainty in the temperature was ±0.05°. We will try two additional temperatures at the uncertainty limits:

$$T = 29.05 \qquad \text{Result} = 4.28 \times 10^{-15}$$

$$T = 29.15 \qquad \text{Result} = 4.33 \times 10^{-15}$$

The relative uncertainty in temperature is $0.05/29.1 = 0.2$ percent. The relative uncertainty in the result is about $0.03/4.30 = 0.7$ percent. Thus the uncertainty in the result is over three times the uncertainty of the input temperature. Also note that the second digit in the calculated result may be a 2 or 3 because of uncertainty of the temperature. Therefore the calculated result has only 2 significant figures, and the final result should be reported as

$$4.3 \times 10^{-15}$$

with 1 certain figure and 1 estimated figure determined by rounding rather than

$$4.30 \times 10^{-15}$$

Be careful with nonlinear functions. The simple rules of multiplication and division cannot be applied. Also, when using well-known constants, such as π, do not limit the significant digits of your result by using a constant value with too few significant digits. On the other hand, it may cause unnecessary extra effort to use too many significant digits.

EXAMPLE 21.3   *Significant Figures and Calculations*

PROBLEM STATEMENT: Let us say that our astronauts want to estimate the gravity of a newly explored planet by measuring the time for a screwdriver to fall from the height of their head to the ground. They then propose to use the following formula to calculate the acceleration due to gravity:

$$g = \frac{2h}{t^2}$$

where t is the time required to reach the ground and h is the height of the tallest astronaut. After several tries they conclude that

$$t = 1.6 \pm 0.1 \text{ s}$$

and

$$h = 73.2 \pm 0.5 \text{ in}$$

Thus,

$$g = \frac{2(73.2)}{(1.6)^2} = 57.2 \; \frac{\text{in}}{\text{s}^2}$$

Determine the uncertainty of this result.

SOLUTION: The maximum value for g is given by

$$g = \frac{2(73.7)}{(1.5)^2} = 65.5 \; \frac{\text{in}}{\text{s}^2}$$

whereas the minimum is

$$g = \frac{2(72.7)}{(1.7)^2} = 50.3 \; \frac{\text{in}}{\text{s}^2}$$

Note that because the function is nonlinear, the error is nonsymmetric. Thus, the maximum value exhibits a $+8.3$ discrepancy, whereas the minimum exhibits a -6.9 deviation from 57.2. Such behavior becomes more pronounced as the errors become larger. Taking the average of these two quantities, we could, after rounding to 2 digits, express the error generally as

$$g = 57 \; \frac{\text{in}}{\text{s}^2} \pm 7.6 \; \frac{\text{in}}{\text{s}^2}$$

Because of the nonsymmetry of the plus and minus errors, such general representations should be used with care. Although they are suitable for providing a rough approximation of uncertainty, a more conservative approach would be to report the bounds directly, as in

$$g = 57 \; \frac{\text{in}}{\text{s}^2} \quad (\text{range: 50 to 66}) \; \frac{\text{in}}{\text{s}^2}$$

Observe that most of the error comes from the inaccurate measurement of time. Therefore any efforts to improve the estimate for g should be directed toward improving the measurement of time.

21.2   ACCURACY AND PRECISION

The errors associated with engineering calculations and measurements can be characterized with regard to their precision and accuracy. *Accuracy* of a numerical value, commonly expressed in terms of inaccuracy, is the agreement of a computed or measured value with the true value. This may be assessed by comparison with a standard.

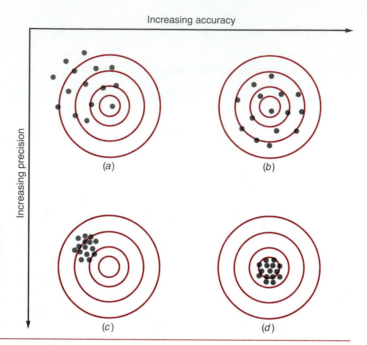

Increasing accuracy

Increasing precision

FIGURE 21.2
Examples from marksmanship
illustrating the concepts of
accuracy and precision:
(*a*) inaccurate and imprecise,
(*b*) accurate and imprecise,
(*c*) inaccurate and precise,
(*d*) accurate and precise.

Precision refers to how closely individual computed or measured values agree with each other. In this sense, it quantifies the extent to which values are free of random error. Precise values may be inaccurate due to bias.

These concepts can be illustrated graphically using an analogy from marksmanship (Fig. 21.2). *Inaccuracy* (also called *bias*) is defined as systematic deviation from the truth. Thus, although the shots in Fig. 21.2*c* are more tightly grouped than in Fig. 21.2*a*, the two cases are equally biased because they are both centered on the upper left quadrant of the target. *Imprecision* (also called *uncertainty*), on the other hand, refers to the magnitude of the scatter. Therefore, although Fig. 21.2*b* and *d* are equally accurate (that is, centered on the bull's-eye), the latter is more precise because the shots are tightly grouped.

21.3   ERROR

The engineer must be aware of sources of error in numerical data which is obtained, especially via experiment. This will limit the reliability of all derived quantities. There are two classes of numerical error in measurement: random and bias. Any error which can be discovered by a suitable investigation and compensated by a simple additive correction is a bias. If a driver measured the actual speed of a car to be consistently 55 mph when the speedometer read 62 mph, he or she would apply a correction of −7 mph, at least at 62 mph. An error which cannot be dependably compensated in this way is an indeterminate or random error. The existence of random error limits the reliability of the measured value; therefore, it is important to have an estimate of the magnitude of this uncertainty.

Now that we have introduced the qualitative notations of accuracy and precision, we can show how they are expressed quantitatively. Engineering calculations and measurements should be sufficiently accurate, or unbiased, to meet the requirements of a particular problem. They should also be precise enough for adequate engineering design. In this book we use the collective term *error* to represent both the inaccuracy and the imprecision of our calculations and measurements. As to the specific components of the error, we employ the term *bias* to represent inaccuracy and *uncertainty* to represent imprecision.

It should be noted that other terms can be used to signify these quantities. For example, some engineers refer to inaccuracy as *systematic error* and imprecision as *random error* (Ang and Tang, 1975). Whatever terminology is employed, it is critical to recognize the fundamental difference between the two types of error and how it relates to the reliability of your work.

21.3.1 Bias

Bias is the discrepancy between an exact, or true, value and an estimate. The relationship between the true value and the estimate can be formulated as

$$\text{True value} = \text{estimate} + \text{bias} \tag{21.1}$$

By rearranging Eq. (21.1), we find that bias is equal to the discrepancy between the truth and the estimate, as in

$$\text{Bias} = \text{true value} - \text{estimate} \tag{21.2}$$

A shortcoming of this quantitative definition is that it takes no account of the magnitude of the value under examination. For example, a bias of a millimeter is much more significant if we are measuring the diameter of a wire rather than its length. One way to account for the magnitude of the quantity being evaluated is to normalize the bias to the true value, as in

$$\text{Fractional relative bias} = \frac{\text{bias}}{\text{true value}} \tag{21.3}$$

where, as in Eq. (21.2), bias = true value − estimate. The fractional relative bias can also be multiplied by 100 percent and expressed as

$$\text{Percent relative bias} = \frac{\text{bias}}{\text{true value}} \times 100\% \tag{21.4}$$

An important advantage of relative bias is that it is independent of units. Whereas a change in units can have a marked effect on Eq. (21.2), it has no effect on

Eqs. (21.3) and (21.4) because both the numerator and denominator of their right-hand sides would be multiplied by the same conversion factors.

The signs of Eqs. (21.2) through (21.4) may be either positive or negative. If the estimate is greater than the true value, the bias will be negative; if the estimate is less than the true value, the bias will be positive. The denominator for Eqs. (21.3) and (21.4) may be less than zero, which can also lead to a negative relative bias. Often we may not be concerned with the sign of the bias but are interested only in its magnitude. Therefore it is often useful to employ the absolute value of Eqs. (21.2) through (21.4).

EXAMPLE 21.4   Calculating Bias

PROBLEM STATEMENT: Suppose that you have the task of measuring the lengths of a bridge and a rivet, and you come up with 9999 and 11 cm, respectively. If the true values are 10,000 and 10 cm, respectively, compute (*a*) the bias and (*b*) the absolute value of the percent relative bias for each case.

SOLUTION: (*a*) The bias for measuring the bridge is [Eq. (21.2)]

$$\text{Bias} = 10,000 - 999 = 1 \text{ cm}$$

and for the rivet is

$$\text{Bias} = 10 - 11 = -1 \text{ cm}$$

Thus both biases have the same magnitude but different signs because one measurement is an underestimate and the other an overestimate.

(*b*) The absolute value of the percent relative bias for the bridge is

$$|\text{Percent relative bias}| = \left| \frac{1}{10,000} \right| 100\% = 0.01\%$$

and for the rivet is

$$|\text{Percent relative bias}| = \left| \frac{-1}{10} \right| 100\% = 10\%$$

Although both measurements represent discrepancies of 1 cm, the absolute value for the percent relative bias for the rivet is much greater. You would, therefore, conclude that you have done an adequate job of measuring the bridge, whereas your estimate for the rivet leaves something to be desired.

When performing actual measurements, the true value is rarely known. If you knew the true value, there would be little point in performing the measurement in the first place. Therefore, you might wonder whether Eqs. (21.1) through (21.4) have more than merely theoretical significance.

One practical application occurs when an equipment manufacturer or user checks out the accuracy of an instrument. For example, an automobile manufacturer could place a car on rollers that can be set at a prespecified velocity. Then the speedometer can be checked to ensure that it provides an accurate reading. If not—that is, if the measurement instrument is biased—it would have to be adjusted to conform to the true value. This adjustment process is called *calibration*. Some instruments are adjusted at the factory, whereas others must be adjusted by the user. In either case, the calibration involves comparison of a measurement with a true value in order to evaluate the bias.

21.3.2 Uncertainty

When using an unbiased instrument, the key question in the measurement process reduces to one of precision. A formulation to represent such cases is

$$\text{True value} = \text{estimate} \pm \text{uncertainty} \tag{21.5}$$

In contrast to Eq. (21.1), the true value is not defined exactly but rather lies somewhere within the range specified by the uncertainty. The estimate is usually expressed with a number of significant figures corresponding to the precision of the measuring device. For example, for the speedometer in Fig. 21.1, we might have concluded that the velocity was 48.5 with an error of 0.5 or, using Eq. (21.5),

$$\text{True value} = 48.5 \pm 0.5$$

In other words, we can say with confidence that the true value lies somewhere between 48 and 49. It must be stressed, however, that our confidence in this result is predicated on the assumption that the graduations on the speedometer are a valid expression of the instrument's precision. Although this is often true, there are certain instances where the precision is poorer than implied by the instrument's gauge. In such cases, Eq. (21.5) would merely be modified to reflect the increase or decrease in uncertainty. For example, if we knew that the speedometer had a precision of 0.7 km/h, the reading would be expressed as

$$\text{True value} = 48.5 \pm 0.7$$

This signifies that the result is between 47.8 and 49.2.

Many equipment manufacturers specify the precision of their instruments. In other cases, you may have to estimate the uncertainty yourself. In the following chapter, we present methods for accomplishing this objective.

Just as we developed a relative fractional bias to remove scale effects from our estimate of accuracy, we can develop a fractional relative uncertainty, as in

$$\text{Fractional relative uncertainty} = \frac{\text{uncertainty}}{\text{estimate}}$$

and a percent relative uncertainty, as in

$$\text{Percent relative uncertainty} = \frac{\text{uncertainty}}{\text{estimate}} \times 100\%$$

Note that in both cases the uncertainty is normalized to the estimate rather than to the true value, as was the case for the relative biases. Also note that absolute values are usually employed in these formulas because the magnitude rather than the sign is typically of interest.

EXAMPLE 21.5   *Calculating Uncertainty*

PROBLEM STATEMENT: A manufacturer places an automobile on rollers that maintain the car's velocity at a steady level of 48.5 km/h. During this test, the car is subjected to various environmental factors such as different temperatures and simulated road surfaces. At various times, the speedometer is read and the results recorded as in Table 21.1. Note that because of difficulty in estimating the third digit, all values are recorded as the midpoint between the smallest division on the scale. Use this information to determine (*a*) the uncertainty and (*b*) the percent relative uncertainty for the speedometer.

TABLE 21.1   *Speedometer Readings under Test Conditions
with True Value of Velocity Held at 48.5 km/h*

48.5	47.5	49.5	48.5
48.5	48.5	47.5	49.5
49.5	50.5	48.5	
47.5	48.5	46.5	

SOLUTION: (*a*) Although the average value for the data in Table 21.1 is 48.5 km/h (which indicates no bias), recorded values range from 46.5 to 50.5. If this range is considered a valid estimate of the instrument's precision, the uncertainty can be calculated as

$$\text{Uncertainty} = \frac{50.5 - 46.5}{2} = 2 \text{ km/h}$$

Therefore, if a value of 48.5 is read, we can state that

$$\text{True value} = 48.5 \pm 2 \text{ km/h}$$

That is, we can state that the true velocity falls somewhere between 46.5 and 50.5 km/h.
(*b*) The percent relative uncertainty is

$$\text{Percent relative uncertainty} = \frac{2}{48.5} 100\% = 4\%$$

21.3.3 Total Error

For certain cases, a measurement will be both inaccurate and imprecise. In these situations the total error consists of both bias and uncertainty, as in

Total error = bias ± uncertainty

Therefore the true value is defined as

True value = estimate + bias ± uncertainty

The above formula provides a comprehensive framework for considering measurement error. We will now take a closer look at the uncertainty term in this formula. As described in the following chapter, the field of statistics has been developed to quantify random errors of this type.

PROBLEMS

21.1 Give examples of the errors in common everyday measuring devices. Define what might introduce bias and uncertainty into the readings obtained from these devices and how you might be able to assess the magnitude of the error.

21.2 Three survey crews are sent out to measure the distance between two immovable landmarks. From prior measurements, it is known that the distance is 74.37 ft, with the last digit uncertain. The three crews measure the distance three times each, with the results:

Crew 1	Crew 2	Crew 3
75.01	74.23	74.30
74.75	74.37	74.75
74.80	74.30	74.29

Determine the bias and uncertainty of each crew.

21.3 You measure the inflow concentration of two reactors and come up with the following estimates (the true values are also shown):

	Estimate	True Value
Reactor A	15	17
Reactor B	278	272

Compute the bias and the percent relative bias for each reactor. On the basis of your results, which measurement is better? Why?

21.4   The number 17150 has 3 certain digits, 1 estimated digit, and 1 zero to place the decimal point properly. Estimate the error in the number.

21.5   The number 0.00152159 has an error of ±1 percent. Express the number using an appropriate number of certain digits and 1 estimated digit. Suppose that the error was ±0.01 percent.

21.6   Round the following numbers to 1, 2, 3, 4, and 5 digits:
(*a*) 855,555   (*f*) 0.0154789
(*b*) 123,456   (*g*) 0.0515151
(*c*) 726,171   (*h*) 0.0771371
(*d*) 571,650   (*i*) 15.6525
(*e*) 589,189   (*j*) 7.25124

21.7   Let us assume that you can measure the radius to within ±5 percent. Estimate the relative percent error of the area of a circle and a spheroid. Use $\pi = 3.14159265$.

21.8   Repeat Prob. 21.7, assuming that $\pi = 3.14$ and that the radius can be measured with an accuracy of ±0.01 percent. What is the most important factor influencing the error of the area of a circle and spheroid?

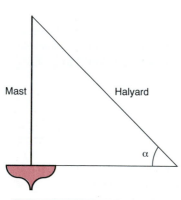

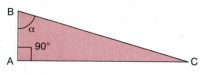

FIGURE P21.9

21.9   It is proposed to estimate the height of a ship mast by measuring the length and angle of the halyard (Fig. P21.9). The halyard length is 91.2 ± 0.1 ft. The angle $\alpha = 39°$ and can be measured to within 1°. Estimate the error in calculating the height of the mast, given

$$\text{height} = \text{halyard length} \times \sin \alpha$$

21.10   A plot of land is shown in Fig. P21.10. Length AB is 9.3 mi ± 0.1 mi, and length AC is 137 mi ± 1 mi. Calculate (*a*) length BC and its error, (*b*) the area and its error, and (*c*) the angle α and its error.

FIGURE P21.10

21.11   A population grows exponentially according to the formula

$$p = a(1 - e^{-bt})$$

where t = time in years. Calculate the error in the calculated population after 5 years if $a = 10,500 \pm 1$ percent and $b = 0.2 \pm 0.05$ percent.

21.12   If $x = 5386$, $y = 0.126$, and $z = 19$, evaluate the following expressions and report your result to the appropriate level of significance:

(*a*) $x + y + z$

(*b*) $x - y - z$

(*c*) $\dfrac{xy}{z}$

(*d*) $\dfrac{x + y}{z}$

STATISTICS

In the previous chapter we introduced the notion of uncertainty, or random error, as a measure of the imprecision of a measurement. As a first example, we quantified the imprecision of an automobile speedometer. To do this, we placed the car on rollers that maintained a constant velocity of 48.5 km/h. Then we took a reading at various times from the speedometer and compiled the results (recall Table 21.1). The 14 results indicated an average of 48.5 km/h and a range from 46.5 to 50.5. This range was used to quantify the uncertainty, or spread of the readings, as ±2 km/h.

We could continue this experiment to obtain additional readings. Table 22.1 contains the original 14 data points along with 10 additional readings. As with the previous case, most of the measurements are close to 48.5. However, notice that the range now stretches from 45.5 to 52.5. Thus the range has grown from a level of 4 km/h for the 14 readings to a level of 7 km/h for the 24 readings. The fact that the range grows as the number of value increases reduces its value as a quantitative measure of uncertainty. For this reason, an alternative measure of spread must be devised.

The field of statistics has been developed to address such problems. *Statistics* can be defined as the science of systematic collection and organization of quantifiable data. It also provides methods for using the data to draw conclusions.

One of the most fundamental objectives of statistics (and the origin of its name) is the summarization of data in one or more well-chosen "statistics" that convey as much

TABLE 22.1   Speedometer Readings from Table 21.1 plus 10 Additional Readings (Shaded)

48.5	47.5	49.5	48.5
48.5	48.5	47.5	49.5
49.5	50.5	48.5	52.5
47.5	48.5	46.5	47.5
45.5	46.5	51.5	48.5
46.5	50.5	50.5	49.5

information as possible about specific characteristics of the data set. These *descriptive statistics* are most often selected to represent (1) the location of the center of the distribution of the data and (2) the degree of spread of the data set. We already took a first step in this direction in the previous chapter when we examined the average value and the range of the speedometer readings. The following sections are devoted to taking a closer look at these and other descriptive statistics. First, however, we will discuss some general concepts related to the statistical characterization of data.

22.1   POPULATIONS AND SAMPLES

A common exercise in engineering is to collect information in order to describe and draw conclusions regarding an entire *population* of individuals, objects, or characteristics. A problem with such an exercise is that it is often impossible or impractical to study each and every individual item of a large population. In such cases, the engineer must attempt to make the characterization on the basis of a limited *sample.*

One example is the industrial engineer who must assess the quality of a population of a particular manufactured product, say an automobile. It would clearly be impractical to test each automobile individually. Consequently, the engineer could randomly select a number of the cars and, on the basis of the sample, attempt to characterize the quality of the entire population.

Another example of a sample of a population was our effort to characterize the uncertainty or imprecision of the speedometer in the previous chapter. For this case, the population is the total number of speedometer readings that could conceivably be taken to measure the car's velocity. This type of population is theoretically infinite in the sense that, assuming the car never malfunctioned, the readings could be taken indefinitely. Thus the total population could *never* be completely characterized. Our only option is to take a limited sample of readings and use this sample to characterize the population.

In certain cases you may have the opportunity to analyze an entire population. A good example is a census. However, more frequently your task will be to characterize an "infinite" population on the basis of a limited sample. With this as background, we can proceed to the first method of descriptive statistics—the histogram.

22.2   THE DISTRIBUTION OF DATA AND HISTOGRAMS

Aside from determining their range, another way to characterize a set of random measurements is to determine their distribution—that is, the shape with which the data is spread out. A histogram is a simple visual way to do this.

A histogram is constructed by sorting the measurements into intervals. The total number of measurements that falls within each interval is called the *frequency.* These frequencies are plotted in the form of a bar diagram. The units of measurement are plotted on the horizontal axis, and the frequency of occurrence of each interval is plotted on the vertical axis.

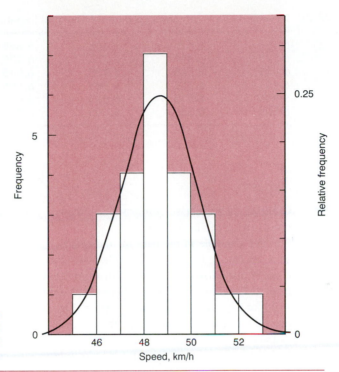

FIGURE 22.1
A histogram for the speedometer data. An ideal bell-shaped or normal distribution is superimposed.

Figure 22.1 is a histogram that was constructed for the speedometer data from Table 22.1. In addition to the frequency, this plot also shows the relative frequency as a right-hand-side ordinate. The *relative frequency* is the frequency divided by the total number of measurements. Thus, the height of each bar is the fraction of the total that falls within a particular interval.

Notice how most of the measurements group around the center of the distribution and fall off in an almost symmetrical fashion to each side. Many distributions have this characteristic shape. If we have a very large set of data and employ narrow intervals, the histogram will often be transformed to a smooth shape. The symmetric bell-shaped curve superimposed on Fig. 22.1 is one such characteristic shape—the *normal distribution*. Given enough additional measurements, the histogram for this particular case could eventually approach the normal distribution. Because of its importance in statistical analysis, we will return to the normal distribution at a later point in this chapter.

It is a relatively easy task to construct Fig. 22.1 because of the manner in which the speedometer readings were taken. Recall from Chap. 21 that the recording of the speedometer readings constituted a sorting process. Whenever the indicator fell between two of the speedometer graduations, we could record the velocity as the midpoint. Thus all the readings would be automatically placed within intervals of 1 km/h. Consequently, one reading fell between 45 and 46, three fell between 46 and 47, and so on. Therefore, for this particular case, the construction of the histogram was easy.

TABLE 22.2   Measurements* of the Coefficient of Thermal Expansion of a Structural Steel

6.555	6.625	6.465	6.655	6.455	6.545
6.675	6.445	6.725	6.485	6.665	6.715
6.405	6.645	6.785	6.645	6.535	6.495
6.495	6.685	6.525	6.585	6.755	6.705
6.745	6.515	6.695	6.685	6.555	6.695
6.505	6.535	6.565	6.705	6.675	6.545

*10^{-6} in/in/°F

Other cases are not so automatic and require judgment on the part of the individual constructing the histogram.

An example is the data for the coefficient of thermal expansion of a structural steel, as reported in Table 22.2. The instrument used to make these measurements has a gauge that allows readings to four significant figures. Because of this, each measurement falls within an interval that is 0.01 unit wide. If each of these is taken as the width of the intervals for the histogram, Fig. 22.2a results. Because of the range of the measurements and the amount of the data, it is difficult to perceive the shape of the data distribution on the basis of this histogram. To obtain a more coherent and smoother-looking histogram we must use broader intervals to characterize the data. As seen in Fig. 22.2b, grouping the data into 0.04-unit intervals reveals that the data seems to be concentrated into two groups, that is, a *bimodal distribution*. Note that if the intervals are made too broad, as shown Fig. 22.2c, the true shape of the data will also be obscured.

A few general guidelines can help to create meaningful histograms. It is good practice to use about 10 intervals when making a first attempt to construct the histogram. In any event, try to avoid using less than 6 or more than 18. Of course, the proper choice will depend on the amount of data. If you have only 10 to 20 items of data, 10 intervals will probably be too many because none of the intervals would have frequencies large enough to make the distribution evident. Conversely, for a very large quantity of data, more than 20 intervals might be used effectively.

Another rule is to ensure that each measurement falls into a single interval. That is, try to make sure that none of the measurements falls exactly on the upper or lower bound of an interval. If this occurs, it will be unclear into which interval the measurement should be placed.

Finally, and most important, be flexible. Try a number of possibilities in order to ensure that you gain significant insight into the true shape of the data distribution. As we will see in the next section, the ability to try a number of different cases is greatly facilitated by computer-generated histograms.

Comparisons of Figs. 22.1 and 22.2 should also make it evident that histograms come in a variety of characteristic shapes. Figure 22.3 illustrates three other types. Figure 22.3a is called a *uniform distribution* because the number of instances within a range occurs with equal likelihood. Figure 22.3b is similar to the normal distribution in that the data is concentrated around one particular location. However, the distribution is not symmetrical for this case but is skewed to one side. The particular distribution in Fig. 22.3b is said to be "skewed to the right" because the right tail is drawn out. Figure 22.3c shows a distribution that is skewed to the left.

FIGURE 22.2 Three histograms constructed with the same data but using different segment widths.

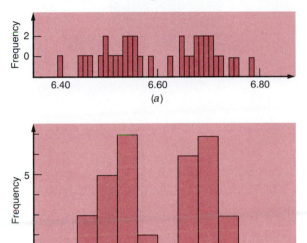

(a)

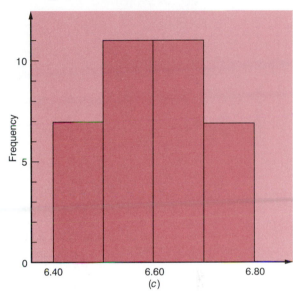

(b)

(c)

Coefficient of thermal expansion, $\times 10^{-6}$ in/in/°F

FIGURE 22.3
Some characteristic shapes of histograms: (a) uniform, (b) skewed right, and (c) skewed left. Idealized shapes for such distributions are superimposed.

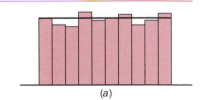

(a)

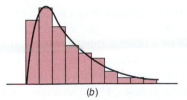

(b)

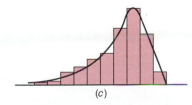

(c)

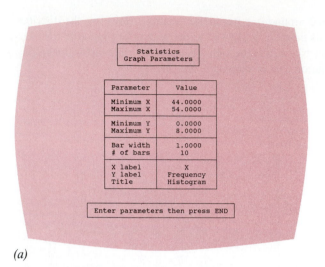

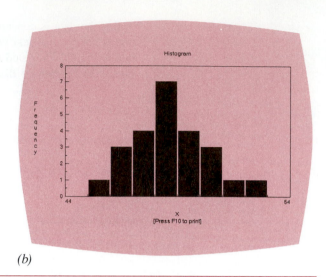

(a)   *(b)*

FIGURE 22.4
Computer screen displaying
the Electronic TOOLKIT
software to create a
histogram for the data from
Table 22.1: (*a*) the way in
which the parameters of the
histogram are entered and
(*b*) the resulting histogram.

22.2.1 Histogram Software

It is a relatively simple matter to develop a computer code to generate a histogram. Problem 22.7 at the end of this chapter is devoted to this task.

A user-friendly computer program to create histograms is contained in the Electronic TOOLKIT software associated with the text. We can use the software to generate a histogram for the data in Tables 22.1 and 22.2.

Figure 22.4*a* shows the screen used to enter the parameters for the histogram. Notice that the parameters allow you to scale the height of the histogram, the limits of the abscissa, and the width of the bars. This allows you to quickly try different values until you obtain an acceptable histogram. The results for the data from Table 22.1 are shown in Fig. 22.4*b*.

An example of three histograms developed for the data from Table 22.2 is shown in Fig. 22.5. Notice how the choice of the bar width results in different shapes. The version in Fig. 22.5*b* is the most satisfactory in that it captures the bimodal nature of this distribution. Because software allows you to construct the histograms in seconds, you have the capability of trying many versions. Such exploratory data analysis is one of the strengths of computerized statistical software.

22.3 MEASURES OF LOCATION

The histogram is more descriptive than a simple listing of data because histograms are presented in an organized manner and convey information that may be apparent only after a careful study of a table of data. There are many times, however, when we may want to compare or summarize sets of data. For example, we might want to determine whether, in a collective sense, the values in one distribution are greater or less than those in another. For such cases, histograms are cumbersome and statistics will be better suited to our needs. As described in this section, the most fundamental statistics for this purpose provide measures of the central location of the data set.

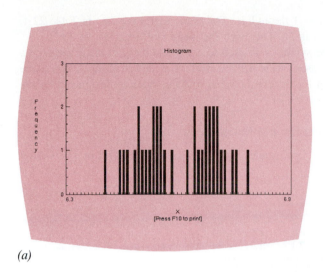

(a)

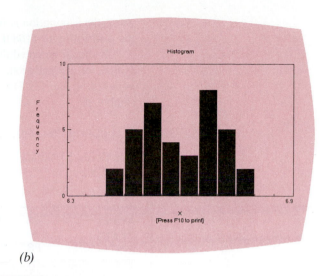

(b)

FIGURE 22.5
Three computer-generated histograms for the data from Table 22.2 using bar widths of (a) 0.01, (b) 0.04, and (c) 0.1. Computerized software such as the Electronic TOOLKIT allows you to generate different versions in an efficient manner. This sort of exploratory data analysis is invaluable for building insight regarding data distributions.

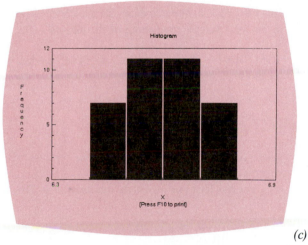

(c)

One of the first steps in analyzing data of the sort found in Table 22.1 is to determine the average value. However, as described in this section, a simple average or mean is but one of several ways to quantify the location of a data set.

22.3.1 The Mean

The most commonly used location statistic is the arithmetic mean. The *arithmetic mean* $\bar{y}$ of a sample is defined as the sum of the individual data points y_i divided by the number of points n, or

$$\bar{y} = \frac{\sum y_i}{n}$$

(22.1)

where the summation is from $i = 1$ through n.[1] For example, for the data in Table 22.1, the summation is 1168.0 and n is 24. Therefore the mean is 1168.0/24, or 48.7.

It should be noted that the symbol $\bar{y}$ is used to designate the arithmetic mean of a *sample* of a population. For cases where we are dealing with the mean of the entire *population,* Eq. (22.1) is used for the computation, but the mean is designated by the symbol μ.

EXAMPLE 22.1   Computing the Mean of Samples and Populations

PROBLEM STATEMENT: The final grades for all the students enrolled in a particular course are

Juniors	67	99	79	71	62
Juniors	96	84	68	77	81
Sophomores	88	71	73	52	86

Determine the arithmetic mean for the entire population. Also, compute the arithmetic mean for a sample consisting of the juniors—that is, the first two rows.

SOLUTION: The summation of all the grades is 1154 and $n = 15$. Therefore the mean is

$$\mu = \frac{1154}{15} = 76.93$$

where μ is used for this particular case because we are characterizing the mean of the entire population.

For the first two rows, the summation is 784 and $n = 10$. The mean is

$$\bar{y} = \frac{784}{10} = 78.40$$

where $\bar{y}$ is used because we are taking a sample from the population to characterize the mean. The distinction between the sample and the population is mainly of theoretical interest. However, when we turn to measures of spread in Sec. 22.4, it will actually have a bearing on the proper choice of formulas.

Although the mean is usually the preferred statistic for defining central location, it can sometimes be misleading in this regard. In particular, the mean can be very sensitive to a few extreme measurements when the sample size is small. Consequently, alternative measures are sometimes employed.

One alternative, a *trimmed mean,* is computed after excluding a fixed percentage

[1]Unless noted otherwise, all summations in the remainder of this chapter are also from 1 through n.

of the smallest and largest numbers. For example, a 5 percent trimmed mean is calculated after excluding the smallest and largest 5 percent of the numbers.

EXAMPLE 22.2   Computing Trimmed Means

PROBLEM STATEMENT: The following data represents the concentration of a chemical entering a reactor:

1.755	1.835	1.785	1.775
1.945	0.115	1.815	
1.555	1.615	1.755	

where all the concentrations are in moles per liter. Construct a histogram and compute the mean and the 10 percent trimmed mean for the readings.

SOLUTION: Because of the small amount of data, we will use intervals of 0.1 mol/L in width. The resulting histogram (Fig. 22.6) indicates that most of the data is grouped around 1.75, while a single extreme point falls in the interval from 0.1 to 0.2. The mean is computed as

$$\bar{y} = \frac{15.950}{10} = 1.595$$

Thus, although the points are clearly clustered around 1.75, the extreme point has a pronounced effect on the mean.

For the trimmed mean, 10 percent of the values (that is, one value) are trimmed from the bottom and from the top of the data set prior to determining the mean. Thus the values of 0.115 and 1.945 are excluded. The 10 percent trimmed mean is calculated as

$$\bar{y}_{t,10} = \frac{13.890}{8} = 1.736$$

Therefore the result is closer to the visual depiction of the distribution's center evident in Fig. 22.6.

FIGURE 22.6
A histogram for concentrations of a chemical feeding into a reactor (from Example 22.2). The concentration off to the side is sometimes called an "outlier."

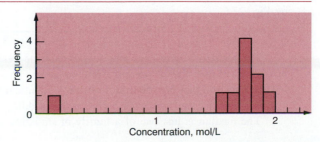

Sometimes the trimmed portion of a distribution will not be an integer. For example, a 10 percent trimmed mean for 24 data points involves trimming 2.4 points from each end. Since this is clearly impossible, the usual strategy is to compute two trimmed means—one excluding two and the other excluding three extreme points. Then linear interpolation (to be discussed in Chap. 24) can be used to determine the 10 percent trimmed mean.

Although trimmed means have their advantages, they should be implemented with caution. They are particularly appropriate for cases when extreme values, or *outliers,* are, in fact, erroneous. For example, the lab technician making the readings for Example 22.2 could have made a blunder when recording the measurement of 0.115. On the other hand, the reading could be true, in which case it is a valid piece of information that should be considered when evaluating the data.

Aside from trimmed means, other location statistics are available that are insensitive to outliers. One of the most common is the median.

22.3.2 Medians and Percentiles

The *median* is the midpoint of a group of data. It is calculated by first putting the data in ascending order. If the number of measurements is odd, the median is the middle value of the measurements. If the number is even, the median is the arithmetic mean of the two middle values.

EXAMPLE 22.3   Computing the Median

PROBLEM STATEMENT: Determine the median of the data from Table 22.1 and from Example 22.2.

SOLUTION: First the data from Table 22.1 is put in ascending order:

45.5	47.5	48.5	(48.5)	49.5	50.5
46.5	47.5	48.5	48.5	49.5	50.5
46.5	47.5	48.5	48.5	49.5	51.5
46.5	47.5	(48.5)	48.5	50.5	52.5

The two middle values are 48.5; therefore the median is 48.5.

For Example 22.2, the two middle values are 1.755 and 1.775, for which the arithmetic mean is $(1.755 + 1.775)/2$, or 1.765. Thus, in contrast to the arithmetic mean calculated previously in Example 22.2 (1.595), the median is insensitive to the outlier value of 0.115 and is much closer to the center of the distribution indicated by Fig. 22.6.

The median is even less sensitive to extreme values than the trimmed mean. The median is, in fact, the 50 percent trimmed mean. This is because the process of ordering the data and picking the middle value is identical with trimming the values one at a time from each end until you arrive at the middle.

The median is just one of a family of statistics, called percentiles, that divide an ordered set of data into portions. For example, the three quartiles divide the data at the 25, 50, and 75 percent levels. Of course, the second quartile is the median. There are also percentiles that divide the data up to a specified percent. For example, the 95 percentile is the value that separates the lower 95 from the upper 5 percent of the data.

Aside from some advanced applications, the median is not as useful as the arithmetic mean for general mathematical statistics. However, for cases with extreme values or asymmetric distributions (that is, strongly skewed), it can be a useful descriptor of central location.

22.3.3 The Mode

The *mode* is the value that occurs most frequently. For example, the value of 48.5 occurs the most often in Table 22.1 and is, therefore, the mode. Often a single value for a mode has little meaning. For example, Fig. 22.2a indicates that nine values occur two times. In such cases the *modal interval* is more meaningful. This refers to the interval with the highest frequency from the histogram that is used to describe the distribution. In Fig. 22.2b, there are two modal intervals from 6.52 to 6.56 and from 6.64 to 6.72. These two peaks are what led us to call this distribution bimodal. Distributions with more than one peak are called multimodal.

22.3.4 Comparison of Central Location Statistics

The arithmetic mean is generally the preferred measure of central location because it has a number of advantages. Among other things, it is much easier to work with mathematically. On the other hand, the median and mode are less sensitive to extreme values. The trimmed mean offers a compromise that somewhat remedies this defect.

For data that is unimodal and symmetric, the mean, median, and mode are identical (Fig. 22.7a). For a unimodal, asymmetric distribution, such as the skewed distributions in Fig. 22.3, the mean will be the closest of the three to the drawn out tail, followed by the median and then the mode. An easy way to remember this is that they fall in alphabetical order away from the long tail (Fig. 22.7b).

22.3.5 The Geometric Mean

Aside from the arithmetic mean, the median, and the mode, there are other measures of central location. Many of these involve transformations. For log-transformed data, a *geometric mean* can be computed as

$$GM = \text{antilog}\left(\frac{1}{n}\sum \log_{10} y_i\right)$$

Because the addition of logarithms is equivalent to the multiplication of their antilogarithms, the geometric mean can be represented alternatively as

$$GM = \sqrt[n]{y_1 y_2 \cdots y_n} = \sqrt[n]{\prod_{i=1}^{n} y_i}$$

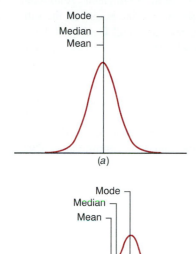

FIGURE 22.7
The relative position of measures of central tendency for (a) a symmetric and (b) an asymmetric distribution.

where Π means "the product of." Note that y_i must be >0.

Many beginners in statistics have difficulty accepting the fact that measures of central tendency other than the arithmetic mean are valid. However, there are cases in engineering where measures such as the geometric mean are even preferable. Additional information on descriptive statistics can be found in the many fine books on the subject—for example, Wonnacott and Wonnacott (1972), Snedecor and Cochran (1967), Box, Hunter, and Hunter (1978), and Milton and Arnold (1990).

22.4   MEASURES OF SPREAD

The three distributions in Fig. 22.8 all have approximately the same mean. However, there are obvious differences in how they are spread out around the central value. Thus additional statistics, beyond those for central location, are required to characterize the spread, or dispersion, of data.

FIGURE 22.8
Three distributions with the same arithmetic mean but different spreads.

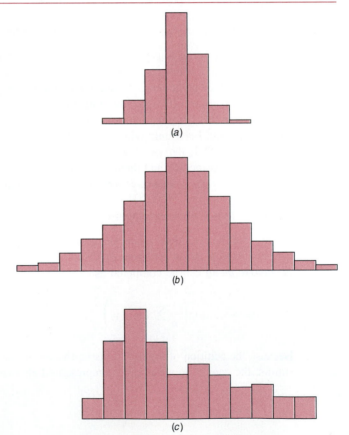

(a)

(b)

(c)

22.4.1   Range and Average Deviation

To this point we have used the range as our first estimate of dispersion. However, as noted previously, the range is inadequate because it is strongly affected by outliers.

An alternative measure that minimizes the effect of extreme values would be to compute an average deviation from the mean, as in

$$\frac{\sum(\bar{y} - y_i)}{n}$$

This statistic is faulty because it is always equal to zero, a result of the fact that the mean marks the spot at which positive and negative deviations are balanced. Consequently, the positive and negative deviations cancel when they are summed.

A possible alternative would be to average the absolute values of the deviations, as in

$$\frac{\sum|\bar{y} - y_i|}{n}$$

Although this statistic can be used under some circumstances, experience as well as theoretical difficulties (that are beyond the scope of this text) have shown it to be generally inadequate. As discussed in the next section, the average of the squared deviations is the preferred statistic for measuring dispersion.

22.4.2   Variance and Standard Deviation

The primary statistic to characterize the dispersion of the data is the *variance,* which is defined as the average of the squared deviations from the mean. The squared deviations for an entire population are represented by

$$S_t = \sum (\mu - y_i)^2 \tag{22.2}$$

where S_t is called the *total sum of the squares of the residuals.* The formula for the variance is

$$\sigma^2 = \frac{\sum (\mu - y_i)^2}{n} \tag{22.3}$$

where σ^2 is the symbol for the variance. Squaring the deviations overcomes the fact that the sum of the deviations is zero.

The units for the variance are the square of the unit of y—for example, square feet, kilograms squared, or kilometers squared per hour squared. In order to have a statistic that has the same units as the quantity being measured, another statistic called the *standard deviation* is defined as the square root of the variance, or

$$\sigma = \sqrt{\frac{\sum (\mu - y_i)^2}{n}}$$ (22.4)

EXAMPLE 22.4 Computing the Variance and Standard Deviation of an Entire Population

PROBLEM STATEMENT: Determine the variance and the standard deviation of the data from Example 22.1.

SOLUTION: The mean for the data from Example 22.1 is 76.93, and therefore the sum of the squared deviations can be calculated as

y_i	$(\mu - y_i)^2$
67	99
96	364
88	123
99	487
84	50
71	35
79	4
68	80
73	15
71	35
77	0
52	622
62	223
81	17
86	82
	$\sum (\mu - y_i)^2 = 2236$

Therefore the variance is ($n = 15$)

$$\sigma^2 = \frac{2236}{15} = 149$$

and the standard deviation is

$$\sigma = \sqrt{149} = 12.2$$

When dealing with samples rather than with the entire population, slightly different formulas are involved. For the total sum of the squared deviations,

$$S_t = \sum (\bar{y} - y_i)^2$$ (22.5)

For the sample variance,

$$s_y^2 = \frac{\sum (\bar{y} - y_i)^2}{n - 1} \tag{22.6}$$

and for the sample standard deviation,

$$s_y = \sqrt{\frac{\sum (\bar{y} - y_i)^2}{n - 1}} \tag{22.7}$$

Observe that for these sample statistics, we divide by $n - 1$ rather than by n. Note that these formulas require two or more samples from the population.

EXAMPLE 22.5   Computing the Sample Variance and Standard Deviation

PROBLEM STATEMENT: Determine the variance and the standard deviation for the first two rows of data in Example 22.1.

SOLUTION: The summation of the squared deviations for the first 10 values in Example 22.1 can be calculated as

y_i	$(\mu - y_i)^2$
67	130
96	310
99	424
84	31
79	0
68	108
71	55
77	2
62	269
81	7
	$\sum (\bar{y} - y_i)^2 = 1336$

Therefore the variance is ($n = 10$)

$$s_y^2 = \frac{1336}{10 - 1} = 148.4$$

and the standard deviation is

$$s_y = \sqrt{148.4} = 12.2$$

We divide by $n - 1$ in Eq. (22.6) so that the sample variance s_y^2 is an unbiased estimator of the population variance σ^2. Mathematical analysts have proved that if n is used as the divisor, then the resulting value of s_y^2 will tend to underestimate σ^2. This is because the y_i's will tend to be closer to $\overline{y}$ than to the true mean μ of the population. Dividing by $n - 1$ corrects for this effect. Of course, when n is large, the resulting correction will be negligible, as in Examples 22.4 and 22.5.

The quantity $n - 1$ is referred to as the *degrees of freedom*. Hence s_y^2 is said to be based on $n - 1$ degrees of freedom. This nomenclature derives from the fact that the sum of the n quantities upon which s_y^2 is based (that is, $\overline{y} - y_1$, $\overline{y} - y_2, \ldots,$ $\overline{y} - y_n$) adds up to zero. Consequently, if any $n - 1$ of the values is specified, the remaining value is fixed. Thus only $n - 1$ of the values are said to be freely determined. Another justification for dividing by $n - 1$ is the fact that there is no such thing as the spread of a single data point. For the case where $n = 1$, Eq. (22.6) yields a meaningless result of infinity.

It should be noted that we almost always deal with statistics for samples rather than for the entire population. Consequently, the standard deviation and variance calculated on a pocket calculator or in most software packages are usually determined by Eqs. (22.6) and (22.7). Practically speaking, the only time that the distinction really matters is when dealing with very small values of n. Otherwise, the difference between the $n - 1$ in the sample variance and the n in the population variance is negligible.

22.4.3   The Coefficient of Variation

Recall that in Sec. 21.3 we introduced relative errors to remove scale effects from error estimates. In the same fashion, we can normalize the standard deviation to the mean, as in (for a sample of a population)

$$CV = \frac{s_y}{\overline{y}} \cdot 100\% \tag{22.8}$$

where CV is referred to as the *coefficient of variation*. For a population, CV would be computed as $(\sigma/\mu) \times 100\%$.

EXAMPLE 22.6   Computing the Coefficient of Variation

PROBLEM STATEMENT: Use the results of Examples 22.1 and 22.5 to determine the coefficient of variation for the first two rows of grades from Example 22.1.

SOLUTION: From Example 22.1, the mean of the first two rows is $\overline{y} = 78.4$; and from Example 22.5, the standard deviation is $s_y = 12.2$. Therefore the coefficient of variation is

$$CV = \frac{12.2}{78.4} \, 100\% = 15.6\%$$

Recall that in Sec. 21.3.2 we used the range to quantify the uncertainty of a group of measurements. For instance, in Example 21.5, we determined that the true value of the speedometer readings in Table 21.1 could be expressed as 48.5 ± 2 km/h. We also indicated that the 2 km/h uncertainty could be expressed in a percent relative uncertainty of 4 percent. The standard deviation and coefficient of variation provide alternative and superior statistics for the same purpose.

EXAMPLE 22.7   Using the Standard Deviation and Coefficient of Variation as Measures of Uncertainty

PROBLEM STATEMENT: Repeat Example 21.5, but use the standard deviation and the coefficient of variation to quantify the uncertainty.

SOLUTION: The mean and standard deviation of the data in Table 21.1 can be determined to be $\bar{y} = 48.5$ and $s_y = 1.04$. Therefore the uncertainty of the readings can be expressed as 48.5 ± 1.04 km/h. The coefficient of variation is

$$CV = \frac{1.04}{48.5} \, 100\% = 2.1\%$$

Notice that, in the foregoing example, the values for the standard deviation ($s_y = 1.04$ km/h) and the coefficient of variation (CV = 2.1 percent) are smaller than the uncertainty (2 km/h) and the percent relative uncertainty (4 percent) based on the range. This is as expected because the former values are averaged deviations rather than the extreme deviation represented by the range. As such, the standard deviation is less sensitive to outlying values and does not tend to grow as the number of measurements increases. Rather, it converges on a constant value—the population standard deviation—as n increases. This makes it preferable to the range. In addition, there are mathematical advantages, discussion of which is beyond the scope of this text, that make it preferable.

Although the standard deviation is preferable, to this point its actual meaning is probably less clear than that of the range. We know that it represents a measure of dispersion, but it may not be evident just how much of the dispersion it measures. Even though the range has its deficiencies, we at least know it represents the entire uncertainty of the data. Thus we can say with confidence that 100 percent of our measurements fall within the range. To this point, a similar statement cannot be made regarding the standard deviation. In Sec. 22.5, we will make the meaning of this statistic more tangible. However, before doing this we will first present some programs for determining descriptive statistics on the computer.

22.4.4   Algorithm for Descriptive Statistics

The computer is an ideal tool for determining descriptive statistics. Figure 22.9 shows pseudocode that has been developed for this purpose. After the data is entered, the al-

```
n ← 0
DO
    INPUT x
    IF x = sentinal THEN EXIT
    n ← n + 1
    yₙ ← x
LOOP

CALL SORT (y, n)

sum ← 0
DO FOR i ← 1 TO n
    sum ← sum + yᵢ
LOOP

mean ← sum / n

j ← INT (n / 2)
IF n / 2 − j = 0 THEN
    median ← yⱼ
Else
    median ←  (yⱼ + yⱼ₊₁) / 2
END IF

sum ← 0
DO FOR i ← 1 TO n
    sum ← sum + (yᵢ − mean)²
LOOP
variance ←  sum / (n − 1)

std dev ← √variance

coeff var ←  std dev / mean

OUTPUT n, mean, median
OUTPUT x₁, xₙ
OUTPUT variance, std dev, coeff var
END
```

FIGURE 22.9
Pseudocode to determine descriptive statistics. Note that the variables i, j, and n should be declared as integers. All other variables should be declared as real.

gorithm implements a sort (recall Chap. 20) to place the numbers in ascending order. This serves two purposes. First, it defines the range of the data. After sorting, the first and the last elements, $X(1)$ and $X(N)$, will contain the lower and upper bounds for the data. Second, it allows us to determine the median in a simple manner.

The algorithm calculates and outputs two measures of central tendency—the arithmetic mean and the median—and four measures of spread—the range, variance, standard deviation, and coefficient of variation. Figure 22.10 shows output that would be produced for the speedometer data from Table 21.1.

NUMBER OF VALUES = 24
MEAN = 48.66667
MEDIAN = 48.5
LOWEST VALUE = 45.5
HIGHEST VALUE = 52.5
VARIANCE = 2.84058
STANDARD DEVIATION = 1.685402
COEFFICIENT OF VARIATION (%) = 3.463155

FIGURE 22.10
Descriptive statistics
produced with a program
developed from the
pseudocode shown in
Fig. 22.9. This application
corresponds to the
speedometer data from
Table 21.1.

22.5 THE NORMAL DISTRIBUTION

As mentioned in Sec. 22.2, the most common distribution in statistics is the normal distribution. It is common because (1) it closely approximates many data sets and (2) it has advantages in mathematical statistics. It can be represented formally by the normal probability density function

$$f(y) = \frac{1}{\sigma\sqrt{2\pi}} e^{(y-\mu)^2/2\sigma^2}$$

(22.9)

where $f(y)$ is the density of a particular value of y. The density is similar to the relative frequency shown in Fig. 22.1. Recall from our discussion of histograms that the relative frequency is the frequency of occurrence divided by the total number of occurrences. Thus the summation of all the relative frequencies should add up to 1. The *density* is a similar quantity that is suited for a *continuous* function (in contrast to the *discrete* histogram) such as the normal distribution. The area under the normal distribution curve between two values of y is equal to the probability that readings will fall between these values. Similarly, the area from $-\infty$ to ∞ is equal to 1, connoting the obvious fact that there will be a 100 percent probability that a reading will fall between these extreme limits.

If various values of y are employed, Eq. (22.9) can be used to calculate a bell-shaped curve. Notice that the population mean μ and the population variance σ appear as parameters in this equation. Consequently, there is not just one normal distribution but an infinite number of such curves, depending on the values of μ and σ that are used in Eq. (22.9).

EXAMPLE 22.8   Calculating the Normal Probability Density Function

PROBLEM STATEMENT: Use Eq. (22.9) to compute values of the normal probability density function for (*a*) $\mu = 5$, $\sigma = 2.2$; (*b*) $\mu = 10$, $\sigma = 2.2$; and (*c*) $\mu = 5$,

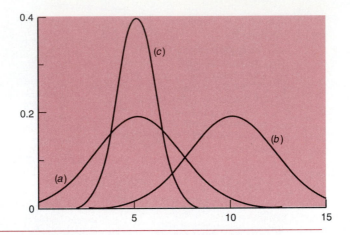

FIGURE 22.11
Three bell-shaped curves with
different means and standard
deviations.

$\sigma = 1.0$. Compute values from $y = 0$ to 20, and plot the results.

SOLUTION: For $\mu = 5$ and $\sigma = 2.2$, Eq. (22.9) can be used to compute

$$f(y) = \frac{1}{2.2\sqrt{2\pi}} e^{-(y-5)^2/2(2.2)^2}$$

Values of $f(y)$ can be determined by substituting different y's into this equation. For example, if $y = 8$,

$$f(8) = \frac{1}{2.2\sqrt{2\pi}} e^{-(8-5)^2/2(2.2)^2} = 0.072$$

The results for other values of y as well as for cases (b) and (c) are

y	(a)	(b)	(c)
1	0.035	0.000	0.000
2	0.072	0.000	0.004
3	0.120	0.001	0.054
4	0.164	0.004	0.242
5	0.181	0.014	0.399
6	0.164	0.035	0.242
7	0.120	0.072	0.054
8	0.072	0.120	0.004
9	0.035	0.164	0.000
10	0.014	0.181	0.000
11	0.004	0.164	0.000
12	0.001	0.120	0.000

These values are plotted in Fig. 22.11. Notice that because they have the same variance, the curves for (a) and (b) have the same spread. However, because they have different means, they are centered on different values of y. On the other hand, because curves (a) and (c) have the same means, they are both centered on y = 5. However, because the variance of (c) is smaller, its spread is narrower than that of (a).

As stated previously, the area under a normal probability density function between two values of y is equal to the probability that readings will fall between those values. For example, in Fig. 22.12, the area between 0.5 and 2.0 is equal to 0.2857. This means that 28.57 percent of the values will fall between these limits. Obviously, the area under the entire function—that is, from $y = \infty$ to $-\infty$—is equal to 1.0, meaning that 100 percent of the values fall within this interval.

A useful property of the normal distribution is that the intervals defined by multiples of the standard deviation contain a fixed percentage of the items. For example, $\mu \pm \sigma$ contains 68.3 percent of the items, $\mu \pm 2\sigma$ contains 95.5 percent of the items, and so on (Fig. 22.13). These percentages hold for any and all normal distributions. Consequently, if the speedometer readings from Table 21.1 are truly normally distributed and if $\overline{y}$ and s_y are adequate measures of μ and σ, then the result of Example 22.7, 48.5 ± 1.04 km/h, implies that 68.3 percent of the values fall between 47.46 and 49.54. Similarly, for two standard deviations, 48.5 ± 2.08 km/h implies that 95.5 percent of the values fall between 46.42 and 50.58.

FIGURE 22.12
A normal probability curve showing how the area under the curve represents the probability that a random event will occur at locations between the abscissa values.

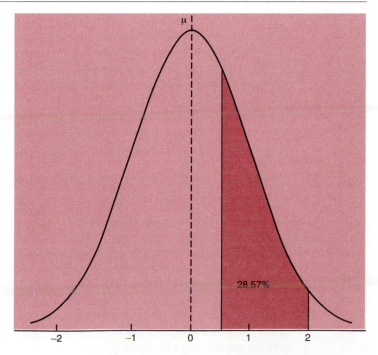

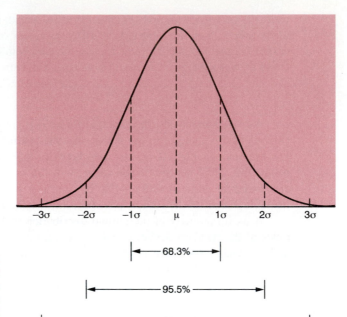

FIGURE 22.13
The percent probabilities for
the normal distribution
encompassed by various
multiples of the standard
deviation.

22.6   PREPACKAGED STATISTICS SOFTWARE

The material in this chapter provides you with the capability of determining descriptive statistics with the computer. In addition to your own programs, prepackaged (or canned) software is also available for statistical analyses. As with other forms of commercially available software, these programs are designed to make your analytical work as effortless as possible.

The Electronic TOOLKIT software that accompanies this text includes a statistics program that is representative of such software. As shown in the following example, it represents a potent tool for analyzing engineering data.

EXAMPLE 22.9   *Using Canned Software to Determine Descriptive Statistics*

PROBLEM STATEMENT: Employ the statistics program from the Electronic TOOL-KIT to analyze the data from Example 22.2.

SOLUTION: A user-friendly computer program to determine descriptive statistics is contained on the Electronic TOOLKIT software associated with the text. The results of applying this program to the data from Example 22.2 are shown in Fig. 22.14. The data to be analyzed is entered in a tabular format that is easy to edit. As each point is entered, the statistics are automatically updated. This feature is included because of its utility for determining the impact of outliers—that is, atypical values that lie far outside the normal range of points.

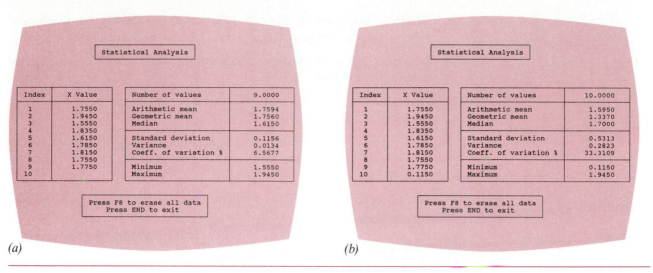

FIGURE 22.14 Screens generated by the statistics package from the Electronic TOOLKIT. The package is used to analyze the data from Example 22.2: (a) shows the analysis for all the points, and (b) shows the results when the outlier value of 0.115 is excluded.

For example, Fig. 22.14a shows the result when all the data is analyzed. Because the software is designed to easily add or delete data, the effect of removing the outlying value of 0.115 can be readily determined.

The result, as shown in Fig. 22.14b, indicates that the outlier has a pronounced effect on the statistics. Thus you might use this result as the basis for reexamining this point to determine whether it was erroneous.

PROBLEMS

22.1 Name at least four measures of central tendency.

22.2 Plot the normal distribution given by Eq. (22.9), and indicate the symmetrical range on the plot where 90 percent of the values would fall. Employ a mean and variance of 1 for this plot.

22.3 The following weights were determined for a group of students in a class:

90	200	150	135	138
175	188	177	142	112
137	115	192	172	160
128	122	134	176	181
140	151	155	166	175
194	220	140	142	126

Develop a histogram for this data. How would you classify the distribution? What reason can you give to explain the shape of the distribution?

22.4 Determine the mean, median, modal interval, and geometric mean of the weights given in Prob. 22.3.

22.5 Determine the range, standard deviation, and variance of the data given in Prob. 22.3.

22.6 Why is the standard deviation for a sample calculated by a different formula from that used for the entire population? It is not necessary to show this mathematically; use qualitative reasons.

22.7 Develop a computer program to determine the histogram of a group of data. Test it by analyzing the data given in Table 22.2.

22.8 Employ the program developed in Prob. 22.7 to solve Prob. 22.3.

22.9 On the basis of Fig. 22.9, develop a computer program to determine descriptive statistics. Test it by analyzing the data from Table 22.2.

22.10 Employ the program developed in Prob. 22.9 to solve Probs. 22.4 and 22.5.

22.11 Combine the programs developed in Probs. 22.7 and 22.9 into a package that is capable of generating both histograms and descriptive statistics. Use a modular approach, and employ structured programming techniques. Document the program both internally and externally, and test it with the data from Table 22.2.

22.12 Employ the program developed in Prob. 22.11 to analyze the data from Table 22.1.

22.13 Employ the Electronic TOOLKIT statistics package to perform Probs. 22.3 through 22.5.

22.14 Develop a program to determine the mean, the median, and the percent trimmed means (from 0 to 100 percent in increments of 10). Apply this program to the data from Table 22.2. Plot the values of the percent trimmed mean versus percent trimmed, and indicate the location of the mean and the median on this plot.

22.15 Perform the same analysis as in Prob. 22.14, but apply it to the data from Example 22.2. Interpret the resulting plot and how it might prove useful as an aid in determining the best measure of central tendency.

22.16 The "drunkard's walk" is a term used to describe the type of random motion found in many diffusion processes. (The name stems from the similarity between this type of motion and the random stumbling around that a drunkard might exhibit.) In one-dimensional space, the random motion can be only to the right or to the left. A random-number generator can be employed to simulate the process by generating random numbers. For example, if the random number generator creates numbers between 0 and 1, we could consider a 0 to 0.5 to represent a step left and a 0.5 to 1 a step right. Write a program that will simulate a

100-step random walk, and run this routine 50 times to obtain 50 different results for the final location of the drunkard (that is, the number of steps away from the starting position, where steps to the right are positive and steps to the left negative). Plot your results as a histogram, and comment on it.

22.17 In engineering it is sometimes necessary to acquire data from many sources and then compare it. For example, suppose that a chemical manufacturing firm wants to assess the efficiency of each of its five manufacturing plants. Efficiency tests are conducted at each of the factories on 5 successive days. Because of unforeseen circumstances, plants 2 and 4 are able to successfully conduct only 3 and 4 of the tests, respectively. Therefore the data is returned as:

Day	Plant 1	Plant 2	Plant 3	Plant 4	Plant 5
1	70	—	53	72	44
2	55	—	48	70	48
3	62	52	52	65	52
4	73	37	61	75	51
5	57	50	56	—	47

Enter this data into your program, and print it out as shown. Determine the mean efficiency for each plant. Which plant exhibits the most variable efficiency?

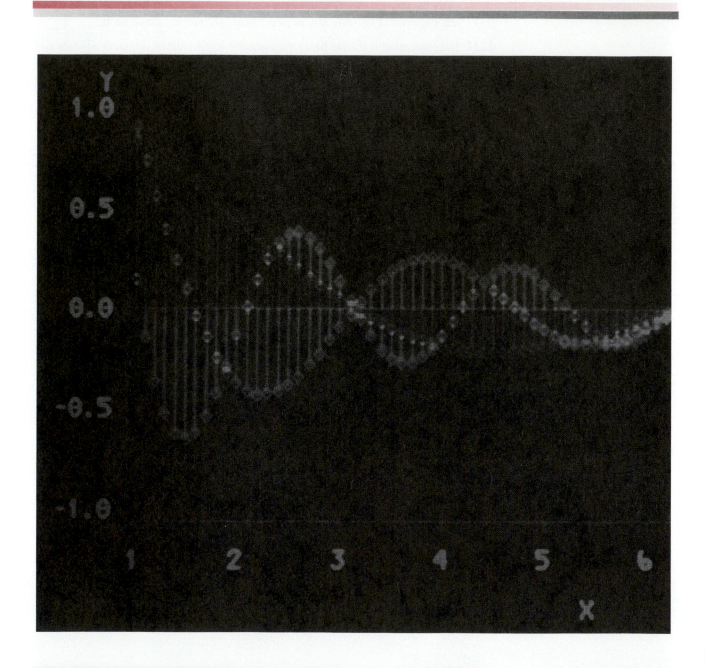

COMPUTER MATHEMATICS

Now that you have been introduced to some aspects of data analysis, the next step in the problem-solving process is to investigate how the computer is used to manipulate these numbers and solve mathematical problems. At the core of computer mathematics is the area of study known as numerical methods.

Numerical methods are techniques whereby mathematical problems are formulated so that they can be solved by arithmetic operations. Because computers are very good at performing arithmetic, they are ideal vehicles for implementing numerical methods. Because of this, the field of numerical methods is sometimes loosely defined as "computer mathematics."

The present part of the book constitutes a brief introduction to some of the elementary ideas and techniques in the field of numerical methods. It is intended to provide you with a feeling for how computers can be effectively utilized to solve mathematical problems. The material is divided into three major groups of techniques. *Chapters 23* and *24* are devoted to methods for fitting curves to data. *Chapters 25* and *26* deal with solving equations for unknowns. Finally, *Chaps. 27, 28,* and *29* introduce techniques for solving calculus problems with the computer.

COMPUTER MATHEMATICS

CURVE FITTING: REGRESSION

A great deal of engineering work is based on cause-effect interactions. One of our routine tasks is to establish the relationship between two variables. A graph provides one means to accomplish this objective. For cartesian coordinate systems, it is conventional to plot the *dependent variable* on the vertical axis, or *ordinate,* and the *independent variable* on the horizontal axis, or *abscissa*. The dependent variable is then said to be plotted "versus" or "as a function of" the independent variable.

By displaying the data on a two-dimensional space, patterns emerge that might not be evident from the tabulated information alone. For example, close inspection of the table in Fig. 23.1a suggests that higher values of y seem to be associated with higher values of x. Although this positive relationship between the variables can be inferred from the table, the graph in Fig. 23.1b provides the additional insight that the relationship curves upward, or accelerates, as x increases. Such insights are one of the great strengths of graphical approaches.

Although graphs have utility for establishing the qualitative relationship between variables, an even more powerful approach is to fit a curve through the points. As depicted in Fig. 23.2, such a curve can then be used to predict values of y as a function of x. The simplest method for deriving such curves is to merely "eyeball" the data and sketch the curve. Although this is certainly valid for "quick and dirty" estimates, it is fundamentally flawed because it is subjective. That is, for most cases, each individual would come up with a slightly different version of a "best" curve.

The following two chapters are devoted to some mathematical techniques that have been developed to remove this subjectivity from the curve-fitting process. These techniques are objective in that they yield consistent results regardless of the individual performing the analysis.

The approaches can be divided into two general categories depending on the amount of error that is associated with the data. First, where the data exhibits a significant degree of error, or uncertainty, the strategy is to derive a curve that represents the general trend of the data and minimizes some measure of error. Because any individual point may be incorrect, no effort is made to intersect every point. Rather, the curve is derived

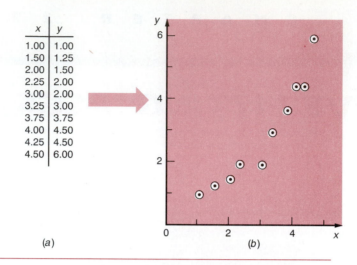

x	y
1.00	1.00
1.50	1.25
2.00	1.50
2.25	2.00
3.00	2.00
3.25	3.00
3.75	3.75
4.00	4.50
4.25	4.50
4.50	6.00

FIGURE 23.1
(a) Tabulated values of a dependent variable y and an independent variable x; (b) a graph of the data.

(a)   (b)

FIGURE 23.2
A curve fit of the data from Fig. 23.1a. This fit was based on an "eyeball" sketch. Such curves can be used to predict values of the dependent variable as a function of the independent variable. Here it is used to predict that at $x = 2.75$, y is approximately equal to 2.25.

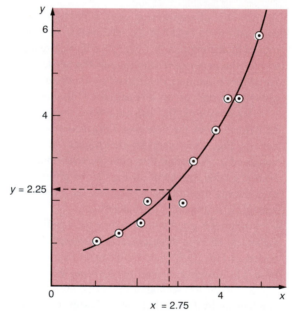

to follow the pattern of the points taken as a group (Fig. 23.3a). This approach, called *least-squares regression,* is the subject of the present chapter.

Second, where the data is known to be very precise, the basic approach is to fit a curve or a series of curves that pass directly through each of the points. Common sources of this data are the tables found in such references as engineering and scientific handbooks. Examples range from engineering economics tables to tables of physical properties. The estimation of values between well-known discrete points is called *interpolation* (Fig. 23.3b). It will be discussed in Chap. 24.

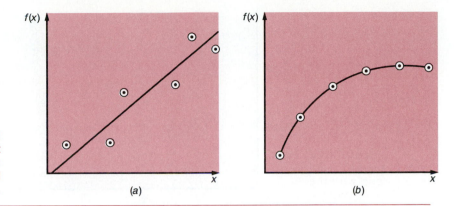

FIGURE 23.3
Two fundamentally different
ways to fit curves to data:
(*a*) regression and
(*b*) interpolation.

23.1 LINEAR REGRESSION

Because of their interest in cause-effect interactions, engineers have frequent occasion to establish the relationship between a set of paired variables: (x_1, y_1), (x_2, y_2), ..., (x_n, y_n). Although examples abound throughout engineering, this task is especially common in experimental studies. In this context, the independent variable x is usually a characteristic that is under the control of the experimenter (that is, x is the cause), whereas the dependent variable y is the resultant behavior that is being measured (that is, y is the effect.).

For a materials engineer, x might be the force per unit area (that is, the stress) imposed on a steel rod, whereas y is the resulting deformation per unit length (that is, the strain). In electrical engineering, x might be the current imposed on a resistor, whereas y is the resulting voltage drop. Numerous other examples exist in all fields of engineering. In any case, because x is controlled and y is measured, the former is usually considered to be error-free, while the latter can have error associated with it.

Table 23.1 contains a set of six paired observations of the type compiled during a typical experimental study. In this hypothetical example, we are interested in characterizing factors influencing the downward velocity of a free-falling parachutist. From physics, it is known that a falling object is subject to two vertical forces: the downward force of gravity and the upward, retarding force of air resistance. Preliminary observations suggest that air resistance is positively correlated with fall velocity. In other words, the higher the vertical velocity is, the greater the air resistance will be. In order to characterize this relationship quantitatively, a wind-tunnel experiment is conducted to measure the air-resistance force as a function of various velocities (see Table 23.1). When this data is plotted, it suggests the linear or straight-line pattern seen in Fig. 23.4.

Linear regression is the technique for determining the "best" straight line through data such as that contained in Table 23.1. The equation for the straight line can be expressed as

$$\hat{y} = a_0 + a_1 x \tag{23.1}$$

TABLE 23.1
Six Values of Air
Resistance (y_i) versus
Velocity (x_i) Collected
during a Wind-Tunnel
Experiment

x_i, m/s	y_i, N
10	110
15	230
20	210
25	350
30	330
35	460

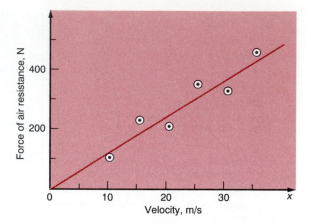

FIGURE 23.4
A plot of the data from
Table 23.1 along with the
best-fit line computed in
Example 23.1.

where $\hat{y}$ (read as y hat or y caret) is the predicted value of the dependent variable, and a_0 and a_1 are the intercept and the slope, respectively, of the straight line. The object of linear regression is to determine the straight line (as defined by a_0 and a_1) that passes as close to as many points as possible. Because most of the error is associated with the y values, "closeness" is defined in terms of the vertical distance between the points and the line. This distance is referred to as the *residual* (Fig. 23.5) and can be computed as

$$\text{Residual}_i = y_i - \hat{y}_i$$

where $\hat{y}_i$ is the y value of the straight line at x_i as calculated by Eq. (23.1).

For a variety of theoretical reasons that are beyond the scope of this text, the "best" straight line is defined as the one that minimizes the sum of the squares of the residuals between the points and the line. Thus the object of linear regression is to determine values of a_0 and a_1 that result in a minimum value of

$$S_r = \sum_{i=1}^{n} (y_i - \hat{y}_i)^2 \tag{23.2}$$

where S_r is the *sum of the squares of the residuals*. Using a derivation that employs calculus [see Chapra and Canale (1988) for details], the values of a_0 and a_1 that minimize Eq. (23.2) can be calculated using the formulas

$$a_1 = \frac{n \sum x_i y_i - \sum x_i \sum y_i}{n \sum x_i^2 - (\sum x_i)^2} \tag{23.3}$$

and

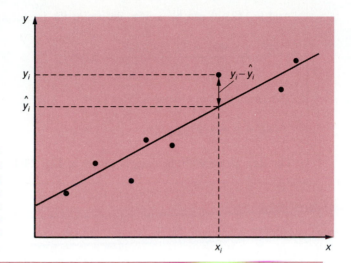

FIGURE 23.5
Plot of dependent versus independent variables along with regression line. The regression line chosen minimizes the sum of the squares of the residuals $y_i - \hat{y}_i$.

$$a_0 = \frac{\sum y_i - a_1 \sum x_i}{n}$$

(23.4)

where all the summations are from $i = 1$ to n.

EXAMPLE 23.1   Linear Regression

PROBLEM STATEMENT: Use linear regression to fit a straight line to the data from Table 23.1.

SOLUTION: The summations needed for Eqs. (23.3) and (23.4) can be computed as

x_i	y_i	$x_i y_i$	x_i^2
10	110	1100	100
15	230	3450	225
20	210	4200	400
25	350	8750	625
30	330	9900	900
35	460	16100	1225
135	1690	43,500	3475

These values can be substituted into Eq. (23.3) to calculate the slope:

$$a_1 = \frac{6(43,500) - 135(1690)}{6(3475) - (135)^2} = 12.5143$$

And Eq. (23.4) can be used to determine the intercept:

$$a_0 = \frac{1690 - 12.5143(135)}{6} = 0.095$$

Therefore, the linear least-squares fit is

$$\hat{y} = 0.095 + 12.5143x$$

The line, along with the data, is shown in Fig. 23.4.

23.2   QUANTIFYING THE "GOODNESS" OF THE LEAST-SQUARES FIT

Any line other than the one determined in Example 23.1 results in a larger sum of the squares of the residuals. Thus the line is unique and, in terms of our chosen criterion, is a "best" line through the points. A number of additional properties of this fit can be elucidated by examining more closely the way in which the residuals were computed.

Recall from statistics that the total sum of the squared deviations of values around a sample mean can be computed as [Eq. (22.5)]

$$S_t = \sum (\bar{y} - y_i)^2 \tag{23.5}$$

where y_i are the individual values and $\bar{y}$ is the mean. This value can in turn be employed to compute the sample standard deviation, s_y, as [Eq. (22.7)]

$$s_y = \sqrt{\frac{\sum (\bar{y} - y_i)^2}{n - 1}} \tag{23.6}$$

Notice the similarity between Eq. (23.5) and the total sum of the squared residuals from Eq. (23.2). In the case of Eq. (23.5), the residuals represent the discrepancy between a measurement and an estimate of central tendency—the mean. For Eq. (23.2), the residuals represent the discrepancy between a measurement and another estimate of central tendency—the straight line.

Consequently, just as the standard deviation can be used to quantify the spread of data around the mean, an estimate of spread around the regression line can be determined with

$$s_{y/x} = \sqrt{\frac{S_r}{n - 2}} \tag{23.7}$$

where $s_{y/x}$ is called the *standard error of the estimate*. We divide by $n - 2$ because two data-derived estimates—a_0 and a_1—were used to compute S_r; thus we have lost two degrees of freedom. Another justification for dividing by $n - 2$ is that there is no such thing as the "spread of data" around a straight line connecting two points. Thus, for the case where $n = 2$, Eq. (23.7) yields a meaningless result of infinity.

Just as was the case with the standard deviation, the standard error of the estimate quantifies the spread of the data. However, $s_{y/x}$ quantifies the spread *around the regression line* as shown in Fig. 23.6*b*, in contrast to the standard deviation s_y that quantifies the spread *around the mean* (Fig. 23.6*a*).

The above concepts can be used to quantify the "goodness" of our fit. This is particularly useful for comparison of several regressions (see Fig. 23.7). To do this, we return to the original data and determine the *total sum of the squares* around the mean for the dependent variable (in our case, y). As was the case for Eq. (23.5), this quantity is designated S_t. This is the uncertainty associated with the dependent variable prior to regression. After performing the regression, we can compute S_r, the sum of the squares of the residuals around the regression line (Eq. 23.2). This represents the uncertainty that remains after the regression. It is, therefore, sometimes called the *unexplained sum of the squares*. The difference between the two sums, $S_t - S_r$, quantifies the improvement or error reduction due to describing the data in terms of a straight line rather than

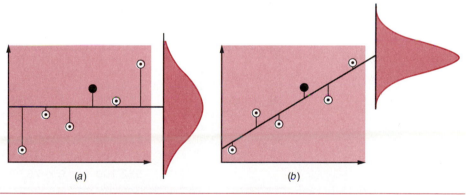

FIGURE 23.6
Regression data showing (a) the spread of the data around the mean of the dependent variable and (b) the spread of the data around the best-fit line. The reduction in the spread in going from (a) to (b), as indicated by the bell-shaped curves at the right, represents the improvement due to linear regression.

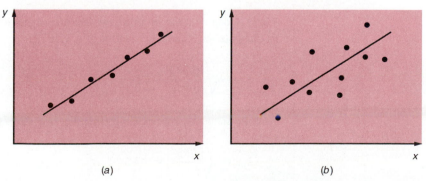

FIGURE 23.7
Examples of linear regression with (a) small and (b) large residual errors.

as an average value. Because the magnitude of this quantity is scale-dependent, the difference is normalized to the total error to yield

$$r^2 = \frac{S_t - S_r}{S_t} \qquad (23.8)$$

where r^2 is called the *coefficient of determination* and r is the *correlation coefficient* ($= \sqrt{r^2}$). For a perfect fit, $S_r = 0$ and $r = r^2 = 1$, signifying that the line explains 100 percent of the variability of the data. For $r = r^2 = 0$, $S_r = S_t$ and the fit represents no improvement. An alternative formulation for r that is more convenient for computer implementation is

$$r = \frac{n \sum x_i y_i - \sum x_i \sum y_i}{\sqrt{n \sum x_i^2 - (\sum x_i)^2} \sqrt{n \sum y_i^2 - (\sum y_i)^2}} \qquad (23.9)$$

EXAMPLE 23.2   "Goodness" of a Linear Least-Squares Fit

PROBLEM STATEMENT: Compute the total standard deviation, the standard error of the estimate, and the correlation coefficient for the data in Example 23.1.

SOLUTION: The mean of the y values from Example 23.1 is $\bar{y} = 1690/6 = 281.67$. Therefore the additional summations needed to solve this example can be computed as

x_i	y_i	y_i^2	$(y_i - \bar{y})^2$	$\hat{y}_i$	$(y_i - \hat{y}_i)^2$
10	110	12,100	29,469	125	225
15	230	52,900	2,669	188	1764
20	210	44,100	5,136	250	1600
25	350	122,500	4,669	313	1369
30	330	108,900	2,336	376	2116
35	460	211,600	31,803	438	484
		552,100	$S_r = 76,082$		$S_t = 7558$

These, along with the summations determined previously in Example 23.1, can be used to compute the standard deviation [Eq. (22.7)]:

$$s_y = \sqrt{\frac{76,082}{6 - 1}} = 123$$

And the standard error of the estimate is [Eq. (23.7)]

$$s_{y/x} = \sqrt{\frac{7558}{6 - 2}} = 43.5$$

Thus, because $s_{y/x} < s_y$, we can see that the linear regression has decreased our uncertainty regarding the measurements. The extent of the improvement can be quantified by the coefficient of determination [Eq. (23.8)]:

$$r^2 = \frac{76,082 - 7558}{76,082} = 0.901$$

or by the correlation coefficient

$$r = \sqrt{0.901} = 0.949$$

An alternative computation of r can be made using Eq. (23.9):

$$r = \frac{6(43,500) - 135(1690)}{\sqrt{6(3475) - (135)^2} \sqrt{6(552,100) - (1690)^2}} = 0.949$$

Before we proceed to the algorithm for linear regression, a word of caution is in order. Although the correlation coefficient provides a handy measure of goodness of fit, you should be careful not to ascribe more meaning to it than is warranted. Just because r is "close" to 1 does not mean that the fit is necessarily "good." For example, it is possible to obtain a relatively high value of r when the underlying relationship between y and x is not even linear! Draper and Smith (1981) provide guidance and additional material regarding assessment of results for linear regression. In addition, you should always, at the minimum, inspect a plot of the data along with your best-fit line whenever you employ regression.

```
INPUT n
DO FOR i ← 1 to n
    INPUT x, y
    sumx ← sumx + x
    sumy ← sumy + y
    sumxy ← sumxy + xy
    sumx2 ← sumx2 + x²
LOOP

x̄ ← sumx
     ─────
       n

ȳ ← sumy
     ─────
       n

a₁ ← n sumxy − sumx sumy
     ────────────────────
      n sumx2 − (sumx)²

a₀ ← ȳ − a₁ x̄
OUTPUT a₀, a₁
```

FIGURE 23.8
Pseudocode to implement linear regression.

23.3 ALGORITHM FOR LINEAR REGRESSION

As seen in Fig. 23.8, it is a relatively trivial matter to develop an algorithm for linear regression. Notice that the algorithm in Fig. 23.8 does not compute the standard error of the estimate, the coefficient of determination, or the correlation coefficient. In Prob. 23.1 you will have the task of including these capabiities in your version of the program.

In addition to the computation outlined in Fig. 23.8, a computer-generated plot is critical for the effective use and interpretation of regression results. If your computer system has plotting capabilities, we recommend that you expand your program to include a plot of y versus x, showing both the data and the regression line. The inclusion of this capability will greatly enhance the program's utility for problem solving.

Figure 23.9 shows some screens generated using the Electronic TOOLKIT. The package is employed to solve the problem from Examples 23.1 and 23.2. The regression results are presented in Fig. 23.9a, and the best-fit line along with the data is plotted in Fig. 23.9b. These screens illustrate how output can be designed in a clear and accessible fashion. You can employ them as models for the software you develop from Fig. 23.8.

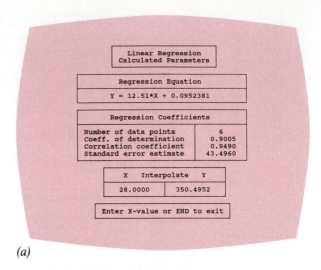

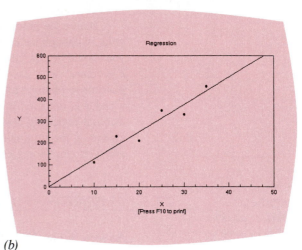

FIGURE 23.9
Computer screens
generated by a linear
regression program from
the Electronic TOOLKIT.
The package is used to
solve the problem
previously analyzed in
Examples 23.1 and 23.2.
The regression results are
shown in (*a*); a plot of the
data along with the
best-fit line is shown in (*b*). (*b*)

23.4 ALTERNATIVE AND ADVANCED METHODS

Linear regression provides a powerful technique for fitting a "best" straight line to
data. However, it is predicated on the fact that the relationship between the dependent
and independent variables is linear. This is not always the case, and the first step in any
regression analysis should be to plot and visually inspect the data to ascertain whether
a linear model applies.

Figure 23.10 shows some data that is obviously curved. A number of options are
available for fitting a "best" curve to such data. First, the data can sometimes be "lin-

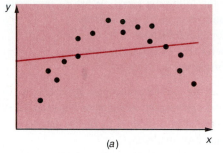

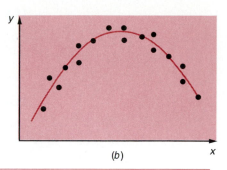

FIGURE 23.10
(a) Data that is ill-suited for linear least-squares regression; (b) indication that a parabola is preferable.

earized" by transforming it logarithmically prior to regression. This is the same strategy used for semilog and log-log plots.

If transformations do not straighten the data out, two additional alternatives are polynomial and nonlinear regression. As the name implies, *polynomial regression* is used to fit a least-squares polynomial through data. Thus a best quadratic or a best cubic equation could be developed with this technique. In fact, linear regression is the simplest form of polynomial regression because a straight line is a first-order polynomial.

Nonlinear regression is employed to fit any equation to data. An introduction to some additional regression methods can be found in Chapra and Canale (1988). Draper and Smith (1981) provide a detailed description of the subject.

PROBLEMS

23.1 Utilize Fig. 23.8 as a starting point for developing your own program for linear regression. Employ structured programming techniques, modular design, and internal and external documentation in your development. Incorporate goodness-of-fit statistics into the program. (*Optional:* Include the capability of plotting both the data and the best-fit line on the same line graph.)

23.2 Develop the program from Prob. 23.1, but incorporate an algorithm that classifies each of the data points in terms of how far away it is from the line. This can be accomplished by using a quantity such as the residual normalized to the standard error of the estimate:

$$\frac{|y_i - \hat{y}_i|}{s_{y/x}}$$

This quantity provides a relative measure of the distance of each point away from the line, and it can help you identify which points are farthest away from the line. It should be noted that although the above quantity is adequate for the present context, there are better and more sophisticated measures and methods for flagging outliers.

23.3   Define the following terms. What do they indicate?
(a) Standard error of the estimate
(b) Standard deviation
(c) Coefficient of determination
(d) Correlation coefficient

23.4   Curve fitting to data points can generally be divided into two categories. Name them, and contrast the differences between the two approaches. For the following situations, select a method to fit the data to a curve and explain the reasons for your choice of method:
(a) A chemical engineer needs to find a very precise property of steam in a table. Unfortunately, the temperature for which the engineer wants this property is not given exactly in the table.
(b) An electrical engineer takes many measurements of a network's intrinsic capacitance for different inputs of current. The engineer wants to fit a curve through this data to be able to predict the effect on capacitance for all values of current.
(c) The average temperature of the ocean at a certain location has been measured very accurately and precisely for the past several years. This temperature has been dropping slightly, and an atmospheric scientist wants to fit a curve through this data to be able to predict the date of the next ice age.

23.5   An experiment is performed to determine the percent elongation of a material as a function of temperature. The resulting data is

Temperature °F	400	500	600	700	800	900	1000	1100
Elongation, %	11	13	13	15	17	19	20	23

Predict the percent elongation for a temperature of 780°F.

23.6   The shear strengths, in kips per square foot (ksf), of nine specimens taken at various depths in a clay stratum are

Depth, m	1.9	3.1	4.2	5.1	5.8	6.9	8.1	9.3	10
Strength, ksf	0.3	0.6	0.5	0.8	0.7	1.1	1.5	1.3	1.6

Estimate the shear stress at a depth of 7.5 m.

23.7   The distance required to stop an automobile is a function of its speed. The following experimental data was collected to quantify this relationship:

Speed, mi/h	15	20	25	30	40	50	60	70
Distance, ft	15	21	45	46	65	90	111	98

Estimate the stopping distance for a car traveling at 50 mi/h.

23.8 You are provided with the following stress/strain data for an aluminum alloy:

Stress	1	2	3	4	5	6	7	8	9	10	11	12
Strain	2	4	6	6	6	7	8	7.5	7	7.5	8	7.5

Use linear regression to determine the strain corresponding to a stress of 7.4.

23.9 A transportation engineering study was conducted to determine the proper design of bike lanes. Data was gathered on bike-lane widths and average distance between bikes and passing cars. The data from 11 streets is

Width x, ft	5	10	7	7.5	7	6	10	9	5	5.5	8
Distance y, ft	3	8	5	8	6	6	10	10	4	5	7

(a) Plot the data.
(b) Fit a straight line to the data with linear regression. Add this line to the plot.
(c) If the minimum, safe average distance between bikes and passing cars is considered to be 6 ft, determine the corresponding minimum lane width.

23.10 An experiment is performed to define the relationship between applied stress and the time to fracture for a stainless steel. Eight different values of stress are applied, and the resulting data is

Stress x, kg/mm²	5	10	15	20	25	30	35	40
Time y, h	40	30	25	40	18	20	22	15

(a) Plot the data.
(b) Fit a straight line to the data with linear regression. Superimpose this line on your plot.
(c) Use the best-fit equation to predict the fracture time for an applied stress of 33 kg/mm².

23.11 In water-resources engineering the sizing of reservoirs depends on accurate estimates of water flow in the river that is being impounded. For some rivers, long-term historical records of such flow data are difficult to obtain. In contrast, meteorological data on precipitation is often available for many years past. Therefore it is often useful to determine a relationship between flow and precipitation. This relationship can then be used to estimate flows for years when only precipitation measurements were made. The following data is available for a river that is to be dammed:

Precipitation x, in	35	40	41	55	52	37	46	48	39	45
Flow y, ft³/s	4050	6075	5540	9500	7290	5700	6210	8440	4590	8000

(a) Plot the data.

(b) Fit a straight line to the data with linear regression. Superimpose this line on your plot.

(c) Use the best-fit line to predict the annual water flow if the precipitation is 50 in.

23.12 The concentration of total phosphorus (p in mg/m^3) and chlorophyll a (c in mg/m^3) for each of the Great Lakes is

	p	c
Lake Superior	4.5	0.8
Lake Michigan	8.0	2.0
Lake Huron	5.5	1.2
Lake Erie:		
West basin	39.0	11.0
Central basin	19.5	4.4
East basin	17.5	3.3
Lake Ontario	21.0	5.5

Chlorophyll a is a parameter that indicates how much plant life is suspended in the water. As such, it indicates how unclear and unsightly the water appears. Use the above data to determine a relationship to predict c as a function of p. Use this equation to predict the level of chlorophyll that can be expected if waste treatment is used to lower the phosphorus concentration of western Lake Erie to 10 mg/m^3.

23.13 It is known that the tensile strength of a plastic increases as a function of the time it is heat-treated. The following data is collected:

Time	10	15	20	30	40	50	55	60	75
Strength	4	20	18	50	33	48	80	105	78

Fit a straight line to this data, and use the equation to determine the tensile strength at a time of 70 min.

23.14 The following data was gathered to determine the relationship between pressure and temperature of a fixed volume of 1 kg of nitrogen. The volume is 10 m^3.

T, °C	−20	0	20	40	50	70	100	120
p, N/m³	7500	8104	8700	9300	9620	10,200	10,500	11,700

Employ the ideal gas law $pV = nRT$ to determine R on the basis of this data. Note that for the law T must be expressed in kelvins.

23.15 The following data was taken from an experiment that measured the current in a wire for various imposed voltages:

Voltage, V	0	2	3	4	5	7	10
Current, A	0	5.2	7.8	10.7	13	19.3	26.5

On the basis of a linear regression of this data, determine current for a voltage of 5 V. Plot the line and the data, and evaluate the fit.

23.16 It is known that the voltage drop across an inductor follows Faraday's law:

$$V_L = L\frac{di}{dt}$$

where V is the voltage drop (in volts), L is inductance (in henries; 1 H = 1 Vs/A), and i is current (in amperes). Employ the following data to estimate L:

di/dt, A/s	1	2	4	6	8	10
V_L, V	5	11	19	31	39	50

What is the meaning, if any, of the intercept of the regression equation derived from this data?

23.17 Semilog plots can be used to fit the following equation:

$$y = y_0 10^{(slope)x}$$

Linear regression can also be employed for the same purpose by linearizing the equation by taking its common logarithm to give

$$\log_{10} y = \log_{10} y_0 + (slope)\, x$$

The population (p) of a small community on the outskirts of a city grows rapidly over a 20-year period:

t	0	5	10	15	20
P	100	212	448	949	2009

As an engineer working for a utility company, you must forecast the population 5 years into the future in order to anticipate the demand for power. Employ linear regression to make this prediction.

23.18 Log-log plots can be used to fit the following equation:

$$y = y_0 x^{\text{slope}} \qquad\qquad\qquad\qquad\qquad\qquad\qquad\qquad\text{(P23.18)}$$

Linear regression can also be employed for the same purpose by linearizing the equation by taking its common logarithm to give

$$\log_{10} y = \log_{10} y_0 + (\text{slope}) \log x$$

Employ this concept and linear regression to fit the following data for viscosity (v) as a function of temperature (T):

T, °F	40	50	60	70	80
v, 10^{-5} ft^2/s	1.7	1.4	1.2	1.05	0.95

Determine the best-fit equation of the form of Eq. (P23.18), and use this relationship to predict v at $T = 63°F$.

CURVE FITTING: INTERPOLATION

TABLE 24.1
Data from an
Engineering
Economics Table*

i, %	F/P
10	6.7275
11	8.0623
12	9.6463
13	11.5231
14	13.7435
15	16.3665
20	38.3376
25	86.7362

*The independent variable is
interest rate i, and the
dependent variable is the
ratio of future to present
worth, F/P, for a 20-year
investment period.

As described in the previous chapter, there are many situations where engineers must use regression to fit a curve to uncertain data. Just as often, however, we must estimate intermediate values between precise data. Common sources of such information are the reference tables found in engineering and scientific handbooks. Examples range from tables of well-known physical and chemical properties to tables giving the values of mathematical functions.

Table 24.1 is an example from engineering economics. Because it is impossible to tabulate every possible bit of information, such tables usually include data spaced at uniform intervals. For example, in Table 24.1, the economic data corresponds to various interest rates from 10 to 25 percent. If we require information for an interest rate of 11 percent, the proper value of 8.0623 can be read directly from the table. However, if we require an intermediate value, say at 11.5 percent, interpolation is required.

The most common method used for this purpose is *linear interpolation*. The rationale and shortcomings of this approach can be best illustrated by displaying the data from Table 24.1 in graphical form. Just as with any set of paired values, the economic data can be plotted on cartesian coordinates. As shown in Fig. 24.1, the interest rate is the independent variable and F/P, the ratio of future to present worth, is the dependent variable. In linear interpolation, intermediate values are estimated to lie on the straight line connecting two adjacent points. As will be shown in this chapter, the straight line can be expressed mathematically as

$$f(x) = a_0 + a_1 x \tag{24.1}$$

where a_0 is the intercept, a_1 is the slope, and $f(x)$ and x are the dependent and the independent variables, respectively. Thus Eq. (24.1) is used to calculate intermediate values between any two points.

The strengths and shortcomings of this approach are both evident from Fig. 24.1. Because of the close spacing of the points between 11 and 12 percent, this approach yields an excellent result when predicting a value, say, at 11.5 percent. However, sup-

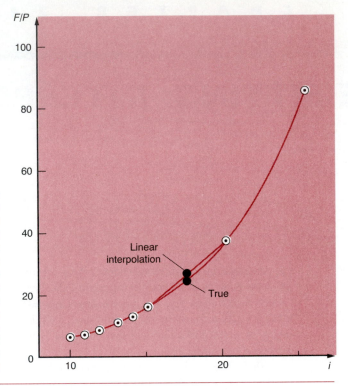

pose that we desired a prediction at 17.5 percent. For this case, the wide spacing makes the curvature of the data more prominent, and as is evident from Fig. 24.1, a substantial error results.

One remedy for such situations is based on a simple extension of the reasoning underlying linear interpolation. As seen in Fig. 24.2*a,* linear interpolation is based on the premise that there is one and only one straight line connecting two points. Similarly, there is also a unique parabola,

$$f(x) = a_0 + a_1x + a_2x^2$$

that connects any three points (Fig. 24.2*b*). An interpolating parabola is capable of capturing some of the curvature of data and, hence, usually yields a prediction that is superior to linear interpolation.

Carrying this reasoning one step farther, we see that there is one and only one cubic equation,

$$f(x) = a_0 + a_1x + a_2x^2 + a_3x^3$$

that fits four points (Fig. 24.2*c*). This leads us to the general conclusion that there is a unique *n*th-order polynomial,

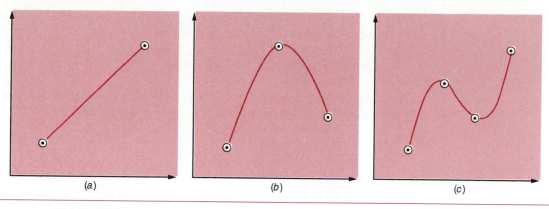

FIGURE 24.2
Examples of interpolating polynomials: (*a*) first-order (linear) connecting two points, (*b*) second-order (parabolic or quadratic) connecting three points, and (*c*) third-order (cubic) connecting four points.

$$f(x) = a_0 + a_1 x + a_2 x^2 + \ldots + a_n x^n$$

that passes exactly through $n + 1$ data points.[1] As described in this chapter, *polynomial interpolation* is the technique used to derive the *n*th-order polynomial equation to fit $n + 1$ points. This polynomial can then be used to predict intermediate values between the points.

Although there is one and only one *n*th-order polynomial that fits $n + 1$ points, there are a variety of mathematical formats in which the polynomial can be expressed. In the present chapter, we will describe one version that is well-suited for implementation on computers—the *Lagrange interpolating polynomial*. Before presenting the general equation, we will introduce the linear and parabolic versions because of their simple visual interpretation.

24.1 LINEAR INTERPOLATION

The simplest form of interpolation is to connect two data points with a straight line. This technique, called *linear interpolation,* is depicted graphically in Fig. 24.3. Using similar triangles,

$$\frac{f(x) - f(x_0)}{x - x_0} = \frac{f(x_1) - f(x_0)}{x_1 - x_0}$$

which can be rearranged to give

$$f(x) = f(x_0) + \frac{f(x_1) - f(x_0)}{x_1 - x_0}(x - x_0) \tag{24.2}$$

[1]This relationship results from the fundamental theorem of algebra.

This particular form is called the *first-order Newton interpolating polynomial*. Notice that the equation is in the form of a straight line:

Prediction = constant + slope × distance

The Lagrange interpolating polynomial can be derived directly from this formulation. In order to derive the Lagrange version, we reformulate the slope as

$$\frac{f(x_1) - f(x_0)}{x_1 - x_0} = \frac{f(x_1)}{x_1 - x_0} + \frac{f(x_0)}{x_0 - x_1} \tag{24.3}$$

which is referred to as the symmetric form. Substituting Eq. (24.3) into Eq. (24.2) and collecting terms yields

$$f_1(x) = \frac{x - x_1}{x_0 - x_1} f(x_0) + \frac{x - x_0}{x_1 - x_0} f(x_1) \tag{24.4}$$

which is the *first-order Lagrange interpolating polynomial*. The prediction, $f_1(x)$, has been subscripted with a 1 to designate that this is the first-order, or linear, version.

FIGURE 24.3
Linear interpolation: The shaded areas are the similar triangles used to derive the linear interpolation formula.

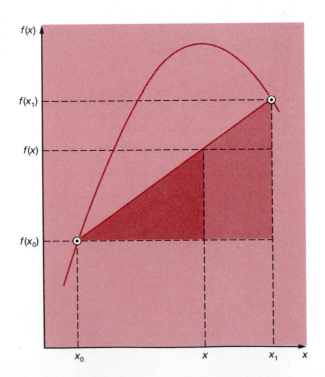

EXAMPLE 24.1   The First-Order Lagrange Interpolating Polynomial

PROBLEM STATEMENT: Use linear interpolation to evaluate $F/P = 17.5$ percent on the basis of the data from Table 24.1 Note that the exact value computed from economic theory is 25.1627.

SOLUTION: From Table 24.1, the following data can be used to interpolate between $i = 15$ and 20 percent:

$$x_0 = 15 \qquad f(x_0) = 16.3665$$
$$x_1 = 20 \qquad f(x_1) = 38.3376$$

Therefore the estimate at $i = 17.5$ percent is [Eq. (24.4)]

$$f_1(17.5) = \frac{17.5 - 20}{15 - 20}\,16.3665 + \frac{17.5 - 15}{20 - 15}\,38.3376 = 27.352$$

This result represents a percent relative error of

$$\frac{25.1627 - 27.352}{25.1627}\,100\% = -8.7\%$$

24.2   PARABOLIC INTERPOLATION

As depicted in Fig. 24.1, the error in Example 24.1 is due to the fact that we used a straight line to approximate a curve. Consequently, a strategy for improving the estimate is to introduce some curvature into the line connecting the points. If three data points are available, this can be accomplished with a second-order polynomial or parabola. The Lagrange polynomial for this purpose is

$$f_2(x) = \frac{(x - x_1)(x - x_2)}{(x_0 - x_1)(x_0 - x_2)}\,f(x_0) + \frac{(x - x_0)(x - x_2)}{(x_1 - x_0)(x_1 - x_2)}\,f(x_1)$$
$$+ \frac{(x - x_0)(x - x_1)}{(x_2 - x_0)(x_2 - x_1)}\,f(x_2) \tag{24.5}$$

EXAMPLE 24.2   The Second-Order Lagrange Interpolating Polynomial

PROBLEM STATEMENT: Estimate F/P at $i = 17.5$ percent from the data in Table 24.1 by fitting a parabola to the points at $i = 15$, 20, and 25 percent.

SOLUTION: From Table 24.1, the following data can be used for the parabolic interpolation:

$$x_0 = 15 \qquad f(x_0) = 16.3665$$
$$x_1 = 20 \qquad f(x_1) = 38.3376$$
$$x_2 = 25 \qquad f(x_2) = 86.7362$$

Equation (24.5) can be used to compute the estimate at $i = 17.5$ percent:

$$f_2(17.5) = \frac{(17.5 - 20)(17.5 - 25)}{(15 - 20)(15 - 25)} 16.3665$$

$$+ \frac{(17.5 - 15)(17.5 - 25)}{(20 - 15)(20 - 25)} 38.3376$$

$$+ \frac{(17.5 - 15)(17.5 - 20)}{(25 - 15)(25 - 20)} 86.7362 = 24.049$$

which represents a percent relative error of 4.43 percent. Thus the error is about half that obtained in Example 24.1 with linear interpolation.

Figure 24.4 shows the parabolic interpolation. In addition, the entire parabola [as computed by substituting other values of x into Eq. (24.5)] is also plotted. Notice that

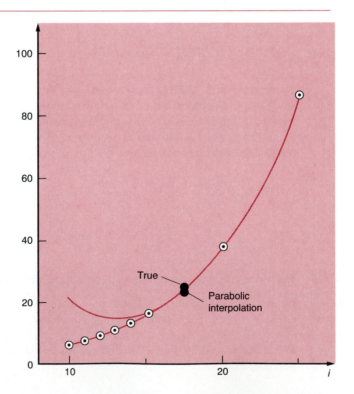

FIGURE 24.4
Parabolic interpolation.

by intersecting the three points at $i = 15, 20,$ and 25 percent, the parabola captures some of the curvature of the data.

24.3 THE GENERAL FORM OF THE LAGRANGE POLYNOMIAL

Just as we have developed first- and second-order Lagrange polynomials, a third-order version can be derived as

$$f_3(x) = \frac{(x - x_1)(x - x_2)(x - x_3)}{(x_0 - x_1)(x_0 - x_2)(x_0 - x_3)} f(x_0) + \frac{(x - x_0)(x - x_2)(x - x_3)}{(x_1 - x_0)(x_1 - x_2)(x_1 - x_3)} f(x_1)$$

$$+ \frac{(x - x_0)(x - x_1)(x - x_3)}{(x_2 - x_0)(x_2 - x_1)(x_2 - x_3)} f(x_2)$$

$$+ \frac{(x - x_0)(x - x_1)(x - x_2)}{(x_3 - x_0)(x_3 - x_1)(x_3 - x_2)} f(x_3) \tag{24.6}$$

Inspection of Eqs. (24.4) through (24.6) leads to the following general representation of the nth-order Lagrange polynomial:

$$f_n(x) = \sum_{i=0}^{n} L_i(x) f(x_i) \tag{24.7}$$

where

$$L_i(x) = \prod_{\substack{j=0 \\ (j \neq i)}}^{n} \frac{x - x_j}{x_i - x_j} \tag{24.8}$$

where the symbol Π stands for "the product of." Equation (24.8) is evaluated for $j = 0$ to n, with the exception of $j = i$.

Although Eqs. (24.7) and (24.8) might seem complicated, they are really quite simple. This can be appreciated by noticing that each term, $L_i(x)$, will be 1 at $x = x_i$ and 0 at all the other data points. Thus each product, $L_i(x)f(x_i)$, will be exactly equal to $f(x_i)$ at x_i and will be zero at all the x values for the other data points (Fig. 24.5). Consequently, the summation of all the products designated by Eq. (24.7) represents the unique nth-order polynomial that passes exactly through all $n + 1$ data points.

An important feature of the Lagrange polynomial is that the points do not have to be equispaced or in any particular order. Such is the case for the following example.

EXAMPLE 24.3   The Third-Order Lagrange Interpolating Polynomial

PROBLEM STATEMENT: Estimate F/P at $i = 17.5$ percent from the data in Table 24.1 by fitting a third-order polynomial to the points at $i = 15, 20, 25,$ and 14 percent.

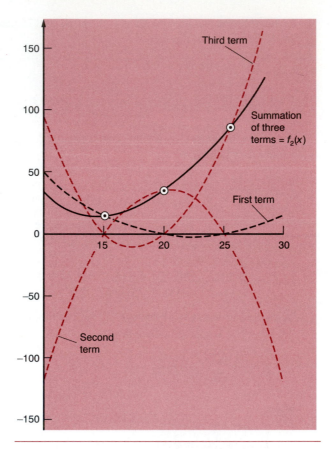

FIGURE 24.5 A visual depiction of the rationale behind the Lagrange polynomial. This figure shows the second-order case from Example 24.2. Each of the three terms in Eq. (24.5) passes through one of the data points and is zero at the other two. The summation of the three terms must, therefore, be the unique second-order polynomial $f_2(x)$ that passes exactly through the three points.

SOLUTION: From Table 24.1, the following data can be used for the parabolic interpolation:

$$x_0 = 15 \qquad f(x_0) = 16.3665$$

$$x_1 = 20 \qquad f(x_1) = 38.3376$$

$$x_2 = 25 \qquad f(x_2) = 86.7362$$

$$x_3 = 14 \qquad f(x_2) = 13.7435$$

Equation (24.6) can be used to compute the estimate at 17.5 percent:

$$f_3(17.5) = \frac{(17.5 - 20)(17.5 - 25)(17.4 - 14)}{(15 - 20)(15 - 25)(15 - 14)}16.3665$$

$$+ \frac{(17.5 - 15)(17.5 - 25)(17.5 - 14)}{(20 - 15)(20 - 25)(20 - 14)}38.3376$$

$$+ \frac{(17.5 - 15)(17.5 - 20)(17.5 - 14)}{(25 - 15)(25 - 20)(25 - 14)}86.7362$$

$$+ \frac{(17.5 - 15)(17.5 - 20)(17.5 - 25)}{(14 - 15)(14 - 20)(14 - 25)}13.7435 = 25.043$$

which represents an error of 0.48 percent. Thus, for this case, cubic interpolation represents a great improvement over the linear and parabolic versions from Examples 24.1 and 24.2.

24.4 ALGORITHM FOR LAGRANGE INTERPOLATION

As seen in Fig. 24.6, it is a relatively trivial matter to develop an algorithm for Lagrange interpolation. Notice how nested loops are used to implement the summations and products from Eqs. (24.7) and (24.8).

In addition to the computation outlined in Fig. 24.6, a computer-generated plot is useful for the effective interpretation of interpolation results. If your computer system has plotting capabilities, we recommend that you expand your program to include a plot of y versus x, showing both the data and the interpolating polynomial. The inclusion of this capability will greatly enhance the program's utility for problem solving.

Figure 24.7 shows some screens generated using the Electronic TOOLKIT. The package is used to solve the problem from Example 24.3. The solution is shown in Fig. 24.7a, and a plot of the interpolating polynomial and the data are shown in Fig. 24.7b. These screens illustrate how output can be designed in a clear and accessible fashion. You can employ them as models for the software you develop from Fig. 24.6.

FIGURE 24.6
Pseudocode for Lagrange interpolation. This algorithm is set up to compute a single nth order prediction where $n + 1$ is the number of data points.

```
y ← 0
DO FOR i ← 0 to n
    product ← f_i
    DO FOR j ← 0 to n
        IF i ≠ j
            product ← product · (x − x_j)/(x_i − x_j)
        END IF
    LOOP
    y ← y + product
LOOP
```

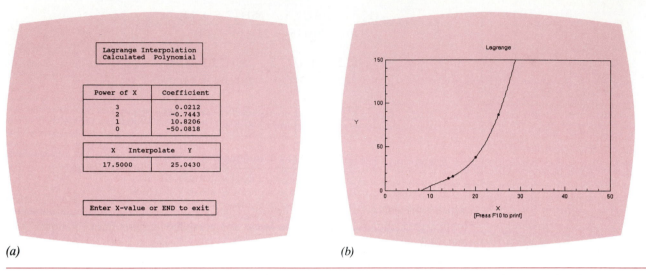

(a)   *(b)*

FIGURE 24.7 Computer screens generated by the interpolation option from the Electronic TOOLKIT. The package is used to solve the problem previously analyzed in Example 24.3.

24.5   PITFALLS AND ADVANCED METHODS

Although there is one and only one *n*th-order polynomial that fits $n + 1$ points, there are a variety of mathematical formats in which this polynomial can be expressed. Aside from the Lagrange equation, an alternative formulation is *Newton's interpolating polynomial.* We have already seen the linear form of the Newton polynomial in Eq. (24.2). Higher-order versions are also available. Newton's polynomial is usually the preferred method when the appropriate order is not known beforehand. This is because Newton's method allows for convenient error analysis that can provide insight into the proper order.

There are situations where both the Lagrange and the Newton formats are inappropriate. Of course, one example is for data that has error associated with it. As you learned in the previous chapter, regression is a more suitable alternative for fitting curves to such data. If you try to fit a higher-order polynomial to uncertain data, the results are often unacceptable (see Fig. 24.8).

Another case for which interpolating polynomials do not work well is for data that is generally smooth but has abrupt local changes (see Fig. 24.9). *Cubic splines* are interpolating polynomials that are better suited for such cases. In this approach, a separate cubic equation is fit between each pair of adjacent data points. The cubics are formulated so that the connections between the separate cubic equations are smooth. By limiting the polynomials to cubics, the wild oscillations of the higher-order versions do not occur. Thus, as seen in Fig. 24.9*d,* the resulting fit is much more acceptable for data with abrupt local changes.

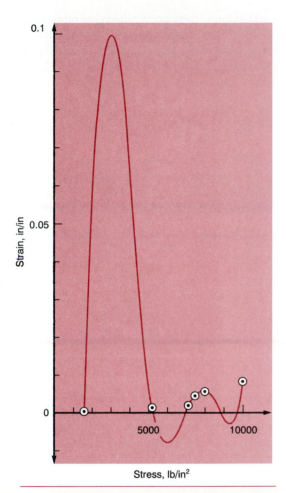

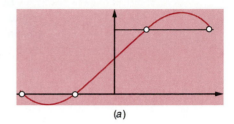

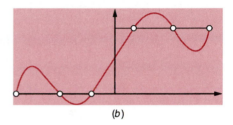

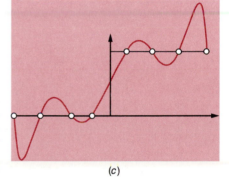

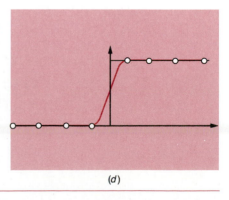

FIGURE 24.8
Fit of a fifth-order Lagrange interpolating polynomial to experimentally derived stress-strain data. Although the curve passes directly through each of the points, the error in the data induces wild oscillations between points.

FIGURE 24.9
A visual representation of a situation where splines are superior to higher-order interpolating polynomials. The function to be fit undergoes an abrupt increase at $x = 0$. Parts (a) through (c) indicate that the abrupt change induces oscillations in interpolating polynomials. In contrast, because it is limited to third-order curves with smooth transitions, the cubic spline (d) provides a much more acceptable approximation.

Beyond interpolating polynomials and splines, there are a variety of other techniques available for determining intermediate values between precisely known points. Chapra and Canale (1988) provide an introduction to the subject. A more detailed treatment can be found in other references (for example, Ralston and Rabinowitz, 1975).

PROBLEMS

24.1 Utilize Fig. 24.6 as a starting point for developing your own program for Lagrange interpolation. Employ structured programming techniques, modular design, and internal and external documentation in your development. (*Optional:* Include the capability to plot both the data and the interpolating polynomial on the same line graph.)

24.2 Show that

$$L_i(x) = \prod_{\substack{j=0 \\ (j \neq i)}}^{n} \frac{x - x_j}{x_i - x_j} = 1$$

at $x = x_i$ and is equal to zero at all other data points. Also, show that

$$\sum_{i=0}^{n} L_i(x) = 1$$

for all the data points.

24.3 An experiment is used to determine the ultimate moment capacity, C, of a concrete beam as a function of cross-sectional area, A. The moment capacity has units of in·kips, and the area has units of m^2. The experiment, which is performed with great precision, yields the following data:

C	932.3	1785.2	2558.6	3252.7	3867.4	4402.6	4858.4	5234.8
A	1	2	3	4	5	6	7	8

Determine the ultimate moment capacity (to one decimal place) for an area of 5.4 in^2.

24.4 The acceleration due to gravity at an altitude y above the surface of the earth is given by

y, m	0	20,000	40,000	60,000	80,000
g, m/s^2	9.8100	9.7487	9.6879	9.6278	9.5682

Compute g at $y = 55,000$ m to four decimal places of accuracy.

24.5 The ratio of the purchase price to annual payments (*P/A*) can be found in economic tables as a function of interest rate. For a 20-year loan, these values are

i	0.10	0.12	0.14	0.16	0.18	0.20	0.25	0.30
P/A	8.514	7.469	6.623	5.929	5.353	4.870	3.954	3.316

Compute *P/A* for an interest rate of 0.195.

24.6 The specific volume of a superheated steam is listed in steam tables for various temperatures. For example, at a pressure of 2950 lb/in², absolute:

T, °F	700	720	740	760	780
v	0.1058	0.1280	0.1462	0.1603	0.1703

Determine *v* at $T = 728°F$.

24.7 The vertical stress σ_z under the corner of a rectangular area subjected to a uniform load of intensity *q* is given by the solution of Boussinesq's equation:

$$\sigma_z = \frac{q}{4\pi}\left[\frac{2mn\sqrt{m^2 + n^2 + 1}}{m^2 + n^2 + 1 + m^2n^2}\frac{m^2 + n^2 + 2}{m^2 + n^2 + 1}\right.$$

$$\left. + \sin^{-1}\left(\frac{2mn\sqrt{m^2 + n^2 + 1}}{m^2 + n^2 + 1 + m^2n^2}\right)\right]$$

Because this equation is inconvenient to solve manually, it has been reformulated as

$$\sigma_z = qf_{z(m,n)}$$

where $f_{z(m,n)}$ is called the *influence value* and *m* and *n* are dimensionless ratios, with $m = a/z$ and $n = b/z$, and *a* and *b* as defined in Fig. P24.7. The influence value is then listed in a table, a portion of which is given here:

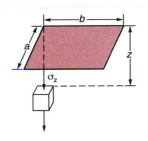

FIGURE P24.7

m	*n* = 1.2	*n* = 1.4	*n* = 1.6
0.1	0.02926	0.03007	0.03058
0.2	0.05733	0.05894	0.05994
0.3	0.08323	0.08561	0.08709
0.4	0.10631	0.10941	0.11135
0.5	0.12626	0.13003	0.13241
0.6	0.14309	0.14749	0.15027
0.7	0.15703	0.16199	0.16515
0.8	0.16843	0.17389	0.17739

If $a = 5.6$ and $b = 14$, compute σ_z at a depth 10 m below the corner of a rectangular footing that is subject to a total load of 100 t (metric tons). Express your answer in tonnes per square meter. Note that *q* is equal to the load per area.

24.8 Bessel functions often arise in advanced engineering analysis. These functions are usually not amenable to straightforward evaluation and, therefore, are often compiled in standard mathematical tables. For example:

x	$J_0(x)$
1.8	0.3400
2.0	0.2239
2.2	0.1104
2.4	0.0025
2.6	−0.0968

Estimate $J_0(2.1)$. Note that the true value is 0.1666.

24.9 The current in a wire is measured with great precision as a function of time:

t	0	0.1250	0.2500	0.3750	0.5000
i	0	6.2402	7.7880	4.8599	0.0000

Determine i at $t = 0.32$.

EQUATION SOLVING: ROOTS OF AN EQUATION

Years ago you learned to use the *quadratic formula*

$$x = \frac{-b \pm \sqrt{b^2 - 4ac}}{2a} \tag{25.1}$$

to solve

$$f(x) = ax^2 + bx + c \tag{25.2}$$

The values calculated with Eq. (25.1) are called the "roots" of Eq. (25.2). They represent the values of x that make Eq. (25.2) equal to zero. Thus we can define the root of an equation as the value of x that makes $f(x) = 0$. For this reason, roots are sometimes called the *zeros* of the equation.

Although the quadratic formula is handy for solving Eq. (25.2), there are many other functions for which the root cannot be determined so easily. For example, the following equation can be used to predict the vertical velocity of an object during free-fall:

$$v = \frac{gm}{c}[1 - e^{-(c/m)t}] \tag{25.3}$$

where velocity v is the dependent variable; time t is the independent variable; and the gravitational constant g, the drag coefficient c, and mass m are parameters. If the parameters are known, Eq. (25.3) can be used to predict the falling object's velocity as a function of time. Such computations can be performed because v is expressed *explicitly* as a function of time. That is, it is isolated on one side of the equal sign.

However, suppose that we had to determine the drag coefficient for a parachutist of a given mass to attain a prescribed velocity in a set time period. Although Eq. (25.3) provides a mathematical representation of the interrelationship among the model variables and parameters, it cannot be solved explicitly for the drag coefficient. Try it; there is no way to rearrange the equation so that c is isolated on one side of the equal sign. In such cases, c is said to be *implicit*.

This represents a real dilemma because many engineering design problems involve determining the properties or composition of a system (as represented by its parameters) in order to ensure that it performs in a desired manner (as represented by its variables). Thus these problems often require the determination of implicit parameters.

The solution to the dilemma is provided by numerical methods for roots of equations. To solve the problem with numerical methods, it is customary to express Eq. (25.3) in an alternative form. This is done by subtracting the dependent variable v from both sides of the equation in order to express the equation in a form similar to Eq. (25.2):

$$f(c) = \frac{gm}{c}\left[1 - e^{-(c/m)t}\right] - v \tag{25.4}$$

The value of c that makes $f(c) = 0$ is, therefore, the root of the equation. This value also represents the drag coefficient that solves the design problem.

Before the advent of digital computers, such equations were usually solved by trial and error. This "technique" consists of guessing a value of c and evaluating whether $f(c)$ is zero. If not (as is almost always the case), another guess is made and $f(c)$ is again evaluated to determine whether the new value provides a better estimate of the root. The process is repeated until a guess is obtained that results in an $f(c)$ that is close to zero.

Such haphazard methods are obviously inefficient and inadequate for the requirements of engineering practice. A numerical method of the sort described in the present chapter represents an alternative that also starts with guesses but then employs a systematic strategy to home in on the true root. In addition, it is ideally suited for computer implementation. As elaborated in the following pages, the combination of such systematic methods and computers makes the solution of most applied-roots-of-equations problems a simple and efficient task. However, before discussing this technique, we will briefly introduce a graphical method for depicting functions and their roots.

25.1   THE GRAPHICAL METHOD

A simple approach for obtaining an estimate of the root of an equation, $f(x) = 0$, is to make a plot of the function and observe where it crosses the x axis. This point, which represents the x value for which $f(x) = 0$, provides a rough approximation of the root.

EXAMPLE 25.1   The Graphical Approach

PROBLEM STATEMENT: Use the graphical approach to determine the drag coefficient c needed for a parachutist of mass $m = 68.1$ kg to have a velocity of 40 m/s after free-falling for time $t = 10$ s. (*Note:* The acceleration due to gravity is 9.8 m/s².)

SOLUTION: This problem can be solved by determining the root of Eq. (25.4) using the parameters $t = 10$, $g = 9.8$, $v = 40$, and $m = 68.1$:

$$f(c) = \frac{9.8(68.1)}{c}[1 - e^{-(c/68.1)10}] - 40$$

or

$$f(c) = \frac{667.38}{c}[1 - e^{-0.146843c}] - 40 \qquad \text{(E25.1.1)}$$

Various values of c can be substituted into this equation to compute

c	$f(c)$
4	34.115
8	17.653
12	6.067
16	−2.269
20	−8.401

These points are plotted in Fig. 25.1. The resulting curve crosses the c axis between 12 and 16. Visual inspection of the plot provides a rough estimate of the root of 14.75. The validity of the graphical estimate can be checked by substituting it into Eq. (E25.1.1) to yield

$$f(c) = \frac{667.38}{14.75}[1 - e^{-0.146843(14.75)}] - 40$$

$$= 0.059$$

which is close to zero. It can also be checked by substituting it into Eq. (25.3) to give

$$v = \frac{9.8(68.1)}{14.75}[1 - e^{-(14.75/68.1)10}] = 40.059$$

which is very close to the desired fall velocity of 40 m/s.

25.2 THE BISECTION METHOD

When applying the graphical method in Example 25.1, you have observed (Fig. 25.1) that the function changed sign on opposite sides of the root. In general, if a function

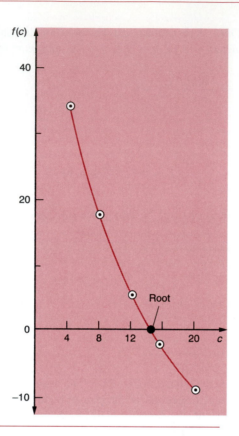

FIGURE 25.1
The graphical approach for
determining the roots of an
equation.

$f(x)$ is real and continuous in an interval from x_l to x_u and $f(x_l)$ and $f(x_u)$ have opposite signs—that is, if

$$f(x_l)f(x_u) < 0$$

then there is at least one real root between x_l (the lower bound of the interval) and x_u (the upper bound).

Bisection—which is alternatively called *binary chopping, interval halving,* or *Bolzano's method*—capitalizes on this observation by locating an interval where the function changes sign. Then the location of the sign change (and, consequently, the root) is identified more precisely by dividing the interval in half—that is, by "bisecting" it into two equal subintervals. Each of these subintervals is searched to locate the sign change. The process is repeated to obtain refined estimates. An algorithm for bisection is listed in Fig. 25.2, and a graphical depiction is provided in Fig. 25.3. The following example goes through the actual computations involved in the method.

Step 1: Choose lower x_l and upper x_u guesses for the root, so that the function changes sign over the interval. This can be checked by ensuring that $f(x_l) f(x_u) < 0$.

Step 2: An estimate of the root x_r is determined by

$$x_r = \frac{x_l + x_u}{2}$$

FIGURE 25.2
An algorithm for bisection.
After step 3, the algorithm
loops back to step 2. This
procedure is continued until
the root estimate is accurate
enough to meet your
requirements.

Step 3: Make the following evaluations to determine in which subinterval the root lies:
(a) If $f(x_l) f(x_r) < 0$, the root lies in the lower subinterval. Therefore set $x_u = x_r$ and return to step 2.
(b) If $f(x_l) f(x_r) > 0$, the root lies in the upper subinterval. Therefore set $x_l = x_r$ and return to step 2.
(c) If $f(x_l) f(x_r) = 0$, the root equals x_r; terminate the computation.

FIGURE 25.3
A graphical depiction of the
bisection method. This plot
conforms to the first three
iterations from Example 25.2.

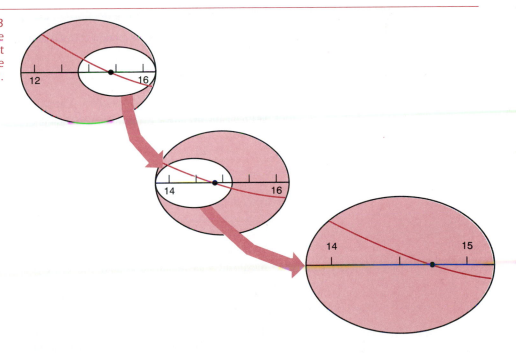

EXAMPLE 25.2   Bisection

PROBLEM STATEMENT: Use bisection to solve the same problem approached graphically in Example 25.1.

SOLUTION: The first step in bisection is to guess two values of the unknown (in the present problem, c) that give values for $f(c)$ with different signs. From Fig. 25.1, we can see that the function changes sign between values of 12 and 16. Therefore the initial estimate of the root lies at the midpoint of the interval:

$$x_r = \frac{12 + 16}{2} = 14$$

Next we compute the product of the function value at the lower bound and at the midpoint:

$$f(12)f(14) = 6.067(1.569) = 9.517$$

which is greater than zero, and hence no sign change occurs between the lower bound and the midpoint. Consequently, the root must be located between 14 and 16. Therefore the lower bound is redefined as 14, and a revised root estimate is calculated as

$$x_r = \frac{14 + 16}{2} = 15$$

The process can be repeated to obtain refined estimates. For example,

$$f(14)f(15) = 1.569(-0.425) = -0.666$$

Therefore the root is between 14 and 15. The upper bound is redefined as 15, and the root estimate for the third iteration is calculated as

$$x_r = \frac{14 + 15}{2} = 14.5$$

The method can be repeated until the result is accurate enough to satisfy your needs.

We ended Example 25.2 with the statement that the method could be continued until the result was "accurate enough" to meet your needs. We must now develop an objective criterion for deciding when enough is enough—that is, when to terminate the method. An initial suggestion might be to end the calculation when the error in the root falls below some prescribed level. This strategy is flawed because determination of the error depends on prior knowledge of the true value of the root. Such would not be the case in an actual situation because there would be no point in using the method if we already knew the root.

Therefore we require an error estimate that needs no foreknowledge of the root. Fortunately, there is a very neat way to accomplish this using bisection. Each time an approximate root is located using bisection as $x_r = (x_l + x_u)/2$, we know that the true root lies somewhere within an interval of $(x_u - x_l)/2 = \Delta x/2$. Therefore the root must lie within $\pm \Delta x/2$ of our estimate (see Fig. 25.4). For instance, when Example 25.2 was terminated, we could make the definitive statement that

$$x_r = 14.5 \pm 0.5$$

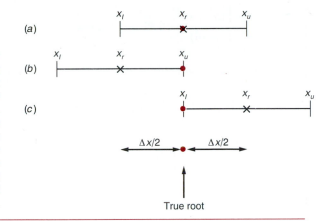

FIGURE 25.4
Three ways in which the interval may bracket the root: In (*a*) the true value lies at the center of the interval, whereas in (*b*) and (*c*) the true value lies near the extreme. Notice that the discrepancy between the true value and the midpoint of the interval never exceeds half the interval length $\Delta x/2$.

For this bound to be exceeded, the true root would have to fall outside the interval, which, by definition, can never occur with the bisection method.

The bisection error estimate can be neatly incorporated into the computer algorithm (Fig. 25.2) as follows. Before starting the method, the user can specify an acceptable error level. Because it takes account of scale effects, the best approach is to specify a percent relative error. This acceptable percent relative error can be designated as ϵ_s. Then after each iteration an approximate percent relative error ϵ_a can be calculated, as in

$$\epsilon_a = \left| \frac{x_u - x_l}{2x_r} \right| 100\% \qquad (25.5)$$

which is the absolute value of the error, $\Delta x/2$, multiplied by 100 percent and divided by the root estimate x_r. Note that the absolute value is used because we are not concerned with the sign of the error but only with its magnitude. When ϵ_a falls below ϵ_s, the computation can be terminated with assurance that the root is known to *at least* the level specified by ϵ_s. Because Eq. (25.5) represents a conservative estimate (that is, an upper bound), the root is always known better.

EXAMPLE 25.3   **Error Estimates for Bisection**

PROBLEM STATEMENT: Continue Example 25.2 until the approximate error ϵ_a falls below $\epsilon_s = 0.5$ percent. After each iteration, compute, along with the approximate error, a true error ϵ_t, in order to confirm that ϵ_t is always less than ϵ_a. In order to calculate ϵ_t, we will provide you with the true value of the root—14.7802.

SOLUTION: For the first iteration, $x_l = 12$, $x_u = 16$, and $x_r = 14$. Therefore Eq. (25.5) can be used to compute

$$\epsilon_a = \left| \frac{16 - 12}{2(14)} \right| 100\% = 14.286\%$$

and the absolute value of the true error is

$$\epsilon_t = \left| \frac{14.7802 - 14}{14.7802} \right| 100\% = 5.279\%$$

Because ϵ_a is greater than $\epsilon_s = 0.5$ percent, the bisection method would be continued. The results for all the iterations are

Iteration	x_t	x_u	x_r	ϵ_a, %	ϵ_t, %
1	12	16	14	14.286	5.279
2	14	16	15	6.667	1.487
3	14	15	14.5	3.448	1.896
4	14.5	15	14.75	1.695	0.204
5	14.75	15	14.875	0.840	0.641
6	14.75	14.875	14.8125	0.422	0.219

Thus after six iterations ϵ_a finally falls below $\epsilon_s = 0.5$ percent, and the computation can be terminated. As anticipated, ϵ_t is less than ϵ_a for every interation. Therefore we can be confident that our final result of 14.8125 is within *at least* 0.5 percent of the true value. As indicated, the true error at this point is actually only 0.219 percent.

25.3 ALGORITHM FOR BISECTION

As seen in Fig. 25.5, it is a relatively trivial matter to develop an algorithm for bisection. In addition to the computation outlined in Fig. 25.5, a computer-generated plot is useful for making effective initial guesses for bisection. If your computer system has plotting capabilities, we recommend that you expand your program to include a plot of the function. The inclusion of this capability will greatly enhance the program's utility for problem solving.

Figure 25.6 shows some screens generated using the Electronic TOOLKIT. The package is used to solve Examples 25.2 and 25.3. A plot of the function is shown in Fig. 25.6a, and the solution is shown in Fig. 25.6b. These screens illustrate how output can be designed in a clear and accessible fashion. You can employ them as models for the software you develop from Fig. 25.5.

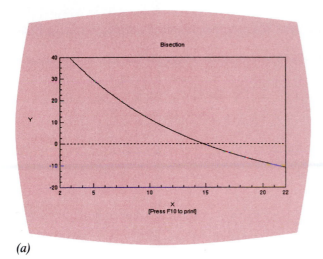

(a)

FIGURE 25.5
Pseudocode for bisection.

$\epsilon_a \leftarrow 100$
$i \leftarrow 0$
DO

$$x_r \leftarrow \frac{x_l + x_u}{2}$$

$i \leftarrow i + I$
IF $x_l + x_u \neq 0$

$$\epsilon_a \leftarrow \left| \frac{x_u - x_l}{2x_r} \right| \cdot 100\%$$

END IF
$test \leftarrow f(x_l) \cdot f(x_r)$
CASE test
   CASE IS < 0
      $x_u \leftarrow x_r$
   CASE IS > 0
      $x_l \leftarrow x_r$
   CASE ELSE
      $\epsilon_a \leftarrow 0$
END CASE
LOOP UNTIL $\epsilon_a \leq \epsilon_s$ OR $i \geq i_{max}$

FIGURE 25.6
Computer screens generated
by the bisection routine from
the Electronic TOOLKIT. The
package is used to solve the
problem previously analyzed
in Example 25.2 and 25.3.
The screen in (*a*) shows a plot
that can be used to make
initial guesses. The screen in
(*b*) shows the results of
bisection.

(b)

Bisection Method

Parameter	Value
Function (1 or 2)	1
Maximum iterations	50
Actual iterations	6
Maximum % error	0.5000
Approximate % error	0.4219
Interval lower bound	12.0000
Interval upper bound	16.0000
x	14.8125
f(x)	-0.0629

Press any key to continue

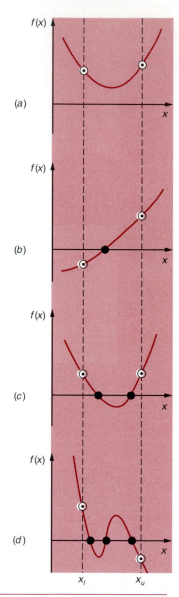

FIGURE 25.7
Illustration of a number of general ways that a real root may occur in an interval prescribed by a lower bound x_l and an upper bound x_u. Parts (a) and (c) indicate that if both $f(x_l)$ and $f(x_u)$ have the same sign, either there will be no real roots or there will be an even number of real roots within the interval. Parts (b) and (d) indicate that if the function has different signs at the end points, there will be an odd number of real roots in the interval.

FIGURE 25.8
Illustration of some exceptions to the general cases depicted in Fig. 25.7. (a) Multiple root that occurs when the function is tangential to the x axis. For this case, although the end points are of opposite signs, there are an even number of real roots for the interval. (b) Discontinuous functions where end points of opposite signs also bracket an even number of real roots. Special strategies are required for determining the roots for these cases.

25.4   PITFALLS AND ADVANCED METHODS

25.4.1   Pitfalls

Aside from providing rough estimates of the root, graphical interpretations are important tools for understanding the properties of functions and anticipating possible pitfalls of root-location methods such as bisection. For example, Fig. 25.7 shows a number of ways in which roots can occur in an interval prescribed by a lower bound x_l and an upper bound x_u. Figure 25.7b depicts the case where a single root is bracketed by negative and positive values of $f(x)$. However, Fig. 25.7d, where $f(x_l)$ and $f(x_u)$ are also on opposite sides of the x axis, shows three roots occurring within the interval. In general, if $f(x_l)$ and $f(x_u)$ have opposite signs, there are an odd number of roots in the interval. As indicated by Fig. 25.7a and c, if $f(x_l)$ and $f(x_u)$ have the same sign, there are no roots or an even number of roots between the values.

Although these generalizations are usually true, there are cases where they do not hold. For example, *multiple roots*—in other words, functions that are tangential to the x axis at the root (Fig. 25.8a)—and discontinuous functions (Fig. 25.8b) can violate these principles. An example of a function that has a multiple root is the cubic equation $f(x) = (x - 2)(x - 2)(x - 4)$. Notice that $x = 2$ makes two items in this polynomial equal to zero. Hence $x = 2$ is called a *double root*. In addition to these cases, complex roots can occur.

The existence of functions of the type depicted in Fig. 25.8 makes it difficult to develop general computer algorithms guaranteed to locate all the roots in an interval. However, when used in conjunction with graphical approaches, numerical methods are extremely useful for solving many roots-of-equations problems confronted routinely by engineers and applied mathematicians.

25.4.2   Advanced Methods

Beyond bisection, there are a variety of other techniques that are available to determine the roots of equations. These can be divided into bracketing and open methods. *Bracketing methods* require two initial guesses that are located on either side of (that is, they "bracket") the root. Bisection is one bracketing method. Others, such as the *method of false position,* use more sophisticated and efficient strategies to systematically home in on the root. One strength of bracketing methods is that they always work.

Open methods also employ an iterative approach to locate the root. Some examples are the *Newton-Raphson* and the *secant methods.* The former requires only one initial guess, whereas the latter requires two. Open methods are usually much more efficient at locating roots than bracketing methods. In addition, they can be used to locate multiple roots, whereas the bracketing methods, by definition, cannot. However, if the initial guess (or guesses) is too distant from the true root, the iterations sometimes *diverge.* That is, as the iterations proceed, they actually move farther away from rather than homing in on the root. Detailed descriptions, along with computer algorithms for all these methods, can be found elsewhere (Chapra and Canale, 1988).

PROBLEMS

25.1 Utilize Fig. 25.5 as a starting point for developing your own program for bisection. Employ structured programming techniques, modular design, and internal and external documentation in your development. Make sure the program checks that the initial guesses are in the proper order ($x_l < x_u$) and bracket the root [$f(x_l) f(x_u) < 0$]. Also, substitute the final result back into the function to verify whether the result is close to zero. (*Optional:* Include the capability to plot the function in order to facilitate making the initial guesses.)

25.2 The number of iterations performed by the bisection method to reduce the error below a predefined value can be calculated before an analysis is undertaken based on the initial interval and the desired error of the final result. Derive a relationship to compute the number of iterations. (*Hint:* The relation between the first, or initial, interval, $\Delta x_1 = x_u - x_1$, and the second interval, Δx_2, is $\Delta x_2 = \Delta x_1/2$.) Additionally, as depicted in Fig. 25.4, the error after any iteration is $\Delta x/2$. Remember to adjust your relationship so that it accounts for the fact that the number of iterations must be a whole number, or integer.

25.3 Why might multiple roots pose problems for a technique such as bisection?

25.4 If the bisection method is employed on an interval where the function changes sign, is it always going to locate a root? Why? Give examples to support your claim.

25.5 The upward velocity of a rocket can be computed by the following formula:

$$v = u \ln \frac{m_0}{m_0 - qt} - gt$$

where v is upward velocity, u is the velocity at which fuel is expelled relative to the rocket, m_0 is the initial mass of the rocket at time $t = 0$, q is the fuel consumption rate, and g is the downward acceleration of gravity (assumed constant $= 9.8$ m/s^2). If $u = 2200$ m/s, $m_0 = 160,000$ kg, and $q = 2680$ kg/s, compute the time at which $v = 1000$ m/s. (*Hint: t* is somewhere between 10 and 50 s.) Determine your result so that it is within 1 percent of the true value. Check your answer.

25.6 The annual worth of an engineering project is characterized by the equation

$$A_t = \frac{-1900(1.18)^n}{1.18^n - 1} + \frac{45n}{1.18^n - 1} + 3000$$

where n is the number of years after the project begins operation. At what value of n will $A_t = 0$? That is, at what n will the project start to yield a profit? (*Hint:* It is less than 10 years.) Solve so that n is within 5 percent of the true value.

25.7 In environmental engineering the following equation can be used to compute the oxygen level in a river downstream from a sewage discharge:

$$c = 10 - 15(e^{-0.1t} - e^{-0.5t})$$

where t is the time of travel downstream in days. Determine the time of travel downstream where the oxygen level first falls to a reading of 4. (*Hint:* It is within 5 d of the discharge.) Determine your answer to a 1 percent error.

25.8 Use bisection to determine the root of

$$f(x) = 1.9 - 0.56x - 0.8x^2 + 0.24x^3$$

Perform two interations with initial guesses of $x_l = 1$ and $x_u = 2$. Compute the approximate error for each iteration.

25.9 Figure P25.9a shows a uniform beam subject to a linearly increasing distributed load. The equation for the resulting elastic curve is (see Fig. P25.9b)

$$y = \frac{w_0}{120EIL}(-x^5 + 2L^2x^3 - L^4x) \tag{P25.9.1}$$

The derivative of the elastic curve is

$$\frac{dy}{dx} = \frac{w_0}{120EIL}(-5x^4 + 6L^2x^2 - L^4)$$

Use bisection to determine the point of maximum deflection (that is, the value of x where $dy/dx = 0$). Then substitute this value into Eq. (P25.9.1) to determine the value of the maximum deflection. Use the following parameter values in your computation: $L = 180$ in, $E = 29 \times 10^6$ lb/in^2, $I = 723$ in^4, and $w_0 = 12$ kips/ft. Express your results in inches.

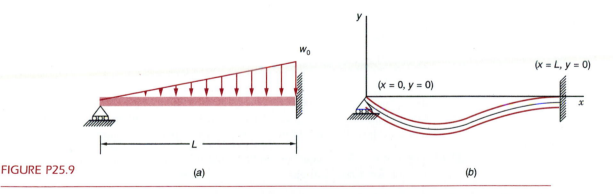

FIGURE P25.9

(a)   (b)

25.10 The height of a hanging cable is given by (Fig. P25.10)

$$y = \frac{H}{w}\cosh\left(\frac{w}{H}x\right) + y_0 - \frac{H}{w}$$

where H is the horizontal force at $x = a$, y_0 is the height at $x = 0$, and w is the

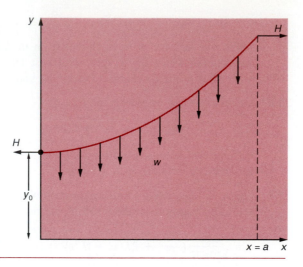

FIGURE P25.10

cable's weight per unit length. Determine the value of H, if $y = 12$ at $x = 50$ for $y_0 = 5$ and $w = 10$. The hyperbolic cosine is defined as

$$\cosh (x) = \frac{e^x + e^{-x}}{2}$$

25.11 The BASIC and FORTRAN intrinsic functions SQR and SQRT employ root-location methods in their routines. Because the roots must be located rapidly, more efficient methods than bisection are used. Nonetheless, it is instructive to formulate a subprogram that employs bisection to determine the square root of any positive number. Include appropriate checks to ensure accuracy and avoid any possible pitfalls.

25.12 Determine the roots of

$$f(x) = \frac{9 - 8x}{(1 - x)^2}$$

using bisection and initial guesses of $x_l = 0$ and $x_u = 3$. Perform the computation by hand to an error level of 1 percent. If any difficulties occur, explain them and state how they might be avoided in a computer program.

25.13 The van der Waals equation provides the relationship between the pressure and volume for a nonideal gas as

$$\left(p + \frac{a}{v^2} \right) (v - b) = RT$$

where p is pressure, a and b are parameters that relate to the type of gas, v is molal volume, R is the universal gas constant (0.082054 L · atm/mol · K), and T is absolute temperature. If $T = 350$ K and $p = 1.5$ atm, compute v for ethyl alcohol ($a = 12.02$ and $b = 0.08407$).

25.14 An oscillating current in an electric circuit is described by

$$I = 10e^{-t} \sin (2\pi t)$$

where t is in seconds. Determine all values of t such that $I = 2$.

25.15 The equation for a reflected standing wave in a harbor is given by

$$h = h_0 \left[\sin \left(\frac{2\pi x}{\lambda} \right) \cos \left(\frac{2\pi tv}{\lambda} \right) + e^{-x} \right]$$

Solve for x if $h = 0.5h_0$, $\lambda = 20$, $t = 10$, and $v = 50$.

EQUATION SOLVING: LINEAR ALGEBRAIC EQUATIONS

In Chap. 25 we determined the value x that satisfied a single equation, $f(x) = 0$. Now we deal with the case of determining the values $x_1, x_2, \ldots, x_n$ that simultaneously satisfy a set of equations:

$$f_1(x_1, x_2, \ldots, x_n) = 0$$
$$f_2(x_1, x_2, \ldots, x_n) = 0$$
$$\cdot \qquad \qquad \cdot$$
$$\cdot \qquad \qquad \cdot$$
$$\cdot \qquad \qquad \cdot$$
$$f_n(x_1, x_2, \ldots, x_n) = 0$$

Such systems can be either linear or nonlinear. In the present chapter, we will deal exclusively with *linear algebraic equations* that are of the general form

$$a_{11}x_1 + a_{12}x_2 + \cdots + a_{1n}x_n = c_1$$
$$a_{21}x_1 + a_{22}x_2 + \cdots + a_{2n}x_n = c_2$$
$$\cdot \qquad \cdot \qquad \qquad \cdot$$
$$\cdot \qquad \cdot \qquad \qquad \cdot$$
$$\cdot \qquad \cdot \qquad \qquad \cdot$$
$$a_{n1}x_1 + a_{n2}x_2 + \cdots + a_{nn}x_n = c_n \qquad (26.1)$$

where the a's are constant coefficients, the c's are constants, and n is the number of equations. Note that the unknowns, $x_1, x_2, \ldots, x_n$, are all raised to the first power and are multiplied by a constant. All other equations are nonlinear.

Linear algebraic equations arise in a variety of problem contexts and in all fields of engineering. In particular, they originate in the mathematical modeling of large systems of interconnected elements such as structures, electric circuits, and networks (see Fig. 26.1). In the previous chapter, you saw how single-component systems result in a single equation that can be solved using root-location techniques. Multicomponent systems of the type depicted in Fig. 26.1 result in a coupled set of mathematical equations that must be solved simultaneously. The equations are coupled because the individual parts of the system are influenced by other parts. For example, in Fig. 26.1a, reactor 4 receives chemical inputs from reactors 2 and 3. Consequently, its response is dependent on the quantity of chemical in these other reactors. When these dependencies are expressed mathematically, the resulting equations are often of the linear algebraic form of Eq. (26.1).

We will use a simple example to illustrate this in the present chapter. Suppose that a team of three parachutists is connected by a weightless cord while free-falling at a velocity of 5 m/s (Fig. 26.2). A simple problem might be to determine the tension in each cord and the acceleration of the team.

To study this problem, free-body diagrams can be developed specifying the various forces acting on each parachutist. As seen in Fig. 26.3, each parachutist is subjected to two major external forces, gravity and air resistance. If a downward force is assigned a positive sign, Newton's second law ($F = ma$) can be used to formulate the force due to gravity as

$$F_D = mg \tag{26.2}$$

where g is the gravitational constant, or the acceleration due to gravity, which is approximately equal to 9.8 m/s^2.

FIGURE 26.1
Some systems of interconnected elements that can be characterized as a system of equations: (a) reactors, (b) a structure, (c) an electric circuit, (d) a fluid pipe network, and (e) a series of blocks and springs.

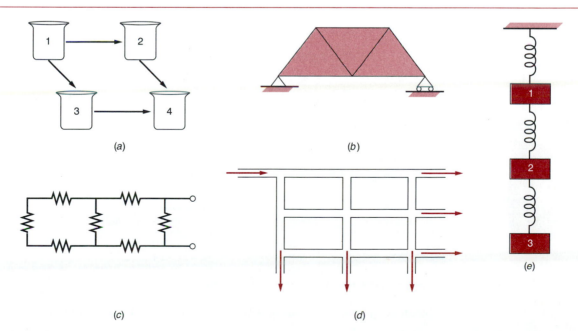

(a)

(b)

(c)

(d)

(e)

FIGURE 26.2
Three parachutists free-falling
while connected by
weightless cords.

Air resistance can be formulated in a number of ways. A simple approach is to assume that it is linearly proportional to velocity, as in

$$F_U = -cv \qquad (26.3)$$

where c is a proportionality constant called the drag coefficient (in kilograms per second). Thus the greater the fall velocity is, the greater will be the upward force due to air resistance.

In addition to these external forces, the cords connecting the parachutists also exert internal forces between the jumpers. The free-body diagrams in Fig. 26.3 specify all the forces for each parachutist. According to Newton's second law, the net force should equal mass times acceleration. Therefore, for the first parachutist,

$$m_1 g - T - c_1 v = m_1 a$$

net force

Similarly, for the second and third parachutists,

$$m_2 g + T - c_2 v - R = m_2 a$$

and

$$m_3 g + R - c_3 v = m_3 a$$

These equations have three unknowns—a, T, and R. By bringing all the terms with unknowns to one side and all the knowns to the other, the following set of equations results:

$$m_1 a + T \qquad\quad = m_1 g - c_1 v \qquad (26.4a)$$

$$m_2 a - T + R = m_2 g - c_2 v \qquad (26.4b)$$

$$m_3 a - R \qquad\quad = m_3 g - c_3 v \qquad (26.4c)$$

Suppose that the parachutists are free-falling at a velocity of 5 m/s; the following values are given for the masses and drag coefficients:

Parachutist	Mass, kg	Drag Coefficient, kg/s
1	70	10
2	60	14
3	40	17

Substituting these values along with $g = 9.8$ m/s^2 into Eq. (26.4) yields

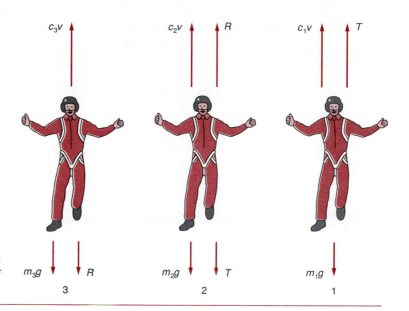

FIGURE 26.3
Free-body diagrams for each of
the three falling parachutists.

$$70a + T \qquad = 636 \qquad\qquad (26.5a)$$

$$60a - T + R = 518 \qquad\qquad (26.5b)$$

$$40a \qquad - R = 307 \qquad\qquad (26.5c)$$

Thus these are three linear algebraic equations with three unknowns that are of the exact form of Eq. (26.1). The method discussed in the present chapter, Gauss elimination, will be used to solve Eq. (26.5) for the unknowns.

It should be noted that a computer is really not required to solve such a small system of equations. Manual algebraic manipulations are quite satisfactory for solving two or three simultaneous equations. However, for four or more equations, manual solutions become arduous, and computers must be utilized. Because many engineering problems involving simultaneous equations can consist of hundreds and even thousands of equations, computers are an absolute necessity.

26.1 GAUSS ELIMINATION

In high school you probably learned to solve two simultaneous equations by eliminating unknowns. The basic strategy of this technique is to multiply one of the equations by a constant in order that one of the unknowns is eliminated when the equations are combined. This results in one equation with one unknown. This single equation can be solved for the unknown and the result substituted back into either of the original equations to determine the other unknown.

(a)
$$a_{11}x_1 + a_{12}x_2 + a_{13}x_3 = c_1$$
$$a_{21}x_1 + a_{22}x_2 + a_{23}x_3 = c_2$$
$$a_{31}x_1 + a_{32}x_2 + a_{33}x_3 = c_3$$

(b)
$$a_{11}x_1 + a_{12}x_2 + a_{13}x_3 = c_1$$
$$a_{22}'x_2 + a_{23}'x_3 = c_2'$$
$$a_{33}''x_3 = c_3''$$

(c)
$$x_3 = c_3'' / a_{33}''$$
$$x_2 = (c_2' - a_{23}'x_3) / a_{22}'$$
$$x_1 = (c_1 - a_{12}x_2 - a_{13}x_3) / a_{11}$$

FIGURE 26.4
The two phases of Gauss elimination: (b) forward-elimination and (c) back-substitution. The primes indicate the number of times that the coefficients and constants have been modified.

The basic approach to eliminating unknowns can be extended to large sets of equations by developing a systematic scheme to eliminate unknowns and then back-substitute. *Gauss elimination* is one of the simplest and most common of these schemes. As depicted in Fig. 26.4, the method consists of two phases. First, a *forward-elimination* is used to eliminate unknowns until the equations are in the form of Fig. 26.4b. Notice that the last equation has one unknown, the next to the last has two unknowns, and so on. Then a *back-substitution phase* is used to solve for the unknowns. The method is best illustrated by an example.

EXAMPLE 26.1   Gauss Elimination

PROBLEM STATEMENT: Solve Eq. (26.5) by Gauss elimination.

SOLUTION: The equations to be solved are

$$70a + T \qquad = 636 \qquad\qquad (E26.1.1a)$$

$$60a - T + R = 518 \qquad\qquad (E26.1.1b)$$

$$40a \qquad - R = 307 \qquad\qquad (E26.1.1c)$$

First we will use forward-elimination to transform the equations into the form of Fig. 26.4b. The initial step will be to eliminate the first unknown, a, from the second and the third equations. To do this, multiply Eq. (E26.1.1a) by 60/70 to give

$$60a + 0.85714T = 545.14$$

Now this equation can be subtracted from Eq. (E26.1.1b) to eliminate the first term from the second equation. The result is

$$-1.8571T + R = -27.14$$

Similarly, the first equation can be multiplied by 40/70 and the result subtracted from the third equation to give

$$-0.57143T - R = -56.429$$

Therefore, the three equations are now

$$70a + \quad T \quad = 636 \tag{E26.1.2a}$$

$$-1.8571T + R = -27.14 \tag{E26.1.2b}$$

$$-0.57143T - R = -56.429 \tag{E26.1.2c}$$

Thus the first unknown is eliminated from the second and the third equations. During these manipulations, the first equation is referred to as the *pivot equation,* and the coefficient of the first unknown in the first equation (70) is called the *pivot element.*

The next step is to eliminate the second unknown from the third equation. For this step, the pivot equation is Eq. (E26.1.2b), and the pivot element is -1.8571. Equation (E26.1.2b) can be multiplied by $-0.57143/-1.8571$ and the result subtracted from Eq. (E26.1.2c) to give

$$-1.3077R = -48.078$$

With this manipulation, the forward-elimination phase is completed, and the three equations have been transformed into the desired form of Fig. 26.4b.

$$70a + \quad T \quad = 636 \tag{E26.1.3a}$$

$$-1.8571T + \quad R = -27.14 \tag{E26.1.3b}$$

$$-1.3077R = -48.078 \tag{E26.1.3c}$$

Now, back-substitution can be used to solve the third equation for

$$R = \frac{-48.078}{-1.3077} = 36.765$$

This result can be back-substituted into Eq. (E26.1.3b) to solve for

$$T = \frac{-27.14 - 36.765}{-1.8571} = 34.41$$

This result can be back-substituted into Eq. (E26.1.3a) to solve for

$$a = \frac{636 - 34.41}{70} = 8.59$$

Therefore, the final result is

$$a = 8.59 \text{ m/s}^2$$

$$T = 34.41 \text{ N}$$

$$R = 36.765 \text{ N}$$

We can check if this result is correct by substituting these values back into the original equations to see if they balance:

$$70(8.59) + 34.41 \qquad\qquad = 636$$

$$60(8.59) - 34.41 + 36.765 = 518$$

$$40(8.59) \qquad\quad - 36.765 = 307$$

Comparing the left- and the right-hand sides shows that the equations balance:

$$635.71 \cong 636$$

$$517.755 \cong 518$$

$$306.835 \cong 307$$

The minor discrepancies are due to round-off error.

26.1.1 Pivoting

The technique illustrated in Example 26.1 is commonly referred to as *naive Gauss elimination*. It is called "naive" because during both the forward-elimination and the back-substitution phases, it is possible that a division by zero could occur. For example, if we use naive Gauss elimination to solve

$$2x_2 + 3x_3 = 8 \tag{26.6a}$$

$$2x_1 + 6x_2 + 7x_3 = 8 \tag{26.6b}$$

$$4x_1 + x_2 + 6x_3 = 5 \tag{26.6c}$$

the initial operation consists of multiplying the first row by 2/0. Thus a division by zero occurs because the pivot element (a_{11}) is equal to zero. Problems may also arise when the pivot element is close to, rather than exactly equal to, zero. For situations where the magnitude of the pivot element is very small compared with the other elements, round-off errors can be introduced.

Therefore before each column of coefficients is eliminated, it is advantageous to determine which of the coefficients in the column is largest. If this coefficient is larger than the pivot coefficient, the rows can be switched so that the largest element becomes the pivot element. For Eq. (26.6), the third equation has the largest coefficient (4) in the column that is to be eliminated. Therefore, Eq. (26.6a) and Eq. (26.6c) are switched to give

$$4x_1 + x_2 + 6x_3 = 5$$
$$2x_1 + 6x_2 + 7x_3 = 8$$
$$2x_2 + 3x_3 = 8$$

If this technique, which is called *partial pivoting,* is implemented prior to each elimination step, division by zero and other problems connected with very small pivot elements can be avoided. For this reason, as described in the next section, computer programs for Gauss elimination must include partial pivoting.

26.2   ALGORITHM FOR GAUSS ELIMINATION

It is a relatively simple matter to develop an algorithm for Gauss elimination. Figure 26.5 presents modules to perform the three fundamental operations involved in the technique: forward-elimination, back-substitution, and partial pivoting. There are two features about these modules that are noteworthy. First, the coefficients are stored in a two-dimensional array, and the right-hand-side constants and the unknowns are stored as one-dimensional arrays. This allows efficient manipulation and storage of the constants and the results. Second, notice the extensive use of loops to perform the methodical steps of the algorithm.

Figure 26.6 shows some screens generated using the Gauss elimination routine from the Electronic TOOLKIT. The package is used to solve Example 26.1. The results are shown in Fig. 26.6a, and an error check is shown in Fig. 26.6b. These screens illustrate how output can be designed in a clear and accessible fashion. You can employ them as models for your own software.

26.3   PITFALLS AND ADVANCED METHODS

26.3.1   Pitfalls

The adequacy of the solution of simultaneous equations depends on the condition of the system. Ill-conditioned systems are those where small changes in the coefficients can result in large changes in the solution. An alternative interpretation is that a wide range of answers can approximately satisfy the equations. Because round-off errors can induce small changes in the coefficients, these artificial changes can lead to large solution errors for ill-conditioned systems. One simple remedy is to use double-precision

$DO\ FOR\ k \leftarrow i\ to\ n-1$
$\quad DO\ FOR\ i \leftarrow k+1\ to\ n$

$$factor \leftarrow \frac{a_{ik}}{a_{kk}}$$

$\quad\quad DO\ FOR\ j \leftarrow k+1\ to\ n$
$\quad\quad\quad a_{ij} \leftarrow a_{ij} - factor \cdot a_{kj}$
$\quad\quad LOOP$
$\quad\quad c_i \leftarrow c_i - factor \cdot c_k$
$\quad LOOP$
$LOOP$

(a) Forward elimination

$x_n \leftarrow c_n / a_{nn}$
$DO\ FOR\ i \leftarrow n-1\ to\ 1\ step -1$
$\quad sum \leftarrow 0$
$\quad DO\ FOR\ j \leftarrow i+1\ to\ n$
$\quad\quad sum \leftarrow sum + a_{ij} \cdot x_j$
$\quad LOOP$

$$x_i \leftarrow \frac{c_i - sum}{a_{ii}}$$

$LOOP$

(b) Back substitution

$pivot \leftarrow k$
$big \leftarrow |a_{kk}|$
$DO\ FOR\ m \leftarrow k+1\ to\ n$
$\quad dummy \leftarrow |a_{mk}|$
$\quad IF\ dummy > big$
$\quad\quad big \leftarrow dummy$
$\quad\quad p \leftarrow m$
$\quad END\ IF$
$LOOP$

$IF\ p \neq k$
$\quad DO\ FOR\ l \leftarrow k\ to\ n$
$\quad\quad dummy \leftarrow a_{pl}$
$\quad\quad a_{pl} \leftarrow a_{kl}$
$\quad\quad a_{kl} \leftarrow dummy$
$\quad LOOP$
$\quad dummy \leftarrow c_p$
$\quad c_p \leftarrow c_k$
$\quad c_k \leftarrow dummy$
$END\ IF$

(c) Partial pivoting

FIGURE 26.5
Algorithms for the major operations of Gauss elimination: (a) forward-elimination, (b) back-substitution, and (c) partial pivoting.

FIGURE 26.6 Computer screens generated by Gauss elimination routine from the Electronic TOOLKIT. The package is used to solve the problem previously analyzed in Example 26.1

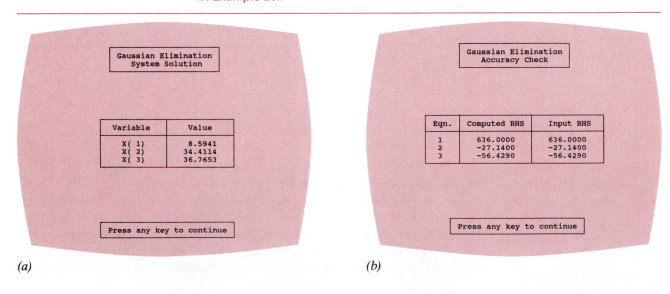

(a)   (b)

variables in your computer program. For a more detailed discussion of the problem, see Chapra and Canale (1988). Fortunately, most linear algebraic equations derived from engineering-problem settings are naturally well conditioned. Therefore, although you should be aware of the problem, it does not occur frequently in practice.

Another potential pitfall is the presence of round-off error for very large systems of equations. This is due to the fact that for a technique such as Gauss elimination, every result is dependent on previous results. Consequently, an error in the early steps will tend to propagate. In general, a rough rule of thumb is that round-off may be important when dealing with greater than 100 equations. For these cases, Gauss elimination sometimes performs poorly, and as discussed in the next section, other methods may be needed.

26.3.2 Advanced Methods

Alternative methods are available for solving simultaneous equations. One approach that is particularly well suited for large systems is the *Gauss-Seidel method*. This technique is fundamentally different from Gauss elimination in that it is an approximate iterative method. That is, it employs initial guesses and then iterates to obtain refined estimates of the solution. The Gauss-Seidel method is particularly well suited for large numbers of equations. In these cases, elimination methods can be subject to round-off errors. Because the error of the Gauss-Seidel method is controlled by the number of iterations, round-off error is usually not an issue of concern. However, there are certain instances where the Gauss-Seidel technique will not converge on the correct answer. These and other trade-offs between elimination and iterative methods are discussed elsewhere (Chapra and Canale, 1988; Ralston and Rabinowitz, 1978).

PROBLEMS

26.1 Utilize Fig. 26.5 as a starting point for developing your own program for Gauss elimination. Employ structured programming techniques, modular design, and internal and external documentation in your development. Include the capability of substituting the answers back into the original equations to substantiate that the results make the equations balance.

26.2 Three parachutists are connected by a weightless cord while free-falling. The team is accelerating at a rate of 5 m/s^2. Calculate the tension in each section of cord and the velocity, given the following:

Parachutist	Mass, kg	Drag Coefficient, kg/s
1 (top)	50	12
2 (middle)	30	10
3 (bottom)	70	14

26.3 Three parachutists are connected by a weightless cord while free-falling. The upward force due to drag is represented by

$$F_U = -cv^2$$

where c is a drag coefficient (in kilograms per meter). If the team has a velocity of 10 m/s, calculate the tension in each rope and the acceleration, given the following:

Parachutist	Mass, kg	Drag Coefficient, kg/m
1 (top)	50	0.6
2 (middle)	30	0.5
3 (bottom)	70	0.7

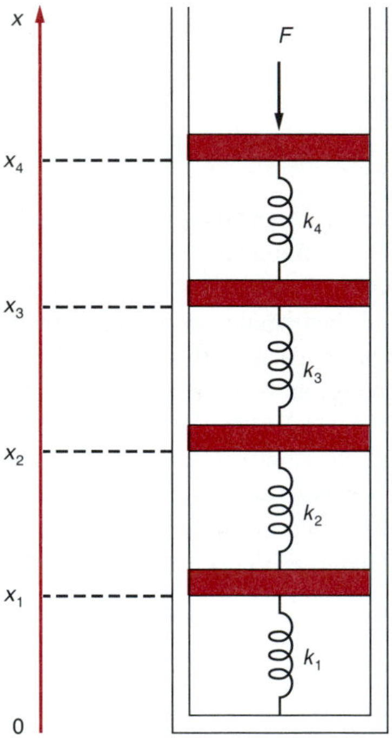

26.4 Solve by Gauss elimination:

$$x_1 + 2x_2 + 3x_3 = -3$$

$$x_1 - x_2 - 2x_3 = 1$$

$$4x_1 - 2x_2 + 5x_3 = -18$$

26.5 Idealized spring-mass systems have numerous applications throughout engineering. Figure P26.5 shows an arrangement of four springs in series being depressed with a force of 2000 lb. At equilibrium, force-balance equations can be developed defining the interrelationships between the springs:

$$k_2(x_2 - x_1) = k_1 x_1$$

$$k_3(x_3 - x_2) = k_2(x_2 - x_1)$$

$$k_4(x_4 - x_3) = k_3(x_3 - x_2)$$

$$F = k_4(x_4 - x_3)$$

where the k's are spring constants. If k_1 through k_4 are 100, 50, 75, and 200 lb/in, respectively, compute the x's.

FIGURE P26.5

26.6 For Prob. 26.5, solve for the force F if x_3 is given as 60 in. All other parameters are as given in Prob. 26.5.

26.7 Three blocks are connected by a weightless cord and rest on an inclined plane (Fig. P26.7a). Employing a procedure similar to the one used in the analysis of the falling parachutists in Example 26.1, the following set of simultaneous equations can be derived (free-body diagrams are shown in Fig. P26.7b):

$$100a + T \qquad = 519.72$$

$$50a - T + R = 216.55$$

$$20a \qquad - R = 86.62$$

Solve for acceleration a and the tensions T and R in the two ropes.

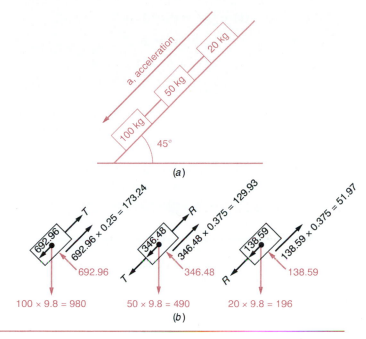

FIGURE P26.7

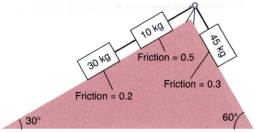

FIGURE P26.8

26.8 Perform a computation similar to that called for in Prob. 26.7, but for the system shown in Fig. P26.8.

26.9 Given two equations, a graphical solution is possible:

$$3x_1 + 2x_2 = 18 \qquad -x_1 + 2x_2 = 2$$

Plot each of these equations on cartesian coordinates, with the ordinate as x_2 and the abscissa as x_1. What is the solution? Compare the graphical result to the analytical solution of the equations obtained by eliminating unknowns.

26.10 Obtain graphical solutions for the following equations (see Prob. 26.9):

$$(a) \ -\tfrac{1}{2}x_1 + x_2 = 1$$

$$-\tfrac{1}{2}x_1 + x_2 = \tfrac{1}{2}$$

(b) $-\frac{1}{2}x_1 + x_2 = 1$

   $-x_1 + 2x_2 = 2$

(c) $-\frac{2.3}{5}x_1 + x_2 = 1.1$

   $-\frac{1}{2}x_1 + x_2 = 1$

Interpret your results for each case, and speculate how these situations might pose problems for a computer algorithm. Such cases are said to be *ill-conditioned*.

26.11 Although the Lagrange polynomial described in Chap. 24 provides a convenient means for determining intermediate values between points, it does not provide a convenient polynomial of the conventional form

$$f(x) = a_0 + a_1 x + a_2 x^2 + \cdots + a_n x^n$$

An interpolating polynomial of this form can be determined using simultaneous equations. For example, suppose that you desired to compute the coefficients of the parabola

$$f(x) = a_0 + a_1 x + a_2 x^2 \qquad\qquad (P26.11.1)$$

Three points are required: $[x_0, f(x_0)]$; $[x_1, f(x_1)]$; and $[x_2, f(x_2)]$. Each can be substituted into Eq. (P26.11.1) to give

$$f(x_0) = a_0 + a_1 x_0 + a_2 x_0^2$$

$$f(x_1) = a_0 + a_1 x_1 + a_2 x_1^2$$

$$f(x_2) = a_0 + a_1 x_2 + a_2 x_2^2$$

Thus, for this case, the x's are the knowns and the a's are the unknowns. Because there are three equations with three unknowns, a technique such as Gauss elimination can be employed to solve for the a's. Test this approach by determining the parabola of the form of Eq. (P26.11.1) to interpolate between (1, 2), (3, 8), and (7, 6). Plot the resulting equation and the points.

COMPUTER CALCULUS: DIFFERENTIATION

The computer's fundamental mathematical capability is limited to arithmetic operations and a handful of intrinsic functions. In the previous chapters, we have seen how this seemingly limited repertoire can be used to develop a variety of powerful problem-solving techniques. Now we will learn how these same simple capabilities can be used to solve calculus problems with the computer. The logical starting point is the mathematical operation that is at the heart of calculus—differentiation.

27.1   THE DERIVATIVE

Calculus is the mathematics of change. Because engineers must continuously deal with systems that move and grow, knowledge of calculus is an essential tool of our profession.

At the heart of calculus lies the mathematical concept of the derivative. It represents the rate of change of a dependent variable with respect to an independent variable. As depicted in Fig. 27.1a, the mathematical definition of the derivative begins with a difference approximation:

$$\frac{\Delta y}{\Delta t} = \frac{f(t_1 + \Delta t) - f(t_1)}{\Delta t} \tag{27.1}$$

where y and $f(t)$ are alternative representations of the dependent variable, and t is the independent variable. If Δt is allowed to approach zero, as occurs in moving from Fig. 27.1a to 27.1c, the difference becomes a derivative:

$$\frac{dy}{dt} = \lim_{\Delta t \to 0} \frac{f(t_1 + \Delta t) - f(t_1)}{\Delta t}$$

where dy/dt [which can also be represented as y' or $f'(t_1)$] is called the *first derivative*

617

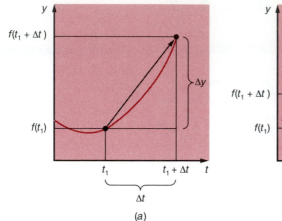

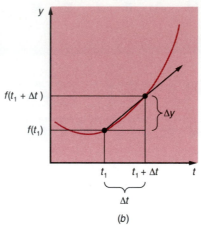

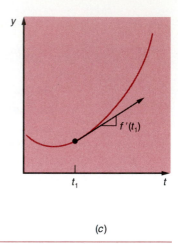

(a)   (b)   (c)

FIGURE 27.1
The graphical definition of a derivative: going from (a) to (c), as Δt approaches zero and the difference approximation becomes a derivative.

of y with respect to t evaluated at t_1. As seen in the visual depiction of Fig. 27.1c, the derivative is the slope of the tangent to the curve at t_1.

Aside from providing the slope at a single point, it is possible to "take the derivative of" or, as it is called in calculus, to "differentiate" a function itself. In this way, a second function results that can be used to compute the derivative for different values of the independent variable. General rules are available for this purpose. In the case of polynomials, the following simple rule applies:

$$\frac{d}{dt}x^n = nx^{n-1} \tag{27.2}$$

For example, suppose that the position of a moving object is described by the function

$$y = 2t^2 \tag{27.3}$$

In this case, y might represent the height of a rocket taking off from a launching pad. By substituting values of t into Eq. (27.3), we can calculate height as a function of time. As seen in Fig. 27.2a and column (b) of Table 27.1, the rocket seems to be accelerating. That is, it travels a progressively larger distance with each passing unit of time. Differentiation can be used to quantify this behavior. Equation (27.3) can be differentiated by applying Eq. (27.2) to yield

$$v = \frac{dy}{dt} = 4t \tag{27.4}$$

where v is velocity. Equation (27.4) provides the slope of Fig. 27.2a, which is equivalent to the rocket's velocity. As plotted in Fig. 27.2b and listed in column (c) of Table 27.1,

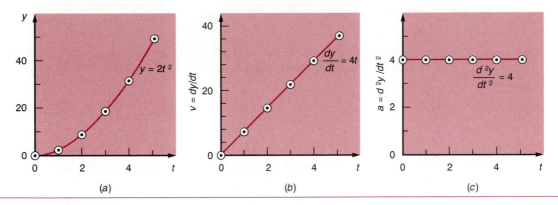

FIGURE 27.2
Graphs of (*a*) distance, in meters; (*b*) velocity, in meters per second; and (*c*) acceleration, in meters per second squared versus time, in seconds, for the flight of a rocket.

this result indicates that velocity is increasing linearly with time. That is, with every unit of time, the velocity increases by 4.

Finally, Eq. (27.4) can also be differentiated to give the second derivative (that is, the derivative of the derivative) of *y* with respect to *t*:

$$a = \frac{d^2y}{dt^2} = 4$$

where d^2y/dt^2 is the second derivative. Thus the rocket has a constant acceleration of 4 [see Fig. 27.2*c* and column (*d*) of Table 27.1].

Differentiation of a function has many applications in engineering. However, conventional programming languages are not capable of the type of analytical differentiation performed above. That is, we cannot input a function such as $y = 2t^2$ and have the computer output the derivative, $y' = 4t$. Nevertheless, the computer can determine slopes of functions at points. It is this sort of differentiation that is described in the present chapter.

TABLE 27.1 Values of Distance (in Meters), Velocity (in Meters per Second), and Acceleration (in Meters per Second Squared) as a Function of Time (in Seconds) for a Rocket

Time, *t* (*a*)	Distance, *y* (*b*)	Velocity, *v* (*c*)	Acceleration, *a* (*d*)
0	0	0	4
1	2	4	4
2	8	8	4
3	18	12	4
4	32	16	4
5	50	20	4

27.2 NUMERICAL DIFFERENTIATION

The computer differentiates in a manner that is similar to the original difference representation (Fig. 27.1a) used in the definition of the derivative [Eq. (27.1)]:

$$\frac{dy}{dt} \cong \frac{\Delta y}{\Delta t} = \frac{f(t + \Delta t) - f(t)}{\Delta t} \tag{27.5}$$

Thus, if we know two points on a curve, the derivative can be approximated by a difference. In anticipation of applying Eq. (27.5) on the computer, we will reexpress it using slightly different nomenclature:

FORWARD)

$$\frac{dy}{dt} \cong \frac{y_{i+1} - y_i}{h} \tag{27.6}$$

where y_i is the value of the dependent variable corresponding to the independent variable t_i at which the derivative is to be estimated, y_{i+1} is a value of the dependent variable at a later value of the independent variable t_{i+1}, and $h = t_{i+1} - t_1$ is called the *step size*. As depicted in Fig. 27.3a, Eq. (27.6) is referred to as a *forward difference*. It is called this because it utilizes data at the base point (that is, at i) and at a forward point (that is, at $i + 1$) to estimate the derivative.

Two similar approximations of the first derivative can also be made. As shown in Fig. 27.3b, a *backward difference* uses the base point and a previous point, as in

BACKWARD)

$$\frac{dy}{dt} \cong \frac{y_i - y_{i-1}}{h} \tag{27.7}$$

FIGURE 27.3
(a) Forward, (b) backward, and (c) central difference approximations of the derivative of y at t_i. Notice how the central difference provides a superior estimate of the derivative at t_i.

Finally, a centered or *central difference* uses the difference between the forward and the backward point to estimate the derivative at the base point (Fig. 27.3c):

$$\frac{dy}{dt} \cong \frac{y_{i+1} - y_{i-1}}{2h} \qquad CENTRAL \tag{27.8}$$

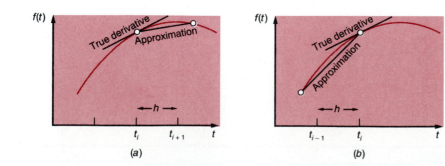

(a)

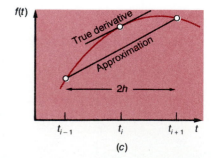

(c)

As is apparent from Fig. 27.3, the central difference is superior to the forward and backward forms. This superiority is also supported by the following example.

EXAMPLE 27.1   Difference Approximations of the First Derivative

PROBLEM STATEMENT: Use Eqs. (27.6) through (27.8) to estimate the velocity of the rocket from column (c) of Table 27.1 based on the times and distances from columns (a) and (b).

SOLUTION: The forward difference [Eq. (27.6)] must be used to estimate the derivative at $t = 0$ because no previous value is available to compute a central or backward difference. The result is

$$\frac{dy}{dt} \cong \frac{2 - 0}{1 - 0} = 2$$

In a similar fashion, the backward difference [Eq. (27.7)] can be employed to estimate the derivative at the end of the table at $t = 5$, as in

$$\frac{dy}{dt} \cong \frac{50 - 32}{5 - 4} = 18$$

Finally, the central difference [Eq. (27.8)] can be used to estimate the derivatives at each of the intermediate points. For example, at $t = 1$,

$$\frac{dy}{dt} \cong \frac{8 - 0}{2 - 0} = 4$$

The results for the entire computation, along with the true velocity [column (c) of Table 27.1] for comparison, are

Time	Distance	True Velocity	Difference Approximation
0	0	0	2
1	2	4	4
2	8	8	8
3	18	12	12
4	32	16	16
5	50	20	18

Note that the forward and backward differences are somewhat erroneous at the end points but that the central difference approximations yield perfect estimates for the intermediate points. The perfect results are not typical and stem from the function used to characterize distance in this problem. However, it *is* generally true that the central difference will perform better than either the forward or backward difference.

27.3   PITFALLS AND ADVANCED METHODS

27.3.1   Pitfalls

A shortcoming of numerical differentiation is that it tends to amplify errors in the data. Figure 27.4*a* and *b* shows the data and results for Example 27.1. Figure 27.4*c* uses the same data, but some points are raised and some are lowered slightly. This minor modification is barely apparent from Fig. 27.4*c*. However, the resulting effect in Fig. 27.4*d* is significant because the process of differentiation amplifies errors. Such amplification is due to the fact that numerical differentiation is a subtraction process. Consequently, random positive and negative errors tend to reinforce each other.

One remedy for this problem is to use polynomial or nonlinear regression to fit a lower-order curve to the data. As discussed in Chap. 23, regression does not attempt to intersect each point but rather traces out the general trend. The regression equation can be differentiated analytically to determine derivative estimates. In addition, a technique such as cubic splines (recall Sec. 24.5) can also be used to fit the data, and the resulting cubic equation can be differentiated to estimate the derivatives. Although cubic splines

FIGURE 27.4
Illustration of how even small data errors introduce large variability into numerical estimates of derivatives: (*a*) data with no error, (*b*) the resulting numerical differentiation, (*c*) data modified slightly, and (*d*) the resulting differentiation manifests significant variability.

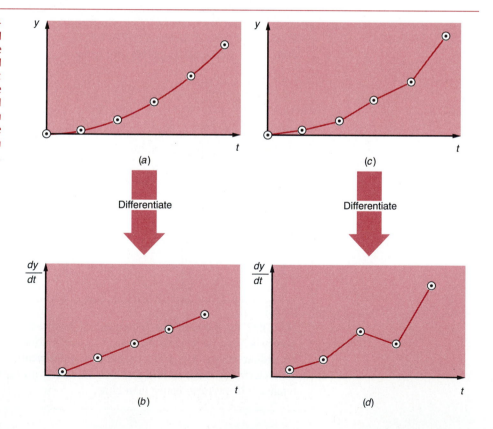

intersect every point, the fact that they are limited to cubic equations usually results in smoother derivatives.

27.3.2   Advanced Methods

Aside from the techniques described in the previous section, there are other advanced methods for numerical differentiation. For equispaced data, higher-order forward-, backward-, and central-difference formulas are available. These formulas are summarized elsewhere (Chapra and Canale, 1988).

Two major difficulties with Eqs. (27.6) through (27.8) are that they are applicable only for equispaced data and the forward and backward differences at the end points are less accurate than the central differences in the middle. One remedy is to fit a second-order Lagrange interpolating polynomial [recall Eq. (24.5)] to each set of three adjacent points. Remember that this polynomial does not require that the points be equispaced. The second-order polynomial can be differentiated analytically to give

$$\frac{dy}{dt} \cong y_{i-1}\frac{2t - t_i - t_{i+1}}{(t_{i-1} - t_i)(t_{i-1} - t_{i+1})} + y_i\frac{2t - t_{i-1} - t_{i+1}}{(t_i - t_{i-1})(t_i - t_{i+1})}$$
$$+ y_{i+1}\frac{2t - t_{i-1} - t_i}{(t_{i+1} - t_{i-1})(t_{i+1} - t_i)} \tag{27.9}$$

where t is the value at which you want to estimate the derivative. Although this equation is certainly more complex than Eqs. (27.6) to (27.8), it has some very important advantages. First, it can be used to estimate the derivatives anywhere within the range prescribed by the three points. Second, the points themselves do not have to be equally spaced. Third, the derivative estimate is of the same accuracy as the centered difference [Eq. (27.8)]. In fact, for equispaced points, Eq. (27.9) evaluated at $t = t_i$ reduces to Eq. (27.8). Because of these properties, Eq. (27.9) is superior to Eqs. (27.6) through (27.8). In Prob. 27.1 you will have the task of incorporating Eq. (27.9) into a computer program. This will provide you with an advanced capability for performing numerical differentiation.

PROBLEMS

27.1   Develop a computer program for numerical differentiation. Employ structured programming techniques, modular design, and internal and external documentation in your development. Incorporate Eq. (27.9) in place of the finite divided differences in order to allow the use of unequally spaced data and to improve the accuracy of the derivative estimates.

27.2   What are the fundamental differences between the forward, backward, and central differences? Which is the most accurate and why?

27.3   Compute the forward-, backward-, and central-difference approximations for

$y = \sin(x)$ at $x = \pi/3$. Note that the derivative as determined by calculus is $dy/dx = \cos x$. Thus the value of the derivative at $\pi/3$ is 0.5. Compute the approximations for (a) $h = \pi/3$, (b) $h = \pi/6$, and (c) $h = \pi/12$. Calculate the percent relative error for each result. What conclusions can be drawn regarding the effect of halving the step size on the accuracy of each type of approximation?

27.4 Prove that for equispaced data points, Eq. (27.9) reduces to Eq. (27.8) at $t = t_i$.

27.5 Apply Eq. (27.9) to recompute the derivative estimates for Example 27.1. Interpret your results.

27.6 Compute the central difference for each of the following functions at the specified location and for the specified step size:

(a) $y = x^3 + 2x - 10$ at $x = 0$, $h = 0.5$

(b) $y = x^3 + 2x - 10$ at $x = 0$, $h = 0.1$

(c) $y = \tan x$ at $x = 4$, $h = 0.2$

(d) $y = \sin(0.5\sqrt{x})$ at $x = 1$, $h = 0.25$

(e) $y = e^x$ at $x = 0$, $h = 0.5$

27.7 A jet fighter's position on an aircraft carrier's runway was timed during landing:

t, s	0	0.51	1.03	1.74	2.36	3.24	3.82
x, m	154	186	209	250	262	272	274

where x is distance from the end of the carrier. Estimate (a) velocity (dx/dt) and (b) acceleration (dv/dt) using numerical differentiation.

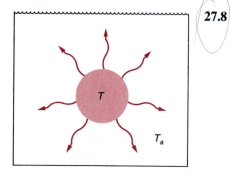

FIGURE P27.8

27.8 The rate of cooling of a body (Fig. P27.8) can be expressed as

$$\frac{dT}{dt} = -k(T - T_a)$$

where T is the temperature of the body (in degrees Celsius), T_a is the temperature of the surrounding medium (in degrees Celsius), and k is a proportionality constant (per minute). Thus this equation (which is called *Newton's law of cooling*) specifies that the rate of cooling is proportional to the difference in the temperatures of the body and of the surrounding medium. If a metal ball heated to 90°C is dropped into water that is held constant at $T_a = 20$°C, the temperature of the ball changes as in

Time, min	0	5	10	15	20	25
Temperature, °C	90	62.5	45.8	35.6	29.5	25.8

Utilize numerical differentiation to determine dT/dt at each value of time. Plot dT/dt versus $T - T_a$, and employ linear regression to evaluate k.

27.9 The following data was collected when a large oil tanker was loading:

t, min	0	15	30	45	60	90	120
V, 10^6 barrels	0.5	0.65	0.73	0.88	1.03	1.14	1.30

Calculate the flow rate Q (that is, dV/dt) for each time.

27.10 The *heat current* H is the quantity of heat flowing through a material per unit time. It can be computed with

$$H = -kA\frac{dT}{dx}$$

where H has units of joules per second; k, a coefficient of thermal conductivity that parameterizes the heat-conducting properties of the material, has units of joules per second per meter per degree Celsius; A is the cross-sectional area perpendicular to the path of heat flow; T is temperature (in degrees Celsius); and x is the distance (in centimeters) along the path of heat flow. If the temperature outside a house is $-20°C$ and inside is $21°C$, what is the rate of heat loss through a 6-in-thick, 1000-ft^2 wall made of

Material	k
Wood	0.08
Red brick	0.60
Concrete	0.80
Insulating brick	0.15

27.11 *Fick's first diffusion law* states that

$$\text{Mass flux} = -D\frac{dc}{dx} \tag{P27.11}$$

where mass flux is the quantity of mass that passes across a unit area per unit time (in grams per square centimeter per second), D is a diffusion coefficient (in square centimeters per second), c is concentration, and x is distance (in centimeters). An environmental engineer measures the following concentration of a pollutant in the sediments underlying a lake ($x = 0$ at the sediment-water interface and increases downward):

x, cm	0	1	2.5
c, 10^{-6} g/cm^3	0.1	0.4	0.75

Use the best numerical differentiation technique available to estimate the derivative at $x = 0$. Employ this estimate in conjunction with Eq. (P27.11) to compute the mass flux of pollutant out of the sediments and into the overlying waters $(D = 10^{-6}$ cm/s$^2)$. For a lake with 10^6 m^2 of sediments, how much pollutant would be transported into the lake over a year's time?

27.12 *Faraday's law* characterizes the voltage drop across an inductor as

$$V_L = L\frac{di}{dt}$$

where V_L is voltage drop (in volts), L is the inductance (in henrys; 1 H = 1 Vs/A), i is current (in amperes), and t is time (in seconds). Determine the voltage drop as a function of time from the following data:

t	0	0.1	0.2	0.4	0.6	0.8
i	0	0.1	0.2	0.6	1.3	2.7

The inductance is equal to 5 H.

SAME AS   27.9

COMPUTER CALCULUS: INTEGRATION

In the previous chapter we learned how the computer is used to determine derivatives of data. Now we will turn to the opposite process in calculus—*integration*.

According to the dictionary definition, to *integrate* means "to bring together, as parts, into a whole; to unite; to indicate the total amount. . . ." Mathematically, integration is represented by

$$I = \int_a^b f(x)\, dx \tag{28.1}$$

which stands for the integral of the function $f(x)$ with respect to the independent variable x, evaluated between the limits $x = a$ and $x = b$.

As suggested by the dictionary definition, the "meaning" of Eq. (28.1) is the *total value,* or *summation,* of $f(x)\, dx$ over the range from $x = a$ to b. In fact, the symbol $\int$ is actually a stylized capital S that is intended to signify the close connection between integration and summation.

Figure 28.1 represents a graphical manifestation of the concept. For functions lying above the x axis, the integral expressed by Eq. (28.1) corresponds to the area under the curve of $f(x)$ between $x = a$ and b.

We will have numerous occasions to refer back to this graphical conception as we develop computer-oriented methods for integration. In fact, some of the common precomputer methods for integration are based on graphs. For example, a simple intuitive approach is to plot the function on a grid (Fig. 28.2) and count the number of boxes that approximate the area. This number multiplied by the area of each box provides a rough estimate of the integral. The estimate can be refined, at the expense of additional effort, by using a finer grid.

In the present chapter we will describe an alternative method for integration that is also based on a graphical approach but is easier to use than the grid method. Also, it is ideal for computer implementation.

Integration has numerous applications in engineering. A number of examples re-

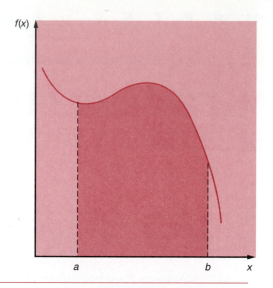

FIGURE 28.1
Graphical representation of the integral of $f(x)$ between the limits $x = a$ to b. The integral is equivalent to the area under the curve.

FIGURE 28.2
The use of a grid to approximate an integral.

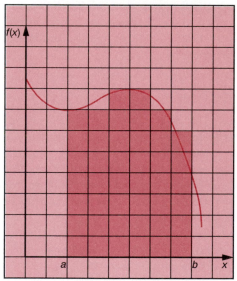

late directly to the idea of the integral as the area under a curve. Figure 28.3 depicts a few cases where integration is used for this purpose.

Other common applications relate to the analogy between integration and summation. For example, the average or mean value of a continuous function between $x = a$ and b can be computed by

$$I = \frac{\displaystyle\int_a^b f(x)\,dx}{b - a} \tag{28.2}$$

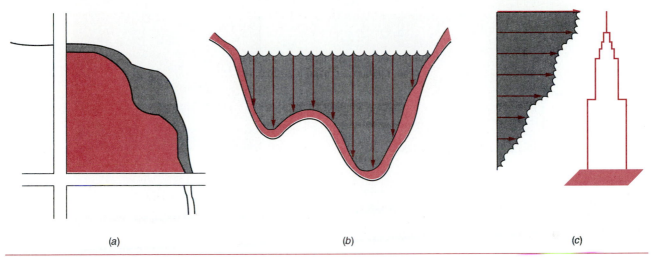

(a) (b) (c)

FIGURE 28.3
Examples of how integration is used to evaluate the areas of surfaces in engineering. (*a*) A surveyor might need to know the area of a field bounded by a meandering stream and two roads. (*b*) A water-resource engineer might need to know the cross-sectional area of a river. (*c*) A structural engineer might need to integrate a nonuniform force due to a wind blowing against the side of a skyscraper.

Thus, the numerator represents the summation of $f(x)\,dx$. Then by dividing by the distance from a to b, we obtain the average, or mean, height of the function (Fig. 28.4). Equation (28.2) can be compared with Eq. (22.1) to contrast the calculation of the mean for continuous functions and discrete data.

Another application, which will serve as an example for this chapter, is to determine a quantity on the basis of its given rate of change. This example most clearly demonstrates the relationship and opposite natures of differentiation and integration. In the previous chapter we used differentiation to determine the velocity of an object, given displacement as a function of time. That is, as shown in Fig. 28.5*a*,

$$v(t) = \frac{d}{dt}\,y(t)$$

In the present chapter we are given velocity and use integration to determine displacement:

FIGURE 28.4
The integral can be used to evaluate the mean value of a function.

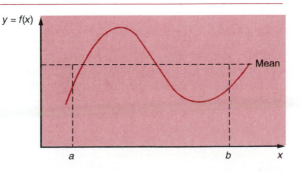

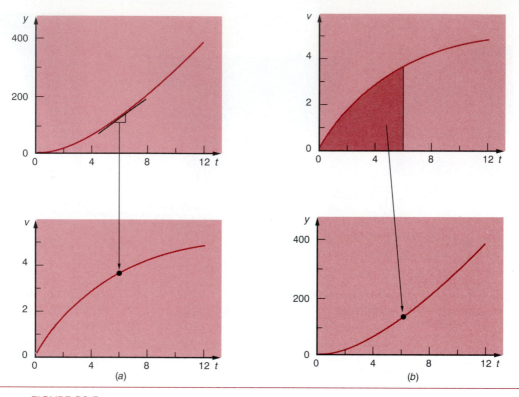

FIGURE 28.5
The contrast between
(*a*) differentiation and
(*b*) integration.

$$y(t) = \int_0^t v(t)\, dt \tag{28.3}$$

Thus by integrating or "summing" the product of velocity multiplied by time, we can determine how far the projectile travels over time *t*. As seen in Fig. 28.5*b*, the area under the velocity-time curve from 0 to *t* gives the distance traveled over that time. In the present chapter we will use the computer to make evaluations of this sort.

28.1 THE TRAPEZOIDAL RULE

As depicted in Fig. 28.6, a simple approach for determining the integral of a function is as the area under the straight line connecting the function values at the ends of the integration interval. The formula to compute this area is

$$I \cong (b - a)\frac{f(a) + f(b)}{2} \tag{28.4}$$

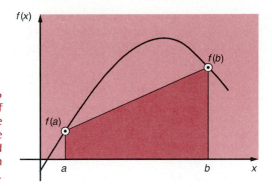

FIGURE 28.6
The trapezoidal rule consists of taking the area under the straight line connecting the function values at the end points of the integration interval.

FIGURE 28.7
(a) The formula for computing the area of a trapezoid: height times the average of the bases. (b) For the trapezoidal rule, the concept is the same but the trapezoid is on its side.

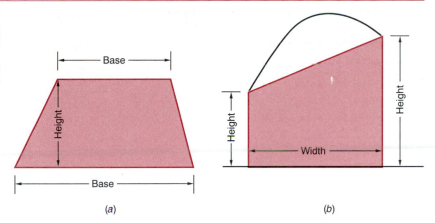

Because the area under the line is a trapezoid, this formula is called the *trapezoidal rule*. Recall from geometry that the formula for computing the area of a trapezoid is the height times the average of the bases (Fig. 28.7a). In the present case, the concept is the same but the trapezoid is on its side (Fig. 28.7b). Therefore the integral estimate can be represented as

$$I \cong \text{width} \times \text{average height}$$

where for the trapezoidal rule, the average height is the average of the function values at the end points, or $[f(a) + f(b)]/2$.

EXAMPLE 28.1   The Trapezoidal Rule

PROBLEM STATEMENT: The velocity of a falling parachutist is given by the following function of time:

$$v(t) = \frac{gm}{c}[1 - e^{-(c/m)t}]$$

where v is velocity in meters per second, g is the gravitational constant of 9.8 m/s^2, m is the mass of the parachutist equal to 68.1 kg, and c is a drag coefficient of 12.5 kg/s. A plot of the function can be developed by substituting the parameters and various values of t into the equation to give

t, s	v, m/s
0	0.00
2	16.40
4	27.77
6	35.64
8	41.10
10	44.87
12	47.49
14	49.30
16	50.56

These points are plotted in Fig. 28.8. Use the trapezoidal rule to determine how far the parachutist has fallen after 12 s.

SOLUTION: According to Eq. (28.3), the distance can be determined by integrating the velocity, as in

$$y(t) = \int_0^t v(t)\, dt$$

or, substituting the function and the given values,

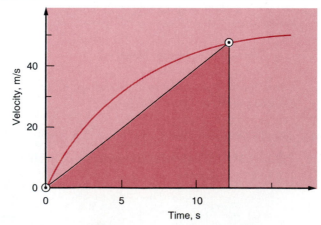

FIGURE 28.8
A plot of the velocity of a falling parachutist versus time. The area under the curve from $t = 0$ to 12 is equal to the distance the parachutist travels over the first 12 s of free-fall. The use of a single application of the trapezoidal rule to approximate this area is shown. The area between the curve and the shaded portion represents the error of the trapezoidal rule approximation.

$$y(t) = \int_0^{12} 53.39[1 - e^{-0.18355t}]\, dt$$

Using calculus, this integral can be evaluated analytically to give an exact result of 381.958 m. This exact value will allow us to assess the accuracy of the trapezoidal-rule approximation.

In order to implement the trapezoidal rule, the values at the end points can be substituted into Eq. (28.4) to give

$$I \cong (12 - 0)\frac{0 + 47.49}{2} = 289.94 \text{ m}$$

which represents a percent relative error of

$$\text{Error} = \frac{381.958 - 284.94}{381.958} 100\% = 25.4\%$$

The reason for this large error is evident from the graphical depiction in Fig. 28.8. Notice that the area under the straight line neglects a significant portion of the integral lying above the line.

One way to improve the accuracy of the trapezoidal rule is to divide the integration interval from a to b into a number of segments and apply the method to each segment. The areas of the individual segments can be added to yield the total integral. Figure 28.9 shows the general format and nomenclature we will employ for the multiple-segment trapezoidal rule. Notice that the integration interval is divided into n segments. If, as shown in Fig. 28.9, a and b are designated as x_0 and x_n, the total integral is

FIGURE 28.9
Nomenclature used for the multiple-segment trapezoidal rule.

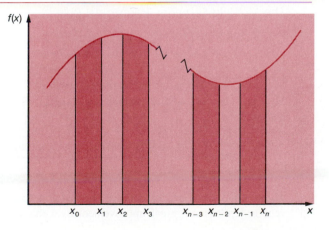

$$I = h_1 \frac{f(x_0) + f(x_1)}{2} + h_2 \frac{f(x_1) + f(x_2)}{2}$$

$$+ \ldots + h_n \frac{f(x_{n-1}) + f(x_n)}{2} \tag{28.5}$$

where h_i is the width of segment $i = x_i - x_{i-1}$. For the case where the segments are of equal width,

$$h = \frac{b - a}{n} \tag{28.6}$$

and the integral can be expressed more concisely by grouping terms, as in

$$I = \frac{h}{2}\left[f(x_0) + 2\sum_{i=1}^{n-1} f(x_i) + f(x_n) \right]$$

or, substituting Eq. (28.6),

$$I = (b - a)\frac{f(x_0) + 2\sum_{i=1}^{n-1} f(x_i) + f(x_n)}{2n} \tag{28.7}$$

Because the summation of the coefficients of $f(x)$ in the numerator divided by $2n$ is equal to 1, the average height is represented by a weighted average of the function values. According to Eq. (28.7), the interior points are given twice the weight of the two end points, $f(x_0)$ and $f(x_n)$.

EXAMPLE 28.2   The Multiple-Segment Trapezoidal Rule

PROBLEM STATEMENT: Solve the same problem given in Example 28.1 but use a three- and a six-segment trapezoidal rule to determine the integral.

SOLUTION: Because the data used to generate the plot in Example 28.1 is equispaced, Eq. (28.7) can be used to evaluate the integral. For the three-segment trapezoidal rule, the result is

$$I = (12 - 0)\frac{0 + 2(27.77 + 41.10) + 47.49}{6} = 370.49$$

which represents a percent relative error of 3 percent. Thus using the multiple-segment version represents a great improvement over the single-segment result of 25.4 percent obtained previously in Example 28.1. The improvement becomes even more pronounced for the six-segment version:

$$I = (12 - 0)\frac{0 + 2(16.40 + 27.77 + 35.64 + 41.10 + 44.87) + 47.49}{12}$$

$$= 379.05$$

which represents an error of 0.76 percent.

28.2 ALGORITHM FOR THE TRAPEZOIDAL RULE

FIGURE 28.10
An algorithm for the multiple-segment trapezoidal rule.

$sum \leftarrow y_1$
$DO\ FOR\ i \leftarrow 2\ to\ n-1$
$\quad sum \leftarrow sum + 2\,y_i$
$LOOP$
$sum \leftarrow sum + y_n$

$height \leftarrow \dfrac{sum}{2n}$

$integral \leftarrow (x_n - x_o) \cdot height$

As seen in Fig. 28.10, it is a relatively trivial matter to develop an algorithm for the trapezoidal rule. This algorithm is set up for the equal-sized, multisegment version of the method. Notice that a loop is employed to perform the computation efficiently.

Figure 28.11 shows some screens generated using the trapezoidal-rule algorithm from the Electronic TOOLKIT. The package is used to solve Example 28.2. The solution is shown in Fig. 28.11a, and a plot is shown in Fig. 28.11b. These screens illustrate how output can be designed in a clear and accessible fashion. You can employ them as models for your own software.

FIGURE 28.11 Computer screens generated by the trapezoidal-rule module of the Electronic TOOLKIT. The package is used to solve the problem previously analyzed in Example 28.2.

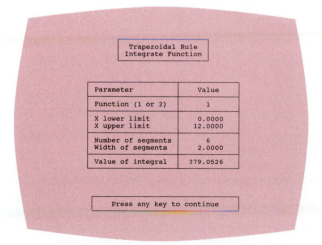

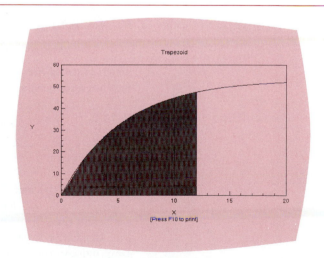

(a)   *(b)*

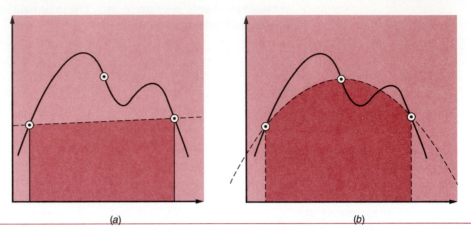

FIGURE 28.12
(*a*) The trapezoidal rule establishes the area under the straight line connecting the function values at the end points. (*b*) Simpson's 1/3 rule establishes the area under the parabola connecting values at each end and one in the middle.

(a)   (b)

28.3   ADVANCED METHODS

In the present chapter we used a simple geometric derivation to develop the trapezoidal-rule formula (see Fig. 28.12*a*). An alternative derivation involves fitting a straight line between the function values at the end points with a Lagrange interpolating polynomial [Eq. (24.4)]. This equation can be integrated analytically, and the result is the same as Eq. (28.4).

Taking this reasoning a step farther, we can develop a more accurate estimate by fitting a Lagrange parabola to three function values at the ends of the integration interval and at the midpoint (Fig. 28.12*b*). When the parabolic equation is integrated, the following formula results:

$$I = (b - a)\frac{f(x_0) + 4f(x_1) + f(x_2)}{6} \tag{28.8}$$

which is called *Simpson's 1/3 rule*. Just as was the case with the trapezoidal rule, a multisegment version can also be developed, as in

$$I = (b - a)\frac{f(x_0) + 4\sum_{i=1,3,5}^{n-1} f(x_i) + 2\sum_{j=2,4,6}^{n-2} f(x_j) + f(x_n)}{3n} \tag{28.9}$$

Equations (28.8) and (28.9) are much more accurate than the trapezoidal rule. In Prob. 28.3 you will have the task of developing a computer program to implement Simpson's rule.

In addition to Simpson's rule, other advanced integration techniques are available. Descriptions of these methods are found elsewhere (Chapra and Canale, 1988).

PROBLEMS

28.1 Utilize Fig. 28.10 as a starting point for developing your own computer program for the multiple-segment trapezoidal rule. Employ structured programming techniques, modular design, and internal and external documentation in your development. Incorporate the capability to integrate a function as well as discrete data points by having the program employ the function to generate a table of values.

28.2 Utilize Fig. 28.10 as a starting point for developing your own computer program for a multiple-segment trapezoidal rule. Design the program so that it can evaluate unequally spaced tabular data [Eq. (28.5)]. Employ structured programming techniques, modular design, and internal and external documentation in your development.

28.3 Repeat Prob. 28.1, but for the multiple-segment Simpson's 1/3 rule.

28.4 If the velocity is known, then the distance that a parachutist falls over a time t is given by

$$d = \int_0^t v(t)\, dt$$

If the velocity is defined by

$$v(t) = 5000\left(\frac{t}{3.75 + t}\right)$$

use numerical integration to determine how far the parachutist falls in 6 s.

28.5 The upward velocity of a rocket can be computed by the following formula:

$$v = u \ln\left(\frac{m_0}{m_0 - qt}\right) - gt$$

where v is upward velocity, u is the velocity at which fuel is expelled relative to the rocket, m_0 is the initial mass of the rocket at time $t = 0$, q is the fuel consumption rate, and g is the downward acceleration of gravity (assumed constant $= 9.8$ m/s^2). If $u = 2200$ m/s, $m_0 = 160{,}000$ kg, and $q = 2680$ kg/s, use a numerical method to determine how high the rocket will fly in 30 s.

28.6 In order to estimate the size of a new dam, you have to determine the total volume of water (in gallons) that flows down a river in a year's time. You have available to you the following long-term average data for the river:

Date	mid Jan	mid Feb	mid Mar	mid Apr	mid June	mid Aug	mid Oct	mid Nov	mid Dec
Flow, ft^3/s	1100	1300	2200	4200	3600	700	750	800	1000

Determine the volume. Be careful of units. Note that there are 7.481 gal/ft^3, and take care to make a proper estimate of flow at the end points.

28.7   You normally jog down a country road and want to determine how far you run. Although the odometer on your car is broken, you can use the speedometer to record the car's velocity at 1-min intervals as you follow your jogging route:

Time, min	0	1	2	3	4	5	6	6.33
Velocity, mi/h	0	35	40	48	50	55	32	36

Use this data to determine how many miles you jog. Be careful of units.

28.8   A transportation engineering study requires the calculation of the total number of cars that pass through an intersection over a 24-h period. An individual visits the intersection at various times over the course of the day and counts the number of cars that pass through the intersection in a minute. Utilize this data to estimate the total number of cars that pass through the intersection per day. (Be careful of units, and use the most accurate numerical method whenever possible.)

Time	Rate, cars/min	Time	Rate, cars/min
12:00 midnight	2	12:30 P.M.	19
2:00 A.M.	2	2:00 P.M.	7
4:00 A.M.	0	4:00 P.M.	9
5:00 A.M.	2	5:00 P.M.	20
6:00 A.M.	5	6:00 P.M.	26
7:00 A.M.	18	7:00 P.M.	12
8:00 A.M.	27	8:00 P.M.	9
9:00 A.M.	12	9:00 P.M.	15
10:30 A.M.	15	10:00 P.M.	8
11:30 A.M.	17	11:00 P.M.	5
		12:00 midnight	3

28.9   A rod subject to an axial load (Fig. P28.9a) will be deformed as shown in the stress-strain curve in Fig. P28.9b. The area under the curve from zero stress out to the point of rupture is called the *modulus of toughness* of the material. It provides a measure of the energy per unit volume required to cause the material to rupture. As such, it is representative of a structure's ability to withstand an impact load. Use numerical integration to compute the modulus of toughness for the stress-strain curve seen in Fig. P28.9b.

28.10 Employ the multiple-segment trapezoidal rule to evaluate the vertical distance traveled by a rocket if the vertical velocity is given by

$$v = 10t^2 \qquad\qquad 0 \le t \le 10$$

$$v = 1000 - 5t \qquad 10 \le t \le 20$$

$$v = 45t + 2(t - 20)^2 \qquad 20 \le t \le 30$$

s	e
0.02	40.0
0.05	37.5
0.10	43.0
0.15	52.0
0.20	60.0
0.25	55.0

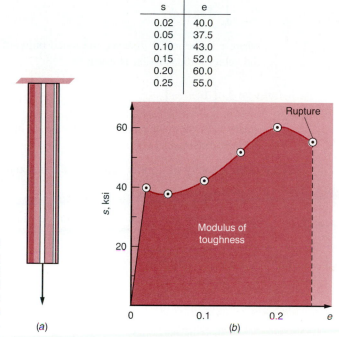

FIGURE P28.9
(a) A rod under axial loading and (b) the resulting stress-strain curve, where stress is in kips per square inch (10^3 lb/in^2) and strain is dimensionless.

28.11 The work done on an object is equal to the force times the distance moved in the direction of the force. The velocity of an object in the direction of a force is given by

$$v = 5t \qquad\qquad 0 \le t \le 7$$

$$v = 35 + 2 * (7 - t)^2 \qquad 7 \le t \le 14$$

Employ the multiple-segment trapezoidal rule to determine the work if a constant force of 200 lb is applied for all t.

28.12 If the velocity distribution of a fluid flowing through a pipe is known (Fig. P28.12), the flow rate Q (that is, the volume of water passing through the pipe per unit time) can be computed by $Q = \int v \, dA$, where v is the velocity and A is the pipe's cross-sectional area. (To grasp the meaning of this relationship physically, recall the close connection between summation and integration.) For a circular pipe, $A = \pi r^2$ and $dA = 2\pi r \, dr$. Therefore

FIGURE P28.12

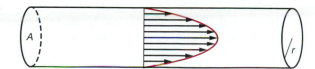

$$Q = \int_0^r v(2\pi r)\, dr$$

where r is the radial distance measured outward from the center of the pipe. If the velocity distribution is given by

$$v = 3.0\left(1 - \frac{r}{r_0}\right)^{1/7}$$

where r_0 is the total radius (in this case, 3 in), compute Q using the multiple-segment trapezoidal rule.

28.13 If the velocity of a roller coaster in the horizontal direction is given by

$$v(t) = (60 - t)^2 + (60 - t)\sin\sqrt{t}$$

determine the horizontal distance traveled in 60 s using the multiple-segment trapezoidal rule.

28.14 The exponential integral

$$E = \int_a^b \frac{e^{-x}}{x}\, dx$$

arises frequently in engineering analysis. Evaluate this integral from $a = 1$ to $b = 5$ by the multiple-segment trapezoidal rule.

28.15 One method for improving integral estimates is to perform the integration twice—once with a certain number of segments (n) and again with twice as many segments ($2n$). It is then possible to combine these two values to obtain a third, improved estimate. This type of approach is known as *Richardson extrapolation*. [See Chapra and Canale (1988) for additional information.] For the trapezoidal rule, the formula to compute the improved estimate I_3 is

$$I_3 = \tfrac{4}{3}I_2 - \tfrac{1}{3}I_1$$

where I_2 is the integral value using $2n$ segments and I_1 is the value using n segments. Employ this approach to obtain the integral of

$$\int_{-3}^3 x^4 + 2x^2 + 4x - 4\, dx$$

by using the multiple-segment trapezoidal rule with $n = 6$ to compute I_1 and $n = 12$ to compute I_2. Determine the percent relative error for I_1, I_2, and I_3 (the exact value of the integral can be computed with calculus as 109.2) in order to substantiate that the Richardson extrapolation does, in fact, lead to superior results.

28.16 The following relationship can be used to compute the height (that is, frequency) of a bell-shaped normal distribution:

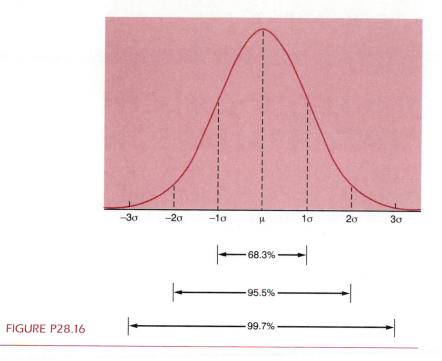

$$f(y) = \frac{1}{\sigma\sqrt{2\pi}} e^{-(y-\mu)^2/2\sigma^2} \tag{P28.16}$$

The area under the curve specified by this equation is equal to the probability that a random event would occur between two values of y (see Fig. P28.16). Furthermore, the area from $y = -\sigma$ to $y = \sigma$ is equal to a 68.3 percent probability. That is,

$$\int_{-\sigma}^{\sigma} \frac{1}{\sigma\sqrt{2\pi}} e^{-(y-\mu)^2/2\sigma^2}\, dy = 0.683$$

Employ the multiple-segment trapezoidal rule to verify that this is correct for $\mu = 0$ and $\sigma = 1$.

COMPUTER CALCULUS: RATE EQUATIONS

The subject of ordinary differential equations is traditionally not taught to freshman engineers. This is not unreasonable because a student usually needs several prerequisite college mathematics courses to fully grasp and appreciate the rich theory associated with the subject. Although it is generally true that ordinary differential equations are usually taken after freshman year, it should be noted that most of the subject's complexity is associated with obtaining exact analytical solutions. In contrast, the simpler numerical solutions for solving ordinary differential equations with the computer are, in fact, quite easy to understand. For this reason, and because of their prominence in computer mathematics, we have chosen to devote the present chapter to solving simple ordinary differential equations. This material will provide you with an introduction to the subject—an introduction that will, we hope, stimulate and motivate your future studies of the theory associated with differential equations. Before discussing how rate equations can be solved with a computer, we will first orient you by describing what they are mathematically and where they originate in engineering.

29.1   WHAT IS A RATE EQUATION?

In order to illustrate what we mean by a rate equation, let us start with a given function:

$$y = -0.5x^4 + 4x^3 - 10x^2 + 8.5x + 1 \tag{29.1}$$

where y is the dependent variable and x is the independent variable. Equation (29.1) is a fourth-order polynomial. As shown in Fig. 29.1a, it can be used to compute values of y for corresponding values of x. Thus the equation characterizes the relationship between y and x. Now if we apply the rule expressed previously in Eq. (27.2), we can differentiate Eq. (29.1) to give

$$\frac{dy}{dx} = -2x^3 + 12x^2 - 20x + 8.5 \tag{29.2}$$

This equation also characterizes the relationship between y and x but in a different manner from that of Eq. (29.1). Rather than explicitly representing the value of y for each value of x, Eq. (29.2) provides the rate of change of y with respect to x (that is, the slope) for each value of x. Figure 29.1 shows plots of both the function and its derivative.

Equations that contain derivatives, such as Eq. (29.2), are called *differential equations*. Although there are various types, the kind represented by Eq. (29.2) is called a *first-order ordinary differential equation*. The adjective "first-order" refers to the fact that the highest derivative in the equation is a first derivative. The adjective "ordinary" refers to the fact that the dependent variable y is a function of only one independent variable x. First-order ordinary differential equations can usually be written in the format of Eq. (29.2)—that is, with the derivative isolated on the left side of the equal sign. Thus the equation can be used to compute the rate of change of the dependent variable with respect to the independent variable at each value of the independent variable. For this reason, we give such equations a special name—*rate equations*.

Before moving on to the place of rate equations in engineering, we will briefly elaborate on what is meant by the "solution" of a rate equation. This is important because

FIGURE 29.1
(a) Plot of y versus x and (b) plot of dy/dx versus x for the function $y = -0.5x^4 + 4x^3 - 10x^2 + 8.5x + 1$. Notice how the zero values of the derivatives correspond to the point at which the original function is flat—that is, has a zero slope. Also notice that the maximum absolute values of the derivatives are at the ends of the interval, where the slopes of the function are greatest.

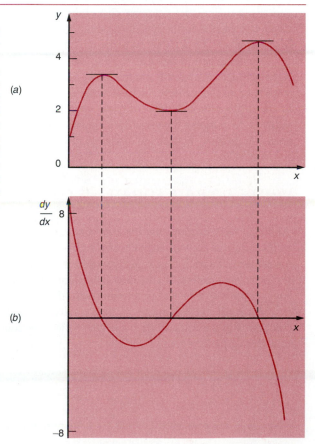

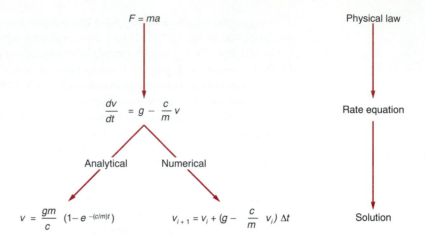

FIGURE 29.2
The sequence of events in the application of rate equations for engineering problem solving. The example shown is for the velocity of a falling parachutist.

the solution is the primary result that we will find valuable in our engineering application of rate equations. As just demonstrated in going from Eq. (29.1) to (29.2), we can generate a differential equation given the function. In most engineering work, we are usually interested in the reverse process; that is, the object is usually to determine the function given the differential equation. The function then represents the solution.

As described in the next section, the differential equation is usually derived from a physical law. It typically characterizes the rate of change of a variable of interest with respect to an independent variable such as distance or time. The object of the exercise or the solution is, therefore, to come up with a nondifferential equation or function to allow us to compute the variable of interest itself as a function of the independent variable. The process is depicted schematically in Fig. 29.2.

29.2   WHERE DO RATE EQUATIONS COME FROM?

Although rate equations are the simplest type of differential equations, they have wide applicability in engineering. This is because many of the fundamental laws of physics, mechanics, electricity, and thermodynamics explain *variations* in physical properties and states of systems. Rather than describing the *state* of physical systems directly, the laws are usually couched in terms of spatial and temporal *changes*.

A simple example relates to the velocity of a falling parachutist. At several previous points in the text, we have used the following equation to characterize the velocity of a falling parachutist:

$$v = \frac{gm}{c}\left[1 - e^{-(c/m)t}\right] \tag{29.3}$$

where m is the parachutist's mass (in kilograms), c is a drag coefficient (in kilograms per second), and t is time (in seconds). We will now demonstrate how this relationship is actually the solution of a rate equation derived from Newton's second law of motion. Although this example is very simple, it is a good illustration of how rate equations are derived from physical laws in engineering.

As stated above, the derivation starts with Newton's second law:

$$F = ma \qquad (29.4)$$

where F is the net force acting on a body (in newtons, or kilogram-meters per second squared) and a is its acceleration (in meters per second squared). This law can be formulated as a differential equation by expressing the acceleration as the time rate of change of the velocity (dv/dt) and dividing by m to yield

$$\frac{dv}{dt} = \frac{F}{m} \qquad (29.5)$$

where v is velocity (in meters per second). Thus by isolating the derivative on the left side of the equal sign, we have expressed Newton's second law in the general format of a rate equation.

Next we must express the net force in terms of measurable variables and parameters. As shown in Fig. 29.3, the net force is composed of two individual forces, gravity and air resistance. We have already developed quantitative relationships for these forces in Chap. 26. We recall [Eq. (26.2)] that the downward force of gravity is represented by

$$F_D = mg$$

whereas, according to Eq. (26.3), the upward force of air resistance can be represented by

$$F_U = -cv$$

Therefore the net force is obtained by adding these individual forces:

$$F = mg - cv$$

This result can then be substituted into Eq. (29.5) to give

$$\frac{dv}{dt} = g - \frac{c}{m}v \qquad (29.6)$$

FIGURE 29.3
Schematic diagram of the forces acting on a falling parachutist. F_D is the downward force due to gravity; F_U is the upward force due to air resistance.

Equation (29.6) is a model that relates the acceleration of a falling object to the forces acting on it. It is also a rate equation that is similar in form to Eq. (29.2). That is, it characterizes the rate of change of a dependent variable v with respect to a single independent variable t. As mentioned previously, the "solution" for such an equation is another equation that characterizes the values of the dependent variable itself with respect to the independent variable. There are two fundamental approaches to obtaining such solutions, analytical and numerical.

An analytical or exact solution for Eq. (29.6) is obtained using the methods of calculus. For example, if the parachutist is originally at rest ($v = 0$ at $t = 0$), calculus can be used to solve Eq. (29.6) for

$$v = \frac{gm}{c}[1 - e^{-(c/m)t}]$$

(29.7)

Thus you can now see that this equation, which we have used on several previous occasions to characterize the parachutist's velocity, is actually the solution of a differential equation.

EXAMPLE 29.1

Analytical Solution for the Falling Parachutist Problem

PROBLEM STATEMENT:　A parachutist with a mass of 68.1 kg jumps out of a stationary hot air balloon. Use Eq. (29.7) to compute velocity prior to opening the chute. The drag coefficient is equal to 12.5 kg/s.

SOLUTION:　Inserting the parameters into Eq. (29.7) yields

$$v = \frac{9.8(68.1)}{12.5}[1 - e^{-(12.5/68.1)t}]$$

$$= 53.39[1 - e^{-0.18355t}]$$

which can be used to compute

t, s	v, m/s
0	0.00
2	16.40
4	27.77
6	35.64
8	41.10
10	44.87
12	47.49
∞	53.39

According to the model, the parachutist accelerates rapidly (Fig. 29.4). A velocity of 44.87 m/s (100.4 mi/h) is attained after 10 s. Note also that after a sufficiently long time, a constant velocity, called the *terminal velocity,* of 53.39 m/s (119.4 mi/h) is reached. This velocity is constant because, after a sufficient time, the force of gravity will be in balance with the air resistance. Thus the net force is zero and acceleration ceases.

Equation (29.7) is called an *analytical* or *exact solution* because it exactly satis-

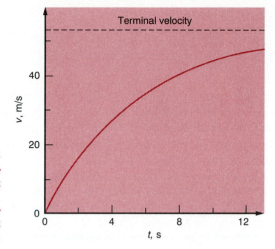

FIGURE 29.4
The analytical solution to the falling parachutist problem as computed in Example 29.1. Velocity increases with time and asymptotically approaches a terminal velocity.

fies the original differential equation. Unfortunately, there are many rate equations that cannot be solved exactly. In many of these cases, the only alternative is to develop a numerical solution that approximates the exact solution. This solution is not itself a continuous function. Rather, it is a table of numerical values that approximates $v(t)$ at various values of t. As described in the next section, the simplest technique available for this purpose is Euler's method.

29.3 EULER'S METHOD

One way to introduce Euler's method is to derive it to solve the falling parachutist problem. As should be obvious by now, the fundamental approach for solving a mathematical problem with the computer is to reformulate the problem so that it can be solved by arithmetic operations. Our one stumbling block for solving Eq. (29.6) in this way is the derivative term dv/dt. However, as we have already shown in Chap. 27, difference approximations can be used to express derivatives in arithmetic terms. For example, using a forward difference [Eq. (27.6)], the first derivative of v with respect to t can be approximated by

$$\frac{dv}{dt} \cong \frac{\Delta v}{\Delta t} = \frac{v_{i+1} - v_i}{h} \tag{29.8}$$

where v_i and v_{i+1} are velocities at a present and a future time, respectively, and h is the step size $\Delta t = t_{i+1} - t_i$. Substituting Eq. (29.8) into Eq. (29.6) yields

$$\frac{v_{i+1} - v_i}{h} = g - \frac{c}{m} v_i$$

which can be solved for

$$v_{i+1} = v_i + \left(g - \frac{c}{m} v_i \right) h \tag{29.9}$$

Notice that the term in parentheses is the rate equation itself [Eq. (29.6)]. That is, it provides a means to compute the rate of change or slope of v. Thus the differential equation has been transformed into an equation that can be used to determine the velocity algebraically at t_{i+1} using the slope and previous values of v. If you are given an initial value for velocity at some time t_i, you can easily compute velocity at a later time t_{i+1}. This new value of v at t_{i+1} can in turn be employed to extend the computation to v at t_{i+2} and so on. Thus at any time along the way,

New value = old value + slope × step size

EXAMPLE 29.2   Numerical Solution to the Falling Parachutist Problem

PROBLEM STATEMENT: Perform the same computation as in Example 29.1, but use Eq. (29.9) to compute velocity. Employ a step size of 2 s for the calculation.

SOLUTION: At the start of the computation ($t_i = 0$), the velocity of the parachutist is zero. Using this information and the parameter values from Example 29.1, Eq. (29.9) can be used to compute velocity at $t_{i+1} = 2$ s:

$$v = 0 + \left[9.8 - \frac{12.5}{68.1} 0 \right] 2 = 19.60 \text{ m/s}$$

For the next interval (from $t = 2$ to 4 s), the computation is repeated, with the result

$$v = 19.6 + \left[9.8 - \frac{12.5}{68.1} 19.6 \right] 2 = 32.00 \text{ m/s}$$

The calculation is continued in a similar fashion to obtain additional values:

t, s	v, m/s
0	0.00
2	19.60
4	32.00
6	39.85
8	44.82
10	47.97
12	49.96
∞	53.39

The results are plotted in Fig. 29.5 along with the exact solution. It can be seen that the

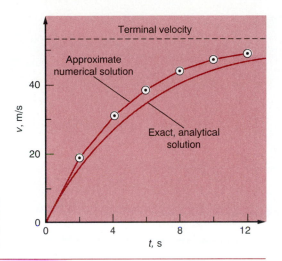

FIGURE 29.5
Comparison of the numerical and analytical solutions for the falling parachutist problem.

numerical method accurately captures the major features of the exact solution. However, because we have employed straight-line segments to approximate a continuously curving function, there is some discrepancy between the two results. One way to minimize such discrepancies is to use a smaller step size. For example, applying Eq. (29.9) at 1-s intervals results in a smaller error, as the straight-line segments track closer to the true solution. Using hand calculations, the effort associated with using smaller and smaller step sizes would make such numerical solutions impractical. However, with the aid of the computer, large numbers of computations can be performed easily. Thus you can accurately model the velocity of the falling parachutist without having to solve the differential equation exactly.

According to Eq. (29.9), the slope estimate provided by the rate equation is used to extrapolate linearly from an old value to a new value over a step size h. This approach can be represented generally as

$$y_{i+1} = y_i + \left(\frac{dy}{dx}\right)_i h \tag{29.10}$$

where $(dy/dx)_i$ represents the value of the rate equation evaluated at x_i. This formula is referred to as *Euler's* (or the *point-slope*) *method* (Fig. 29.6). As described in the next section, it is very easy to implement this method with the computer.

29.4   COMPUTER PROGRAMS FOR EULER'S METHOD

As seen in Fig. 29.7, it is a relatively simple matter to develop an algorithm for Euler's method. Notice that the heart of the algorithm is two nested loops. The inner loop performs the actual computation. The outer loop is used so that answers do not have to

```
OUTPUT  t₀, y₀
t ← t₀ ; y ← y₀
DO
  IF tf − t ≥ tp THEN
    te ← t + tp
  ELSE
    te ← tf
  END IF
  h ← tc
  DO
    IF te − t < h THEN h = te − t
    y ← y + dy/dt
    t ← t + h
  LOOP UNTIL t ≥ te
  OUTPUT t, y
LOOP UNTIL t ≥ tf
```

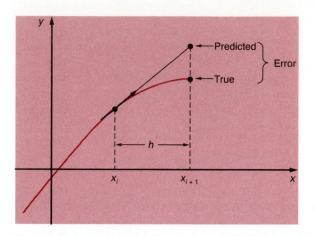

FIGURE 29.6
Graphical depiction of Euler's method.

FIGURE 29.7
Algorithm for Euler's method. Note that t_0 = the initial time, t_f = the final time, t_p = the print interval, t_c = the computation interval (equivalent to Δt), and y_0 = the initial value of the dependent variable.

be printed out for each computation step. This is important because for applications with a very small step size, the quantity of output would be enormous. Thus the algorithm in Fig. 29.7 allows you to control the output so that a reasonable number of values are printed.

A plotting option is a useful enhancement to the algorithm. If your computer system has plotting capabilities, we recommend that you expand your program to include a plot of the solution. The inclusion of this capability will greatly enhance the program's utility for problem solving.

Figure 29.8 shows some screens generated with the Electronic TOOLKIT. The package is used to solve Example 29.2. The solution is shown in Fig. 29.8a, and a plot of the results is shown in Fig. 29.8b. These screens illustrate how output can be designed in a clear and accessible fashion. You can employ them as models for your own software.

29.5 ADVANCED METHODS

Euler's method is called a first-order approach because it employs straight-line projections to approximate the trajectory of the rate equation. Higher-order methods are available that are capable of more faithfully capturing the curvature of the solution; hence they are more accurate than Euler's approach. The most common of these are called *Runge-Kutta methods*. In addition, Euler's and the higher-order techniques can be used to determine solutions of simultaneous rate equations. These are very important in engineering because they provide a way to predict the dynamics of systems of interconnected elements such as the ones shown in Fig. 26.1. Whereas the techniques in Chap. 26 focused on predicting the steady state of such systems, simultaneous differential equations are used to predict how the individual elements of such systems vary dynamically. Additional information on these and other advanced topics related to rate equations can be found elsewhere (Chapra and Canale, 1988).

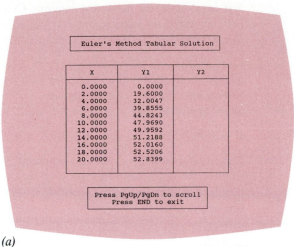

(a)

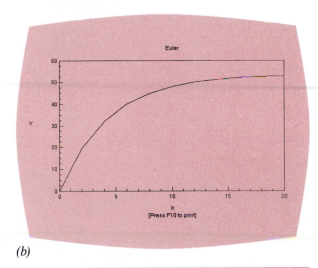

FIGURE 29.8
Computer screens
generated by the Euler's
method module of the
Electronic TOOLKIT. The
package is used to solve
the problem previously
analyzed in Example 29.2.

(b)

PROBLEMS

29.1 Utilize Fig. 29.7 as a starting point for developing your own computer program for Euler's method. Employ structured programming techniques, modular design, and internal and external documentation in your development. Incorporate the capability to plot the solution.

29.2 What is meant by a first-order ordinary differential equation? Give an example. Provide an example of a second-order ordinary differential equation.

29.3 What type of finite-divided difference is employed in Eq. (29.8)?

29.4 Repeat Example 29.2 using 1-s intervals, and compare it with the exact solution and the 2-s interval numerical solution. Compute the percent relative error for both numerical solutions at $t = 2, 4, 6,$ and 8 relative to the analytical solution. What conclusion can be drawn regarding halving the step size?

29.5 Employ Newton's law of cooling (Prob. 27.8) to calculate the temperature of a metal ball dropped into water that is held constant at $T_a = 20°C$. Compute the ball's temperature at 1-s intervals from $t = 0$ to 25 min. The ball's temperature at $t = 0$ is 90°C. Use a value of 0.1 min^{-1} for k.

29.6 Repeat Prob. 29.5, but employ a value of 0.2 per minute for k.

29.7 Population-growth dynamics are important in a variety of engineering planning studies. One of the simplest models of such growth incorporates the assumption that the rate of change of the population p is proportional to the existing population at any time t:

$$\frac{dp}{dt} = Gp \tag{P29.7}$$

where G is a growth rate (per year). This model makes intuitive sense because the greater the population is, the greater the number of potential parents will be.

At time $t = 0$ an island has a population of 10,000 people. If $G = 0.075$ per year, employ Euler's method to predict the population at $t = 20$ years, using a step size of 0.5 years. Plot p versus t on standard and semilog graph paper. Determine the slope of the line on the semilog plot. Discuss your results.

29.8 Although the model in Prob. 29.7 works adequately when population growth is unlimited, it breaks down when factors such as food shortages, pollution, and lack of space inhibit growth. In such cases, the growth rate itself can be thought of as being inversely proportional to population. One model of this relationship is

$$G = G'(p_{max} - p) \tag{P29.8}$$

where G' is a population-dependent growth rate (per people-year) and p_{max} is the maximum sustainable population. Thus when population is small ($p \ll p_{max}$), the growth rate will be at a high, constant rate of $G'p_{max}$. For such cases, growth is unlimited and Eq. (P29.8) is essentially identical with Eq. (P29.7). However, as population grows (that is, p approaches p_{max}), G decreases until at $p = p_{max}$ it is zero. Thus the model predicts that when the population reaches the maximum sustainable level, growth is nonexistent and the system is at a steady state. Substituting Eq. (P29.8) into Eq. (P29.7) yields

$$\frac{dp}{dt} = G'(p_{max} - p)p$$

For the same island studied in Prob. 29.7, employ Euler's method to predict the population at $t = 20$ years, using a step size of 0.5 years. Employ values of $G' = 10^{-5}$ per people-year and $p_{max} = 20,000$ people. At time $t = 0$ the island has a population of 10,000 people. Plot p versus t, and interpret the shape of the curve.

29.9 For a simple ideal resistor-inductor circuit, the following rate equation holds:

$$L\frac{di}{dt} + iR = 0$$

where i is current, L is inductance, and R is resistance. Solve for i, if $L = R = 1$ and i at $t = 0$ equals 0.1 A. Employ Euler's method with a step size of 0.1, and solve for i from $t = 0$ to 1. Plot i versus t.

29.10 Real circuits often do not behave in the idealized fashion represented by Prob. 29.9. For example, circuit dynamics might be described by a relationship such as

$$L\frac{di}{dt} + (-i + i^3)R = 0$$

where all parameters are as defined in Prob. 29.9. Solve for i as a function of time under the same conditions specified in Prob. 29.9. Plot the results.

29.11 Suppose that, after falling for 10 s, the parachutist from Examples 29.1 and 29.2 pulls the rip cord. At this point, assume that the drag coefficient is instantaneously increased to a constant value of 50 kg/s. Compute the parachutist's velocity from $t = 10$ to 30 s with Euler's method. Employ all the parameters from Examples 29.1 and 29.2, and take the initial condition from the analytical solution. Plot v versus t from $t = 0$ to 30 s. Use the analytical solution for $t = 0$ to 10 s and the numerical solution obtained above for $t = 10$ to 30 s.

29.12 The rate of heat flow (conduction) between two points on a cylinder heated at one end is given by

$$\frac{dQ}{dt} = \lambda A\frac{dT}{dx}$$

where λ is a constant, A is the cylinder's cross-sectional area, Q is heat flow, T is temperature, t is time, and x is the distance from the heated end. Because the equation involves two derivatives, we will simplify it by letting

$$\frac{dT}{dx} = \frac{100(L - x)(20 - t)}{100 - xt}$$

where L is the length of the rod. Combining the two equations gives

$$\frac{dQ}{dt} = \lambda A\frac{100(L - x)(20 - t)}{100 - xt}$$

Employ Euler's method to compute the heat flow for $t = 0$ to 15 s, if $\lambda = 0.3$ cal $\cdot$ cm/s $\cdot$ °C, $A = 10$ cm^2, $L = 20$ cm, and $x = 2.5$ cm. The initial condition is that $Q = 0$ at $t = 0$. Plot your results.

29.13 Recognizing that $dx/dt = v$, solve Prob. 28.10 with Euler's method.

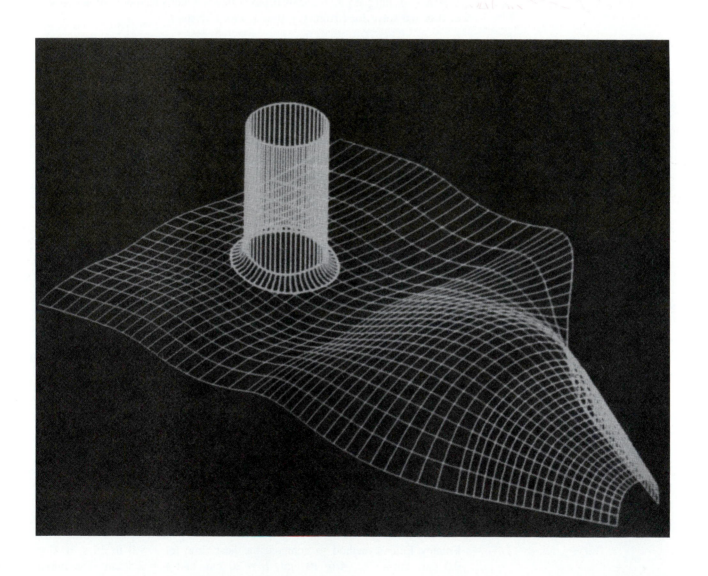

ENGINEERING APPLICATIONS

In the previous part of the book, we reviewed numerical methods for solving mathematical problems with the computer. Along with the methods, we included examples and end-of-the-chapter problems drawn from engineering practice. The present part of the book is intended to examine such computer-oriented applications in more depth.

All of the following chapters focus on mathematical modeling. Rather than using a general example, such as the falling parachutist from Part Five, each chapter here focuses on specific applications from the four major areas of engineering: chemical, civil, mechanical, and electrical. In each case we present an organizing principle that lies at the heart of computer modeling for the area.

Chapter 30 offers an overview of the major organizing principles of engineering. *Chapter 31* is devoted to mass-balance models for reactors in chemical engineering. In *Chap. 32*, which focuses on civil engineering, force balances are used to analyze a structure. *Chapter 33* also applies a force balance, but uses it to predict the dynamics of a machine. *Chapter 34* employs Kirchhoff's laws to perform the electrical engineering problem of network analysis.

ORGANIZING PRINCIPLES IN ENGINEERING

Knowledge and understanding lie at the heart of the effective implementation of any tool. No matter how impressive your tool chest, you will be hard-pressed to repair a car if you do not understand how it works.

The same holds true for the tools that are used for engineering problem solving. To this point, you have been introduced to several of these tools: computer hardware and software, graphics, sorting, statistics, and numerical methods. Although each of the tools has great potential utility, they are practically useless without a fundamental understanding of how engineering systems work.

This understanding is gained by empirical means—that is, by observation and experiment. The section of the book on data analysis (Part Four) provides techniques and concepts that play an important role in empirical studies.

Although careful observations and experiments are at the heart of our efforts to understand engineering systems, they are only half the story. Over years and years of observation and experiment, engineers and scientists have noticed that certain aspects of their empirical studies occur repeatedly. Such general behavior can then be expressed as fundamental laws, or organizing principles that essentially embody the cumulative wisdom of past experience. Thus most engineering problem solving employs the two-pronged approach of empiricism and theoretical analysis.

It must be stressed that the two prongs are closely coupled. As new measurements are taken, the generalizations may be modified or new ones developed. Similarly, the generalizations can have a strong influence on the experiments and observations. In particular, generalizations can serve as organizing principles that can be employed to synthesize observations and experimental results into a coherent and comprehensive framework from which conclusions can be drawn. Among the most important of these organizing principles are the conservation laws.

30.1 CONSERVATION LAWS

Although they form the basis for a variety of complicated and powerful mathematical models, the great conservation laws of science and engineering are very simple to understand. They all boil down to

$$\text{Change} = \text{increases} - \text{decreases} \tag{30.1}$$

This equation can be understood using an example from everyday life—your bank account.

Suppose that you apply Eq. (30.1) to your checking account over the course of a month. If you deposit more money than you withdraw, your balance at the end of the month will be greater than at the beginning. According to Eq. (30.1), your account will have a positive change because the increases (deposits) are greater than decreases (withdrawals). Conversely, if you withdraw more than you deposit, the change will be negative, and your account will be less at the month's end. In either case, the change, as calculated by Eq. (30.1), can be used to determine the outcome.

EXAMPLE 30.1   *Conservation of Cash*

PROBLEM STATEMENT: Suppose you have $150 in your account at the beginning of the month. Use Eq. (30.1) to compute your balance at the end of the month. During this time interval, you make deposits of $40, $30, and $150 and withdraw $75 and $15.

SOLUTION: First we can reformulate Eq. (30.1) in terms of the present problem context:

Change in dollars = total deposits − total withdrawals

This equation is now an expression of the conservation of cash. It provides an organizing principle for computing the change in dollars in your account:

Change in dollars = (40 + 30 + 150) − (75 + 15) = 130

This result can then be used to compute the dollars in the account at the end of the month using the formula

Dollars at end of month = dollars at start of month + change in dollars

or

Dollars at end of month = 150 + 130 = 280

Therefore, we have calculated that there is $280 in the account at the end of the month.

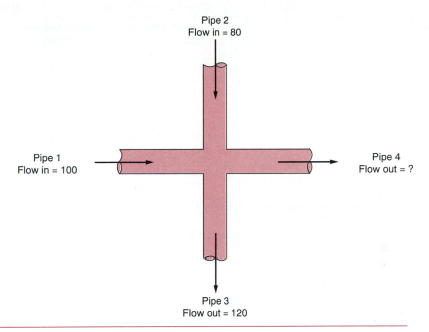

Pipe 2
Flow in = 80

Pipe 1
Flow in = 100

Pipe 4
Flow out = ?

Pipe 3
Flow out = 120

FIGURE 30.1
A flow balance for incompressible fluid flow at the junction of pipes.

Although simple, the foregoing example embodies one of the most fundamental ways in which conservation laws are used in engineering—that is, to predict changes. We will give it a special name—the *transient* or *time-variable computation.*

Aside from predicting changes, another way in which conservation laws are applied is for cases where change is nonexistent. If change is zero, Eq. (30.1) becomes

Change = 0 = increases − decreases

or

Increases = decreases

Thus, if no change occurs, the increases and decreases must be in balance. This case, which is also given a special name—the *steady-state computation*—has many applications in engineering. For example, for incompressible fluid flow in pipes, the flow into a junction must be balanced by flow going out, as in

Flow in = flow out   (30.2)

For the junction in Fig. 30.1, Eq. (30.2) can be used to compute that the flow out of the fourth pipe must be 60.

Although Fig. 30.1 and Example 30.1 might appear trivially simple, they embody

TABLE 30.1 Devices and Types of Balances That Are Commonly Used in the Four Areas of Engineering*

Field	Device	Organizing Principle	Mathematical Expression
Chemical Engineering		Conservation of mass	Mass balance: Input Over a unit time period $\Delta\text{mass} = \text{inputs} - \text{outputs}$
Civil engineering	Structure	Conservation of momentum	Force balance: At each node Σ horizontal forces $(F_H) = 0$ Σ vertical forces $(F_V) = 0$
Mechanical engineering	Machine	Conservation of momentum	Force balance: $F = ma$ $m\dfrac{d^2x}{dt^2} = \text{downward force} - \text{upward force}$
Electrical engineering	Circuit	Conservation of charge	Charge balance: For each node Σ current $(i) = 0$
		Conservation of energy	Voltage balance: Around each loop Σ emf's $- \Sigma$ voltage drops for resistors $= 0$ $\Sigma \xi - \Sigma iR = 0$

*For each case, the conservation law upon which the balance is based is specified.

the two fundamental ways in which conservation laws are applied in engineering. A major objective of the present part of the book is to develop some examples of how this is done.

30.2   BALANCES IN ENGINEERING

In the previous section we showed how conservation laws can be used to develop balances. These, in turn, can be applied to compute both transient and steady-state solutions for engineering problems. In the following chapters, we will develop four such balances for the major areas of engineering: chemical, civil, mechanical, and electrical. For each case we will focus our discussion on a particular device that is fundamental to that area of engineering. The devices and types of balances are summarized in Table 30.1.

Chapter 31, which relates to chemical engineering, focuses on mass balances of reactors. The mass balance derives from conservation of mass. It specifies that the change of mass in the reactor depends on the amount of mass flowing in minus the mass flowing out. We will develop equations from this principle that can be employed for both transient and steady-state computations.

Chapters 32 and 33 both employ force balances that are essentially derived from the conservation of momentum. In Chap. 32, a force balance is utilized for the analysis of a truss. For this case, we assume that the truss is not in motion, and therefore the summation of vertical and horizontal forces at each node of the truss must equal zero. Thus this example focuses on steady-state behavior of the structure. In contrast, Chap. 33 applies the same basic principle but uses it to analyze the up-and-down motion or vibrations of an automobile. For this case, the rate of change of the vertical motion is equal to the difference between downward and upward forces. Thus this example focuses on the transient behavior of the vehicle.

Finally, in Chap. 34 both charge and energy balances are used to model an electric circuit. The charge balance, which derives from conservation of charge, is similar in spirit to the flow balance depicted in Fig. 30.1. Just as flow must balance at the junction of pipes, electric current must balance at the junction of electric wires. The energy balance specifies that the change of energy around any loop of the circuit must add up to zero. These balances will be used to compute both transient and steady-state solutions for electric circuits.

Although there are many other examples of the application of balances, the four outlined in Table 30.1 are fundamental enough to introduce you to the way in which conservation laws serve as organizing principles in engineering problem solving.

30.3   THE ENGINEERING PROBLEM-SOLVING PROCESS

While laws and idealizations are important organizing concepts, they are but one facet of the total scheme of engineering problem solving (Fig. 30.2). Therefore, although the

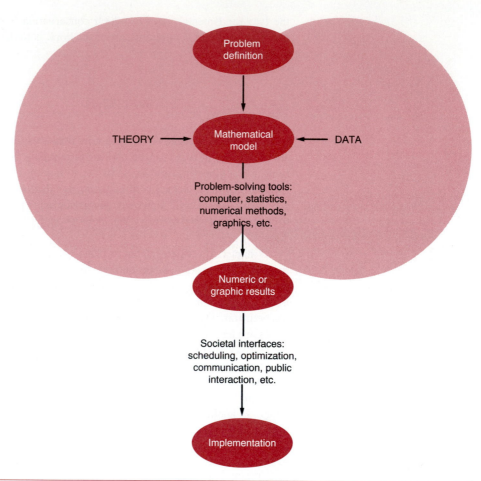

balances serve as the focus of the upcoming chapters, we will also attempt to demonstrate how other tools such as data analysis, numerical methods, and graphics figure in problem solving. Of course, our primary intent will be to illustrate how computers enhance the entire process.

PROBLEMS

30.1 What is the two-pronged approach to engineering problem solving? Into what category should the conservation laws be placed?

30.2 What is the form of the transient conservation law? What is it for steady state?

30.3 The following information is available for a bank account:

Date	Deposits	Withdrawals	Balance
5/1			512.33
	200.13	427.26	
6/1			
	206.80	328.61	
7/1			
	450.25	206.80	
8/1			
	127.31	380.61	
9/1			

Use the conservation of cash to compute the balance on 6/1, 7/1, 8/1, and 9/1. Show each step in the computation. Is this a steady-state or a transient computation?

30.4 Give examples of conservation laws in engineering and in everyday life.

CHAPTER 31

CHEMICAL ENGINEERING: MASS BALANCE

Chemical engineers use reactors to concoct a variety of products, ranging from synthetic detergents to drugs. In the present chapter we will use numerical methods to analyze the behavior of systems of reactors.

It should be noted that other areas of engineering also employ reactor models. For example, environmental engineers use them for a variety of tasks including waste treatment and the prediction of the fate and transport of pollutants in the water, soil, and atmosphere. Biomedical engineers employ reactor models for a number of purposes including the study of how medicines are absorbed and purged by the body. Thus, aside from introducing you to chemical engineering, the present chapter will acquaint you with the use of computers in these other exciting areas.

31.1 USING LINEAR ALGEBRAIC EQUATIONS IN THE STEADY-STATE ANALYSIS OF A SERIES OF REACTORS

One of the most important organizing principles in chemical engineering is the *conservation of mass*. In quantitative terms, the principle is expressed as a mass balance that accounts for all sources and sinks of a material that pass in and out of a unit volume (Fig. 31.1). Over a finite period of time, this can be expressed as

$$\text{Accumulation} = \text{inputs} - \text{outputs} \qquad (31.1)$$

The mass balance represents a bookkeeping exercise for the particular substance being modeled. For the period of the computation, if the inputs are greater than the outputs, the mass of the substance within the volume increases. If the outputs are greater than the inputs, the mass decreases. If inputs are equal to the outputs, accumulation is zero and mass remains constant. For this stable condition, or *steady state,* Eq. (31.1) can be expressed as

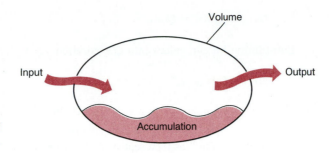

FIGURE 31.1
A schematic representation of mass balance.

Inputs = outputs (31.2)

The mass balance can be used for engineering problem solving by expressing the inputs and outputs in terms of measurable variables and parameters. For example, if we were performing a mass balance for a conservative substance (that is, one that does not increase or decrease due to chemical transformations) in a reactor (Fig. 31.2), we would have to quantify the rate at which mass flows into the reactor through the two inflow pipes and out of the reactor through the outflow pipe. This can be done by taking the product of the flow rate, Q (in cubic meters per minute), and the concentration c (in milligrams per cubic meter) for each pipe. For example, for pipe 1 in Fig. 31.2, $Q_1 = 2$ m^3/min and $c_1 = 25$ mg/m^3; therefore the rate at which mass flows into the reactor through pipe 1 is $Q_1c_1 = (2$ m^3/min$)(25$ mg/m$^3) = 50$ mg/min. Thus, 50 mg of chemical flows into the reactor through this pipe each minute. Similarly, for pipe 2 the mass inflow rate can be calculated as $Q_2c_2 = (1.5$ m^3/min$)(10$ mg/m$^3) = 15$ mg/min.

Notice that the concentration out of the reactor through pipe 3 is not specified by Fig. 31.2. This is because we already have sufficient information to calculate it on the basis of the conservation of mass. Because the reactor is at steady state, Eq. (31.2) holds and the inputs should be in balance with the outputs, as in

FIGURE 31.2
A steady-state, completely mixed reactor with two inflow pipes and one outflow pipe. The flows Q are in cubic meters per minute, and the concentrations c are in milligrams per cubic meter.

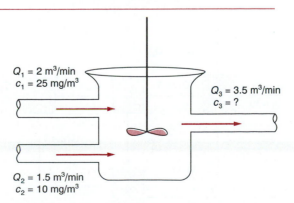

$Q_1 = 2$ m^3/min
$c_1 = 25$ mg/m^3

$Q_3 = 3.5$ m^3/min
$c_3 = ?$

$Q_2 = 1.5$ m^3/min
$c_2 = 10$ mg/m^3

$$Q_1 c_1 + Q_2 c_2 = Q_3 c_3$$

Substituting known values into this equation yields

$$50 + 15 = 3.5 c_3$$

which can be solved for $c_3 = 18.6$ mg/m³. Thus we have determined the concentration in the third pipe. However, the computation yields an additional bonus. Because the reactor is well-mixed (as represented by the propeller in Fig. 31.2), the concentration will be uniform, or homogeneous, throughout the tank. Therefore the concentration in pipe 3 should be identical with the concentration throughout the reactor. Consequently, the mass balance has allowed us to compute the concentration in both the reactor and the outflow pipe. Such information is of great utility to chemical engineers who must design reactors to yield mixtures of a specified concentration.

Because simple algebra was used to determine the concentration for the single reactor in Fig. 31.2, it might not be obvious how computers figure in mass-balance calculations. Figure 31.3 shows a problem setting where computers are not only useful but a practical necessity. Because there are five interconnected, or coupled, reactors, five simultaneous mass-balance equations are needed to characterize the system. For reactor 1, the rate of mass flow in is

$$5(10) + Q_{31} c_3$$

and the rate of mass flow out is

$$Q_{12} c_1 + Q_{15} c_1$$

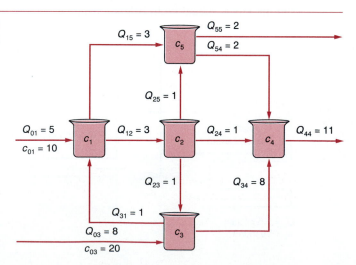

FIGURE 31.3
Five reactors linked by pipes.

Because the system is at steady state, the inflows and outflows must be equal:

$$5(10) + Q_{31}c_3 = Q_{12}c_1 + Q_{15}c_1$$

or, substituting the values for flow from Fig. 31.3,

$$6c_1 - c_3 = 50$$

Similar equations can be developed for the other reactors:

$$-3c_1 + 3c_2 = 0$$

$$-c_2 + 9c_3 = 160$$

$$-c_2 - 8c_3 + 11c_4 - 2c_5 = 0$$

$$-3c - c_2 + 4c_5 = 0$$

A numerical method such as Gauss elimination and a computer can be used to solve these five equations for the five unknown concentrations:

$$c_1 = 11.51$$

$$c_2 = 11.51$$

$$c_3 = 19.06$$

$$c_4 = 17.00$$

$$c_5 = 11.51$$

31.2 USING RATE EQUATIONS IN THE TRANSIENT ANALYSIS OF A REACTOR

In addition to steady-state computations, we might also be interested in the transient response of a completely mixed reactor. To do this, we have to develop a mathematical expression for the accumulation term in Eq. (31.1). Because accumulation represents the change in mass in the reactor per change in time, accumulation can be simply formulated as

$$\text{Accumulation} = \frac{\Delta M}{\Delta t} \tag{31.3}$$

where M is the mass of chemical in the reactor. By definition, concentration is equal to the mass per unit volume, or

$$c = \frac{M}{V}$$

This equation can be solved for mass, and the result ($M = cV$) can be substituted into Eq. (31.3) to reexpress accumulation in terms of concentration:

$$\text{Accumulation} = \frac{\Delta cV}{\Delta t}$$

If we assume that the volume of liquid in the reactor is constant, V can be moved outside the difference:

$$\text{Accumulation} = V\frac{\Delta c}{\Delta t}$$

Finally, as t approaches zero, the difference can be reexpressed as a derivative:

$$\text{Accumulation} = V\frac{dc}{dt} \tag{31.4}$$

Thus a mathematical formulation for accumulation is volume times the derivative of c with respect to t.

Now Eqs. (31.1) and (31.4) can be used to represent the mass balance for a single reactor such as the one shown in Fig. 31.4:

$$V\frac{dc}{dt} = Qc_{\text{in}} - Qc \tag{31.5}$$

This rate equation can be used to determine transient or time-variable solutions for the reactor. For example, if $c = c_0$ at $t = 0$, calculus can be employed to analytically solve Eq. (31.5) for

$$c = c_{\text{in}}[1 - e^{-(Q/V)t}] + c_0 e^{-(Q/V)t}$$

FIGURE 31.4
A single, completely mixed reactor with an inflow and an outflow.

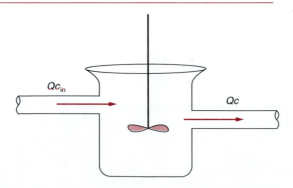

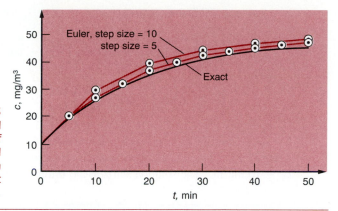

FIGURE 31.5
Plot of analytical and numerical solutions of Eq. (31.5). The numerical solutions are obtained with Euler's method using different step sizes.

If $c_{in} = 50$ mg/m³, $Q = 5$ m³/min, $V = 100$ m³, and $c_0 = 10$ mg/m³, the equation is

$$c = 50(1 - e^{-0.05t}) + 10e^{-0.05t}$$

Figure 31.5 shows this exact, analytical solution.

Euler's method provides an alternative approach for solving Eq. (31.5). Figure 31.5 includes two solutions with different step sizes. As the step size is decreased, the numerical solution converges on the analytical solution. Thus, for this case, the numerical method can be used to check the analytical result. Although the analytical solution for this case is fairly simple and straightforward, other solutions to rate equations may be quite complicated or impossible to obtain. In these cases, a numerical method that can be implemented quickly can be quite valuable in verifying the analytical solution or in providing the only viable solution.

Besides checking the results of an analytical solution, numerical solutions have added value in those situations where analytical solutions are impossible or so difficult that they are impractical. Suppose, for example, that the concentration of the inflow to the reactor is not constant but rather varies sinusoidally with time, as in (see Fig. 31.6a)

$$c_{in} = \bar{c}_{in} + c_a \sin \frac{2\pi t}{T} \tag{31.6}$$

where $\bar{c}_{in}$ is the base-inflow concentration, c_a is the amplitude of the oscillation, and T is its period (that is, the time required for a complete cycle). For this case, the mass-balance equation is

$$V\frac{dc}{dt} = Q\left(\bar{c}_{in} + c_a \sin \frac{2\pi t}{T}\right) - Qc \tag{31.7}$$

It is possible to solve this equation by calculus, but the solution is time-consuming and complicated:

$$c = c_0 e^{-(Q/V)t} + \bar{c}_{in}(1 - e^{-(Q/V)t})$$

$$+ \frac{Qc_0}{V\sqrt{(Q/V)^2 + (2\pi/T)^2}}\left[\sin\left(\frac{2\pi t}{T} - \tan^{-1}\frac{2\pi V}{TQ}\right)\right]$$

$$+ \left[e^{-(Q/V)t}\sin\left(-\tan^{-1}\frac{2\pi V}{TQ}\right)\right] \tag{31.8}$$

For such cases, Euler's approach offers a much easier means of obtaining the solution. Table 31.1 and Fig. 31.6b show the result corresponding to the situation in Fig. 31.6a for $c_0 = 10$ mg/m^3. The solution starts at the initial condition and then oscillates upward.

TABLE 31.1
Values of Concentration Calculated Using Euler's Method to Solve Eq. (31.7) with an Initial Condition of $c_0 = 10$ at $t = 0$*

t, min	c, mg/m³
0	10.00
5	21.62
10	35.00
15	47.05
20	54.73
25	56.31
30	52.16
35	44.59
40	37.07
45	32.91
50	34.06

*A step size of 0.1 min was used for the computation, and results were output every 5 min.

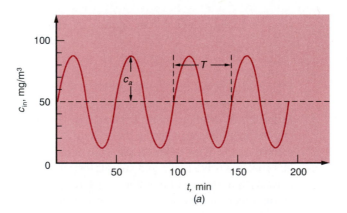

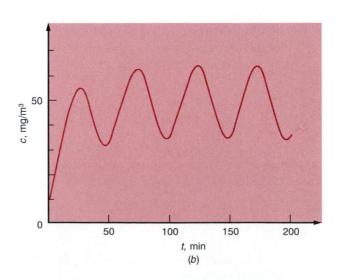

FIGURE 31.6
(a) Sinusoidal inflow concentration and (b) the resulting concentration in the reactor, as computed with Euler's method.

31.3 USING ROOTS OF EQUATIONS IN THE DETERMINATION OF A REACTOR'S RESPONSE TIME

Aside from the direct solution of the rate equation, other numerical methods can be used to investigate additional aspects of transient behavior. For example, suppose that you were interested in determining the time required for the concentration to reach 50 mg/m^3. This type of information might be needed to establish the approximate start-up time for the reactor. Computation of reactor response time can be formulated as a roots-of-equation problem. From the plot (Fig. 31.6b) we can make an approximate estimate of about 17 min. To obtain a more accurate estimate, bisection can be used to find the root of [Eq. (31.8) with numerical values substituted for the parameters]

$$f(t) = 10e^{-0.05t} + 50(1 - e^{-0.05t})$$
$$+ 14.7879[\sin{(0.1257t - 1.1921)} + 0.929e^{-0.05t}] - 50 = 0$$

Using bisection with initial guesses of $x_l = 10$, $x_u = 20$, and $\epsilon_s = 0.01$ percent yields a root of $t = 16.55$ min after 13 iterations, with an estimated error of $\epsilon_a = 7.4 \times 10^{-3}$ percent.

31.4 USING INTEGRATION IN THE DETERMINATION OF TOTAL MASS INPUT OR OUTPUT

A final application of a numerical method to the present problem context would be to compute how much mass entered or left the reactor over a specified time period. Integration provides a means to make this computation, as in

$$M = \int_{t_1}^{t_2} Qc \, dt$$

where t_1 and t_2 are the initial and final times. This formula makes intuitive sense if you recall the analogy between integration and summation. Thus the integral represents the summation of the product of flow times concentration to give the total mass entering or leaving from t_1 to t_2.

Because the flow rate for our example is constant, Q can be moved outside the integral:

$$M = Q \int_{t_1}^{t_2} c \, dt \tag{31.9}$$

Suppose that you want to compute the total mass flowing in from $t_1 = 0$ to $t_2 = 50$ min. To do this, Eq. (31.6) is substituted into Eq. (31.9) to yield (with proper parameter values)

$$M = 5 \int_0^{50} (50 + 40 \sin 0.1257t) \, dt$$

Using calculus, this integral is easy to evaluate, with the result that $M = 12,500$ mg. This result can be visually verified by examining Fig. 31.6a. Because the integral is equivalent to the area between the curve and the t axis, it is clear that the integral equals 50 mg/m^3 × 50 min, or 2500 (mg/m^3) min. Multiplying this result by the flow of 5 m^3/min yields the result of 12,500 mg.

The evaluation of the outflow of mass is not quite as easy as for the inflow. This involves substituting Eq. (31.8) into Eq. (31.9). The result is tedious to evaluate analytically. However, using the values of c computed previously with Euler's method [Table 31.1, which is the numerical solution of Eq. (31.7)] and a method such as the multiple-segment trapezoidal rule yields an estimate for the integral of 2017.4 (mg/m^3) min. This can be multiplied by the flow of 5 m^3/min to give 10,087 mg. Therefore, over the time from $t = 0$ to 50, 12,500 mg was input to the reactor, while 10,087 mg was output. Thus 2413 mg accumulated in the reactor during this time. Dividing this amount by the reactor's volume gives $c = 2413$ mg/100 m$^3 = 24.13$ mg/m^3. Adding this to the 10 mg/m^3 that was originally in the reactor gives a total of 34.13 mg/m^3. This value agrees closely with the value of 34.06 computed with Euler's method (see Table 31.1 at $t = 50$). Thus not only does the integration give us information regarding how much mass enters and leaves the reactor, but it also provides us with an independent check of the validity of our numerical solution to the rate equation.

PROBLEMS

31.1 Write the conservation-of-mass equation. What is its form at steady state?

31.2 Because the system shown in Fig. 31.3 is at steady state, what can be said regarding the four flows: Q_{01}, Q_{03}, Q_{44}, and Q_{55}?

31.3 Recompute the concentrations for the five reactors shown in Fig. 31.3, if the flows are changed to:

$$Q_{01} = 5 \qquad Q_{31} = 2 \qquad Q_{25} = 3 \qquad Q_{23} = 1$$
$$Q_{15} = 3 \qquad Q_{55} = 4 \qquad Q_{54} = 2 \qquad Q_{34} = 5$$
$$Q_{12} = 4 \qquad Q_{03} = 6 \qquad Q_{24} = 0 \qquad Q_{44} = 7$$

31.4 Solve the same system as specified in Prob. 31.3, but set Q_{12} and Q_{54} equal to zero. Use conservation of flow to recompute the values for the other flows. What does the answer indicate to you regarding the physical system?

31.5 If $c_{in} = \bar{c}(1 - e^{-t})$, calculate the outflow concentration of a single completely mixed reactor as a function of time. Use Euler's method to perform the computation. Employ values of $\bar{c} = 50$ mg/m^3, $Q = 5$ m^3/min, $V = 100$ m^3, and $c_0 = 10$ mg/m^3. Perform the computation from $t = 0$ to 50 min.

31.6 The following equation pertains for a concentration of a chemical in a completely mixed reactor:

$$c = c_{in}(1 - e^{-0.05t}) + c_0 e^{-0.05t}$$

If $c_0 = 5$ and $c_{in} = 20$, compute the time required for c to become 95 percent of c_{in}. Employ bisection.

31.7 The outflow chemical concentration from a completely mixed reactor is measured as:

t, min	0	5	10	15	20	30	40	50	60
c, mg/m^3	10	20	30	40	60	80	70	50	60

For an outflow of $Q = 10$ m^3/min, estimate the mass of chemical that exits the reactor from $t = 0$ to 60 min.

31.8 Aside from inflow and outflow, another way by which mass can enter or leave a reactor is by a chemical reaction. For example, if the chemical decays, the reaction can sometimes be characterized as a first-order reaction:

$$\text{Reaction} = -kVc$$

where k is a reaction rate (in minutes^{-1}), which can generally be interpreted as the fraction of the chemical that goes away per unit time. For example, $k = 0.1$ min^{-1} can be thought of as meaning that roughly 10 percent of the chemical in the reactor decays in a minute. The reaction can be substituted into the mass-balance equation [Eq. (31.5)] to give

$$V\frac{dc}{dt} = Qc_{in} - Qc - kVc$$

If $k = 0.1$ min^{-1}, $c_{in} = 50$ mg/m^3, $Q = 5$ m^3/min, and $V = 100$ m^3, what is the steady-state concentration of the reactor?

31.9 Repeat Prob. 31.8, but compute the transient concentration response if $c_0 = 10$ mg/m^3. Compute the response with Euler's method from $t = 0$ to 20 min.

31.10 Duplicate the computation in Secs. 31.2 and 31.4, but change c_0 to 100 mg/m^3, Q to 10 m^3/min, and V to 50 m^3.

31.11 In environmental engineering, mass-balance models have been employed to calculate the fate and transport of pollutants. These models can then be utilized to effectively manage the reduction of these contaminants. Use mass balances to determine the concentration of chloride in each of the Great Lakes using the information shown in Fig. P31.11. Employ your model to determine what would happen if the loadings to Lakes Michigan and Huron were reduced by 50 percent.

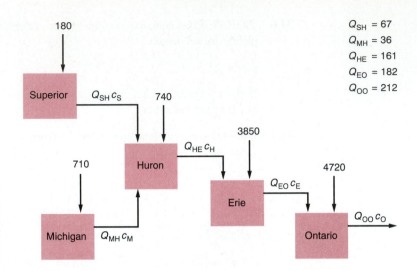

$Q_{SH} = 67$
$Q_{MH} = 36$
$Q_{HE} = 161$
$Q_{EO} = 182$
$Q_{OO} = 212$

FIGURE P31.11
A chloride balance for the Great Lakes. Numbered arrows are direct inputs (that is, flow times concentration) from waste sources.

FIGURE P31.12
A schematic showing the transfers of a drug between the blood, tissue, and an organ of an individual organism.

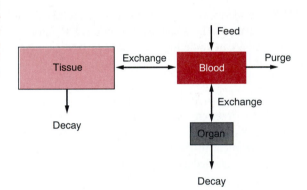

31.12 In biomedical engineering, mass-balance models have been employed to calculate the distribution of drugs within an organism's body. This area of study is called *pharmacokinetics.* As depicted in Fig. P31.12, a simple model can be set up to predict the distribution of a drug between an organism's blood, tissue, and an organ. Mass balances can be utilized to develop the following set of differential equations:

$$M_B \frac{dv_B}{dt} = F_B - k_B M_B v_B + D_{TB}(v_T - v_B) + D_{OB}(v_O - v_B)$$

$$M_T \frac{dv_T}{dt} = D_{TB}(v_B - v_T) - k_T M_T v_T$$

$$M_O \frac{dv_O}{dt} = D_{OB}(v_B - v_O) - k_O M_O v_O$$

where the subscripts B, T, and O represent the blood, tissue, and organ, respectively; t = time [min]; M = the mass [kg]; ν = the mass-specific concentration of the drug in each of the compartments [μg kg^{-1}]; F_B = the feed rate of the drug into the blood [μg h^{-1}]; D_{TB} = the mass-exchange rate of the drug between the tissue and the blood [kg h^{-1}]; D_{OB} = the mass-exchange rate of the drug between the organ and the blood [kg h^{-1}]; k_O = the decay rate of the drug in the organ [h^{-1}]; k_T = the decay rate of the drug in the tissue [h^{-1}]; and k_B = the purging rate of the drug from the blood [h^{-1}]. Determine the steady-state distribution of the drug, given the following parameter values:

$M_B = 10$ kg $\qquad\qquad\qquad\qquad D_{TB} = 0.3$ kg h^{-1}

$M_T = 50$ kg $\qquad\qquad\qquad\qquad D_{OB} = 0.1$ kg h^{-1}

$M_O = \;\; 1$ kg $\qquad\qquad\qquad\qquad k_T = 0.004$ h^{-1}

$F_B \; = 10 \; \mu$g h^{-1} $\qquad\qquad\qquad k_O = 0.01$ h^{-1}

$\qquad\qquad\qquad\qquad\qquad\qquad\quad k_B = 0.1$ h^{-1}

31.13 Repeat Prob. 31.12, but suppose that at $t = 0$, the intravenous feed of the drug is shut off ($F_B = 0$). Predict the levels of the drug in each of the compartments over the ensuing 24-h period if the initial concentrations are $\nu_B = 0.8$, $\nu_T = 0.05$, and $\nu_O = 0.1 \; \mu$g kg^{-1}.

CIVIL ENGINEERING: STRUCTURAL ANALYSIS

Civil engineering encompasses a broad variety of specialties, including structural, water resources, transportation, geotechnical, and environmental engineering. Any one of these specialties could be employed to illustrate the application of physical laws, numerical methods, and computers for engineering problem solving. However, because of its fundamental place in the education of all civil engineers, we have chosen to focus this chapter on structural analysis.

It should also be noted that other areas of engineering deal with this topic. For example, aerospace engineers must examine the effect of forces on structures such as airplanes and spacecraft. In addition, chemical engineers must be cognizant of the structural reliability of large tanks and reactors.

32.1 USING LINEAR ALGEBRAIC EQUATIONS IN THE ANALYSIS OF A STATISTICALLY DETERMINANT TRUSS

An important problem in structural engineering is that of finding the forces and reactions associated with a statically determinant truss. Figure 32.1 is an example of such a truss.

The forces (F) represent either tension or compression on the internal members of the truss. External reactions (H_2, V_2, and V_3) are forces that characterize how the truss interacts with the supporting surface. The hinge at node 2 can transmit both horizontal and vertical forces to the surface, whereas the roller at node 3 transmits only vertical forces. It is observed that an external loading of 1000 lb is applied downward at node 1. The object of this problem is to determine how the structure's members and its supports are influenced by this force. To do this, we can describe the structure as a system of coupled linear algebraic equations. Free-body-force diagrams are shown for each node in Fig. 32.2. The sum of the forces in both horizontal and vertical directions must be zero at each node because the system is at rest. Therefore for node 1,

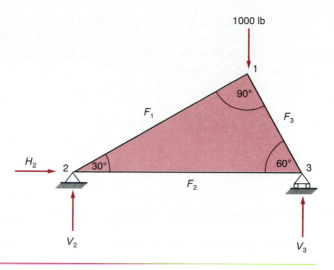

$$\sum F_H = 0 = -F_1 \cos 30° + F_3 \cos 60° + F_{1,h}$$

$$\sum F_V = 0 = -F_1 \sin 30° + F_3 \sin 60° + F_{1,v}$$

For node 2,

$$\sum F_H = 0 = F_2 + F_1 \cos 30° + F_{2,h} + H_2$$

$$\sum F_V = 0 = F_1 \sin 30° + F_{2,v} + V_2$$

For node 3,

$$\sum F_H = 0 = -F_2 - F_3 \cos 60° + F_{3,h}$$

$$\sum F_V = 0 = F_3 \sin 60° + F_{3,v} + V_3$$

where $F_{i,h}$ is the external horizontal force applied to node i (where a positive force is from left to right) and $F_{i,v}$ is the external vertical force applied to node i (where a positive force is upward). Thus in this problem, the 1000-lb downward force on node 1 corresponds to $F_{1,v} = -1000$. For this case, all other $F_{i,v}$'s and $F_{i,h}$'s are zero. Note that the directions of the forces are unknown, but proper application of Newton's laws requires only that consistent assumptions regarding direction be made. Solutions are negative if the directions are assumed incorrectly. Also note that in this problem, the forces in all members are assumed to be in tension and acting to pull adjoining nodes together. This problem can be written as the following system of six equations and six unknowns by expressing the force balances as:

$$
\begin{aligned}
0.866F_1 \qquad\quad -\ 0.5F_3 \qquad\qquad\qquad &= 0 \\
0.5F_1 \qquad\quad +\ 8.66F_3 \qquad\qquad\qquad &= -1000 \\
-0.866F_1 -\ F_2 \qquad\qquad -\ H_2 \qquad\quad &= 0 \\
-0.5F_1 \qquad\qquad\qquad\qquad -\ V_2 \quad &= 0 \\
F_2 +\ 0.5F_3 \qquad\qquad\qquad &= 0 \\
-\ 0.866F_3 \qquad -\ V_3 &= 0
\end{aligned}
\tag{32.1}
$$

Notice in Eq. (32.1) that partial pivoting is required to avoid division by zero diagonal elements. If we employ a pivot strategy, the system can be solved using Gauss elimination. The result is

$$
\begin{aligned}
F_1 &= -500 \\
F_2 &= 433 \\
F_3 &= -866 \\
H_2 &= 0 \\
V_2 &= 250 \\
V_3 &= 750
\end{aligned}
$$

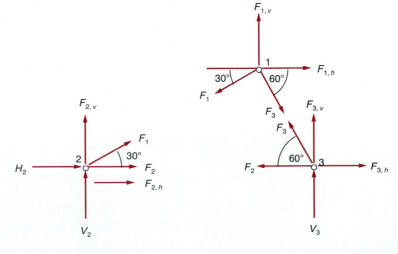

FIGURE 32.2
Free-body-force diagram for the nodes.

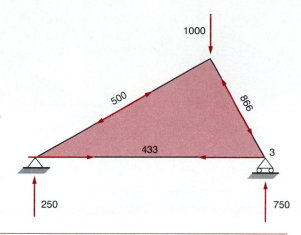

FIGURE 32.3
Forces and reactions due to a
1000-lb force acting
downward at node 1.

The negative signs indicate that members 1 and 3 are in compression. The results are shown in Fig. 32.3. This approach can be used to study the effect of different external forces on the truss. For example, we might want to study the effect of horizontal forces induced by a wind blowing from left to right. If the wind can be idealized as two horizontal point forces of 1000 lb on nodes 1 and 2, the system of equations can be solved for

$$F_1 = 866$$

$$F_2 = 250$$

$$F_3 = -500$$

$$H_2 = -2000$$

$$V_2 = -433$$

$$V_3 = 433$$

For a wind from the right, $F_{1,h} = -1000$, $F_{3,h} = -1000$, and all other external forces are zero, with the result that

$$F_1 = -866$$

$$F_2 = -1250$$

$$F_3 = 500$$

$$H_2 = 2000$$

$$V_2 = 433$$

$$V_3 = -433$$

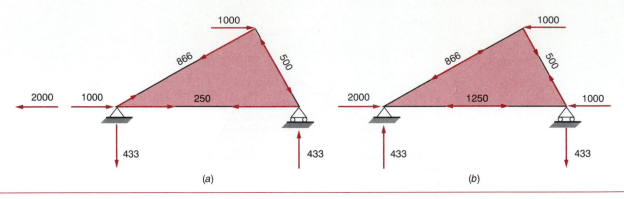

FIGURE 32.4
Two test cases showing
(a) winds from the left and
(b) winds from the right.

The results indicate that the winds have markedly different effects on the structure. Both cases are depicted in Fig. 32.4.

The foregoing method becomes particularly useful when applied to large, complicated structures. In engineering practice it may be necessary to solve trusses with hundreds or even thousands of structural members. Linear equations provide a powerful approach for gaining insight into the behavior of these structures.

32.2 USING LINEAR REGRESSION IN THE CALCULATION OF ELONGATION OF MEMBERS

Once the forces in the members of a structure are determined, a related question involves calculating the resulting elongation of an individual beam. This is based on the relationship of stress and strain. Stress s is defined as force per cross-sectional area:

$$s = \frac{F}{A} \tag{32.2}$$

whereas strain e is defined as change in length per total length:

$$e = \frac{\Delta L}{L} \tag{32.3}$$

For small deformations, Hooke's law can be used to relate stress and strain:

$$s = Ee \tag{32.4}$$

where E is a proportionality constant known as the *modulus of elasticity,* or *Young's modulus.* The region where this equation holds is called the *elastic range.* Table 32.1 and Fig. 32.5 show some strain and stress data that applies to the members in the structure in Fig. 32.1. Linear regression can be used to fit this data with the straight line:

$$s = 29866e + 0.072$$

Thus an estimate of the modulus of elasticity is $E = 29,866$ ksi. Note that the intercept is approximately zero, as would be expected from Eq. (32.4).

In order to compute a value for the change in length of a member, Eqs. (32.2) and (32.3) are substituted into Eq. (32.4):

$$\frac{F}{A} = E\frac{\Delta L}{L}$$

which can be solved for

$$\Delta L = \frac{LF}{AE} \tag{32.5}$$

This equation can be used to determine elongation for any of the members in the structure, provided the elongation remains in the elastic range. For example, for the case depicted in Fig. 32.4*b*, member 3 is in tension with a force of 500 lb. Thus, because $L = 60$ ft, $E = 29.9 \times 10^6$ lb/in^2, and $A = 0.5$ in^2, Eq. (32.5) can be used to determine that the resulting change in length is

$$\Delta L = \frac{(60 \times 12 \text{ in})(500 \text{ lb})}{(0.5 \text{ in}^2)29.9 \times 10^6 \text{ lb/in}^2} = 0.024 \text{ in}$$

TABLE 32.1
Experimentally Derived Strain and Stress Data for Beams of the Type Used to Construct the Structure in Fig. 32.1

Strain e, in/in	Stress s, ksi
0.0002	5.0
0.0002	7.0
0.0002	6.0
0.0004	11.0
0.0004	13.0
0.0004	12.5
0.0006	19.0
0.0006	17.5
0.0006	17.0
0.0008	24.6
0.0008	23.5

FIGURE 32.5
Plot of stress versus strain along with the best-fit straight line derived with linear regression.

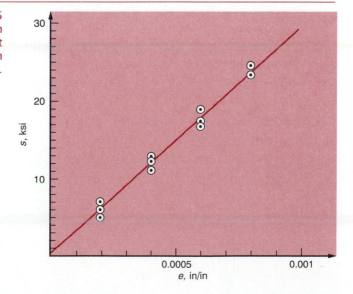

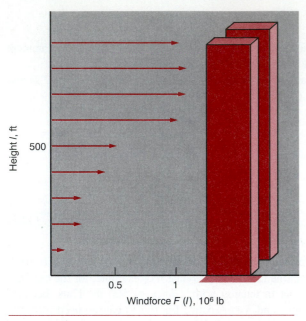

FIGURE 32.6 A distributed wind blowing against the side of a skyscraper.

TABLE 32.2
Wind Force $F(l)$ and Height Times Wind Force $lF(l)$ as a Function of Height above Street Level l for the Skyscraper in Fig. 32.6

l, ft	$F(l)$, 10^6 lb/ft	$lF(l)$, 10^6 lb
0	0.00	0
100	0.10	10
200	0.20	40
300	0.20	60
400	0.40	160
500	0.50	250
600	1.00	600
700	1.05	735
800	1.05	840
900	1.00	900

32.3   USING INTEGRATION IN THE DETERMINATION OF TOTAL FORCE AND LINE OF ACTION FOR DISTRIBUTED LOADS

Another aspect of structural analysis is that forces are often distributed rather than isolated at a joint. Figure 32.6 and Table 32.2 show the distribution of wind force along the side of a skyscraper.

Analyses of such cases can be greatly simplified if the whole distributed force can be represented as a force acting at a single point on the structure. The location of this single point is called the *line of action*. Integration can be used to determine both the total point force (f_{total}) and the line of action (d) by the following formulas:

$$f_{\text{total}} = \int_0^L F(l)\, dl \tag{32.6}$$

and

$$d = \frac{\displaystyle\int_0^L lF(l)\, dl}{\displaystyle\int_0^L F(l)\, dl} \tag{32.7}$$

where L is the total height. With the trapezoidal rule and the data from Table 32.2, Eq. (32.6) can be used to compute $f_{total} = 500 \times 10^6$ lb. Similarly, $\int lF(l)\, dl = 314{,}500$ ft·lb; and Eq. (32.7) can be used to compute $d = 314{,}500/500 = 629$ ft. Therefore, the distributed force can be represented as a point force of 500×10^6 lb acting at a point 629 ft above street level.

PROBLEMS

32.1 Calculate the forces and reactions for the truss in Fig. 32.1 if a downward force of 2000 lb and a horizontal force to the right of 1500 lb are applied at node 1.

32.2 What is Hooke's law? Give the equation, and state its meaning in your own words.

32.3 In the example for Fig. 32.1, where a 1000-lb downward force is applied at node 1, the external reactions, V_2 and V_3, were calculated. But if the lengths of the truss members had been given, we could have calculated V_2 and V_3 by utilizing the fact that $V_2 + V_3$ must equal 1000 and by summing moments around node 2. However, because we do know V_2 and V_3, we can work backward to solve for the lengths of the truss members. Note that because there are three unknown lengths and only two equations, we can solve only for the relationship between lengths. Do so.

32.4 In the free-body diagrams of Fig. 32.2, all member forces are assumed to act away from the nodes. Is this an assumption of tension or of compression of the members? If the solution to a member's force is negative, what does it indicate about the original assumption?

32.5 Why must the summation of vertical or horizontal forces at a node for a structure such as Fig. 32.1 be equal to zero?

32.6 Employing the same methods that were used to analyze Fig. 32.1, determine the forces and reactions for the truss shown in Fig. P32.6.

FIGURE P32.6

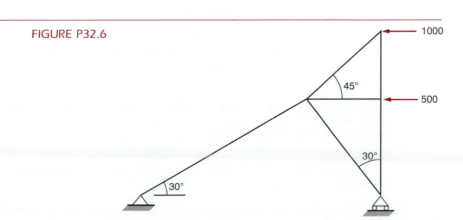

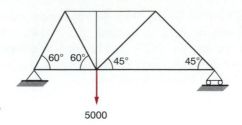

FIGURE P32.7

5000

32.7 Solve for the forces and reaction for the truss in Fig. P32.7. Does the vertical-member force in the middle member seem reasonable? Why?

32.8 If the wind loads in Table 32.2 are changed to

l	0	100	200	300	400	500	600	700	800	900
$F(l)$	0	0.3	0.2	0.4	0.5	0.6	0.9	1.0	1.0	1.2

compute the total force and the line of action.

32.9 In addition to the forces given in Prob. 32.8, a swirling wind produces a small force acting on the opposite side of the structure:

l	0	100	200	300	400
$F(l)$	0	−0.1	−0.2	−0.1	0

where the minus signs indicate that the forces act in opposition to the forces given in Prob. 32.8. Compute the net total force and the line of action due to the combined effect of both forces.

32.10 Verify the results of the linear regression for the data in Table 32.1.

32.11 Perform a linear regression for the following data obtained by subjecting a 2-in-diameter concrete cylinder to various loads and measuring the deflection. The original length of the cylinder is 18 in.

Force	0	100	200	300	400	500	600
ΔL, 10^{-2} in	0	0.011	0.020	0.029	0.035	0.040	0.042

Plot this data and the regression line, and suggest improvements that can be made in this analysis.

32.12 A wind force distributed against the side of a skyscraper is measured as

Height l, ft	Force $F(l)$, 10^3 lb/ft
0	0
100	50
200	155
300	200
400	220
500	400
600	450
700	475
750	490

Compute the net force and the line of action due to this distributed wind.

32.13 Water exerts pressure on the upstream face of a dam as shown in Fig. P32.13. The pressure can be characterized by

$$p(z) = \rho g(D - z) \tag{P32.13}$$

where $p(z)$ is pressure in pascals (or newtons per square meter) exerted at an elevation z meters above the reservoir bottom; ρ is the density of water, which for the present problem is assumed to be a constant value of 10^3 kg/m^3; g is the acceleration due to gravity (9.8 m/s^2); and D is the elevation (in meters) of the water surface above the reservoir bottom. According to Eq. (P32.13), pressure increases linearly with depth, as depicted in Fig. P32.13a. Omitting atmospheric pressure (because it works against both sides of the dam face and essentially cancels out), the total force f_t can be determined by multiplying pressure times the area of the dam face (as shown in Fig. P32.13b). Because both pressure and area vary with elevation, the total force is obtained by evaluating

FIGURE P32.13
Water exerting pressure on the upsteam face of a dam: (a) side view showing force increasing linearly with depth; (b) front view showing width of the dam in meters.

$$f_t = \int_0^D \rho g w(z)\,(D - z)\,dz$$

where $w(z)$ is the width of the dam face in meters at elevation z (Fig. P32.13b). The line of action can also be obtained by evaluating

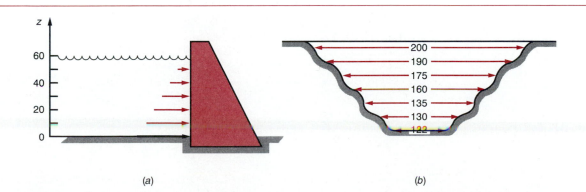

(a)   (b)

$$d = \frac{\displaystyle\int_0^D \rho g z w(z)\,(D - z)\,dz}{\displaystyle\int_0^D \rho g w(z)\,(D - z)\,dz}$$

Use Simpson's rule to compute f_t and d. Check the results with your computer program for the trapezoidal rule.

32.14 Using Gauss elimination, evaluate the forces and reactions for the structure shown in Fig. P32.14.

FIGURE P32.14

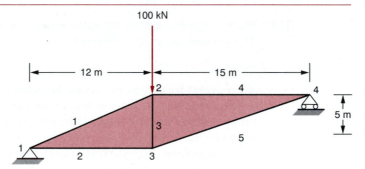

MECHANICAL ENGINEERING: VIBRATION ANALYSIS

Differential equations are often used to model the behavior of engineering systems. Harmonic oscillators form one class of such models that are broadly applicable to most fields of engineering. Some basic examples of harmonic oscillators are a simple pendulum, a mass on a spring, and an inductance-capacitance electrical circuit (Fig. 33.1). Although these are very different physical systems, their oscillations can all be described by similar mathematical models. Thus, although the present problem deals with the mechanical engineering design of an automobile shock absorber, the general approach is applicable to a variety of other problem contexts in all fields of engineering.

33.1 USING ROOTS OF EQUATIONS IN THE DESIGN OF AN AUTOMOBILE SHOCK ABSORBER

As depicted in Fig. 33.2, a sports car of mass M is supported by springs. Shock absorbers offer resistance to motion of the car that is proportional to the vertical speed (up-and-down motion) of the car. Disturbance of the car from equilibrium causes the system to move with an oscillating motion $x(t)$. At any instant, the net forces acting on M are the resistance of the spring and the damping force of the shock absorber. The resistance of the spring is proportional to a spring constant k and the distance x from equilibrium:

$$\text{Spring force} = -kx \tag{33.1}$$

where the negative sign indicates that the restoring force acts to return the car toward the position of equilibrium. Equation (33.1) is another form of *Hooke's law,* which was introduced in the previous chapter.

The damping force of the shock absorber is given by

$$\text{Damping force} = -c\frac{dx}{dt} \tag{33.2}$$

(a)

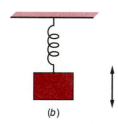

(b)

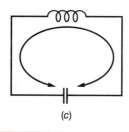

(c)

FIGURE 33.1
Three examples of simple
harmonic oscillators:
(a) pendulum,
(b) spring/mass, and (c) LC
circuit. The two-way arrows
illustrate the oscillations for
each system.

where c is a damping coefficient and dx/dt is the vertical velocity. The negative sign indicates that the damping force acts in the opposite direction against the velocity.

The equation of motion for the system is given by Newton's second law ($F = ma$), which for the present problem is expressed as the following force balance:

$$M \times \frac{d^2x}{dx^2} = -c\frac{dx}{dt} - kx \tag{33.3}$$

Mass $\times$ acceleration = damping force + spring force

or

$$\frac{d^2x}{dx^2} + \frac{c}{M}\frac{dx}{dt} + \frac{k}{M}x = 0 \tag{33.4}$$

This is a second-order linear differential equation that can be solved using the methods of calculus. For example, if the car hits a hole in the road at $t = 0$, such that it is displaced from equilibrium a distance $x = x_0$ and $dx/dt = 0$, then the solution is

$$x(t) = e^{-nt}\left(x_0 \cos pt + x_0\frac{n}{p} \sin pt\right) \tag{33.5}$$

where $n = c/(2M)$, $p = \sqrt{(k/M) - (c/2M)^2}$, and $k/M > c^2/4M^2$. Equation (33.5) gives the vertical position of the car as a function of time. The parameter values are $c = 1.4 \times 10^7$ g/s, $M = 1.2 \times 10^6$ g, $k = 1.25 \times 10^9$ g/s^2, and $x_0 = 0.3$. Substitution of these values into Eq. (33.5) gives

$$x(t) = e^{-5.8333t}[0.3 \cos (31.7433t) + 0.05513 \sin (31.7433t)]$$

Figure 33.3 is a plot of $x(t)$ versus t, as calculated from this equation. Notice that after hitting the bump, the spring's vibrations are damped out. Mechanical-engineering design considerations require that estimates be provided for the first three times the car passes through the equilibrium point.

FIGURE 33.2
A sports car of mass M.

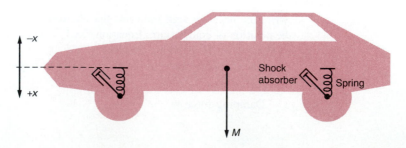

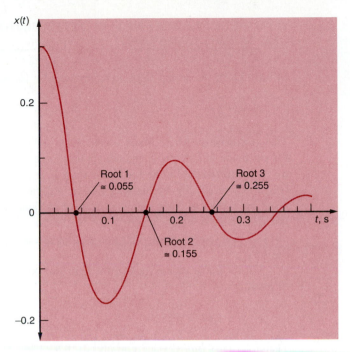

FIGURE 33.3
Plot of the position versus time for a sports car after the wheel hits a hole in the road.

This design problem must be solved using the numerical method of bisection. Estimates of the initial guesses are easily obtained by reference to Fig. 33.3. This example illustrates how graphical methods often provide information that is essential for the successful application of the numerical techniques. The plot indicates that this problem is complicated by the existence of several roots. Thus, in this case, rather narrow bracketing intervals must be used to avoid overlap. Table 33.1 lists the results of using bisection, given a stopping criterion of 0.1 percent.

TABLE 33.1 Results of Using Bisection to Locate the First Three Roots for Vibrations of a Shock Absorber*

Lower Guess	Upper Guess	Root Estimate	Number of Iterations	Approx. Relative Error (%)	True Relative Error (%)
0.0	0.1	0.055322	10	0.088	0.059
0.1	0.2	0.154395	9	0.063	0.001
0.2	0.3	0.253320	8	0.077	0.070

*A stopping criterion of 0.1 percent was used to obtain these results. Note that the exact values of the roots are 0.055289, 0.154393, and 0.253497.

33.2 USING REGRESSION IN THE DETERMINATION OF THE SPRING CONSTANT

In Eq. (33.3), the spring force is represented by Hooke's law,

$$F = -kx \qquad\qquad (33.6)$$

This relationship, which holds when the spring is not stretched too far, signifies that the extension of the spring and the applied force are linearly related. The proportionality is parameterized by the spring constant k. A value for this parameter can be established experimentally by hanging known weights from the spring and measuring the resulting extension. Such data is contained in Table 33.2 and plotted in Fig. 33.4. Notice that above a weight of 40×10^4 N, the linear relationship between the force and displacement breaks down. This sort of behavior is typical of what is termed a "hardening spring."

For the linear region (that is, the first four points), linear regression can be used to fit a straight line[1]:

$$F = -1.67 + 125.5x$$

Therefore, the value of k according to the regression is 125.5×10^4 N/m, which can be converted to 1.26×10^9 g/s^2. This is similar to the value of 1.25×10^9 g/s^2 used in Sec. 33.1. Note that the intercept for the regression result is not exactly zero, as in Eq. (33.6). However, the value of 1.67 is small enough to be disregarded. Be aware of the fact that in the field of statistics, there are rigorous methods available to test the hypothesis that the intercept is zero.

33.3 USING INTEGRATION IN THE CALCULATION OF WORK WITH A VARIABLE FORCE

Many mechanical engineering problems involve the calculation of work, for which the general formula is

Work = force $\times$ distance

When you were introduced to this concept in high school physics, simple examples were presented using forces that remained constant throughout the displacement. For example, if a force of 10 lb was used to pull a block a distance of 15 ft, the work was calculated as 150 ft·lb.

---

[1]Strictly speaking, we should have regressed displacement versus force because the latter is actually the independent variable (in other words, the one that is controlled and precise) in this experiment. However, the end result is not measurably affected by our choosing to regress force versus displacement. For cases with significant scatter, it might make a difference.

TABLE 33.2 Experimental Values for Elongation (x) and Force (F) for the Spring on an Automobile Suspension System

Elongation, m	Force, 10^4 N
0.10	10
0.17	20
0.24	30
0.34	40
0.39	50
0.42	60
0.43	70
0.44	80

FIGURE 33.4
Plot of force (in 10^4 newtons) versus displacement (in meters) for the spring from an automobile suspension system.

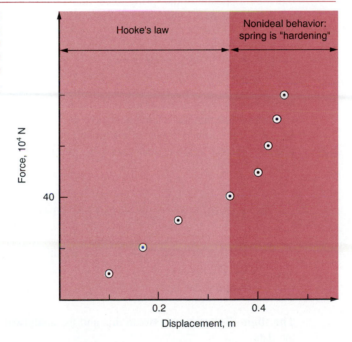

Although such a simple computation is useful for introducing the concept, realistic problem settings are usually more complex. For example, suppose that the force varies during the course of the calculation. In such cases, the work equation is expressed as

$$W = \int_{x_0}^{x_n} F(x)\ dx \tag{33.7}$$

where W is work in foot-pounds; x_0 and x_n are the initial and final positions, respectively; and $F(x)$ is a force that varies as a function of position. If $F(x)$ is easy to integrate, Eq. (33.7) can be evaluated with calculus. However, in a realistic problem setting the force might not be expressed in such a manner. In fact, when we are analyzing measured data, the force might be available in tabular form only. For such cases, numerical integration is the only viable option for the evaluation.

This can be illustrated by calculating the work done in stretching the spring from the experiment in the last section. For the region for which Hooke's law applies (that is, from $x_0 = 0$ to $x_n = 0.34$ m), Hooke's law can be substituted into Eq. (33.7) to give

$$W = \int_{x_0}^{x_n} kx \, dx$$

The sign is positive because in order to stretch the spring we must exert a force that is equal but opposite to the spring force. Substituting $k = 125.5 \times 10^4$ N/m, $x_0 = 0$, and $x_n = 0.34$ yields

$$W = \int_{0}^{0.34} 125.5 \times 10^4 \, x \, dx$$

which can be determined easily with calculus to give $W = 7.25 \times 10^4$ J. We can verify this by independently integrating the data from Table 33.2 with the trapezoidal rule. Because the points are spaced unequally, an individual trapezoidal rule has to be applied to find the area between each pair, and these can be summed to develop the total integral:

$$W = (0.10 - 0)\frac{0 + 10}{2} + (0.17 - 0.10)\frac{10 + 20}{2}$$

$$+ (0.24 - 0.17)\frac{20 + 30}{2} + (0.34 - 0.24)\frac{30 + 40}{2}$$

$$= 6.8 \times 10^4 \text{ J}$$

The slight discrepancy between this and the analytical solution is due to the scatter in the data.

Now suppose that we desire to compute the work to elongate the spring from $x_0 = 0.34$ to $x_n = 0.44$. A simple option is to again apply the trapezoidal rule:

$$W = (0.39 - 0.34)\frac{40 + 50}{2} + (0.42 - 0.39)\frac{50 + 60}{2}$$

$$+ (0.44 - 0.42)\frac{60 + 2(70) + 80}{4} = 5.3 \times 10^4 \text{ J}$$

Note that because the last three points are equally spaced, we have used the multiple-segment trapezoidal rule to compute their integral. Therefore the total work to stretch the spring the entire distance from 0 to 0.44 is

$$W = 6.8 \times 10^4 + 5.3 \times 10^4 = 12.1 \times 10^4 \text{ J}$$

PROBLEMS

33.1 What is Hooke's law for a spring? The spring constant in Eq. (33.1) corresponds to what constant in Eq. (32.4)?

33.2 If all the parameters given in the text for Eq. (33.5) remain the same but $c = 0.3 \times 10^7$ g/s, calculate the first three zeros of the equation to an error level of 0.1 percent. (*Note:* Graph the equation first before calculating the roots.)

33.3 If c decreases to zero in Eq. (33.5), what is the resulting equation? How would this fact affect the car's motion?

33.4 Because the data points in Fig. 33.4 in the nonideal region resemble a power law relationship, suggest what might be done to perform a linear regression for the data in that region.

33.5 In Eq. (33.4), what happens if the mass of the car M is very much greater than c or k?

33.6 If the mass were zero in Eq. (33.4), the following equation would hold:

$$c\frac{dx}{dt} + kx = 0$$

This equation corresponds to the hypothetical situation where the shock absorber and spring are connected to a body with zero mass. If the spring and shock absorber were displaced $x = 0.5$ initially, solve, using Euler's method, for the resulting motion (x versus t) for $c = 5 \times 10^7$ g/s and $k = 2 \times 10^7$ g/s^2.

33.7 Using the following data, calculate the work done by stretching a spring that has a spring constant of $k \cong 3 \times 10^2$ N/m to $x = 0.4$ m.

F, 10^3 N	0	0.01	0.028	0.046	0.063	0.082	0.11	0.13	0.18
x, m	0	0.05	0.1	0.15	0.2	0.25	0.3	0.35	0.4

Compute the work using the trapezoidal rule, and then repeat the analysis using Simpson's 1/3 rule.

33.8 At what point (x distance) does the spring in Prob. 33.7 deviate from Hooke's law?

33.9 For the data points given in Prob. 33.7 that do follow Hooke's law, determine the spring constant by linear regression and compare that result with the approximation in Prob. 33.7.

33.10 The velocity of the spring can be formulated as

$$\frac{dx}{dt} = v \qquad\qquad\text{(P33.10.1)}$$

where v = velocity (m s^{-1}). This equation can be substituted into Eq. (33.4) to yield

$$\frac{dv}{dt} + \frac{c}{M}v + \frac{k}{M}x = 0$$

or

$$\frac{dv}{dt} = -\frac{c}{M}v - \frac{k}{M}x \qquad\qquad\text{(P33.10.2)}$$

Equations (P33.10.1) and (P.33.10.2) are a pair of rate equations that can be used to compute displacement and velocity as a function of time. Employ Euler's method to solve these equations simultaneously for the case depicted in Fig. 33.3.

ELECTRICAL ENGINEERING: CIRCUIT ANALYSIS

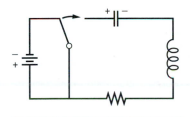

FIGURE 34.1
A passive circuit.

A common device or problem setting in electrical engineering is the electric circuit (Fig. 34.1), which is composed of wires and elements interconnected to form a closed path. Electricity is conducted through the circuit in order to convey information or, when the circuit is linked to electromechanical devices, to perform work.

In the present chapter we will study simple passive circuits of the type shown in Fig. 34.1. Such circuits are called "passive" because they consist of elements such as resistors, capacitors, and inductors that store or dissipate electric energy. This is in contrast to active circuits containing elements such as transducers, amplifiers, and transistors that can serve as energy sources.

As is true with systems in all fields of engineering, organizing principles are required to analyze electrical systems. Kirchhoff's laws, along with mechanisms of the sort outlined in Table 34.1, provide the framework for predicting the behavior of passive electric circuits.

34.1 ANALYZING A RESISTOR NETWORK WITH LINEAR ALGEBRAIC EQUATIONS

A common problem in electrical engineering involves determining the currents and voltages at various locations in resistor circuits. These problems are solved using Kirchhoff's current and voltage rules. The current (or point) rule states that the algebraic sum of all currents entering a node must be zero (see Fig. 34.2), or

$$\Sigma i = 0$$

where all current entering the node is considered positive in sign. The current rule is an application of the principle of conservation of charge.

The voltage (or loop) rule specifies that the algebraic sum of the potential differences (that is, voltage changes) in any loop must equal zero. For a resistor circuit, this is expressed as

TABLE 34.1   Some Relationships or Mechanisms Defining the Ideal Behavior of the Elements of a Passive Circuit

	Resistor	Inductor	Capacitor
Function:	Dissipates energy	Stores energy in a magnetic field	Stores energy in an electric field by storing charges on plates separated by a nonconducting medium
Schematic representation:	—⌇⌇⌇—	—〰〰〰—	—⊣⊢—
Ideal relationship or mechanism:	Ohm's law	Henry's law	Faraday's law
Mathematical relationship:	$V = iR$	$V = L\dfrac{di}{dt}$	$V = \dfrac{q}{C}$

$$\Sigma \xi - \Sigma iR = 0$$

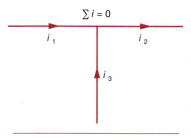

$\Sigma i = 0$

i_1 i_2

i_3

FIGURE 34.2
Kirchhoff's current rule.

where ξ is the emf (electromotive force) of the voltage sources and R is the resistance of any resistors on the loop. Note that the second term derives from Ohm's law (Table 34.1), which states that the voltage drop across an ideal resistor is equal to the product of the current and the resistance. Kirchhoff's voltage rule is an expression of the conservation of energy.

Application of these rules results in systems of simultaneous linear algebraic equations because the various loops within a circuit are coupled. For example, consider the circuit shown in Fig. 34.3. The currents associated with this circuit are unknown in both magnitude and direction. This presents no great difficulty because one simply assumes a direction for each current. If the resultant solution from Kirchhoff's laws is negative, then the assumed direction was incorrect. For example, Fig. 34.4 shows some assumed currents.

Given these assumptions, Kirchhoff's current rule is applied at each node to yield

$$i_{12} + i_{52} + i_{32} = 0$$

$$i_{65} - i_{52} - i_{54} = 0$$

$$i_{43} - i_{32} = 0$$

$$i_{54} - i_{43} = 0$$

Application of the voltage rule to each of the two loops gives

$$-i_{54}R_{54} - i_{43}R_{43} - i_{32}R_{32} + i_{52}R_{52} = 0$$

$$-i_{65}R_{65} - i_{52}R_{52} + i_{12}R_{12} - 200 = 0$$

or, substituting the resistances from Fig. 34.3 and bringing constants to the right-hand side,

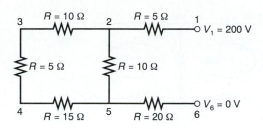

FIGURE 34.3
A resistor circuit to be solved
using simultaneous linear
algebraic equations.

$$-15i_{54} - 5i_{43} - 10i_{32} + 10i_{52} = 0$$

$$-20i_{65} - 10i_{52} + 5i_{12} = 200$$

Therefore the problem amounts to solving the following set of six equations with six unknown currents:

$$i_{12} + i_{52} + i_{32} = 0$$

$$i_{52} + i_{65} - i_{54} = 0$$

$$i_{32} + i_{43} = 0$$

$$i_{54} - i_{43} = 0$$

$$10i_{52} - 10i_{32} - 15i_{54} - 5i_{43} = 0$$

$$5i_{12} - 10i_{52} - 20i_{65} = 200$$

Although impractical to solve by hand, this system is easily handled using the Gauss elimination method discussed in Chap. 26. Proceeding in this manner, the solution is

$$i_{12} = 6.1538$$

$$i_{52} = -4.6154$$

$$i_{32} = -1.5385$$

$$i_{65} = -6.1538$$

$$i_{54} = -1.5385$$

$$i_{43} = -1.5385$$

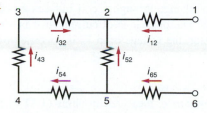

FIGURE 34.4
Assumed currents.

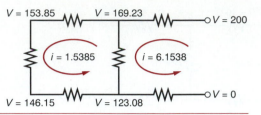

Thus, with proper interpretation of the signs of the result, the circuit currents and voltages are as shown in Fig. 34.5. The advantages of using numerical algorithms and computers for problems of this type should be evident.

34.2 DESIGNING AN *RLC* CIRCUIT USING BISECTION

Aside from steady-state computations, electrical engineers regularly analyze the transient, or time-variable, behavior of circuits. Such a situation occurs when the switch in the circuit shown in Fig. 34.1 is closed. In this case there will be a period of adjustment following the closing of the switch as a new steady state is approached. The length of this adjustment period is closely related to the energy-storing properties of the capacitor and the inductor. Energy storage may oscillate between these two elements during a transient period. However, resistance in the circuit will dissipate or dampen the magnitude of the oscillations.

According to Kirchhoff's voltage law, the sum of the voltage drops around a closed circuit is zero, or

$$V_L + V_R + V_C = 0 \tag{34.1}$$

where V_L, V_R, and V_C are the voltage drops across the inductor, resistor, and capacitor, respectively. Mathematical formulations for each of these voltage drops are summarized in Table 34.1. For the inductor, the voltage drop V_L is specified by *Henry's law:*

$$V_L = L\frac{di}{dt} \tag{34.2}$$

where L is a proportionality constant called the *inductance,* which has units of henrys (1 H = 1 V · s/A). Equation (34.2) is a mathematical expression of Henry's observation that a current change through a coil induces an opposing voltage that is directly proportional to the rate of change of current (that is, di/dt).

For the resistor, the voltage drop V_R is specified by *Ohm's law:*

$$V_R = iR \tag{34.3}$$

where R is the *resistance,* measured in ohms (1 Ω = 1 V/A).

Finally, the voltage drop for the capacitor V_C is specified by *Faraday's law:*

$$V_C = \frac{q}{C} \tag{34.4}$$

where C is a proportionality constant called the *capacitance,* which has units of farads ($1 \text{ F} = 1 \text{ A} \cdot \text{s/V}$). Equation (34.4) is a mathematical expression of the fact that the voltage drop across a capacitor is directly proportional to its charge.

Equations (34.2) through (34.4) can be substituted into Eq. (34.1) to yield

$$L\frac{di}{dt} + Ri + \frac{q}{C} = 0 \tag{34.5}$$

However, the current is equivalent to the rate of change of charge per time:

$$i = \frac{dq}{dt} \tag{34.6}$$

Therefore

$$L\frac{d^2q}{dt^2} + R\frac{dq}{dt} + \frac{q}{C} = 0 \tag{34.7}$$

This is a second-order linear ordinary differential equation that can be solved using the methods of calculus. If $q = q_0 = V_0 C$ at $t = 0$ (where V_0 is the voltage from a battery), the solution is

$$q(t) = q_0 e^{-Rt/2L} \cos\left[\sqrt{\frac{1}{LC} - \left(\frac{R}{2L}\right)^2} \, t\right] \tag{34.8}$$

Equation (34.8) describes the time rate of change of charge on the capacitor as a function of time. The solution is plotted on Fig. 34.6.

A typical electrical engineering design problem might involve determining the proper resistor to dissipate energy at a specified rate, given known values of L and C. For the present case, assume that charge must be dissipated to 1 percent of its original value ($q/q_0 = 0.01$) in $t = 0.05$ s, with $L = 5$ H and $C = 10^{-4}$ F.

This problem involves determining the root of Eq. (34.8). To do this, move q to the right-hand side and divide by q_0 to yield

$$f(R) = 0 = e^{-Rt/2L} \cos\left[\sqrt{\frac{1}{LC} - \left(\frac{R}{2L}\right)^2} \, t\right] - \frac{q}{q_0}$$

or, using the numerical values given, the problem reduces to determining the root of

$$f(R) = e^{-0.005R} \cos\left[\sqrt{2000 - 0.01R^2}\,(0.05)\right] - 0.01 \tag{34.9}$$

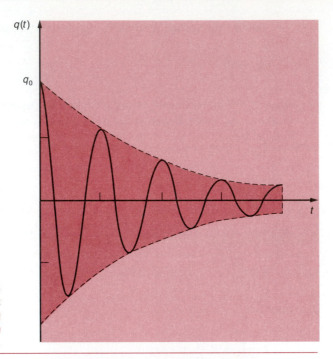

FIGURE 34.6
The charge on a capacitor as a function of time following the closing of the switch in Fig. 34.1.

FIGURE 34.7
Plot of Eq. (34.9) used to obtain initial guesses for R that bracket the root.

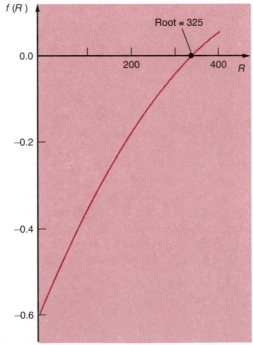

Examination of this equation suggests that a reasonable range for R is 0 to 400 Ω (because $2000 - 0.01R^2$ must be greater than zero to avoid a complex number). Figure 34.7—a plot of Eq. (34.9)—confirms this. Twenty-one iterations of bisection (Chap. 25) give $R = 328.1515$ Ω, with an error of less than 0.0001 percent.

Thus you can specify a resistor with this rating for the circuit shown in Fig. 34.1 and expect to achieve energy dissipation that is consistent with the requirements of the problem. This design problem could not be solved efficiently without a numerical method such as bisection.

34.3 ANALYZING A NONIDEAL *RL* CIRCUIT USING INTERPOLATION AND RATE EQUATIONS

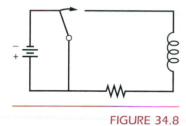

FIGURE 34.8
An *RL* circuit.

Aside from using calculus as in the previous section, we can use numerical techniques such as Euler's method to calculate the transient response of a circuit. Consider the simple *RL* circuit seen in Fig. 34.8. Kirchhoff's voltage law can be applied in conjunction with Eqs. (34.2) and (34.3) to give

$$L\frac{di}{dt} + Ri = 0$$

or

$$\frac{di}{dt} = -\frac{R}{L}i \tag{34.10}$$

This simple linear rate equation can be solved easily by calculus or Euler's method for current as a function of time. However, real resistors may not always obey Ohm's law, and the R in Eq. (34.10) is not necessarily constant.

Suppose that you performed some very precise experiments to measure the voltage drop and corresponding current for a resistor. The results, as listed in Table 34.2 and Fig. 34.9, suggest a curvilinear relationship rather than the straight line represented by Ohm's law. In order to quantify this relationship, a curve must be fit to the data.

Because of measurement error, regression would typically be the preferred method of curve fitting for analyzing such experimental data. However, the smoothness of the relationship suggested by the points as well as the precision of the experimental methods leads you to fit an interpolating polynomial as a first try. Using Lagrange interpolation (Chap. 24), you discover that a third-order polynomial yields almost perfect results. The polynomial, which is developed from the Lagrange form by grouping terms, is

$$V = 40i + 160i^3$$

Kirchhoff's voltage law can be applied in conjunction with Eq. (34.2) to give

$$L\frac{di}{dt} + 40i + 160i^3 = 0$$

TABLE 34.2

Experimental Data for Voltage Drop across a Resistor Subjected to Various Levels of Current

i	V
-1.00	-200.0
-0.50	-40.0
-0.25	-12.5
0.25	12.5
0.50	40.0
1.00	200.0

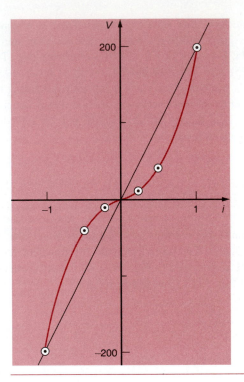

FIGURE 34.9
Plot of voltage versus current for a resistor. The curve is the actual relationship, whereas the straight line is an idealized version following Ohm's law.

or

$$\frac{di}{dt} = -\frac{40i + 160i^3}{L} \tag{34.11}$$

Contrast this equation with the rate equation for the ideal RL circuit [Eq. (34.10)]. Euler's method can be used to solve both; the results are displayed in Fig. 34.10. Notice how the nonideal case differs significantly from the solution for Eq. (34.10).

PROBLEMS

34.1 State Kirchhoff's current and voltage rules for electric circuits.

34.2 Solve the resistor circuit in Fig. 34.3, using Gauss elimination, if $V_1 = 110$ V and $V_6 = -110$ V. Notice that the voltages do not cancel each other out.

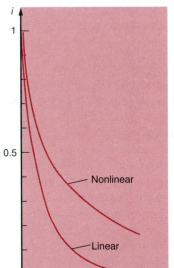

FIGURE 34.10
Numerical solutions for both the ideal [Eq. (34.10) with $R = 200\ \Omega$] and the nonideal [Eq. (34.11)] resistors.

34.3 Solve the circuit in Fig. P34.3 for the currents in each wire. Use Gauss elimination with pivoting.

34.4 Form the mathematical relationship with Kirchhoff's rules for the circuit shown in Fig. P34.4.

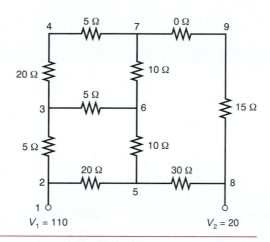

FIGURE P34.3 $V_1 = 110$ $V_2 = 20$

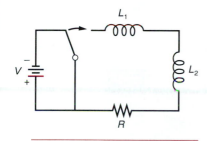

FIGURE P34.4

34.5 Solve for the current variation with time in the circuit shown in Fig. P34.4 if $L_1 = 5$ H, $L_2 = 20$ H, $R = 5\ \Omega$, and $V = 20$ V. Use Euler's method, and determine the time when the current is 50 percent of its maximum value.

34.6 Repeat Prob. 34.5 if R is given by

$$R = 20 + 100i^2$$

34.7 Solve Eq. (34.9) to determine when the charge is 10 percent of its original value. Employ bisection.

34.8 Graph Eq. (34.8) for $R = 200\ \Omega$, and let t vary. (That is, let t be the independent variable.) How many seconds elapse before the ratio of q/q_0 drops permanently below 0.10? Use a graphical approach for a coarse estimate, and use bisection to refine it.

34.9 Note that the pair of first-order rate equations (34.5) and (34.6)

$$\frac{di}{dt} = -\frac{R}{L}i - \frac{1}{LC}q = 0 \tag{P34.9.1}$$

and

$$\frac{dq}{dt} = i \tag{P34.9.2}$$

are equivalent to the single second-order equation (34.7)

$$L\frac{d^2q}{dt^2} + R\frac{dq}{dt} + \frac{q}{C} = 0$$

Equations (P34.9.1) and (P34.9.2) are a pair of rate equations that can be used to compute charge and current as a function of time. Employ Euler's method to solve these equations simultaneously for the case depicted in Fig. 34.6.

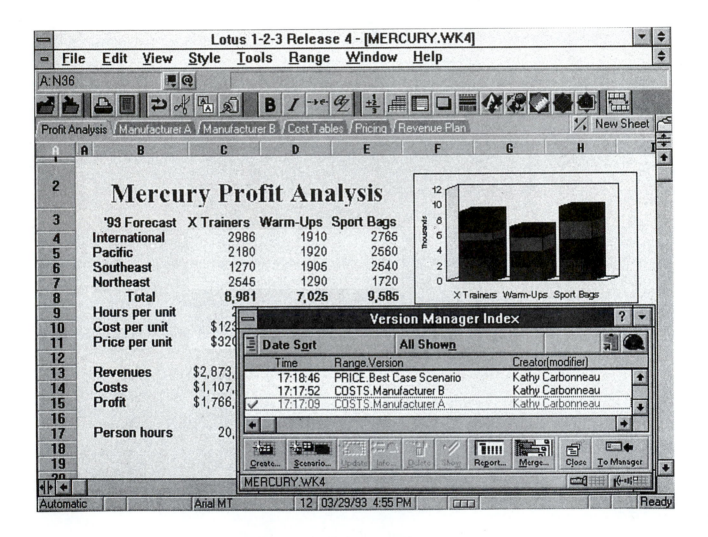

SPREADSHEETS

To this point you have been introduced to a number of ways in which personal computers can serve your engineering career. We have presented material on BASIC and FORTRAN and have shown you how these languages can be used to develop your own "homemade" computer programs to solve engineering problems. Although the capability to write your own programs will serve you well, there will be many times throughout your career when you will use software created by others.

Because of the sheer number of choices and their expense, the selection of the proper software pack-ages might at first seem overwhelming. In the earlier days of the computer era, separate pieces of soft-ware were often required for each individual applica-tion. Today, there are a number of multipurpose or generic programs that can be applied in numerous problem contexts. Noteworthy examples include word processors, and "canned" software packages for graphics, statistics, and numerical methods. In the present part of the book, we will provide an introduc-tion to another major type of generic software: the electronic spreadsheet.

FUNDAMENTALS OF SPREADSHEETS

As an engineer, you will have many occasions to perform long, interconnected computations. Often such calculations will be recorded on a large sheet of paper called a *spreadsheet,* or *worksheet* (Fig. 35.1). Whether or not they are formally called spreadsheets, large tabulated computations are used in all fields that deal with numbers. The areas where they were first recognized are those related to business and economics. For this reason, the original electronic spreadsheets were created in a business context. However, they also have great utility in engineering. As described in the next section, they are one of the most popular software packages ever invented.

35.1   THE ORIGIN OF SPREADSHEETS

While a student at the Harvard Business School in 1978, Daniel Bricklin was suffering through the task of preparing spreadsheets for some of his courses. Not only were these spreadsheets long and involved, but most of the numbers were interdependent. Therefore, if he made a mistake or wanted to change one of the numbers, numerous other computations had to be recalculated and many results modified to correct the entire sheet.

Bricklin realized that a way out of his dilemma was to write software so that the spreadsheet could be implemented on a microcomputer. Then, if a number was changed, the computer could modify the entire sheet automatically. In this way, the electronic spreadsheet was born. Together with Robert Frankston, a programmer, and Daniel Flystra, a businessman, Bricklin formed a company called Software Arts and in 1979 began selling the program under the trademark VisiCalc.

Soon other companies began marketing spreadsheets. For years, the most successful was *1-2-3,* offered by the Lotus Development Corporation. Today other spreadsheets, such as Borland Corporation's *Quattro Pro* and Microsoft Corporation's *Excel,* are becoming increasingly popular. To date, millions of spreadsheet programs have been sold. Along with word-processing programs, they are the best-selling software packages of all time.

There are three reasons why spreadsheets have been so popular. First, they are extremely flexible. That is, although they are simple in concept, they can be applied in a wide number of problem contexts. In this sense they are quintessential examples of generic software. Second, they are usually inexpensive to purchase and easy to apply. Thus a businessperson or engineer can obtain practical benefits from the software with a reasonable investment of time, money, and effort. Third, they allow rapid, interactive evaluation of computations. This capability, called "what if?," is their most innovative attribute and probably the key reason for their popularity.

FIGURE 35.1
An engineering student's homework problem is an example of a handwritten preelectronic spreadsheet.

3 – 22 – 92	ENGINEERING 101 EXAMPLES 23.1 & 23.2	ARELLA, ALISON

LINEAR REGRESSION TO FIT A STRAIGHT LINE TO THE DATA FROM TABLE 23.1

x_i	y_i	$x_i y_i$	x_i^2	y_i^2
10	110	1100	100	12100
15	230	3450	225	52900
20	210	4200	400	44100
25	350	8750	625	122500
30	330	9900	900	108900
35	460	16100	1225	211600
135	1690	43500	3475	552100

SLOPE [EQ. (23.3)]

$$a_i = \frac{6\,(43500) - 135\,(1690)}{6\,(3475) - (135)^2} = 12.5143$$

INTERCEPT [EQ. (23.4)]

$$a_0 = \frac{1690 - 12.5143\,(135)}{6} = 0.095$$

CORRELATION COEFFICIENT [EQ. (23.8)]

$$r = \frac{6\,(43500) - 135\,(1690)}{\sqrt{6\,(3475) - (135)^2}\ \sqrt{6\,(552100) - (1690)^2}} = 0.949$$

COEFFICIENT OF DETERMINATION

$$r^2 = (0.949)^2 = 0.901$$

35.2 THE SPREADSHEET ENVIRONMENT

As stated previously, many spreadsheet products are presently available. Just as was the case with the dialects of BASIC and FORTRAN, each version has its own special characteristics and idiosyncrasies. In order to make the following material broadly applicable, we will attempt to keep our descriptions as generic as possible. In particular, we will limit ourselves to the general features that are typical of two of the most popular spreadsheets—Lotus 1-2-3 and Quattro Pro. Wherever specific examples are required, we will usually opt for 1-2-3. However, because the two products are very similar, this choice should not prove to be a serious impediment for those employing Quattro Pro. Much of the material and all the general concepts and applications will also be compatible with other products such as Excel.

Figure 35.2 displays a new blank spreadsheet screen for the non-Windows version of 1-2-3. Although the layout of each product may differ slightly, all the popular spreadsheets include three major sectors: the spreadsheet area, the input line, and the menu bar.

FIGURE 35.2
The "anatomy" of a spreadsheet showing its major components. Note in particular how a column letter and a row number are used to locate the active cell. Also, on this particular spreadsheet (the non-Windows version of 1-2-3), the menu bar is hidden if it is not consciously activated by the user.

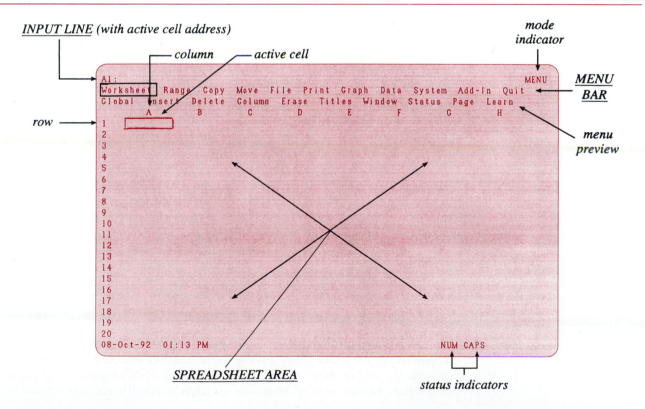

The *spreadsheet area* consists of a two-dimensional space divided into rows and columns. Such two-dimensional characterizations should be very familiar to you by now. In graphing, cartesian (*x, y*) coordinates are employed to locate a point on a two-dimensional space. For computer programming languages, dimensioned variables employ two subscripts to store values in a two-dimensional array.

For spreadsheets, the same thinking applies. As shown in Fig. 35.2, the vertical columns are identified by letters and the horizontal rows by numbers. As with cartesian coordinates or two-dimensional arrays, the combination of two values—a letter and a number—can be used to define each location. In spreadsheet jargon, each location is called a *cell.* Thus the highlighted, or *active, cell* in Fig. 35.2 is designated as A1 because it is located in column A and row 1. The designation A1 is referred to as the *cell address.*

The *input line* is located at the top of the spreadsheet. It is used to display the cell address, its contents, or a message. Because we have yet to enter information, only the *active cell address* (A1) is shown on Fig. 35.2. This serves to indicate your location on the spreadsheet. If you move to another cell, this label changes automatically to reflect the new position.

The *menu bar,* also located at the top of the screen, lists commands that can be used to manipulate data on the spreadsheet area. We will learn more about these commands as we proceed. It should be noted that on some spreadsheets, such as the non-Windows version of 1-2-3 shown in Fig. 35.2, the menu bar appears only when the user strikes the [/] key. On most other popular spreadsheets, it is always visible.

Aside from the three major areas, the spreadsheet includes additional standard information. The *status indicators* display information related to the state of the activities of the spreadsheet. For example, the status indicators for Fig. 35.2 show that the [Caps Lock] and [Num Lock] keys are on. *Mode indicators* reflect the mode of the spreadsheet. For example, the MENU mode in Fig. 35.2 indicates that the [/] key has been struck and the user can make a menu selection. Finally, miscellaneous information such as the time and date may also be displayed.

Before moving on, we should note that modern spreadsheets are among the easiest software to learn and to use. In particular, most have context-sensitive help utilities as a standard part of their environments, which can usually be accessed with the [F1] key. Consequently, you can learn a lot about spreadsheets by simply playing with them and using the help facility when you get stuck. For this reason, the examples in this chapter are designed somewhat differently from those in previous parts of this book. In fact, we include a different type of example called a "hands-on exercise." We dub them that because they are designed to be performed while you are sitting at a computer.

HANDS-ON EXERCISE 35.1   Getting Help and Moving around the Spreadsheet

GOAL: To explore how you can get help, move around, and exit the spreadsheet.

EXERCISES:

1. Before starting this exercise, you should turn on your computer and access your software so that the screen shows a blank spreadsheet like Fig. 35.2. Before proceeding, strike the [F1] or "Help" key. At this point, the spreadsheet should be replaced by a help screen. Because each of the popular spreadsheets has different help utilities, we

will not attempt to describe their capabilities. However, they are self-explanatory, and you should take a few moments to explore your spreadsheet's help environment. When you are finished probing, you can get back to the blank spreadsheet by striking the [Esc] key. Remember that whenever you get stuck, the [F1] key can provide helpful information to get you going again.

2. After you return to your blank spreadsheet screen, you can use the four *arrow cursor keys* on the number pad at the right-hand side of your keyboard to move to any other cell. Experiment with the arrow keys. Do not hold them down too long because they will repeat and take you too far. Or worse, if you try to advance beyond the spreadsheet boundary, this will cause repeated, and annoying, beeping.

3. Try the [Home] key. It should return the active cell to A1. Use the [PgDn] and [PgUp] keys to move the window down and up 20 rows at a time. Set [Scroll Lock] on, and move the active cell down and up with the arrow keys. Note the difference between this case and when the [Scroll Lock] key was off. Before moving on, make sure that the [Scroll Lock] is off (it is a toggle switch[1]).

4. There are two ways to move right or left by windows. First, use the [Tab] and [Shift-Tab] keys. Then try [Ctrl →] and [Ctrl ←] keys. Note that when we enclose two key commands in brackets, it means that the first key must be held down when pressing the following key. This should be familiar to you from typing uppercase letters by holding down the [Shift] key when the [Caps Lock] key is off.

5. Set the END mode by depressing the [End] key (again this is a toggle). After making sure that it is on (check the status line), strike the [↓] key. With an empty spreadsheet, the active cell should move to the lower bound (on most spreadsheets, row 8192). Also observe that once the END mode is "used," it is automatically toggled off. Now, set END and try the [→] key. On the most common spreadsheets, the active cell should now be in column IV. Note that this does not refer to the Roman numeral IV (see Prob. 35.2). Go [Home].

6. Strike the [F5] function key. This is sometimes referred to as the "Go To" key. Observe how the input and the status line have been modified. Type the cell address G15, strike return, and the active cell jumps to that location. Strike [F5] again, enter BT7777, and the active cell jumps there, placing the cell in the upper-left-hand corner of the new window. Go [Home].

35.3 MENU OPERATION

Now that you know how to move around the spreadsheet, the next step is to familiarize yourself with the general operation of the spreadsheet's menu system. Before launching into an exercise, we should mention that different spreadsheets exhibit variations in their menu structure. That is, each has somewhat different nomenclature and options. However, they are fundamentally similar. In the following exercise, we will emphasize these common attributes.

In addition, how you use menus will be partly dictated by the operating system you employ. In particular, if you are in a Windows-type environment, you may employ the

[1]A toggle switch alternately turns on or off every time it is operated. A light switch is a common example.

mouse as your primary means of interaction. Otherwise, you will rely on keystrokes. In all the following exercises, we will emphasize keystroke interactions because they work on all types of systems. As such, keystrokes represent the "least common denominator." In addition, as you become more familiar with your spreadsheet menu, you will learn that keystrokes are actually much faster than pointing and clicking with the mouse.

HANDS-ON EXERCISE 35.2   Navigating the Menu System and Saving and Retrieving Files

GOAL: To explore how you can move around the menu system in order to initiate commands to manipulate data on the spreadsheet.

EXERCISES:

1. If you do not have a mouse-oriented system, you will activate the main menu with the forward slash [/] key. You can move the [→] or [←] keys back and forth, and each option will be highlighted. Note that the menu "wraps around" when you go off the end in either direction. Strike the [F1] key, and you will automatically be provided with help. Hitting the [Esc] key will get you back to the main menu. You can select a highlighted menu item by striking the [Enter] key.

2. If you have a mouse, you do not have to strike the [/] key. You merely have to point to the option and click. In addition to the arrow/Enter or point/click actions, you can also select by typing a *key character*. This relates to the fact that on most spreadsheets, each menu item starts with a unique character. Some products do not use this scheme but rather employ a special **highlight** or underscore to indicate the character that can be employed to make the selection. Either way, it is faster (and, therefore, usually preferable) to select a menu item by merely typing this key character.

Make a selection of the farthest left choice on the menu by typing its key character. Once you make the menu selection, two things typically happen. On some products, a new line will appear which is usually a new menu or a query (Fig. 35.3*a*). On others, new information will appear, but in pull-down format (Fig. 35.3*b*). If you make a selection from this new menu, another group of options will appear. For either type, moving down this menu structure is like moving along the branches of a tree. Eventually, you will come to some terminal point where an action will be invoked.

If you move down the wrong branch, you can always "back out" by striking the [Esc] key. When you have struck enough [Esc] keys, you will return to the spreadsheet. You can then return to the menu system with the [/] key. Take some time now to traverse your spreadsheet's menu tree. If something seems unclear to you, strike the [F1] key to get help. If you get to a point where the computer asks for something you do not understand, just strike [Esc] to back out.

Note that you can also jump immediately out of the menu system by striking the [Ctrl-Break] combination. Also, certain menus have a **q**uit option. For these cases, typing a **q** will be equivalent to either [Esc] or [Ctrl-Break] depending on the menu or the software product. Finally, certain menu selections will automatically jump back up the tree an intermediate number of steps. This occurs in instances where the software developers have felt that in most cases, this is what the user would most often want to do.

3. After you have finished exploring, return to the main menu. Select the file option by

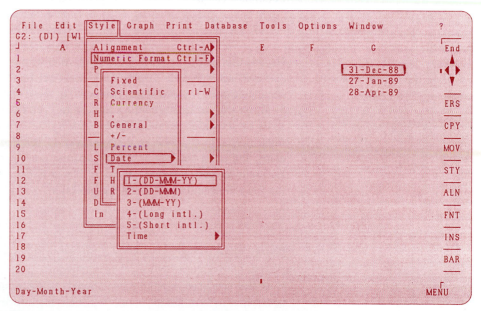

```
G2: (D1) [W11] 33500                                                  MENU
Worksheet  Range  Copy  Move  File  Print  Graph  Data  System  Add-In  Quit
Format  Label  Erase  Name  Justify  Prot  Unprot  Input  Value  Trans  Search
        A          B        C        D        E      F        G
1
2                                                            19-Sep-91
3                                                            28-Dec-91
4                                                            06-Apr-92
5
6
7
8
9
10
11
12
13
14
15
16
17
18
19
20
08-Oct-92  01:23 PM
```

(a)

```
   File   Edit   Style   Graph   Print   Database   Tools   Options   Window        ?
G2: (D1) [W1
   J        A      Alignment        Ctrl-A         E        F           G         End
   1               Numeric Format Ctrl-F
   2            P                               ▶                                    ◀▶
   3            ─      Fixed                            31-Dec-88                     ▼
   4            C      Scientific       rl-W            27-Jan-89
   5            R      Currency                         28-Apr-89                    ERS
   6            H      ,                        ▶
   7            B      General                  ▶                                    CPY
   8            ─      +/-
   9            L      Percent                                                       MOV
   10           S      Date                     ▶
   11           F      T                                                             STY
   12           F      H      1-(DD-MMM-YY)
   13           U      R      2-(DD-MMM)                                             ALN
   14           D      ─      3-(MMM-YY)
   15           In            4-(Long intl.)                                         FNT
   16                         5-(Short intl.)
   17                         Time                     ▶                             INS
   18
   19                                                                               BAR
   20
   Day-Month-Year                                                                   MENU
```

(b)

FIGURE 35.3

(a) The row menu system used on the non-Windows version of 1-2-3. Notice how the next menu options are previewed in the second row of the menu line. (b) The pull-down menu system used on Quattro Pro. This interface makes the treelike structure of the menu system visually obvious. For this case, the commands / S N D are used to instruct the spreadsheet that you would like to specify the style of a numeric format for a range of cells so that they can hold dates. Note that the 1-2-3 commands / R F D perform the same action. Once this is done, if the first selection from the final menu is chosen [1-(DD-MMM-YY)], a screen appears to query you for the range of cells you would like to format. When you enter the range, the values in that range would appear in the desired style. Both screens indicate that this was done previously to the cells G2..G4.

typing an **f**. Then, type **s** for save. The series of keystrokes can be represented in shorthand by

```
/ f s
```

At this point, the spreadsheet will prompt you for a file name for your spreadsheet. If the file has already been saved, the computer will automatically answer its own prompt by filling in the file name. It will also display and allow you to conveniently select any other spreadsheet files that may be on your computer's default drive. Otherwise, you can type in a typical legal DOS file name without an extension. (Remember to use a unique name or you will destroy the file with the same name on your disk.) Once the file name is typed (by you or the computer) striking the [Enter] key will either write the file to your storage disk (if it is a new file) or give you the option of changing your mind (if the computer senses that the file name already exists).

It should be noted that most spreadsheets automatically afix special extensions to spreadsheet files when they are saved. For instance, versions 2.01, 2.2, and 2.3 of 1-2-3 employ the extension .WK1. Version 3 uses .WK3. Other spreadsheets use different extensions (.WQ1 for Quattro Pro and .XLS for Excel).

4. Retrieving the file from the external storage involves a similar reverse process. Again, the file option is selected by typing / **f**. Then, type **r** for **r**etrieve. The series of keystrokes can be represented in shorthand by

```
/ f r
```

Beyond this point, the process is very similar to saving the file. For example, the products will automatically display the spreadsheet files that are on your system's external storage device. You can select from among these files to retrieve their contents.

5. When you are finished with your session, you can exit the spreadsheet. This process starts by activating the menu bar by striking the forward slash key [/]. The exit is then implemented in a number of different ways depending on the software product you are using. For some (such as 1-2-3), the exiting process starts from the main menu by selecting **q** for **q**uit. For others (Quattro Pro or Excel), there is an **x** for e**x**it selection off the **f**ile submenu. In any event, the procedure is straightforward.

35.4 NUMBERS, LABELS, AND FORMULAS

Now that you know how to move around the spreadsheet and use its menu system, the next step is to familiarize yourself with entering information into a cell. On most spreadsheets, each cell may contain three fundamental types of information—a label, a numeric value, or an expression (Fig. 35.4).

Before proceeding we should mention some fundamental aspects of the ways that numbers and formulas are represented on spreadsheets. For example:

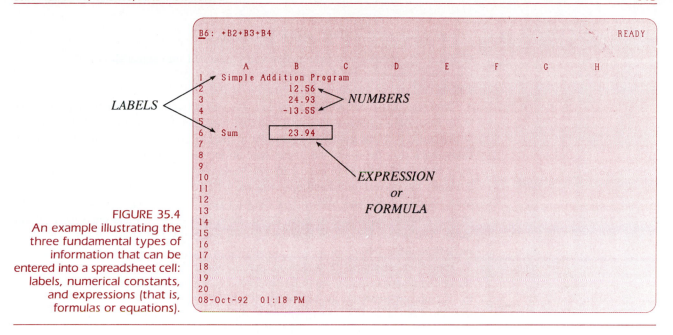

B6: +B2+B3+B4   READY

	A	B	C	D	E	F	G	H
1	Simple Addition Program							
2		12.56						
3		24.93						
4		-13.55						
5								
6	Sum	23.94						

08-Oct-92   01:18 PM

LABELS

NUMBERS

EXPRESSION
or
FORMULA

FIGURE 35.4
An example illustrating the three fundamental types of information that can be entered into a spreadsheet cell: labels, numerical constants, and expressions (that is, formulas or equations).

• You need not enter a decimal point if there is no fractional part in a number.

• It is possible to enter (and display) numbers in scientific notation (for example, 6.023E23) just like in QuickBASIC or FORTRAN.

• Spreadsheets store all numerical quantities in a real format with 15 significant digits, the same as double precision in QuickBASIC and FORTRAN.

• Spreadsheet formulas are written in a style that is similar to most high-level computer languages. The arithmetic operators and their associated priorities are listed in Table 35.1. Operators with equal priorities are implemented from left to right.

We will now explore how numerical quantities, along with labels and formulas, can be entered into a spreadsheet.

TABLE 35.1 The Priorities of Parentheses and Arithmetic Operators Used for Spreadsheet Formulas

Symbol	Meaning	Priority
( )	Parentheses	1
^	Exponentiation	2
−	Negation	3
*	Multiplication	4
/	Division	4
+	Addition	5
−	Subtraction	5

HANDS-ON EXERCISE 35.3   *Entry of Numbers, Labels, and Expressions*

GOAL: To explore how information can be entered into spreadsheet cells.

EXERCISES:

1. Move the active cell to B2. Type 12.34, and notice how the letters appear on the input line. Also glance at the mode indicator, and see that we are in VALUE mode. While we are in this mode, it is possible to use the [Backspace] key to make corrections. Back over the 4 and the 3, and type 56 instead. Now press the [Enter] key. The input line should change to B2: 12.56 and the number 12.56 should appear in cell B2.

2. Move the current cell down one with the [↓] key to B3, and type 24.93. However, instead of pressing [Enter], try the [↓] key. This should cause the value to be entered *and* the active cell to move down. Now, type -13.55 and press the [↓] key twice. The active cell should now be B6 with one blank cell above. Move one cell to the left, to A6.

3. Type Sum, and press [Enter]. Note that the word "Sum" is entered *left-justified* in the cell and appears in the input line preceded by an apostrophe ('). The apostrophe is the default justification code and means left-justified. Now type "Sum, and press [Enter]. The label should now be *right-justified* (or almost—one blank space will be to the right of it). Try ^Sum, and the label will be centered in the cell. Try \Sum, and you should observe that the word is repeated as many times as necessary to fill the cell.

Now enter a label that is too long for the cell, such as The sum is equal to. (This is an example of an instance where a long label works well.) Move the active cell to B6, and enter a number. Observe what has happened to your long label.

Return to A6, and enter Sum. Go [Home], and enter Simple Addition Program. Then move the active cell back to B6.

4. We will now enter a formula into cell B6. Spreadsheets recognize a cell entry as a formula when it begins with +, −, (, @, or a numeric digit. Enter the following formula:

    + b2 + b3 + b4

The formula, as typed, should appear in the input line, but the result of the formula (in this case, the sum of the three cells) should appear in the active cell (see Fig. 35.4).

We will now reenter the formula but use *pointing* this time. As we go through this, do not strike the [Enter] key until you are instructed to do so. Start with a + again, and then hit the [↑] key so that the active cell goes to cell B2. The mode indicator should denote that we are now in POINT mode, and the address B2 should appear automatically in your expression on the input line. Now, type another +. Notice how the highlight returns to the active cell and the mode indicator is now VALUE. Point again, but to the B3 value this time, enter a +, point to B4, and now finish the formula by striking the [Enter] key. The correct sum should appear in B6. Realize that pointing is often more convenient than direct typing of expressions.

5. Spreadsheets have many special functions (called "at functions" in spreadsheet lingo) which are similar in spirit to the intrinsic or library functions available in QuickBASIC, FORTRAN, and other high-level programming languages. These functions always start with the character @. Note that they can be used alone or as a part of a larger expression.

Move the active cell to B6 again, and enter the formula

@sum(b2,b3,b4)

You should get the correct answer. Try the same "at function," but use a range of cells instead of a list of individual cells:

@sum(b2..b4)

Again, the summation should be correct.

6. Now, we will enter the same formula, but using *pointing, anchoring,* and *painting.* Start by typing @sum(, and then point to cell B2. *Anchor* that cell (creating a one-cell range) by pressing the period [.] key, and the B2 in the input line should change to B2..B2. Thus, the anchoring has created a one-cell range. Now, press the [↓] key twice to *paint* the range down to B4. This range should be highlighted and denoted as B2..B4. Finish the expression by typing the right parenthesis) and pressing the [Enter] key.

7. Make a few changes in the numbers in cells B2, B3, and B4; and see how the results of the summation are updated.

As illustrated at the end of the previous exercise, once a computational scheme has been established, the power and convenience of the spreadsheet can be exploited by changing values and watching how the "bottom line" changes. This type of interactive or exploratory calculation is referred to as a *"what if?"* computation. Such automatic updating can be extremely useful and enlightening in many different kinds of complex engineering computations. As such, the "what if?" capability lies at the heart of the spreadsheet's growing popularity in our field.

35.5 EDITING CELL QUANTITIES

The final aspect of spreadsheet fundamentals relates to the way in which cell entries can be edited. To this point, we have changed cells by retyping and reentering information. This is not the most efficient method, especially if you have long formulas or labels. In such cases, you will want to use the "Edit" key or [F2].

HANDS-ON EXERCISE 35.4

Editing Cell Entries

GOAL: To explore how cell entries can be edited conveniently with the [F2] key.

EXERCISES:

1. If you are continuing directly from the end of Hands-On Exercise 35.3, skip this step. Otherwise, enter the spreadsheet in Fig. 35.4.

2. Move the active cell to B6, and press the [F2] key. The mode indicator should change to EDIT, and the formula should appear on the input line with the cursor at the end of the line. Try the [←] key; the cursor should move to the left along the line. When you are in EDIT mode, the [←] and the [→] keys do not move the active cell. Try the [Home] and the [End] keys.

3. Pressing the [Ins] key will put you into "overtype" mode (designated as [OVR] on 1-2-3 or [OT] on Quattro Pro). This key is a toggle switch. Press it again, and you will return to insert mode. Notice how these operations affect the spreadsheet's mode indicator. Now, make sure that you are in insert mode. Move the cursor to the character B in the B2 entry, and press the [Del] key. Enter B again. The cursor should be under the 2. Press the [Backspace] key and then again enter the B. This is typical editing/word-processing protocol. After having made changes, you can enter the revised formula with the [Enter] key. The cursor can be anywhere on the line when you press the [Enter] key. In the event that you are unhappy with your editing changes, you can bail out of the EDIT mode by striking the [Esc] key twice.

4. Another common need is to erase the contents of a cell, a range of cells, or the entire spreadsheet. Move the current cell to A6 where we will now erase the label Sum. Access the menu with the [/] key. Choose the **r**ange option by typing **r** and the **e**rase option by typing **e** again. The range to be erased, in this case a single cell (A6..A6), should be displayed on the input line at the top of the screen. To cause the erase to occur, press [Enter].

For Quattro Pro, the sequence is quite similar. First, an **e** is struck to select **e**dit, followed by an **e** for **e**rase range. To cause the erase to occur, the [Enter] key is struck.

Note that the sequence of keys to erase the single cell can be summarized as

123: / r e [Enter]   QP: / e e [Enter]

In the following chapters, we will present 1-2-3 and Quattro Pro commands in this side-by-side format wherever they differ.

Frequent users of spreadsheets remember sequences like this as a kind of menu language. You can enter these keys as fast as you can, without even looking at the associated menus. However, be careful. If you move too fast and make a typographical error, it may be hard to figure out what happened to your spreadsheet!

5. Now, move to cell B7 and invoke the menu again with the [/] key. Type **r** for **r**ange (or **e** for **e**dit on Quattro Pro) and **e** for **e**rase block, but now, extend the erase range using the [↓] key to paint out the range B7..B11. Then, strike [Enter] to erase that range.

Once you have erased a cell, a range of cells, or the entire spreadsheet, the former contents cannot be retrieved without loading an earlier version of the spreadsheet.[2] Therefore, be *careful* when erasing single cells, *very cautious* when erasing ranges of cells, and *painstakingly deliberate* when erasing entire spreadsheets!

---

[2]At the time of this book's printing, this statement was true of the most popular spreadsheet, Lotus 1-2-3. However, Quattro Pro and Excel have "undo" options that allow the user to retrieve the spreadsheet prior to the last command. Also note that Quattro Pro has shortcut keys to perform some of the actions in this exercise. For example, [Ctrl-e] erases the contents of a cell.

6. You will now erase the entire spreadsheet by the menu commands

123: / w   QP: / f

Then select **e** for **e**rase and **y** for **y**es. The entire spreadsheet should be blank.

35.6   ENGINEERING APPLICATIONS OF SPREADSHEETS

Engineering problem solving usually involves considerable computational effort. In the precomputer era, this computational burden had a limiting effect on the problem-solving process. As shown in Fig. 35.5a, significant amounts of energy were expended on obtaining solutions. This had a number of effects. First, solution techniques were often simplified to make them feasible within the time constraints of a particular problem setting. Second, the number and extent of the computations were limited. Finally, as seen in Fig. 35.5a, because the techniques were so time-consuming, other aspects of the problem-solving process suffered.

As you are well aware, the computer's speed and memory greatly extend our analytical capabilities. By removing some of our computational burden, the computer allows us to place more emphasis on the creative aspects of problem formulation and interpretation of results (see Fig. 35.5b). Nowhere is this benefit more obvious than with the spreadsheet. Although the following example is taken from an academic setting, it illustrates how the spreadsheet can enhance and broaden engineering computations.

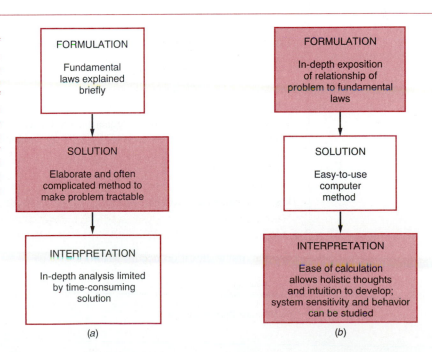

FIGURE 35.5
The three phases of engineering problem solving in (a) the precomputer and (b) the computer era. The sizes of the boxes indicate the level of emphasis directed toward each phase. Computers facilitate the implementation of solution techniques and thus allow more emphasis to be placed on the creative aspects of problem formulation and interpretation of results.

FORMULATION

Fundamental laws explained briefly

SOLUTION

Elaborate and often complicated method to make problem tractable

INTERPRETATION

In-depth analysis limited by time-consuming solution

(a)

FORMULATION

In-depth exposition of relationship of problem to fundamental laws

SOLUTION

Easy-to-use computer method

INTERPRETATION

Ease of calculation allows holistic thoughts and intuition to develop; system sensitivity and behavior can be studied

(b)

EXAMPLE 35.1   *Computing Projectile Motion with a Spreadsheet*

PROBLEM STATEMENT: Figure 35.6 is a sample problem taken from an engineering textbook. Solve this problem with a spreadsheet.

SOLUTION: Before turning to the spreadsheet, we can first solve the problem by hand in a conventional way. To begin, the given variables and constants are

- Height of cliff $= y_0 = -150$ m
- Initial velocity $= v_0 = 180$ m/s
- Angle $= 30°$
- Acceleration of gravity $= a = -9.8$ m/s

The equations used for the solutions are as follows. First, the vertical and horizontal components of the initial velocity are

$$(v_y)_0 = v_0 \sin \theta$$
$$(v_x)_0 = v_0 \cos \theta$$

The equation for vertical motion is [Eq. (2) from Fig. 35.6]

$$y = (v_y)_0 t + \tfrac{1}{2}at^2$$

The projectile hits the ground when y equals the cliff height. Substituting y_0 for y and rearranging the equation gives

$$\tfrac{1}{2}at^2 + (v_y)_0 t - y_0 = 0$$

The root of this quadratic is the value of time that the projectile is in flight:

$$t = \frac{-(v_y)_0 - \sqrt{(v_y)_0^2 + 2ay_0}}{a}$$

Substituting this result into the equation for uniform motion [Eq. (4) from Fig. 35.6] provides the horizontal distance traveled by the projectile:

$$x = (v_x)_0 t$$

The maximum elevation can be computed on the basis of Eq. (3) from Fig. 35.6,

$$v_y^2 = (v_y)_0^2 + 2ay$$

The maximum elevation occurs when $v_y = 0$, or

$$0 = (v_y)_0^2 + 2ay_{max}$$

which can be solved for

SAMPLE PROBLEM 11.7

A projectile is fired from the edge of a 150-m cliff with an initial velocity of 180 m/s, at an angle of 30° with the horizontal. Neglecting air resistance, find (*a*) the horizontal distance from the gun to the point where the projectile strikes the ground, (*b*) the greatest elevation above the ground reached by the projectile.

    Solution. We shall consider separately the vertical and the horizontal motion.

    Vertical Motion. Uniformly accelerated motion. Choosing the positive sense of the *y* axis upward and placing the origin *O* at the gun, we have

$$(v_y)_0 = (180 \text{ m/s}) \sin 30° = +90 \text{ m/s}$$
$$a = -9.81 \text{ m/s}^2$$

Substituting into the equations of uniformly accelerated motion, we have

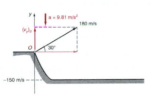

$$v_y = (v_y)_0 + at \qquad v_y = 90 - 9.81t \qquad (1)$$
$$y = (v_y)_0 t + \tfrac{1}{2}at^2 \qquad y = 90t - 4.90t^2 \qquad (2)$$
$$v_y^2 = (v_y)_0^2 + 2ay \qquad v_y^2 = 8100 - 19.62y \qquad (3)$$

    Horizontal Motion. Uniform motion. Choosing the positive sense of the *x* axis to the right, we have

$$(v_x)_0 = (180 \text{ m/s}) \cos 30° = +155.9 \text{ m/s}$$

Substituting into the equation of uniform motion, we obtain

$$x = (v_x)_0 t \qquad x = 155.9t \qquad (4)$$

    a. Horizontal Distance. When the projectile strikes the ground, we have

$$y = -150 \text{ m}$$

Carrying this value into Eq. (2) for the vertical motion, we write

$$-150 = 90t - 4.90t^2 \qquad t^2 - 18.37t - 30.6 = 0 \qquad t = 19.91 \text{ s}$$

Carrying $t = 19.91$ s into Eq. (4) for the horizontal motion, we obtain

$$x = 155.9(19.91) \qquad x = 3100 \text{ m}$$

    b. Greatest Elevation. When the projectile reaches its greatest elevation, we have $v_y = 0$; carrying this value into Eq. (3) for the vertical motion, we write

$$0 = 8100 - 19.62y \qquad y = 413 \text{ m}$$
$$\text{Greatest elevation above ground} = 150 \text{ m} + 413 \text{ m}$$
$$= 563 \text{ m}$$

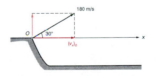

FIGURE 35.6
A problem dealing with projectile motion. (*From Beer and Johnston, Vector Mechanics for Engineers: Dynamics, 1984.*)

$$y_{max} = -\frac{(v_y)_0^2}{2a}$$

This result can be used to compute the maximum elevation above the ground by subtracting the cliff height:

$$y'_{max} = -\frac{(v_y)_0^2}{2a} - y_0$$

where the prime is used to indicate that this is the height above the ground.

The boxed formulas above are the relationships that must be solved in sequence to obtain the answers. For example, using the given variables and constants results in

$$(v_y)_0 = 180\sin 30 = 90 \text{ m/s}$$

$$(v_x)_0 = 180\cos 30 = 155.88 \text{ m/s}$$

$$t = \frac{-90 - \sqrt{90^2 + 2(-9.81)(-150)}}{-9.81} = 19.886 \text{ s}$$

$$x = 155.88(19.886) = 3100.0 \text{ m}$$

$$y'_{max} = -\frac{90}{2(-9.81)} - (-150) = 562.84 \text{ m}$$

Aside from minor discrepancies due to significant figure differences, these results are consistent with those of Fig. 35.6.

The same computation can be set up as a spreadsheet (Fig. 35.7). The commands for accomplishing this are listed in Fig. 35.8.

To this point, the manual and the spreadsheet computations demand similar levels of effort. A little more work is necessary to set up the spreadsheet, but this effort is compensated for by the fact that a neat report of the computation has been developed.

In addition, once the spreadsheet is set up, it can be easily employed to investigate how changes in the given variables will affect the outcome. For example, suppose that we desired to increase the range of the cannon and had to decide whether to build a platform in order to increase the initial height or to purchase a higher-powered cannon in order to increase the initial velocity.

To gain insight into this question, we determine how much the maximum range is lengthened by a 10 percent increase in either the initial height or the initial velocity. First we determine the maximum range under present conditions. This can be done by trying a number of different angles with the following results:

Angle	Distance
30	3099.98
35	3304.75
40	3422.46
45	3446.49
50	3373.91

The computations suggest a maximum at about 45°. We try other angles in this vicinity, with the final result of 3449.48 m for an angle of 43.8° determined as the maximum.

Now we can analyze how increasing the height or the initial velocity will increase the range. For example, increasing the height to 165 m increases the maximum range to 3463.82 m for an angle of 43.6°. This means that a 10 percent increase in height yields about a 0.4 percent increase in distance.

For the 10 percent increase in initial velocity, a much greater range is obtained. For an initial velocity of 198 m/s, the maximum distance is 4143.61 m, with an angle of 44°. This represents a 20.1 percent increase in the maximum range. Therefore, all other things (such as cost) being equal, the increase in firepower is a much more effective option.

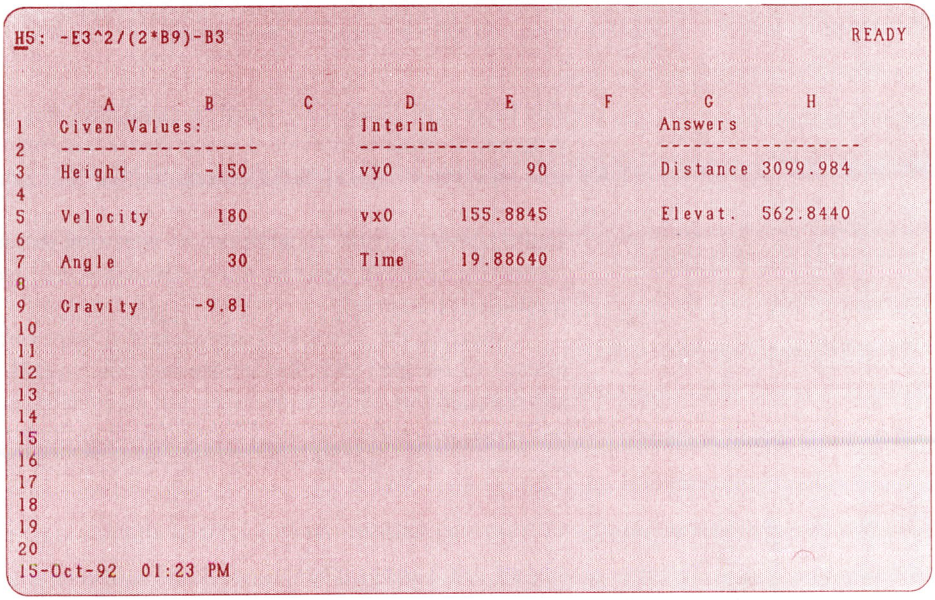

H5: -E3^2/(2*B9)-B3   READY

	A	B	C	D	E	F	G	H
1	Given Values:			Interim			Answers	
2	-------------			-------			-------	
3	Height	-150		vy0	90		Distance	3099.984
4								
5	Velocity	180		vx0	155.8845		Elevat.	562.8440
6								
7	Angle	30		Time	19.88640			
8								
9	Gravity	-9.81						
10								
11								
12								
13								
14								
15								
16								
17								
18								
19								
20								

15-Oct-92   01:23 PM

FIGURE 35.7
The spreadsheet for the projectile motion problem.

FIGURE 35.8
Commands needed to program the projectile problem on a spreadsheet.

```
A1:  'Given values
D1:  'Interim
G1:  'Answers
A2:  \-
B2:  \-
D2:  \-
E2:  \-
G2:  \-
H2:  \-
A3:  'Height
B3:  -150
D3:  'vy0
E3:  +B5*@SIN(B7*@PI/180)
G3:  'Distance
H3:  +E5*E7
A5:  'Velocity
B5:  180
D5:  'vx0
E5:  +B5*@COS(B7*@PI/180)
G5:  'Elevation
H5:  -E3^2/(2*B9)-B3
A7:  'Angle
B7:  30
D7:  'Time
E7:  (-E3-@SQRT(E3^2+2*B9*B3))/B9
A9:  'Gravity
B9:  -9.81
```

The above is called a *sensitivity analysis.* Its intent is to build up insight regarding the behavior of the system being studied. Once the spreadsheet is set up and entered into the personal computer, the entire analysis takes less than 5 min. The ease with which such analyses can be implemented suggests the great potential of spreadsheets for enhancing and broadening the scope of our analytical work.

35.7   THE ELECTRONIC TOOLKIT

The Electronic TOOLKIT, which accompanies this text, includes a simple spreadsheet program. In order to provide this software at no cost, some advanced capabilities available on other spreadsheets have been omitted. For example, it is small, consisting of a

single screen of cells. In addition, its operation differs slightly from the trade spreadsheets. For example, its input line is located at the bottom of the screen, and there is no menu-tree structure.

Although the TOOLKIT spreadsheet is somewhat limited, it is very easy to learn to use and interpret (full documentation is included in Appendix D). It is offered to you as *freeware* that is quite adequate for many engineering applications and classroom problems. In addition, as we will mention after the following example, it has the advantage that it is integrated with the other elements of the TOOLKIT.

EXAMPLE 35.2   *Computing Grades with a Spreadsheet*

PROBLEM STATEMENT: Use the spreadsheet from the Electronic TOOLKIT to compute the average grade in a course. Have the spreadsheet determine the average quiz and homework grades. Then compute the final grade FG by

$$FG = \frac{WQ * AQ + WH * AH + WF * FE}{100}$$

where WQ, WH, and WF are the weighting factors for quizzes, homework, and the final exam, respectively; AQ and AH are the average quiz and homework grades, respectively; and FE is the final-exam grade. Use the following data: WQ = 35; WH = 25; WF = 40; quizzes = 96, 94, and 85; and homework = 85, 88, 82, 100, 100, 95, and 96. Use the spreadsheet to determine the final-exam grade needed to obtain an A in the course—that is, to receive a final grade of 90 or better.

SOLUTION: The commands and data to set up the spreadsheet are shown in Fig. 35.9. After you have entered all these commands and data, the only missing variable is the final-exam grade. As a first try, you assign a grade of 80 for the final exam. When this value is entered into cell B15, the computer instantaneously calculates a "bottom line" or final grade of 87.155 (Fig. 35.10). Therefore, an 80 on the final exam would not earn an A.

```
[B1] quizzes           [A13] average
[C1] homeworks         [A15] fin exam=
[B2] - - - - - - - -   [G1] weight
[C2] - - - - - - -     [F2] - - - - - - - -
[A3] number            [G2] - - - - - - - -
[B4] - - - - - - - -   [F3] quizzes=
[C4] - - - - - - - -   [F4] homework=
[A5] grades            [F5] fin exam=
[B12] - - - - - - - -  [A18] fin grade
[C12] - - - - - - - -
[B13] = (B5 + B6 + B7 + B8 + B9 + B10+ B11)/B3
[C13] = (C5 + C6 + C7 + C8 + C9 + C10 + C11)/C3
[B18] = (G3 * B13 + G4 * C13 + G5 * B15)/100
```

(a)

FIGURE 35.9
(a) The commands to set up a spreadsheet to determine the grade in a course. The dashes are used to make the spreadsheet more coherent and visually appealing. (b) The data for the spreadsheet to determine a final grade in a course.

```
[B3] 3       [C3] 7       [C8] 100     [G3] 35
[B5] 96      [C5] 85      [C9] 100     [G4] 25
[B6] 94      [C6] 88      [C10] 95     [G5] 40
[B7] 85      [C7] 82      [C11] 96
```

(b)

FIGURE 35.10
The spreadsheet to determine the final grade for a course.

```
                              ─────Spreadsheet─────
CELL │  A        B        C        D      E        F        G        H

 1            quizzes  homeworks                        weight
 2            ----------------                      ------------------
 3  │number   3.0000   7.0000                       quizzes   35.0000
 4            ----------------                      homework= 25.0000
 5  │grades   96.0000  85.0000                      fin exam= 40.0000
 6            94.0000  88.0000
 7            85.0000  82.0000
 8                     100.0000
 9                     100.0000
10                      95.0000
11                      96.0000
12            ----------------
13  │average  91.6667  92.2857
14
15  │fin exam= 80.0000
16
17
18  │fin grade 87.1548
19
20
    └─F1help─F2copy─F3edit─F4calc─F5auto/man─F6row/col─F7trans─F8clear─F9keys─┘
[B15] 80
```

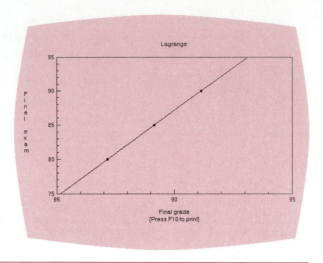

FIGURE 35.11
Plot to determine the
final-exam grade needed to
attain a final grade of 90.

Now, at this point, the real power of the spreadsheet becomes evident. If you were performing this computation by hand, you would have to recalculate the weighted average to determine the effect of changing the final-exam grade. In contrast, because the computation is already set up on the spreadsheet, all that must be done is to enter another final grade to cell B15, and the bottom line is instantaneously modified. For example, if you enter an 85, the final grade is automatically upgraded to 89.155. You could then try another value, say 90, and again the outcome of 91.155 would be immediate. Thus, within seconds, you have obtained the following results:

Final Exam	Final Grade
80	87.155
85	89.155
90	91.155

You could plot the final exam versus the final grade (Fig. 35.11) and use an "eyeball" estimate or linear interpolation to determine that a final exam of 87.1 is needed to obtain a final grade of 90. This can be verified by entering 87.1 to cell B15, with the result that cell B18 shows approximately 90.

It should be noted that the plot is not even necessary. Because the results are almost instantaneous, you can simply try a number of different values in a trial-and-error fashion until you come up with the desired result.

Although the Electronic TOOLKIT spreadsheet has limited capabilities when compared with some of the large commercial software, as in the previous example, it is easy to use and perfectly adequate for many routine classroom engineering problems (see Appendix D and Chap. 37 for additional examples). In addition, it has the advantage that it is integrated with the other programs on the TOOLKIT. For example, a column of numbers can be generated within the spreadsheet and then conveniently analyzed with some of the numerical methods options from the TOOLKIT. It is this integrated nature that makes the TOOLKIT unique and useful.

For example, Fig. 35.11 was created in exactly this manner by transferring data generated in the TOOLKIT spreadsheet to a program that performs Lagrange interpolation. The plot shows the three data points along with the line that matches the data.

PROBLEMS

35.1 Define or identify the following terms:

block	cursor	menu
cell	label	value
cell address	VisiCalc	1-2-3
active cell	Quattro Pro	Borland

35.2 Determine the number of columns on the spreadsheet you are using for this course. Describe how the letter designation scheme for column location works.

35.3 Duplicate Fig. 35.4 by performing Hands-On Exercise 35.3. Then perform the following operations:

(*a*) Move to cell B7, and enter the formula

```
@avg(B2..B4)
```

Do this by pointing, anchoring, and painting. What is the answer? Is it correct?

(*b*) Enter the following formulas in the cells below B7, and report your answers:

```
@var(B2..B4)
@std(B2..B4)
@min(B2..B4)
@max(B2..B4)
```

(*c*) Make a few changes in the numbers in cells B2, B3, and B4; see how the results of all the formulas are updated.

35.4 Use a spreadsheet to solve for the real roots of the parabola

$$ax^2 + bx + c = 0$$

where the roots are given by

$$\text{Roots} = \frac{-b \pm \sqrt{b^2 - 4ac}}{2a}$$

Input various values for a, b, and c; and compute the roots. Note the results of the computations when $a = 0$ and when $4ac > b^2$.

35.5 A boom of length L supports a weight W of 5000 lb, as shown in Fig. P35.5. Solve for the stress in the cable and the boom as a function of θ and W. Set the solution up on a spreadsheet. Vary θ and the length of the boom in a manner such that the cable maintains an angle of 90° with the wall. Output values for the length of the boom, the length of the cable, and the stresses in the boom and cable as θ varies from 0 to 75°.

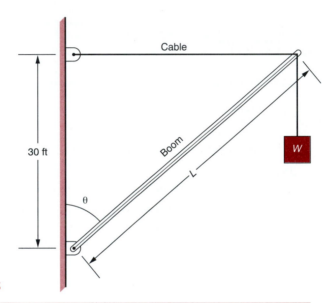

FIGURE P35.5

35.6 A block of mass M is at rest (Fig. P35.6). The coefficient of static friction between the block and the horizontal surface is θ. Investigate the forces between the block and the surface and the motion of the block as the angle between the horizontal surface and the plane varies between 0 and 75°. Try various values of M and θ, using a spreadsheet.

FIGURE P35.6

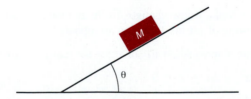

35.7 A current I divides into two parallel resistances of R_1 and R_2 (Fig. P35.7). Determine the voltage across the parallel combination and the current in each resistor for the following values:

$I = 2, 5, 10$

$R_1 = 2, 4, 6$

$R_2 = 4, 7, 9$

Use a spreadsheet to perform your analysis.

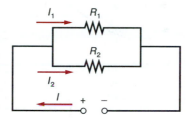

FIGURE P35.7

35.8 Use a spreadsheet to monitor your weekly expenses and income. Adjust the data, if necessary, to show a balanced monthly budget.

35.9 A triangular field is defined by three points A, B, and C with coordinates (ax, ay), (bx, by), and (cx, cy), as seen in Fig. P35.9. Determine how much fence is required to enclose the field for various values of (ax, ay), (bx, by), and (cx, cy). Use a spreadsheet to facilitate your computations.

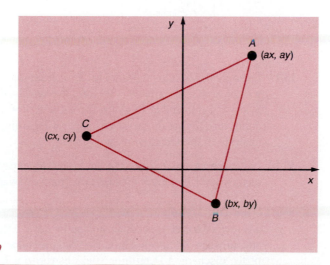

FIGURE P35.9

35.10 A chemical reactor converts a toxic organic compound dissolved in water to carbon dioxide, as shown in Fig. P35.10. The efficiency of the system depends on the flow rate of the water (Q), the volume of the reactor (V), and the decomposition rate of the toxic material (K):

$$\text{Eff} = \frac{Q}{Q + KV}$$

Use a spreadsheet to complete the following table:

Efficiency	Volume, in^3	Flow Rate, m^3/day	Decomposition Rate, 1/day
0.90	100	?	0.10
0.95	?	90	0.08
0.98	150	50	?
?	200	120	0.08
0.99	150	?	0.08
0.99	150	20	?

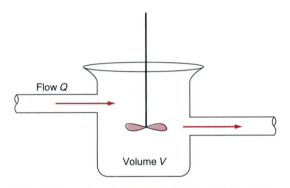

Flow Q

Volume V

FIGURE P35.10

35.11 Use a spreadsheet to program Fig. 35.1. Examine how a 10 percent increase of the y value at $x = 35$ affects r. Perform the same sensitivity analysis for the y value at $x = 25$ (remember to return the value at $x = 35$ to its original value prior to doing this). What do the results suggest regarding the sensitivity of a regression analysis to the location of the points?

35.12 In environmental engineering the following equation can be used to compute the oxygen level in a river downstream from a sewage discharge:

$$c = 10 - 15(e^{-0.1t} - e^{-0.5t})$$

where t is travel time downstream in days. Determine the travel time downstream where the oxygen level first falls to a reading of 4. (*Hint:* It is within 5 d of the discharge.) Determine your answer to a 1 percent error.

35.13 The van der Waals equation provides the relationship between the pressure and volume for a nonideal gas as

$$\left(p + \frac{a}{v^2}\right)(v - b) = RT$$

where p is pressure, a and b are parameters that relate to the type of gas, v is molal volume, R is the universal gas constant ($0.082054 \text{ L} \cdot \text{atm/mol} \cdot °\text{K}$), and T is absolute temperature. If $T = 350$ K and $p = 1.5$ atm, employ a spreadsheet to compute v for ethyl alcohol ($a = 12.02$ and $b = 0.08407$).

35.14 An oscillating current in an electric circuit is described by

$$I = 10e^{-t} \sin(2\pi t)$$

where t is in seconds. Employ a spreadsheet to determine all values of t such that $I = 2$.

35.15 The equation for a reflected standing wave in a harbor is given by

$$h = h_0 \left[\sin\left(\frac{2\pi x}{\lambda}\right) \cos\left(\frac{2\pi t v}{\lambda}\right) + e^{-x} \right]$$

Use a spreadsheet to solve for x if $h = 0.5h_0$, $\lambda = 20$, $t = 10$, and $v = 50$.

TABLE OPERATIONS

In the present chapter, we illustrate how spreadsheets can be used to produce and manipulate tables of data. Before showing how such tables can be created, we will first review the idea of how a table can act as a database. This includes an illustration of how rows and columns of sequential numbers can be conveniently generated with a data fill. We will also show how such tables can be sorted.

Next, we show how static tables can be produced in order to perform sensitivity analyses. Then, we will present an overview on the important concepts of cell moving and copying. Finally, we present information on developing dynamic tables by copying cell formulas and capitalizing on absolute cell references.

36.1 TABLES AS DATABASES

Engineers are continually organizing information and must be capable of evaluating it in a systematic fashion. The most direct example is the compilation of the technical data that is typically amassed during an engineering project. Beyond such technical information, other examples abound on the business side of engineering. The spreadsheet provides one means to organize such information and analyze it in an effective and efficient manner.

A *database* can be generally defined as a collection of data organized in a fashion that permits access and use of that data. In general there are a variety of ways that a database can be organized. One straightforward approach, referred to as a *relational database,* is organized in a tabular format that is very much like the structure of a spreadsheet.

A relational database consists of a collection of individual pieces of information called *items* or *elements.* Examples might be a name or social security number. These, in turn, are organized into larger entities called *records.*

For example, in your university's mainframe computer there is undoubtedly a record referring to your academic standing. It probably consists of a number of items,

TABLE 36.1   A Database for Toxic Substance Concentrations in Fish
for Lakes Near a Superfund Site

Field
↓

Sample	Lake	Fish	PCB	Toxaphene
080884	Green	Perch	25	100
270491	Portage	Bass	2	4
010990	Green	Bass	50	50	← *Record*
082185	Adams	Bass	10	8
101087	Adams	Perch	15	12
240789	Adams	Perch	10	23
050886	Portage	Perch	2	10
120987	Green	Bass	15	78

Item or element

including your name, social security or identification number, class, department, grades, and so on. This record is in turn part of the database that contains the academic records of all the students at your school.

Note that the columns in the relational database are referred to as *fields*. Thus, for your academic record, there is a name field, an identification number field, and so on. Table 36.1 shows an example of a database taken from engineering.

High-level programming languages (such as FORTRAN 90, Pascal, C, or modern BASIC) provide statements that can be employed to develop computerized relational databases. In addition, software manufacturers have designed specialized computer programs for the express purpose of effectively managing databases. Called *database management systems* (or *DBMS*), these programs serve as an interface between the user and the database. As such, they are designed to expedite and simplify the process of creating, accessing, and maintaining databases.

Although they were not originally developed for that purpose, spreadsheets provide a tool for database management that is intermediate between programming languages and a full-blown DBMS. In particular, they exhibit the tabular organization that is at the heart of the relational database. Understanding this, spreadsheet developers have included a variety of capabilities that allow spreadsheets to act as databases. One purpose of the present chapter will be to describe a few of these features.

36.1.1   Sorting

The sort is among the most fundamental database operations. Sorting the database consists of ordering the records (that is, the rows) according to the values in one or more *key fields* (in other words, particular columns). For example, for the database shown in Table 36.1, you could sort the records alphabetically in terms of the lake name or numerically using the PCB concentration.

Most spreadsheets allow you to choose up to five key fields. For example, if you choose lake name as the primary key field and PCB concentration as the secondary, the records are sorted first according to the primary field, then by the secondary field, and so on. The following exercise is designed to show how this is done.

HANDS-ON EXERCISE 36.1   Entering and Sorting a Database

GOAL: To set up a spreadsheet as a relational database and to demonstrate how the information in this database can be sorted.

BACKGROUND: Table 36.1 shows a database that might be used by an environmental engineer to evaluate toxic substance contamination in fish. Each record consists of a number of fields, including a sample number, the lake where the fish was collected, the type of fish, and the level or concentration of two toxic substances in the fish. Enter this database into a spreadsheet, and then sort the data using the lake name as the primary key and the PCB concentration as the secondary key.

EXERCISES:

1. Start up your spreadsheet, and enter the database shown in Table 36.1 Enter the headings commencing in cell A1 and the data commencing in cell A2. Aside from column F, the final result should appear as in Fig. 36.1. Note that some of the heading labels are left-justified (') and others are right-justified (") in order to make the table visually appealing. Your version should look the same way.

2. Before commencing the sorting process, you will first implement a *data fill* in order to generate a sequence of evenly spaced integers in column F. This is done so that you can easily re-sort the data back to its original order at a later time.

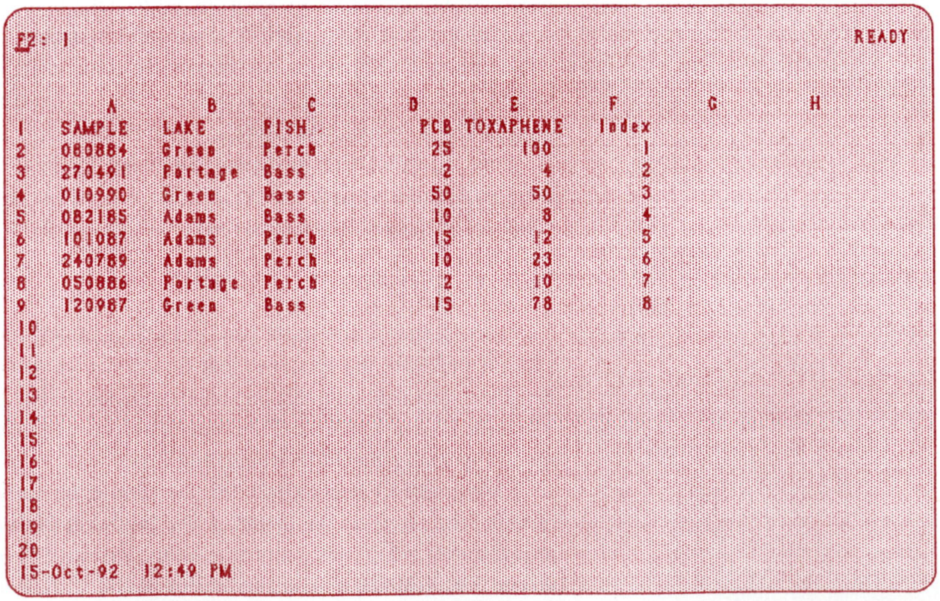

FIGURE 36.1
Spreadsheet set up for the
database from Table 36.1.

F2: 1							READY
	A	B	C	D	E	F	G	H
1	SAMPLE	LAKE	FISH	PCB	TOXAPHENE	Index		
2	080884	Green	Perch	25	100	1		
3	270491	Portage	Bass	2	4	2		
4	010990	Green	Bass	50	50	3		
5	082185	Adams	Bass	10	8	4		
6	101087	Adams	Perch	15	12	5		
7	240789	Adams	Perch	10	23	6		
8	050886	Portage	Perch	2	10	7		
9	120987	Green	Bass	15	78	8		
10								
11								
12								
13								
14								
15								
16								
17								
18								
19								
20								

15-Oct-92 12:49 PM

Enter the label "Index in cell F1. Move the active cell to F2. Initiate the fill by the commands

    123: / d f   QP: / e f

At this point, you will be asked to enter the destination block for the values. Anchor with the [.] key, and paint downward to the bottom of the block: F2..F9. Press the [Enter] key. A dialogue box now appears where you are prompted for a start value, with 0 as the default. Type a 1, and strike [Enter]. Next, for the step value, accept the default value of 1. Finally, the default stop value is 8191 (that is, the maximum number of rows in the spreadsheet). However, the fill will reach the end of your range long before it gets to 8191, so just hit [Enter]. You should observe that column F becomes "filled" with numbers from 1 to 8.

3. Now we will show how this table can be sorted. In the present case, lake name (column B) is the primary key field and PCB concentration (column D) is the secondary.

First, position the highlight in cell A2. A sort can then be initiated by the menu command sequence

    / d s

At this point, you will be presented with the sort menu. Select

    123: **d** (for **d**ata range)   QP: **b** (for **b**lock)

You will then be prompted for a cell range. Either point to or type in the cell block you wish to sort (in the present case: A2..F9), and strike [Enter].

Next, you must specify the primary key field. Do this by selecting

    123: **p** (for **p**rimary key)   QP: **1** (for **1**st key)

Again you will be prompted for a cell range. Enter B2..B9 by either pointing or typing. For this case, you will then be queried whether you would desire to sort in ascending or descending order. Because we want to sort the names in alphabetical order, select **a** for **a**scending.

Specify that the PCB field is to be the secondary key field by selecting

    123: **s** (for **s**econdary key)   QP: **2** (for **2**nd key)

and enter D2..D9. Because we want the highest concentrations toward the top of the list, select **d** for **d**escending.

Finally, select **g** for **g**o, and the spreadsheet will be sorted. The results are revealing (Fig. 36.2). Among other things, the sort makes it clear that Green Lake is the most polluted of the three waterbodies.

A2: '101087 READY

	A	B	C	D	E	F	G	H
1	SAMPLE	LAKE	FISH	PCB	TOXAPHENE	Index		
2	101087	Adams	Perch	15	12	5		
3	240789	Adams	Perch	10	23	6		
4	082185	Adams	Bass	10	8	4		
5	010990	Green	Bass	50	50	3		
6	080884	Green	Perch	25	100	1		
7	120987	Green	Bass	15	78	8		
8	270491	Portage	Bass	2	4	2		
9	050886	Portage	Perch	2	10	7		
10								
11								
12								
13								
14								
15								
16								
17								
18								
19								
20								

15-Oct-92 12:52 PM

FIGURE 36.2 Spreadsheet set up for the database from Table 36.1, but sorted according to lake name and PCB concentration.

36.2 SENSITIVITY ANALYSES WITH STATIC TABLES

Suppose that you have a formula that is a function of several variables and parameters. You might want to vary one or more of these parameters and see how the result changes. Because such numerical experiments provide a way to detect the "sensitivity" of the formula to the parameter variations, it is called a *sensitivity analysis.*

There are two basic approaches for developing formula sensitivity analyses on a spreadsheet. Both involve tables, but do so in fundamentally different ways. In the present section, we will illustrate static sensitivity (or "what if?") tables. In a later section, we will present an alternative approach involving dynamic tables.

36.2.1 One-Way Sensitivity Tables

We will now show how a one-way sensitivity table can be generated. The terminology refers to the fact that we will examine how a formula varies in response to changes of a single parameter.

HANDS-ON EXERCISE 36.2 One-Way Sensitivity Table

GOAL: To create a one-way sensitivity table to analyze the ideal gas law.

BACKGROUND: Recall that the ideal gas law is

$$pV = nRT$$

where p = absolute pressure [atm], V = volume [L], n = the number of moles [mole], R = the universal gas constant [= 0.082054 L atm/(mole °K)], and T = absolute temperature [°K]. For the present case, you will employ the law to solve for the molal volume, v [L/mole], by rearranging it to yield

$$v = \frac{V}{n} = \frac{RT}{p} \qquad\qquad (X36.2.1)$$

Employ this formula to compute values of the molal volume for $T = 300$ °K and pressures corresponding to $p = 20, 40, \ldots, 100$ atm.

EXERCISES:

1. Start up a new spreadsheet, and make the following entries:

    [A1]: 'One-way sensitivity table

    [A3]: 'R =   [B3]: 0.082054   [C3]: 'L atm/(mol K)

    [A4]: 'T =   [B4]: 300   [C4]: 'K

    [A6]: "v (L/mol): [A10]: "p(atm):

2. Now move the cursor to B8, and perform a data fill to create a column of pressures. Recall from Exercise 36.1 that the data fill begins with the commands

    123: / d f   QP: / e f

Start with 20, use a step of 20, and stop with 100.

3. When you have completed these entries, move to cell C7 and enter Eq. (X36.2.1). A good way to do this is by pointing (recall Hands-On Exercise 35.3). Type a plus sign + and then point (that is, use the arrow keys to move) to the universal gas constant in cell B3. Type a multiplication operator * and point to the temperature in cell B4. Finally, type a division operator / and point to the blank cell B7. This blank cell is referred to as the *input cell* because it is the cell referenced by the formula. The formula should now look like

    [C8]: +B3 * B4/B7

When it is entered, the value ERR will be displayed due to a division by zero.

4. One way to make this result look nicer is to change the format of the cell. Make sure that you are still in cell C7, and then strike the menu sequence

    123: / r f   QP: / s n

Select **t** for **t**ext. Enter cell range C7..C7. The result will be that the formula rather than the result will be displayed.

5. Notice that the column is not wide enough to fit the entire formula in cell C7. This can be rectified by using the following commands to expand the column width:

123: / w c s   QP: / s c

After these keystrokes have been implemented, you will be prompted to enter the desired column width. Enter 10; the column will be expanded, and the entire formula will be displayed.

6. Although the appearance of cell C7 has been improved, it is still not very communicative. This is because it is written in terms of cell references rather than in terms of variable names as in Eq. (X36.2.1). This deficiency can be rectified by naming the cells that hold the variables. To do this, move the highlight to B3 and implement the following commands:

123: / r n c   QP: / e n c

You will be prompted for a name. Enter R, and then strike [Enter] again to select the cell range B3..B3. Repeat this procedure, and name cell B4 as T and B7 as V. At this point, the formula in cell C7 should now appear as +R*T/V (note that on certain spreadsheets, you may have to strike the recalc key [F9] once to see the entire formula in its new format).

7. Now we will create the sensitivity table. To do this, press the menu sequence

123: / d t   QP: / t w

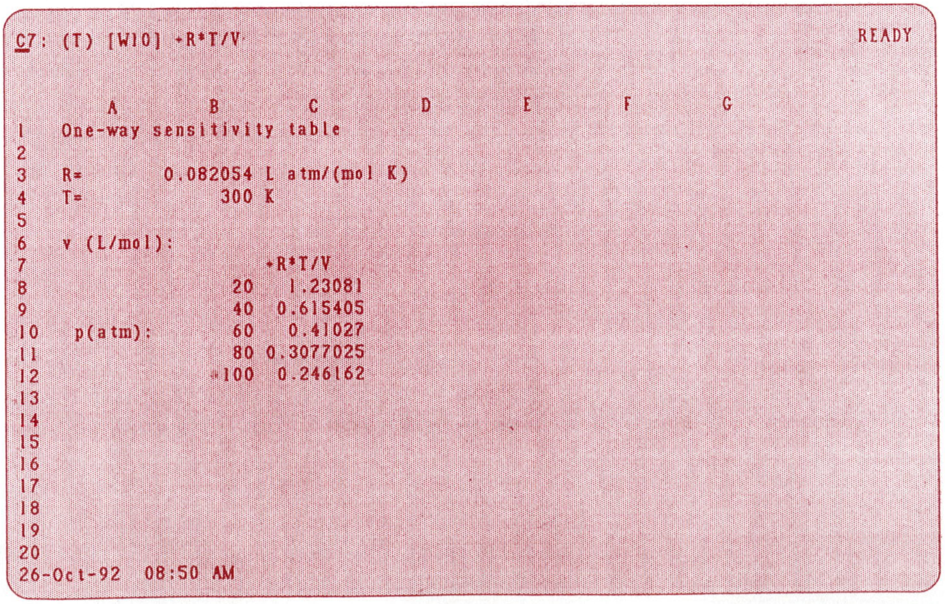

FIGURE 36.3
Spreadsheet screen showing molal volumes calculated as a function of pressure with a one-way sensitivity table.

```
C7: (T) [W10] +R*T/V                                                    READY

         A        B        C        D        E        F        G
 1   One-way sensitivity table
 2
 3   R=        0.082054 L atm/(mol K)
 4   T=            300 K
 5
 6   v (L/mol):
 7                         +R*T/V
 8                 20    1.23081
 9                 40   0.615405
10   p(atm):      60    0.41027
11                 80  0.3077025
12                100   0.246162
13
14
15
16
17
18
19
20
26-Oct-92  08:50 AM
```

Next select **1** to indicate that you want to perform a **1**-variable sensitivity computation. You will then be queried to enter a table range. For our case, paint out and enter the range B7..C12. Finally, you will be queried for an input cell; enter B7.

The one-way sensitivity table will be displayed. As in Fig. 36.3, it shows how the molar volume changes as a function of variations in the partial pressure.

36.2.2 Two-Way Sensitivity Tables

Although the one-way tables are useful, there will be many occasions where you might want to determine how a formula varies as a function of changes of two parameters. We will now show how a two-way sensitivity table can be generated for this purpose.

HANDS-ON EXERCISE 36.3   Two-Way Sensitivity Table

GOAL: To create a two-way sensitivity table to analyze the ideal gas law.

BACKGROUND: As in Exercise 36.2, you will employ the law to solve for the molal volume, v [L/mole],

$$v = \frac{RT}{p} \qquad\qquad (X36.3.1)$$

where p = absolute pressure [atm], R = the universal gas constant [= 0.082054 L atm/ (mole °K)], and T = absolute temperature [°K]. Employ this formula to compute values of the molal volume for all combinations of temperature and pressure corresponding to T = 300, 500, and 700 °K, and p = 20, 40, . . . , 100 atm.

EXERCISES:
1. Start up a new spreadsheet, and make the following entries:

> [A1]: 'Two-way sensitivity table
>
> [A3]: 'R =   [B3]: 0.082054   [C3]: 'L atm/(mol K)
>
> [A10]: "p(atm): [A6]: "v (L/mol):   [D6]: ^T(K)

2. Now move the cursor to B8, and perform a data fill to create a column of pressures. Recall from Exercise 36.1 that the data fill begins with the commands

> 123: / d f   QP: / e f

Create the fill in the range B8..B12. Start with 20, use a step of 20, and stop with 100.
3. Move the cursor to C7, and perform a data fill to create a row of temperatures. Create the fill in the range C7..E7. Start with 300, use a step of 200, and stop with 700.

4. When you have completed these entries, move to cell B7 and enter Eq. (X36.3.1) by pointing. Type a plus sign + and then point (that is, use the arrow keys to move) to the universal gas constant in cell B3. Type a multiplication operator ⋆ and point to the blank cell, A8. Finally, type a division operator / and point to the blank cell A7. The formula should now look like

    [C8]: +B3 ⋆ A8/A7

When it is entered, the value ERR will be displayed due to a division by zero. Make this result look nicer by changing the format of the cell to text and naming the cells that hold the variables: R, T, and v (recall Exercise 36.2).
5. Now we will create the sensitivity table. To do this, press the menu sequence

    123: / d t   QP: / t w

Select **2** to indicate that you want to perform a **2**-variable sensitivity computation. You will then be queried to enter a table range. For our case, paint out and enter the range B7..E12. Next, you will be queried for the first input cell; enter A7. Finally, you will be queried for the second input cell; enter A8.

At this point, the two-way sensitivity table will be displayed. As shown in Fig. 36.4, the table indicates how the molar volume changes as a function of variations in both the partial pressure and the temperature.

FIGURE 36.4
Spreadsheet screen showing molal volumes calculated as a function of pressure and temperature with a two-way sensitivity table.

```
A7:                                                                           READY

         A          B          C          D          E          F          G          H
 1   Two-way sensitivity table
 2
 3   R=        0.082054
 4
 5                           L atm/(mol K)
 6   v (L/mol):                        T(K)
 7          +R*T/V          300        500        700
 8              20    1.23081    2.05135    2.87189
 9              40   0.615405   1.025675   1.435945
10   p(atm):    60    0.41027   0.683783   0.957296
11              80   0.307702   0.512837   0.717972
12             100   0.246162    0.41027   0.574378
13
14
15
16
17
18
19
20
26-Oct-92  09:11 AM
```

In the previous exercise, the values within the table are "dead." That is, they are merely constants that will not change unless you modify the formula or some of the temperatures and/or pressures, and recompute the sensitivity table. This is why we refer to this type of table as *static*.

Now that we have seen how this is done, we will concoct a table that is very similar but dynamic. However, before doing this, we must review some extremely significant aspects related to cell moving and copying.

36.3   CELL MOVING AND COPYING

All spreadsheets have the capability to move and copy cells. As described in the following exercise, there are some fundamental concepts connected with these operations.

HANDS-ON EXERCISE 36.4   Moving and Copying Cells

GOAL: To learn how cells can be moved and copied, and to appreciate the subtleties associated with these operations.

EXERCISES:

1. Start up your spreadsheet, and enter the following information into the cells:

[B1]: 1   [C1]: 2   [A2]: 6   [A3]: 5

[B2]: +B1 + A2

The result should be as shown in Fig. 36.5*a*. The value in cell B2 (7) is equal to the summation of the values above it (1) and to its left (6).

FIGURE 36.5
The contrast between "moving" and "copying" cells on an electronic spreadsheet.

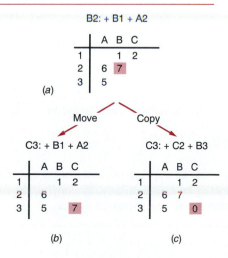

2. Now you will move the information in B2 to another cell, C3. First, go to A1 by striking the [Home] key. A cell move is initiated by the menu command sequence

123: / m   QP: / e m

At this point, you will be asked for the source block of cells that are to be moved. In addition, the spreadsheet will automatically fill in a cell range corresponding to the present active cell. In this case, it should display A1..A1. Now move the highlight to B2, and observe how the cell range changes. Move back to cell A1 either by using the arrow keys or by striking [Home]. Unanchor the cell range by pressing [Esc], and again move to B2. Then, hit [Enter] to specify that the source block is B2.

The spreadsheet now prompts you for the destination to which you are moving. Again, the active cell A1 appears. However, this time, notice that it is not represented as a range and, therefore, does not need to be unanchored. Move the highlight to cell C3, and strike [Enter]. The 7 should leave cell B2 and reappear in cell C3.

Now move the highlight to C3, and examine the contents of the cell (see Fig. 36.5b) As expected, the formula, +B1 + A2, has moved intact to the new location.

3. Move the formula back to B2. The spreadsheet should again resemble Fig. 36.5a. Now you will copy the formula in B2 to C3. First, initiate the copy by the menu sequence

123: / c   QP: / e c

At this point, you will be asked to enter the source and destination blocks in a similar fashion to the cell move. Perform these operations. As you are doing this, notice how the steps are a little easier to implement because you started with the highlight on the source cell, B2. The finished result should look just like Fig. 36.5c.

In contrast to the move operation, notice some significant differences. First, the source block B2 is unchanged. This makes sense because you "copied" the cell, not "moved" it. Second, the formula that has been copied to C3 has been modified to +C2 + B3. Therefore a value of 0 shows up in C3.

In summary, the original contents of B2,

+B1 + A2

when copied to C3 have been transformed to

+C2 + B3

Thus, you see that by copying the formula one column to the right, each column reference is increased by one. Similarly, by copying the formula to a location down a row, each row reference is also increased by one.

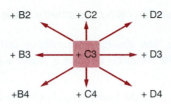

FIGURE 36.6
A simple illustration of how the formula cell references are modified in a relative fashion when the formula is copied to other cells.

As illustrated in the foregoing exercise, the cell copy is a *relative* operation. In other words, if a formula is copied to a cell that is a certain number of rows and columns away, the row and column references for each term in the formula are modified accordingly. Figure 36.6 illustrates this behavior. Notice in particular that when formulas are copied horizontally (that is, from column to column), only the column references change (the letters), whereas when they are copied vertically (that is, from row to row), only the row references change (the numbers).

Now, suppose that you want to copy a formula but you do not want these relative changes to occur. Most spreadsheets provide a simple way to do this by having you enter a dollar sign ($) prior to the cell reference you would like to remain fixed. For example, suppose that you wanted to perform the cell copy shown in Fig. 36.5 but you wanted to keep the column references fixed. To do this, you would edit the formula in B2 so that it looked like

    [B2]: +$B1 + $A2

When this formula was copied, the resulting value in C3 would be 12, which corresponds to the formula

    [C3]: +$B2 + $A3

Such fixed cell references are referred to as *absolute cell references*. This is in contrast to the *relative cell references* without dollar signs. As illustrated in the following section, their existence is a critical ingredient in the effective use of spreadsheet formulas.

36.4 DYNAMIC TABLES

In contrast to the static tables previously presented in Sec. 36.2, we will now use absolute and relative cell references to construct a dynamic sensitivity table. By *dynamic*, we mean that the table will be composed of live formulas rather than numbers (as was the case for the static tables). Consequently, the values displayed in the table will change immediately when you modify the values of parameters that are included in their underlying formulas.

HANDS-ON EXERCISE 36.5   Creating a Dynamic Two-Way Sensitivity Table

GOAL: To explore how absolute and relative cell references can be effectively utilized to copy scientific formulas in order to create a dynamic two-way sensitivity table.

BACKGROUND: As in the earlier Hands-On Exercises 36.2 and 36.3, you will employ the ideal gas law to solve for the molal volume, v [L/mole], as in

$$v = \frac{RT}{p} \tag{X36.5.1}$$

where p = absolute pressure [atm], R = the universal gas constant [= 0.082054 L atm/
(mole °K)], and T = absolute temperature [°K]. Employ this formula to compute values
of the molal volume for all combinations of temperature and pressure corresponding to
T = 300, 500, and 700 °K, and p = 20, 40, ..., 100 atm.

EXERCISES:
1. Start up a new spreadsheet, and make the following entries:

```
[A1]:  'Two-way sensitivity table (Dynamic version)
[A3]:  'R =       [B3]: 0.082054      [C3]: 'L atm/(mol K)
[A10]: "p(atm):  [A6]: "v (L/mol):  [D6]: ^T(K)
```

2. Now move the cursor to B8, and perform a data fill to create a column of pressures.
Recall from Exercise 36.1 that the data fill begins with the commands

```
123: / d f                          QP: / e f
```

Create the fill in the range B8..B12. Start with 20, use a step of 20, and stop
with 100.
3. Move the cursor to C7, and perform a data fill to create a row of temperatures. Create the fill in the range C7..E7. Start with 300, use a step of 200, and stop with 700.
4. When you have completed these entries, move to cell C8 and enter Eq. (X36.5.1). A
good way to do this is by pointing (recall Hands-On Exercise 35.3). Type a plus
sign + and then point (that is, use the arrow keys to move) to the universal gas constant
in cell B3. Type a multiplication operator * and point to the temperature in cell C7.
Finally, type a division operator / and point to the proper pressure in cell B8. The formula should look like

```
[C8]: +B3 * C7/B8
```

which when entered will result in the correct answer, 1.23081, being displayed in C8.
5. Copy cell C8 to the range C8..E12 (if you forget how to do this, review the necessary information from Hands-On Exercise 36.4). The result should look like

	A	**B**	**C**	**D**	**E**
1	Two-way sensitivity table (Dynamic version)				
2					
3	R=	0.082054	L atm/(mol K)		
4					
5					
6	v (L/mol):			T(K)	
7			300	500	700
8		20	1.23081	0	ERR
9		40	0	ERR	ERR
10	p(atm):	60	0	ERR	ERR
11		80	0	ERR	ERR
12		100	0	ERR	ERR

Clearly, the relative nature of the copy operation has resulted in an erroneous outcome. This can be seen clearly by inspecting the individual formulas. For example,

    [D9]: +C4 * D8/C9

You no longer have the ideal gas law, but rather some strange amalgam of cells, including a division by zero which leads to the result ERR!

6. To remedy the situation, return to cell C8 and go into EDIT mode [F2]. Use the arrow keys to position the cursor anywhere under the B3 term in the formula. Press the [F4] key, and note that the term is changed to $B$3, meaning that both the row and the column references are made absolute. Now hit the [F4] key several more times, and observe that it changes from $B$3 to B$3 to $B3 and finally back to B3, whereupon the sequence repeats itself. Thus, the [F4] represents a toggle switch that cycles through all the possible absolute and relative cell reference combinations.

Now press [F4] to get back to $B$3. Then move the cursor under the C7 term which represents temperature. When this term is copied, you would like every cell to have the temperature that is aligned above it. Thus, you would like the row reference to be absolute and the column to be relative. Therefore, strike [F4] two times so that C7 becomes C$7.

Finally, move the cursor under the B8 term which represents pressure. When this term is copied, you would like every cell to have the pressure that is aligned to the left of it. Thus, you would like the row reference to be relative and the column to be absolute. Therefore, press [F4] three times so that B8 becomes $B8. At this point, the final formula should look like

    [C8]: +$B$3 * C$7/$B8

Press [Enter], and the correct result will appear in C8.

```
C8:  +$B$3*C$7/$B8                                                    READY


        A          B          C          D          E      F      G      H
1  Two-way sensitivity table (Dynamic version)
2
3  R=           0.082054 L atm/(mol K)
4
5
6  v (L/mol) :                          T(K)
7                         300        500        700
8                    20  1.23081    2.05135    2.87189
9                    40  0.615405  1.025675   1.435945
10 p(atm) :          60  0.41027   0.683783   0.957296
11                   80  0.307702  0.512837   0.717972
12                  100  0.246162   0.41027   0.574378
13
14
15
16
17
18
19
20
15-Oct-92  01:12 PM
```

FIGURE 36.7
Spreadsheet screen showing molal volumes calculated as a function of pressure and temperature with a dynamic table.

7. Now, copy cell C8 to the range C8..E12. The result is shown in Fig. 36.7. Because of the use of relative and absolute cell references, the molal volumes have been calculated correctly. You can verify this by computing some of the values with your pocket calculator. Also, move the highlight from cell to cell, and study the formula entries to make sure that you understand how the absolute and relative cell references operate.

8. Move the highlight to cell C7, and change the first temperature to 100. Observe that the first column immediately recalculates. Similarly, move to B12, change the pressure to 200, and see that the last row instantly adjusts. This is what we mean when we say that this type of table is "dynamic."

Note that the Electronic TOOLKIT software provided with the text copies and moves cell contents on a relative basis. Absolute copies or moves are not possible. On the other hand, the TOOLKIT is set up so that spreadsheet data can be conveniently transferred into tables and matrices for subsequent numerical methods operations. We will review these capabilities at the end of the next chapter.

PROBLEMS

36.1 Consider the spreadsheet shown on the following page:

	A	B	C	D
1	1	3		
2	2	4		

(*a*) If the formula entered in C1 is +B2 ^ 2, what is the value that results in C1?

    a) 1   b) 4   c) 9   d) 16   e) None of the above

(*b*) If that formula is copied to cell D2, what value will appear in D2?

    a) 1   b) 4   c) 9   d) 16   e) None of the above

(*c*) If the formula $A1 ^ 2 is entered in cell C1 and copied to cell D2, what value will appear in D2?

    a) 1   b) 4   c) 9   d) 16   e) None of the above

(*d*) If the formula $A$1 ^ 2 is entered in cell C1 and copied to cell D2, what value will appear in D2?

    a) 1   b) 4   c) 9   d) 16   e) None of the above

36.2 Consider the spreadsheet shown below:

	A	B	C	D	E	F
1					x	
2			1	2	3	4
3		2				
4		4				
5	y	6				
6		8				
7		10				

You want to complete a table in the range C3 . . F7 which evaluates the function

$$|x^2 - y^2|$$

for the respective row values of *y* and column values of *x*. Write down precisely what formula you would enter in cell B2, which could then be copied throughout the range C3..F7 to create the desired format.

36.3 Consider the spreadsheet shown below:

	A	B	C
1	1	5	10
2	2	6	11
3	3	7	12
4	4	8	13

If the cursor is in cell A1, what happens if the following keys are struck

123: / r e [Enter]   QP: / e e [Enter]

followed by [End] ↓ ?

a) Cell A1 will be erased, and the cursor will end up in B2.
b) Cell A1 will be erased, and the cursor will end up in A1.
c) The entire range of data will be erased.
d) The first column of data will be erased.
e) None of the above

36.4 Enter the following Student Aptitude Test (SAT) data into a spreadsheet:

ID Number	Name	Math	Verbal
551556681	Reckhow, Sarah	736	710
337653878	Chen, David	650	680
263748294	Morales, Juan	690	550
876770183	Hornacek, Heather	620	640
627839874	Kelly, Carlos	720	480
863454637	Jones, Isaac	780	715

Develop a fifth field to hold the total (that is, math plus verbal) score. Sort the table in descending order by the final score.

36.5 Enter the following information into a spreadsheet:

Name	Team	AB (At Bats)	H (Hits)
Kruk	Phil	355	121
Van Slyke	Pitt	378	126
Sheffield	SD	390	127
Grace	Chi	380	121
Butler	LA	376	119
DeShields	Mon	408	128

Add a fifth field for the batting average (= H/AB) for each player. Note that batting averages are typically expressed to three decimal places (for example, 0.305). Format your spreadsheet to display it in this manner. Sort the table in descending order by batting average.

36.6 A modification of the ideal gas law, which can handle nonideal conditions, is the Redlich and Kwong equation of state:

$$P = \frac{RT}{V - b} - \frac{a}{\sqrt{T}V(V + b)}$$

where

$$a = \frac{0.4278R^2T_c^{2.5}}{P_c}$$

$$b = \frac{0.0867RT_c}{P_c}$$

where P_c = the critical pressure of the substance [atm] and T_c = the critical temperature [K]. These parameters have been compiled for numerous gases. For example,

Compound	T_c	P_c
Methane	190.6	45.4
Ethylene	282.4	49.7
Nitrogen	126.2	33.5
Water	647.1	217.6

Develop a spreadsheet program to compute pressure for methane for a particular temperature and a range of volumes. Test it for $T = 400$ K and V ranging from 100 to 900 L. Employ a one-way static table to investigate the sensitivity.

36.7 Repeat Prob. 36.6, but employ a two-way static sensitivity table to investigate the behavior of pressure as a function of a range of temperatures (100 to 500 K) and volumes (100 to 900 L).

36.8 Using the information from Prob. 36.6, develop a two-way dynamic sensitivity table to investigate the behavior of pressure as a function of a range of temperatures (100 to 500 K) and volumes (100 to 900 L). Set up the spreadsheet so that it is general (that is, not set up for a specific compound). Then test it for all the compounds in Prob. 36.6.

36.9 In engineering economics the annual payment, A, on a debt that increases linearly with time at a rate G for n years is given by the following equation:

$$A = G\left[\frac{1}{n} - \frac{n}{(1 + i)^n - 1}\right]$$

Develop a one-way static sensitivity table to study how A varies as a function of interest rates ranging from 0.05 to 0.25.

36.10 Repeat Prob. 36.9, but use a two-way static sensitivity table to investigate the behavior of A as a function of a range of interest rates (0.05 to 0.25) and n (5 to 30 years).

36.11 In environmental engineering the following formula is employed to calculate the dissolved oxygen concentration in a river at a location x downstream from a sewage treatment plant effluent:

$$c = \frac{k_1 D_0}{k_1 - k_2}(e^{-k_2(x/u)} - e^{-k_1(x/u)})$$

Develop a one-way static sensitivity table to study how c varies as a function of distance downstream ranging from 0 to 20 mi. Enter all the equation parameters as named cells. Employ the following values as a test case: $D_0 = 10$ mg/L, $k_1 = 0.2$ d^{-1}, $k_2 = 0.1$ d^{-1}, and $U = 0.1$ ft s^{-1}. Be careful that your units are consistent.

36.12 The saturation value of dissolved oxygen in fresh water is calculated by the equation

$$c_s = 14.652 - 0.41022T_c + 7.9910 \times 10^{-3}T_c^2 - 7.7774 \times 10^{-5}T_c^3$$

where T_c is temperature in degrees Celsius. Saltwater values can be approximated by multiplying the freshwater result by $1 - (9 \times 10^{-6})n$, where n is salinity in milligrams per liter. Finally, temperature in degrees Celsius is related to temperature in degrees Fahrenheit, T_F, by

$$T_c = \tfrac{5}{9}(T_F - 32)$$

Develop a spreadsheet to calculate c_s, given a range of n and T_F. Employ a two-way sensitivity table for your computation. Employ values for n of 0 to 30,000 and for T_F of 32 to 100.

36.13 Repeat Prob. 36.12, but employ a dynamic table.

36.14 Population growth in an industrial area is given by the following formula:

$$p(t) = P(1 - e^{-at})$$

where $p(t)$ = population at any time ≥ 0, P = maximum population, a = constant [yr], and t = time [yr]. Develop a one-way sensitivity table to compute the population at $t = 0$ to 20 years with $P = 100,000$ for $a = 0.05$.

36.15 Repeat Prob. 36.14, but develop a static two-way sensitivity table to compute the population at $t = 0$ to 20 years with $P = 100,000$ for $a = 0.05$ to 0.5 yr^{-1}.

36.16 Repeat Prob. 36.15, but use a dynamic table. Enter the value of P as a cell.

DATA ANALYSIS

In the present chapter, we will illustrate how spreadsheets can be used to manipulate, view, and analyze information. First, we will demonstrate how graphs can be produced. Then, we will explore the statistical capabilities with special emphasis on regression. Next, we will show how spreadsheets can be employed to solve equations. Finally, we will review some of the data analysis capabilities of the Electronic TOOLKIT that comes with the text.

37.1   GRAPHICS

One of the really convenient and powerful aspects of spreadsheets is the ability to concoct graphs and charts. Before making such plots, we must first enter values into the spreadsheet. In the following exercise, we will employ an equation to generate a table of values.

HANDS-ON EXERCISE 37.1   Creating Values from a Function

GOAL: To create a table of (x, y) coordinates by entering and copying formulas.

BACKGROUND: A wide variety of engineering processes exhibit behavior that can be modeled with the exponential and the sinusoidal functions. As in Fig. 37.1a and b, the former decays whereas the latter oscillates. Oftentimes a system's behavior consists of a combination of the two shapes called a *damped sinusoid* (Fig. 37.1c).

$$y = e^{-at} \cos (bt)$$

where y and t are the dependent and independent variables, respectively, and a and b are constants that dictate the temporal scaling of the two functions.

As an example of an engineering application of the damped sinusoid, y might be the vertical displacement of a spring (such as an automobile shock absorber) and t would be time. Thus, as in Fig. 37.1c, the damped sinusoid provides a simple model of how the shock absorber moves vertically after a vehicle hits a bump in the road.

In the present exercise, you will generate a table of values consisting of time, the two component parts of the damped sinusoid, and the total equation.

EXERCISES:

1. Start up a new spreadsheet, and make the following entries:

```
[A1]: 'Analysis of the damped-sinusoid equation
[A3]: 'a=            [A4]: 'b=            [B3]: 0.5
[B4]: 3             [A6]: "time         [B6]: "exp
[C6]: "cos          [D6]: "y
```

Notice how different justification codes (' and ") have been employed in order to make some of the labels left-justified and others right-justified.

2. You will now implement a *data fill* in order to generate a sequence of evenly spaced times in column A. Move the active cell to A8. Initiate the fill by the commands

FIGURE 37.1
Graphs of (*a*) the decaying exponential function, (*b*) the cosine sinusoidal function, and (*c*) a composite function called the damped sinusoid. Notice how different values of the parameters influence the way in which the functions are scaled in time.

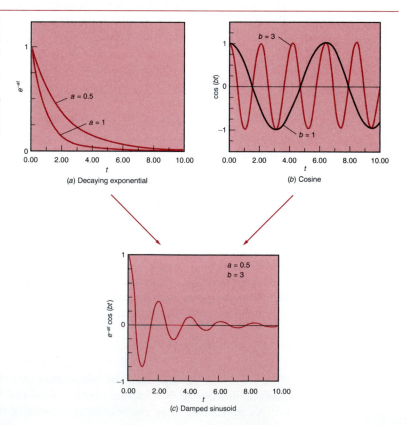

(a) Decaying exponential

(b) Cosine

(c) Damped sinusoid

123: / d f   QP: / e f

At this point, you will be asked to enter the fill range or destination block for the values. Anchor with the [.] key, and paint downward (use [PgDn] to paint 20 cells at a time and the arrow keys to adjust to the exact range of 101 cells: A8..A108). Press the [Enter] key. A dialogue box now appears where you are prompted for a start value with 0 as the default. Accept this with the [Enter] key. Next, for the step value, enter 0.1. The default stop value is 8191 (that is, the maximum number of rows in the spreadsheet), but the fill will reach the end of your range long before it gets to 8191, so just hit [Enter]. You should observe that column A becomes "filled" with numbers. Check the bottom of the column by switching into the END mode by striking the [End] key and then pressing the [↓] key. The value 10 should be in the last cell A108.

3. Notice that the various values are displayed in column A with or without decimal fractions as required, even though they are stored internally to 15 significant figures. To improve the appearance, it is possible to have all the numbers in a column appear in the same format. To do this, move back to the top of the column of numbers with the [End] followed by the [↑] key. Then, execute the following key sequence:

123: / r f   QP: / s n

Now select **f** for **fixed**, and note that the default number of fractional digits is 2. Type **1** to change this, followed by the [Enter] key. The display should now request the cell range that is to be formatted, which is automatically set to the active cell: A8..A8. Now you need to paint out the entire column again. Use [End] and [↓] to do this efficiently. Finish off with the [Enter] key, and your column should be reformatted nicely.

4. Now you will enter and copy the formulas. Move the active cell to B8. Type the first part of the exponential function (remember not to strike [Enter] until the formula is complete),

@exp(-

Then, point with the arrow keys to cell B3 (note how B3 is entered automatically into the formula). Press [F4] once to make the cell reference absolute: $B$3. (Actually, for the present case, you really only need to make the row reference absolute, as in $B3; however, either way will work.) Type the multiplication operator ∗ and then point left to A8. End by typing the closed parenthesis) and then finish by pressing [Enter]. The formula should be displayed in the input line as

[B8]: @exp(-$B$3∗A8)

and the result displayed in cell B8 should be 1.

5. You will now copy this formula to the cells directly below. With the active cell in B8, enter the copy commands and proceed until you reach the prompt that asks for the "where to?" or destination block. It should be set up for cell B8. Anchor the B8 cell by pressing the [.] key. You will now paint down through the range of cells by using a

technique called "hanging on another column." Use the [←] key to paint one cell left to A8. Then hit [End] and [↓], and finally move back one cell to the right with the [→] key. The destination block range should read B8..B108. This procedure took four keystrokes and is generally faster than other methods. Now finish the copy by pressing [Enter]. As a final touch, set the format of the B column to display 3 digits to the right of the decimal point.

6. Enter the cosine function into column C. Start the process by making the following entry:

[C8]: @cos($B$4*A8)

Then copy this formula into the other cells in column C.

7. Enter the product of columns B and C into column D. For example,

[D8]: +B8 * C8

Then copy this formula into the other cells in column D. To improve the appearance of the spreadsheet, set the format of the C and D columns to display 3 digits to the right of the decimal point.

Now that a table of values has been generated, we can display these values on a graph. Commercial spreadsheets offer a number of ways to do this. From an engineering perspective, the most useful of these plots are *XY* graphs.

HANDS-ON EXERCISE 37.2   Creating an *XY* Graph of a Function

GOAL: To graph a table of (x, y) coordinates generated from an equation.

BACKGROUND: In the present exercise, you will create a plot of the values previously generated in Exercise 37.1. These are the two component parts of the damped sinusoid and the total equation.

EXERCISES:

1. After setting up the spreadsheet previously generated in Exercise 37.1, you are ready to produce a graph. First, we have to get into GRAPH mode and specify the graph type. This can be done with the commands

123: / g t   QP: / g g

For the present example, you will choose **x** for an **xy** plot. Note that spreadsheets often use the variables x and y to refer to the independent and dependent variables you are plotting. Of course, in our present example, although you are using the same nomenclature for the dependent variable (y), your independent variable is time (t).

2. Next, you must indicate the ranges of the data that you wish to plot. You can start by

specifying the x range. To do this in 1-2-3, you merely select **x**. For Quattro Pro, select **s** (for **s**eries) and **x** (for **x**-axis range). At this point, you will be queried regarding the range of the x values. Move the active cell to A8. Anchor and paint out the range A8..A108.

3. Now you should specify the ranges for the three y values we have generated in columns B, C, and D. Note that 1-2-3 and Quattro Pro allow you to plot several dependent variables. Lotus 1-2-3 designates these variables using letters (A through F), whereas Quattro Pro employs numbers (the first through the sixth series). In either case, the process is identical with setting the independent variable.

If you are using Quattro Pro, you must exit the **s**eries menu by either striking [Esc] or typing **q**. In 1-2-3 this is unnecessary. Finally, type **v** for **v**iew, and the graph will be displayed.

At this point, the graph may include both symbols and lines. Because we have plotted a function, the symbols detract from the graph's appearance. To rectify this, 1-2-3 and Quattro Pro take somewhat different approaches. If you are using 1-2-3, proceed to step 4. For Quattro Pro go to step 5.

4. *For 123 users:* Starting from the graph submenu, the following 1-2-3 commands can be executed to employ lines for your entire graph:

123: **o**ptions **f**ormat **g**raph **l**ines

After implementing these commands, you would return to the graph submenu by striking **q** twice. Then you can press **v** to view the new version of the graph. As displayed in Fig. 37.2, it should now consist solely of lines tracing out the three functions.

5. *For Quattro Pro:* Starting from the graph submenu, type **c** for **c**ustomize series. The screen will then be filled with a large table that can be employed to customize your graph. Notice that the screen consists of a series of menus which themselves consist of

FIGURE 37.2
Spreadsheet-generated graph of the damped sinusoid along with its component parts.

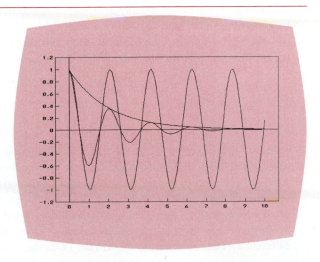

a number of options. The arrangement is particularly well suited for mouse-oriented systems. However, you can also move around among the menus by striking the [Tab] key or the space bar, by typing key letters, or by using the arrow keys. Within each of the menus, you can make individual selections in a similar fashion. You should experiment at this point to gain familiarity with these modes of movement.

Notice that each dependent variable can be customized individually, by selecting the individual series from the series menu at the top of the screen. For the present case, make sure that each graph has a format of lines. When you have done this, select **q** for quit to return to the graph menu. Then you can press **v** to view the new version of the graph. As displayed in Fig. 37.2, it should now consist solely of lines tracing out the three functions.

Aside from XY graphs, spreadsheets also provide the means to develop a variety of other styles of plots. These include bar and pie charts, which are extremely useful for engineering presentations. Some of the problems at the end of the chapter deal with these types of plots.

37.2　STATISTICS AND REGRESSION

Aside from graphics, spreadsheets also provide the means to summarize data with statistics and regression. The following sections provide an introduction to these capabilities. Additional background information can be found in Chaps. 22 and 23 of this book, or in any good statistics text.

37.2.1　Spreadsheet Statistics

Spreadsheet statistics are usually available in the form of @ functions. The most commonly available are summarized in Table 37.1.

Note that both the standard deviation and the variance represent population estimates. For example, the variance is calculated as

$$\sigma^2 = \frac{\sum (\mu - y_i)^2}{n} \tag{37.1}$$

where σ^2 is the symbol for the variance, μ is the population mean, y_i are the individual values, n is the number of values, and the summation is from $i = 1$ through n. The standard deviation is merely the square root of the variance.

Recall from our discussion of statistics in Chap. 22 that such formulas apply only when we are dealing with a complete population of values. If we are analyzing a subset or sample drawn from a population (which is often the case), we must use a slightly different formula to compute the sample variance,

TABLE 37.1   Statistical @ Functions Available on Most Commercial Spreadsheets

Function	Description
@AVG	The mean of all numeric values in the list.
@COUNT	The number of nonblank cells in the list.
@MAX	The largest numeric or last date value in the list.
@MIN	The smallest numeric or earliest date value in the list.
@STD	The population standard deviation of all the values in the list.
@SUM	The total of all numeric values in the list.
@VAR	The population variance of all the values in the list.

$$s_y^2 = \frac{\sum (\bar{y} - y_i)^2}{n - 1} \qquad (37.2)$$

where s_y^2 is the special symbol for the sample variance. Although some commercial spreadsheets provide special @ functions to compute both population and sample statistics, some of the commonly available ones do not. The following exercise shows how the sample standard deviation can be computed for such cases.

HANDS-ON EXERCISE 37.3   Applying Statistics to Summarize Data

GOAL: To develop statistics for a table of data points.

BACKGROUND: Aside from equations, engineers also employ experimental data. Such data often comes in the form of pairs of observations that embody a cause-effect relationship.

Such data is listed in Table 37.2. For this case, the dependent variable x (that is, the effect) is the temperature at which the reaction is conducted, whereas the independent variable y (the cause) is the percentage yield of a chemical reaction.

In the present exercise, you will enter these values into a spreadsheet and then determine a number of statistics for both the temperature and the percentage yield.

TABLE 37.2   Data for Percentage Yield of a Chemical Reaction as a Function of the Temperature at Which the Reaction Occurred

Temperature (°C)	Yield (%)
35	14
45	16
55	19
65	18
75	21

EXERCISES:

1. Start up a new spreadsheet, and make the following entries:

```
[A1]: 'Analysis of the percentage yield of a chemical
[A2]: 'reaction as a function of temperature
[B4]: "temp          [C4]: "%yield        [D4]: "predicted
[B5]: "x             [C5]: "y             [D5]: "y
[A13]: 'number       [A14]: 'minimum      [A15]: 'maximum
[A16]: 'average      [A17]: 'std dev
```

Notice how different justification codes (' and ") have been employed in order to make some of the labels left-justified and others right-justified.

Then from Table 37.2 enter the values for temperature in column B starting with cell B7, and for percentage yield in column C starting with cell C7.

2. Now enter the following formulas:

```
[B13]: @COUNT(B7..B11)
[B14]: @MIN(B7..B11)
[B15]: @MAX(B7..B11)
[B16]: @AVG(B7..B11)
[B17]: @STD(B7..B11)
```

Note that using pointing and painting can facilitate their entry.

3. Recognize that the result in cell B17 is the population standard deviation. For our case, this would be incorrect because clearly we have gathered only a sample of data. In order to rectify this error, move the highlight to cell B17 and enter the following formula based on Eqs. (37.1) and (37.2):

```
[B17]: @STD(B7..B11)*@SQRT(B13/(B13-1))
```

When this modification is entered, the sample standard deviation will appear in cell B17.

4. Move the cursor to cell B13. Copy the formulas in the range B13..B17 to the range C13..C17. When you are finished, the spreadsheets should resemble the left side of Fig. 37.3.

37.2.2 Spreadsheet Regression

Aside from graphics, spreadsheets also provide the means to summarize data with regression. Most packages provide the capability to perform linear regression. This involves specifying a range of cells containing independent and dependent variables. The linear regression algorithm will then determine the slope and intercept of the straight line that fits the data in an optimal sense. That is, the line minimizes the sum of the squares of the residuals between the line and the points.

```
B17: (F3) @STD(B7..B11)*@SQRT(B13/(B13-1))                          READY

          A        B        C        D        E        F        G        H
  1   Analysis of the percentage yield of a chemical
  2   reaction as a function of temperature
  3
  4                temp    %yield predicted         Regression Output:
  5                 x        y        y Constant                         8.8
  6                                     Std Err of Y Est            1.095445
  7                35       14      14.4 R Squared                  0.876712
  8                45       16        16 No. of Observations               5
  9                55       19      17.6 Degrees of Freedom               3
 10                65       18      19.2
 11                75       21      20.8 X Coefficient(s)      0.16
 12                              ------------- Std Err of Coef.  0.034641
 13   number        5        5
 14   minimum      35       14
 15   maximum      75       21
 16   average      55     17.6
 17   std dev   15.811    2.702
 18
 19
 20
 20-Oct-92   07:50 AM
```

FIGURE 37.3
Spreadsheet showing statistics and linear regression results for the percentage yield-temperature data from Table 37.2.

HANDS-ON EXERCISE 37.4   Applying Linear Regression to Summarize Data

GOAL: To develop a linear least-squares fit of tabulated xy data.

BACKGROUND: In the present exercise, you will develop a linear least-squares fit of tabulated xy data. In addition, you will plot the fit along with the original data.

EXERCISES:
1. After setting up the spreadsheet previously generated in Exercise 37.3, you are ready to fit a straight line to the data. First, we have to access the regression capability via the command sequence

    123: / d r   QP: / t a r

At this point, you will be queried for the ranges of the independent variable (x) and the dependent variable (y). For x, paint out B7..B11; and for y paint out C7..C11. Then select the location of your output range as E4. Type go, and the regression summary should appear.
2. Look over the regression summary. Note that the X coefficient is the slope and the constant is the intercept of the best-fit line. These values can be employed to generate predictions. Position the cursor in cell D7, and enter the formula

    +$H$5+$G$11*B7

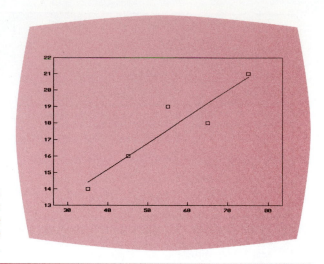

FIGURE 37.4
Spreadsheet-generated graph
of a least-squares fit of the
percentage yield-temperature
data from Table 37.2.

Copy this formula to the range D7..D11. Column D should show the best-fit values as calculated by the linear model. The completed spreadsheet is displayed in Fig. 37.3.

3. Now you can develop a plot of the original data along with the best-fit line (review Exercise 37.2 for details on plotting). Develop a plot with the independent variable (x) as the range B7..B11. Enter the first dependent variable as the range C7..C11 and the second dependent variable as the range D7..D11. Before viewing, make sure that symbols are used for the first range and lines for the second. The completed plot can be displayed by selecting view (or alternatively with the [F10] key). The result, as shown in Fig. 37.4, indicates that the linear model does an adequate job of capturing the trend of the data.

Note that all spreadsheets provide the means to add descriptive labels and titles to plots. You should explore these capabilities by consulting your user's manual or with the "Help" [F1] key. Also, versions of some of the popular commercially available spreadsheets provide additional capabilities beyond simple statistics and regression. For example, certain products provide an optimization capability that can be employed in a number of contexts including nonlinear regression. You can consult your users manual or "Help" facility [F1] to explore the capabilities of your own spreadsheet.

37.3   EQUATION SOLVING

Another standard capability available on most popular spreadsheets is linear equation solving. An example of a linear system of equations is

$$a_{11}x_1 + a_{12}x_2 + a_{13}x_3 = c_1$$

$$a_{21}x_1 + a_{22}x_2 + a_{23}x_3 = c_2 \qquad (37.3)$$

$$a_{31}x_1 + a_{32}x_2 + a_{33}x_3 = c_3$$

where the a's are constant coefficients, the c's are constants, and x's are the unknowns. Note that the unknowns are all raised to the first power and are multiplied by a constant. This is why such equations are called "linear."

Linear algebraic equations arise in a variety of problem contexts and in all fields of engineering. In particular, they originate in the mathematical modeling of large systems of interconnected elements such as structures, electric circuits, and networks.

In Chap. 26, we described a technique for solving such equations called Gauss elimination. Commercial spreadsheets employ a somewhat different approach involving matrices. To understand this approach we must introduce you to some fundamentals of matrix algebra.

37.3.1 Matrix Algebra

A *matrix* consists of a rectangular array of elements represented by a single symbol. For example,

$$A = \begin{bmatrix} a_{11} & a_{12} & a_{13} \\ a_{21} & a_{22} & a_{23} \\ a_{31} & a_{32} & a_{33} \end{bmatrix} \qquad (37.4)$$

where [A] is the shorthand notation for the entire matrix and a_{ij} designates an individual *element* of the matrix.

A horizontal set of elements is called a *row,* and a vertical set is called a *column.* The first subscript i always designates the number of the row in which the element lies. The second subscript j designates the column. For example, element a_{23} is in row 2 and column 3.

A matrix can be characterized by its dimensions. It has n rows and m columns and is said to have a *dimension* of n by m (or $n \times m$). Such a matrix is commonly referred to as an n by m matrix.

Matrices with column dimension $m = 1$, such as

$$\{C\} = \begin{Bmatrix} c_1 \\ c_2 \\ c_3 \end{Bmatrix}$$

are called *column vectors.* Note that for simplicity, the second subscript of each element is dropped. Also, it should be mentioned that it is often desirable to distinguish a column vector from other types of matrices. Consequently, we have employed special brackets, {}, to enclose the values.

Matrices such as the one in Eq. (37.4), where $n = m$, are called *square matrices*. Square matrices are particularly important when solving sets of simultaneous linear equations. For such systems, the number of equations (corresponding to the number of rows) and the number of unknowns (corresponding to the number of columns) must be equal in order for a unique solution to be possible. Consequently square matrices of coefficients are encountered when dealing with such systems.

There are a number of special types of square matrices. In the present context, the most important is the *identity matrix*,

$$[I] = \begin{bmatrix} 1 & 0 & 0 \\ 0 & 1 & 0 \\ 0 & 0 & 1 \end{bmatrix}$$

This matrix has properties similar to unity. Just as $a \times 1 = a$ in simple algebra, so also

$$[A] \times [I] = [A]$$

Matrix algebra must follow rules in the same fashion as algebraic manipulations of simple variables. In the present context, the two most important rules involve matrix multiplication and matrix inversion.

Matrix multiplication.   The *product* of two matrices is represented as $[C] = [A][B]$, where the elements of $[C]$ are defined as

$$c_{ij} = \sum_{k=1}^{n} a_{ik} b_{kj}$$

where $n =$ the column dimension of $[A]$ and the row dimension of $[B]$. That is, the c_{ij} element is obtained by adding the product of individual elements from the ith row of the first matrix, in this case $[A]$, by the product of individual elements from the jth column of the second matrix $[B]$.

It should be clear now that matrices provide a succinct notation for representing simultaneous linear algebraic equations. For example, Eq. (37.3) can be expressed concisely as

$$[A]\{X\} = \{C\} \tag{37.5}$$

where the matrix $[A]$ contains the coefficients

$$[A] = \begin{bmatrix} a_{11} & a_{12} & a_{13} \\ a_{21} & a_{22} & a_{23} \\ a_{31} & a_{32} & a_{33} \end{bmatrix}$$

the matrix $\{x\}$ contains the unknowns

$$\{X\} = \begin{Bmatrix} x_1 \\ x_2 \\ x_3 \end{Bmatrix}$$

and the matrix $\{C\}$ contains the constants

$$\{C\} = \begin{Bmatrix} c_1 \\ c_2 \\ c_3 \end{Bmatrix}$$

At this point you should apply the rule for matrix multiplication to convince yourself that Eqs. (37.3) and (37.5) are equivalent.

Matrix inversion.   Although multiplication is possible, matrix division is not a defined operation. However, if a matrix $[A]$ is square, there is usually another matrix $[A]^{-1}$ called the inverse of $[A]$, for which

$$[A][A]^{-1} = [A]^{-1}[A] = [I]$$

Thus, the multiplication of a matrix by its inverse is analogous to division, in the sense that a number divided by itself is equal to 1. That is, multiplication of a matrix by its inverse leads to the identity matrix.

37.3.2   Spreadsheet Matrix Manipulations

In high school, you undoubtedly solved simultaneous equations by elimination of unknowns. Computer techniques, such as Gauss elimination, use similar algebraic manipulations to obtain solutions. In contrast, commercial spreadsheets employ a somewhat different approach involving matrix inversion and multiplication.

The spreadsheet approach can be understood by formulating the simultaneous equations in matrix form,

$$[A]\{X\} = \{C\}$$

A formal way to obtain a solution using matrix algebra is to multiply each side of the equation by the inverse of $[A]$ to yield

$$[A]^{-1}[A]\{X\} = [A]^{-1}\{C\}$$

Because $[A]^{-1}[A]$ equals the identity matrix, the equation becomes

$$\{X\} = [A]^{-1}\{C\} \tag{37.6}$$

Therefore, if we multiply the inverse of the coefficient matrix, $[A]^{-1}$, times the matrix of constants, $\{C\}$, we will obtain the solution for the unknowns, $\{X\}$. This is another example of how the inverse plays a role in matrix algebra similar to division.

Spreadsheets include the capability to invert and multiply matrices. Thus, they can be employed to solve simultaneous equations in the manner depicted in Eq. (37.6). The following exercise illustrates how this is done.

HANDS-ON EXERCISE 37.5   **Solving Systems of Equations with Spreadsheet Matrix Operations**

GOAL: To illustrate how matrix operations can be employed to solve systems of linear algebraic equations with spreadsheets.

BACKGROUND: There are a variety of ways to determine the coefficients of interpolating polynomials of the form

$$f(x) = a_0 + a_1 x + a_2 x^2 + \cdots + a_n x^n \tag{X37.5.1}$$

For example, in Chap. 24, we developed an algorithm for Lagrange interpolation. Although the Lagrange polynomial provides a means for determining intermediate values between points, it does not provide a convenient polynomial of the form of Eq. (X37.5.1).

An alternative, which does result in an interpolating polynomial of this form, can be determined using simultaneous equations. For example, suppose that you desired to compute the coefficients of the parabola

$$f(x) = a_0 + a_1 x + a_2 x^2 \tag{X37.5.2}$$

Three points are required: $[x_0, f(x_0)]$; $[x_1, f(x_1)]$; and $[x_2, f(x_2)]$. Each can be substituted into Eq. (X37.5.2) to give

$$f(x_0) = a_0 + a_1 x_0 + a_2 x_0^2$$

$$f(x_1) = a_0 + a_1 x_1 + a_2 x_1^2$$

$$f(x_2) = a_0 + a_1 x_2 + a_2 x_2^2$$

Thus, for this case, the x's are the knowns and the a's are the unknowns. Because there are three equations with three unknowns, the equations can be solved for the a's.

Per capita water use in a city varies according to the following measurements:

Time	Hours after Midnight	Water Usage, million gallons per day (MGD)
4:00 A.M.	4	21
9:00 A.M.	9	86
2:00 P.M.	14	55
5:00 P.M.	17	71

Fit an interpolating polynomial to this data, and use it to estimate the water use at 12:00 noon.

SOLUTION:

1. Because we have four data points, a third-degree polynomial can be fit where y = water use and x = time,

$$y = a_0 + a_1 x + a_2 x^2 + a_3 x^3$$

This equation must be satisfied at 4:00 A.M., 9:00 A.M., 2:00 P.M., and 5:00 P.M. Substituting the values for time and water use yields the simultaneous equations

$$a_0 + 4a_1 + 16a_2 + 64a_3 = 21$$

$$a_0 + 9a_1 + 81a_2 + 729a_3 = 86$$

$$a_0 + 14a_1 + 196a_2 + 2744a_3 = 55$$

$$a_0 + 17a_1 + 289a_2 + 4913a_3 = 71$$

or in matrix form

$$\begin{bmatrix} 1 & 4 & 16 & 64 \\ 1 & 9 & 81 & 729 \\ 1 & 14 & 196 & 2744 \\ 1 & 17 & 289 & 4913 \end{bmatrix}$$

2. Start up your spreadsheet, and enter the values for labels and the matrices as shown in Fig. 37.5.

3. Position the cursor in cell [B11]. Strike the following menu sequence to begin the process of inverting the matrix

        123: / d m i   QP: / t a i

You will then be queried for the range to invert (or the source block of the cells). Paint out and enter the range B6..E9. Next, you will be asked to specify the output range or destination block for the inverse. Move the cursor to B11, and press [Enter]. The following values for the matrix inverse should appear in the block B11..F14:

3.295385	−4.76	4.08	−1.61538
−0.79538	1.81	−1.71333	0.698718
0.061538	−0.175	0.2	−0.08654
−0.00154	0.005	−0.00667	0.003205

4. Strike the following menu sequence to begin the process of matrix multiplication:

        123: / d m m   QP: / t a m

You will then be queried for the first range to multiply. Paint out and enter the range B11..E14. Next, you will be asked to identify the second matrix. Paint out and enter the range F6..F9. Finally, you will be asked to specify the destination block for the result. Move the cursor to B16, and press [Enter]. The values for the solution should appear in the block B16..B19. The final spreadsheet is shown in Fig. 37.6.

FIGURE 37.5
Spreadsheet set up to
evaluate the coefficients of a
cubic polynomial to fit the
data from Exercise 37.5.

```
B11:                                                                        READY

          A           B         C          D          E          F          G          H
   1  Evaluation of polynomial coefficients
   2
   3                          terms of polynomial
   4                  zero     first     second      third        y
   5
   6                    1         4        16         64         21
   7  original          1         9        81        729         86
   8  matrix:           1        14       196       2744         55
   9                    1        17       289       4913         71
  10
  11
  12  matrix
  13  inverse:
  14
  15
  16      a0=
  17      a1=
  18      a2=
  19      a3=
  20
20-Oct-92  08:01 AM
```

FIGURE 37.6
Final version of a spreadsheet
set up to evaluate the
coefficients of a cubic
polynomial to fit the data
from Exercise 37.5.

```
B19: 0.2585897436                                                           READY

          A           B         C          D          E          F          G          H
   1  Evaluation of polynomial coefficients
   2
   3                          terms of polynomial
   4                  zero     first     second      third        y
   5
   6                    1         4        16         64         21
   7  original          1         9        81        729         86
   8  matrix:           1        14       196       2744         55
   9                    1        17       289       4913         71
  10
  11              3.295384    -4.76      4.08   -1.61538
  12  matrix     -0.79538     1.81   -1.71333  0.698717
  13  inverse: 0.061538     -0.175      0.2   -0.08653
  14             -0.00153    0.005  -0.00666  0.003205
  15
  16      a0= -230.449
  17      a1= 94.33256
  18      a2= -8.90192
  19      a3= 0.258589
  20
27-Oct-92  02:54 PM
```

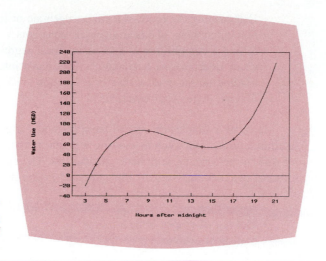

FIGURE 37.7
Spreadsheet-generated graph
of a cubic polynomial fit of
the data from Exercise 37.5.

The interpolating polynomial is, therefore,

$$y = -230.449 + 94.33256x - 8.90192x^2 + 0.25859x^3$$

This polynomial along with the data is displayed in Fig. 37.7. The polynomial can then be employed to predict the water usage at 12:00 noon as in

$$y = -230.449 + 94.33256(12) - 8.90192(12)^2 + 0.25859(12)^3 = 66.51 \text{ MGD}$$

37.4   ELECTRONIC TOOLKIT

Today's commercial spreadsheet was born because Daniel Bricklin and Robert Frankston found that conventional approaches to solving their business applications were inconvenient. With time, new users have expanded the application of spreadsheets into engineering and science. However, as you may have observed, sometimes these applications can be cumbersome and awkward.

The Electronic TOOLKIT software that comes with this text was specifically designed to solve the types of engineering and science problems you will confront routinely as a student or as a practicing professional engineer. Consequently, in contrast to the commercial spreadsheet products, its design makes it especially convenient for attaining solutions to such problems.

As you have seen in previous sections of this book, the heart of the TOOLKIT is a collection of numerical methods programs that solve the most common mathematical problems confronted routinely by engineers and scientists. The programs to solve these problems involve:

- Determining the roots of a single algebraic or transcendental equation
- Solving systems of linear algebraic equations
- Fitting a straight line to data using linear regression
- Using interpolating polynomials to estimate intermediate values between data
- Finding the integral of data or a function
- Solving differential equations

In order to implement any of these methods, engineers must often perform some preliminary analyses such as plotting a graph, statistically characterizing a collection of data, or performing arithmetic calculations. Therefore, the TOOLKIT has three additional modules that *support* the numerical programs. These are:

- Spreadsheet
- Plotting of graphs or data points
- Statistical analysis

In order for the package to be really useful, the eight components must be integrated or linked together. Thus, if we perform some calculations on the spreadsheet, we might want to transfer the results to other parts of the package for further analysis. For example, you might develop some columns of x and y data on the spreadsheet. It would be very convenient to perform linear regression on this data without retyping the points. Furthermore, you might then want to try polynomial interpolation on the same data set. Thus, you would like the individual modules to be able to share data tables.

The TOOLKIT has such capabilities along with some other features that make it uniquely suited to solve the kinds of problems you will routinely encounter. The exercises in this section are designed to introduce you to some of these features. In particular, we will focus on data analysis features that involve the TOOLKIT's spreadsheet module.

HANDS-ON EXERCISE 37.6   Regression, Plotting, and the Electronic TOOLKIT Spreadsheet

GOAL: To illustrate how the Electronic TOOLKIT spreadsheet can be employed to fit a power-law model to data with its integrated plotting and regression routines.

BACKGROUND: Engineers often use power relationships to perform fits of data with models. The following data represents the pressure exerted on a roof as a function of wind speed.

Wind Speed (km/h)	Pressure (kg/m^2)
2	7
3	15
5	35
7	67
9	105

It is proposed to fit this data with an empirical model of the form

$$y = ax^b \tag{X37.6.1}$$

where y = pressure [kg m^{-2}], x = wind speed [km h^{-1}], and a and b = empirical constants.

Use the Electronic TOOLKIT to analyze this data, and employ regression to find the best values for a and b.

SOLUTION: The first step is to see whether the data actually follows a power relationship. This can be accomplished by taking the logarithm of Eq. (X37.6.1) to give

$$\log y = \log a + b \log x$$

Consequently, if the power relationship holds, the data should lie on a straight line when plotted on a log-log scale. A plot of this type is very easy to generate using the TOOLKIT.

The data can be entered into the spreadsheet and transferred to the plot routines. From the TOOLKIT main menu, make selection [1] and enter data into the spreadsheet as shown in columns A and D from Fig. 37.8. Move the highlight to position B5, and type

```
=log(A5)/2.3
```

FIGURE 37.8
Electronic TOOLKIT spreadsheet of force and wind speed data from Exercise 37.6.

Spreadsheet

CELL	A	B	C	D	E	F	G	H
1	Test Data							
2								
3	wind	log-wind		force	log-force			
4								
5	2.0000	0.3014		7.0000	0.8460			
6	3.0000	0.4777		15.0000	1.1774			
7	5.0000	0.6998		35.0000	1.5458			
8	7.0000	0.8460		67.0000	1.8281			
9	9.0000	0.9553		105.0000	2.0235			
10								
11								
12								
13								
14								
15								
16								
17								
18								
19								
20								

F1help-F2copy-F3edit-F4calc-F5auto/man-F6row/col-F7trans-F8clear-F9keys

[A1] Test Data

The resulting value, 0.30137, is the common logarithm of 2. Note that the TOOLKIT's log function is the base e or natural logarithm and that the following formula can be used to compute common logarithms:

$$\log_{10} x = \frac{\log_e x}{2.3}$$

Next, use the [F1] key to copy the contents of cell B5 to the range B6 to B9. Note that the relative copy feature of the TOOLKIT is convenient here. Next move the highlight to E5 and enter

```
=log(D5)/2.3
```

The resulting value is the common logarithm of 7. Use the [F1] key to copy the entry in E5 to the range E6 to E9.

The data in columns A and D of the spreadsheet can be transferred to the X and Y columns of the data tables by using the [F6] key. Simply strike the [F6] key and specify the A5 to A9 range and the X column destination. Strike the [F6] key again, and transfer the D5 to D9 range to the Y column.

Next go to the main menu and select the plot program. Because of the previous steps that transferred the spreadsheet data to the XY data tables, the plotting submenu and selection [3] can be used immediately to plot the data. Note that the program requests the type of plot (linear, semilog, or log-log), the type of information to be plotted (function, data, or both), along with the plot scales and titles. Figure 37.9 shows the data plotted on a log-log scale. The results seem to indicate a linear relationship, which supports the contention that the power model is appropriate.

The relationship can be quantified by applying linear regression between log y and

FIGURE 37.9
Plot of force versus wind speed data on a log-log scale using the Electronic TOOLKIT.

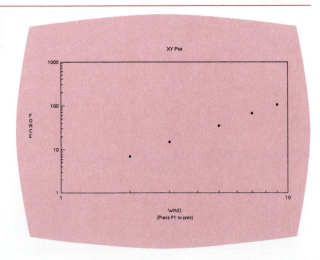

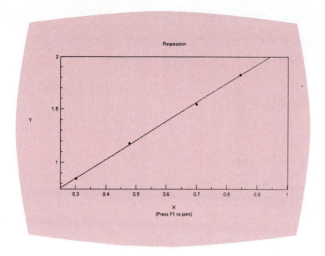

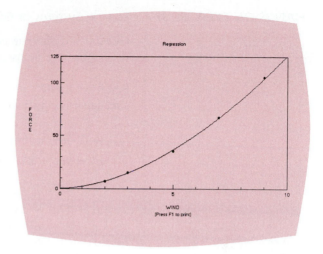

FIGURE 37.10 Plot of least-squares linear fit of log transformed force and wind speed data. Both the fit and the plot were generated with the TOOLKIT.

FIGURE 37.11 Plot of force and wind speed data along with power function fit from Exercise 37.6.

log x. Return to the spreadsheet, and use the [F6] key to transfer the data in columns B and E to the TOOLKIT's linear regression program. Then move to the regression module.

Regression is performed giving

$$\log y = 0.3099 + 1.79 \log x$$

with a correlation coefficient of 0.999. A plot of the regression curve can be easily generated using selection [3] from the regression submenu. This curve is shown in Fig. 37.10.

Taking the antilog of the equation gives

$$y = 2.04x^{1.79} \tag{X37.6.2}$$

Note that $10^{0.3099} = 2.04$. Return to the spreadsheet, and use the [F6] key to transfer the A and D columns to the TOOLKIT's graphics program. Then Eq. (X37.6.2) can be entered into the plotting program as $f_1(x)$ using selection [1] on the plotting submenu. The equation along with the untransformed data is shown in Fig. 37.11. Note that the TOOLKIT is easy to use. This is especially true because there are no special keystroke codes that need to be remembered.

The previous exercise used the spreadsheet, plotting, and regression capabilities. The following example illustrates some of the other TOOLKIT features to solve a problem previously encountered in Exercise 37.5.

HANDS-ON EXERCISE 37.7   Solving Systems of Equations with the Electronic TOOLKIT

GOAL: To illustrate how the Electronic TOOLKIT's spreadsheet and Gauss elimination routine can be used to solve systems of linear algebraic equations.

BACKGROUND: In Exercise 37.5, we were provided with the following set of simultaneous equations to fit a polynomial to data:

$$a_0 + 4a_1 + 16a_2 + 64a_3 = 21$$

$$a_0 + 9a_1 + 81a_2 + 729a_3 = 86$$

$$a_0 + 14a_1 + 196a_2 + 2744a_3 = 55$$

$$a_0 + 17a_1 + 289a_2 + 4913a_3 = 71$$

Employ the Electronic TOOLKIT to perform the same analysis.

SOLUTION: Use selection [5] from the TOOLKIT main menu, and enter the Gauss elimination submenu. Enter the number of equations (in this case, 4), the system coefficients, and the right-hand-side constants for the four equations. Note that the [PgUp] and [PgDn] keys are employed to move from equation to equation. Figure 37.12 shows the screen appearance for the first equation. Similar screens are generated for the others.

You should note that the equation coefficients and right-hand-side constants could have been determined with the spreadsheet and then transferred into the Gauss elimination module with the [F6] key (see Prob. 37.11 at the end of the chapter). Either way is acceptable and is dependent on the type of problem you are solving.

Following entry of all equations, use selection [2] from the submenu to solve the system with Gauss elimination. The results appear in Fig. 37.13. These results can

FIGURE 37.12
First linear equation from
Exercise 37.7 as entered into
the Gauss elimination routine
from the Electronic TOOLKIT.

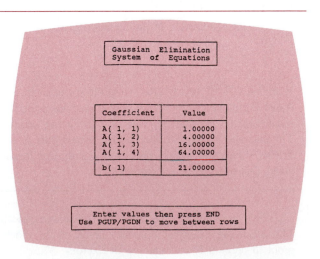

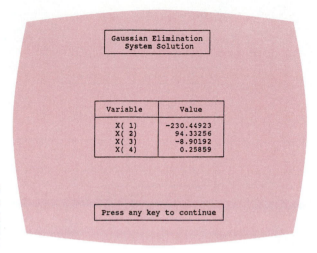

FIGURE 37.13
Solution for the coefficients of a cubic interpolating polynomial to fit the data from Exercise 37.7.

be employed in a manner similar to Exercise 37.5 to obtain the water use rate at 12:00 noon.

Note that the results obtained in the previous exercise are the same as those obtained using matrix inversion and multiplication with 1-2-3 or Quattro Pro in Exercise 37.5. However, recognize that the solution obtained with the TOOLKIT involved less effort (that is, fewer keystrokes). In fact, the largest task consisted of correctly typing in the coefficients and constants. This economy of effort is a direct result of the TOOLKIT's integrated design.

Despite the very straightforward solution presented in the previous exercise, there is an even more direct way to obtain the result with the TOOLKIT. In the next exercise, you will employ the Electronic TOOLKIT to perform the same analysis using Lagrange interpolation.

HANDS-ON EXERCISE 37.8   Interpolation with the Electronic TOOLKIT

GOAL: To learn how to use the TOOLKIT's Lagrange interpolation program to fit a cubic polynomial to data.

BACKGROUND: In Exercise 37.5, you used a commercial spreadsheet to perform interpolation by determining the coefficients of a polynomial by solving a 4×4 system of linear algebraic equations using matrix inversion and multiplication. Then, in Exercise 37.7, you employed the TOOLKIT's Gauss elimination module to evaluate the system of equations. Now we will simplify the analysis further by employing the TOOLKIT's Lagrange interpolation routine to come up with the polynomial coefficients directly.

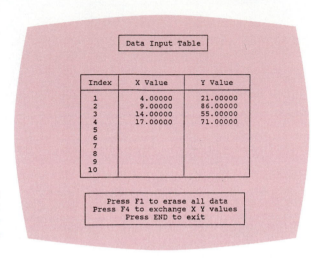

FIGURE 37.14
Values of water use versus
time of day for a city as
entered into the Lagrange
interpolation module of the
Electronic TOOLKIT.

SOLUTION: Make selection [7] from the TOOLKIT main menu to access Lagrange
interpolation. Enter new data using selection [1] on the Lagrange submenu. As shown
in Fig. 37.14, these are the times and flow rates discussed in Exercise 37.5.

Next, use selection [2] from the submenu, input a third-degree polynomial, and
use the first data point as the starting point of the interpolation. Fig. 37.15 is immediately displayed showing the coefficients of the interpolating polynomial. The estimated
value of flow can be obtained by entering an X value of 12.

Finally, a plot can be generated of the curve and the interpolated data points by
making selection [3] from the Lagrange submenu. All that needs to be done to design

FIGURE 37.15
Values of cubic Lagrange
interpolating polynomial
generated with the Electronic
TOOLKIT.

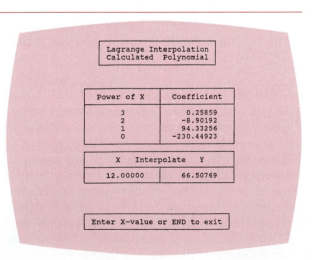

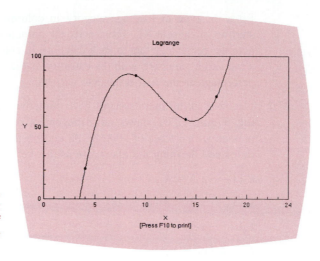

FIGURE 37.16
Electronic
TOOLKIT–generated graph of
a cubic polynomial fit of the
data from Exercise 37.5.

the plot is to supply the upper and lower limits of X and Y, and titles. This plot is shown in Fig. 37.16.

In summary, the exercises in this section have demonstrated how the Electronic TOOLKIT is particularly well suited to solve problems involving spreadsheets and numerical methods. Because of its integrated design, it facilitates movement and transfer of data among the techniques. In this regard, along with the fact that it also solves differential equations and numerically integrates data and functions, it is superior to most commercial spreadsheet packages when dealing with problems of this type.

PROBLEMS

37.1 Phosphorus is the critical nutrient governing the overfertilization of most lakes. As an environmental engineer, you develop an inventory of the phosphorus inputs into an important reservoir in your state. The results are as follows:

Loading Category	Amount (tonnes/yr)
Sewage treatment plants	33
Industrial sources	9
Animal feedlots	7
Forest runoff	12
Agricultural runoff	13
Urban storm runoff	20
Residential runoff	15

Enter the type of loading category in column A of a commercial spreadsheet starting in element A1. Then, enter the amount in the adjacent column B. Compute the percentage due to each category in column C. Graph the values as both a bar and a pie chart, with the category as the independent variable (X) and the percentage as the first dependent variable.

37.2 Idealized spring-mass systems have numerous applications throughout engineering. Figure P37.2 shows an arrangement of four springs in series being depressed with a force of 2000 lb. At equilibrium, force-balance equations can be developed defining the interrelationships between the springs:

$$k_2(x_2 - x_1) = k_1 x_1$$

$$k_3(x_3 - x_2) = k_2(x_2 - x_1)$$

$$k_4(x_4 - x_3) = k_3(x_3 - x_2)$$

$$F = k_4(x_4 - x_3)$$

where the k's are spring constants. If k_1 through k_4 are 100, 50, 75, and 200 lb/in, respectively, use a spreadsheet to compute the x's.

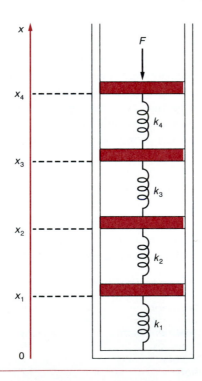

FIGURE P37.2

37.3 For Prob. 37.2, solve for the force F if x_3 is given as 60 in. All other parameters are as given in Prob. 37.2. Employ a spreadsheet to obtain your solution.

37.4 Three blocks are connected by a weightless cord and rest on an inclined plane (Fig. P37.4*a*). On the basis of the free-body diagrams in Fig. P37.4*b*, the following set of simultaneous equations can be derived:

$$100a + T \qquad = 519.72$$

$$50a - T + R = 216.55$$

$$20a \qquad - R = 86.62$$

Employ a spreadsheet to solve for acceleration *a* and the tensions *T* and *R* in the two ropes.

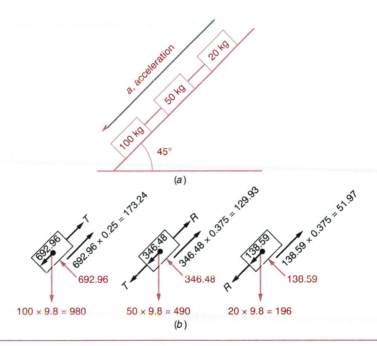

FIGURE P37.4

37.5 The decay of a radioactive waste is governed by the equation

$$\frac{c}{c_0} = e^{-kt}$$

where t = time [yr], c = concentration of the waste at t [pCi L^{-1}], c_0 = concentration of the waste at $t = 0$ [pCi L^{-1}], and k = the decay rate [yr^{-1}]. Note that

$$\ln\left(\frac{c}{c_0}\right) = -kt$$

The following data has been collected from a Superfund site:

t	0	1	2	3	4	5	6	8	10
c	10	7	6.2	4.5	3.0	3.0	2.0	2.0	1.0

Use the Electronic TOOLKIT's semilog plotting capability to evaluate the appropriateness of the model. Employ the linear regression capability to evaluate k. Plot the final curve and the data on a linear scale.

37.6 Repeat Prob. 37.5, but employ a commercial spreadsheet such as 1-2-3, Quattro Pro, or Excel. Discuss the advantages and disadvantages of commercial spreadsheets versus integrated software such as the TOOLKIT.

37.7 The following data represents the flow in a stream:

Month	Feb	Apr	Jul	Aug	Sep	Dec
Flow, cfs	100	800	400	500	700	200

Use the Electronic TOOLKIT to estimate the flow in May using a third-order polynomial based on Gauss elimination. Verify your results with the Lagrange interpolating polynomial. Plot the results.

37.8 Employ the data in Prob. 37.7 and Lagrange interpolation to estimate the flow in May using linear through fifth-order polynomials. Which order polynomial would you recommend?

37.9 At a particular university, the freshman class of engineering students is comprised of the following categories:

Type	Percentage
Aerospace	9
Architectural	6
Chemical	15
Civil	18
Electrical	27
Industrial	5
Mechanical	20

Enter the type of engineer in column A of a commercial spreadsheet starting in element A1. Then, enter the percentages in the adjacent column B. Graph the values as a pie chart, with the type as the independent variable X and the percentage as the first dependent variable.

37.10 Repeat Prob. 37.9, but display the information as a bar-type graph.

37.11 Repeat Exercise 37.7, but calculate the coefficients on the Electronic TOOLKIT's spreadsheet. Then, use the [F6] key to copy the relevant cells to the Gauss elimination module. Use this module to solve for the coefficients.

37.12 The following data was generated by the equation $y = x^2$:

x	1	2	6	7	8	9	10
y	1	4	36	49	64	81	100

Enter the x values in the A column of a commercial spreadsheet starting in element A1. Then, enter the equation in element B1. Copy the formula to fill out the remainder of the y values. Plot the values as (a) an XY plot and (b) a line plot. Discuss the difference between the XY and the line plot options.

37.13 The growth rate of bacteria that can convert sugars and starches to alcohol for beer production is given by

$$r = r_m \frac{S}{k_s + S}$$

where r = growth rate [d^{-1}], r_m = maximum growth rate [d^{-1}], S = substrate (that is, sugar and starch) concentration [$mg\ L^{-1}$], and k_s = the half-saturation constant [$mg\ L^{-1}$]. This relationship can be linearized by inverting it to yield

$$\frac{1}{r} = \frac{k_s}{r_m} \frac{1}{S} + \frac{1}{r_m}$$

Thus, the equation will plot as a straight line on a graph of $1/r$ versus $1/S$. The following data has been collected from a laboratory experiment:

r	0.8	1.6	1.8	2.5	2.5	2.8
S	1.0	2.0	3.0	5.0	7.0	9.0

Test to see whether the model is valid by plotting the data using a spreadsheet. Employ linear regression to evaluate the model coefficients. Plot the final curve and the data on a linear scale.

37.14 A laser-based device mounted on an aircraft is employed to automatically scan elevations as it passes over terrain. As the aircraft passes over two mountain peaks separated by a saddle, it records the following discrete information:

distance, mi	1	3	5	7	9	11
altitude, ft	1010	5100	7250	3950	6050	8125

Use the Electronic TOOLKIT to plot the altitudes versus the distances. Employ fifth-order polynomial interpolation to estimate the maximum height of both peaks and the lowest point in the saddle.

37.15 Use a commercial spreadsheet to solve the following simultaneous equations:

$$x_1 + 2x_2 + 3x_3 = -3$$

$$x_1 - x_2 - 2x_3 = 1$$

$$4x_1 - 2x_2 + 5x_3 = -18$$

37.16 Use the Electronic TOOLKIT to solve Prob. 37.15.

MACROS

Hopefully, your experience to this point should leave you with the feeling that spreadsheets are extremely easy to use and to understand. Because of their simple design, spreadsheets are among the most accessible types of software packages. Armed with the [F1], [/], and [Esc] keys, a novice user can rapidly scale the spreadsheet learning curve and produce useful results.

However, anyone who has ever climbed a major mountain recognizes that after a long gradual climb, there usually comes a point where the situation becomes technical. At this point, the novices admire the view and watch the expert climbers break out the ropes and pitons and proceed to the summit.

Unfortunately, many spreadsheet users tend to view macros in this way, that is, as the advanced facet of spreadsheets that separates the casual user from the expert. The present chapter is intended to help dispel this notion by providing an introduction to this subject.

A *macro* is merely a sequence of recorded keystrokes or commands that the spreadsheet executes automatically. There are two stages to employing a macro. The first stage involves creating the macro by typing in the instructions you want the spreadsheet to perform and then giving the macro a name. The second stage involves executing the macro so that the spreadsheet follows the instructions. Once you have created a macro, you can save it with the rest of your spreadsheet and reuse it over and over again.

38.1   KEYBOARD MACROS

When using spreadsheets over an extended period of time, you will undoubtedly find yourself repeating certain sequences of keystrokes over and over again. A *keyboard macro* is a list of keystrokes that are written down once and are then invoked repeatedly by a much shorter keystroke combination.

HANDS-ON EXERCISE 38.1   A Simple Keyboard Macro to Facilitate Spreadsheet Entry

GOAL: To illustrate how keyboard macros can be employed to economize keystrokes involved in typing recurring labels.

BACKGROUND: When entering a spreadsheet, there will often be occasions when you repeatedly enter a particular string of characters. For example, in a business, it might be your company name. The following exercise illustrates how a macro can be employed to economize such effort. In addition, it introduces the notion of setting up a definition area to keep track of your macros.

EXERCISE:
1. First, we will set up a macro definition area where we will cluster our macros. Type in the following lines:

```
[C12]: 'Macro Definitions
[C14]: 'Name        [A4]: 'Macro        [B3]: 'Description
```

Then move to cell D16 and type the label

```
[D16]: 'Computing
```

and strike [Enter]. Now name the cell with the following commands:

    123: / r n c   QP: / e n c

At this point you will be prompted for a name. Type in \A and [Enter]. A backslash preceding a one-letter name connotes that this is a macro name. You will then be prompted for a range or block of cells. Type D16, and [Enter].
2. Now move to cell A1, begin typing Introduction to, and press [Alt-A]. Notice how the label Computing is inserted. Continue typing for Engineers. Strike [Enter], and the label in A1 should read:

```
[A1]: 'Introduction to Computing for Engineers
```

Complete the process by moving to cell F16 and typing a description of the macro:

```
[F16]: 'Reprints the label
```

The final spreadsheet appears in Fig. 38.1.

In Fig. 38.1, the macro definition area is located close to the active part of the spreadsheet. A preferable practice would be to situate it in a remote location so that you do not have to work around the area when using the spreadsheet. Some macro users always position the beginning cell (that is, the upper-left-hand corner) of the macro defi-

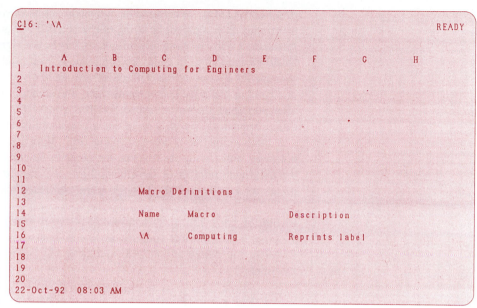

```
C16: '\A                                                              READY

        A         B         C         D         E         F         G         H
 1    Introduction to Computing for Engineers
 2
 3
 4
 5
 6
 7
 8
 9
10
11
12                    Macro Definitions
13
14                    Name      Macro               Description
15
16                    \A        Computing           Reprints label
17
18
19
20
22-Oct-92   08:03 AM
```

FIGURE 38.1
Spreadsheet screen showing
a simple keyboard macro to
print a label.

nition area in the same cell on all their spreadsheets. For example, something memorable like cell AA1000 might be chosen. This way when you want to inspect or modify your macros, you can employ the [F5] key to quickly jump to the area.

A neater alternative is to give an easily remembered name to the beginning cell. This is done with the commands:

123: / r n c   QP: / e n c

When you are prompted for a name, enter something like macros. Then when you want to jump to your macro definition area, you would merely strike [F5] and then enter macros. When you have completed this process, you would be transported to the macro definition area.

38.1.1 Command Macros

Just as you might notice that you repeatedly type certain labels, you will find that certain menu command sequences recur regularly. *Command macros* provide the means to condense these sequences into a single keystroke. The most important type of keyboard macros deal with menu commands.

HANDS-ON EXERCISE 38.2   *Some Command Macros to Economize Repeated Keystrokes*

GOAL: To illustrate how keyboard command macros can be employed to economize keystrokes.

BACKGROUND: When entering a spreadsheet, there will often be occasions when you recurrently enter a particular string of characters. For example, when typed naturally, the spreadsheet will automatically left-justify labels and right-justify numbers. In order that headings and values in tables line up properly, you might find that you would desire to override these conventions. Of course you could always precede the entry with a quote or a caret to make the cell contents right- and center-justified. However, often it is only after you enter the entire spreadsheet that you might notice that you want to change things around. It is at this point that a command macro can come in handy.

Such a situation is depicted in Fig. 38.2. Notice how the labels and the numbers do not align well. Also, observe that the numbers in columns B and C are printed out to the maximum number of significant digits that can fit in the cell. Thus, the decimal points do not align nicely. Develop some command macros to rectify these problems.

EXERCISE:

1. First, make sure that your spreadsheet is entered as shown in Fig. 38.2.
2. Move the cursor to cell A3, and enter the following commands to right-justify the label:

   123: / r l r [Enter]   QP: / s a r [Enter]

At this point the t should be moved to the right side of cell A3.

Now move the highlight to D16 and enter the following macro

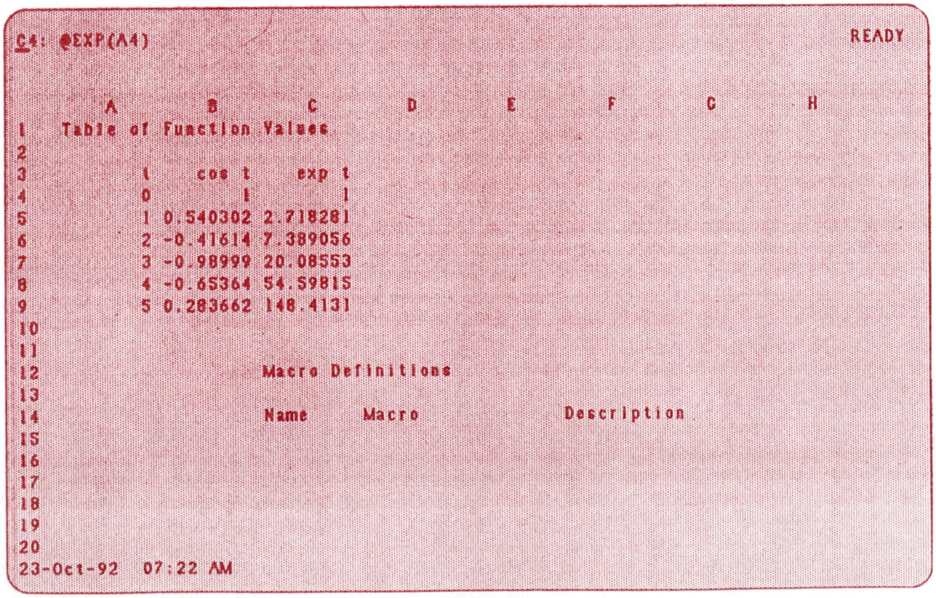

FIGURE 38.2
Spreadsheet screen showing a short table. The labels are left-justified and the numbers right-justified. In addition, the numbers are displayed to the maximum number of significant digits. Both characteristics detract from the presentation. In Exercise 38.2, some keyboard command macros are concocted to change the justification and number format.

```
C4: @EXP(A4)                                                                      READY


        A       B       C       D       E       F       G       H
 1   Table of Function Values
 2
 3           t      cos t    exp t
 4           0        1        1
 5           1  0.540302  2.718281
 6           2  -0.41614  7.389056
 7           3  -0.98999  20.08553
 8           4  -0.65364  54.59815
 9           5  0.283662  148.4131
10
11
12                   Macro Definitions
13
14                   Name     Macro          Description
15
16
17
18
19
20
23-Oct-92   07:22 AM
```

```
123: '/rlr~                    QP: '/sar~
```

The apostrophe is required so that the spreadsheet perceives that the entry is a label. Otherwise, as soon as you typed /, the menu would be activated. The remainder of the entry corresponds to the individual keystrokes you would enter to right-justify the cell's contents. The character at the end of the macro, ~, is called a *tilde* (rhymes with Matilda). It is the macro symbol for striking the [Enter] key.

After entering the macro, name the cell D16: \R. We have chosen this name because it abbreviates the purpose of the macro, "right justification." Move left to cell C16, and enter the label: '\R. Finally, move right to cell F16, and enter a short description of the macro's function: Right-justifies cell.

Now, move the cursor to cell B3, and strike the key combination [Alt-R]. The label should become right-justified. Move to cell C3, and repeat the key combination. 3. Move the cursor to cell B4, and enter the following commands to begin to change the numeric format:

```
123: / r f f                   QP: / s n f
```

At this point you will be prompted to enter the number of decimal places. Enter a 3. Then you will be asked to enter the range of cells to be formatted. This range should originally be B4..B4. Type [End] followed by [↓] to paint the range B4..B9. Finally, strike the [Enter] key, and column B should be formatted to three decimal places.

Now move the highlight to D17, and enter the following macro:

```
123: '/rff3~{END}{D}~        QP: '/snf3~{END}{D}~
```

This entry duplicates the individual keystrokes you would enter to format a column of numbers to three decimal places. Notice that special words are enclosed in curly brackets to signify keys such as [End] and down arrow [↓].

After entering the macro, give a name to cell D17: \F. We have chosen this name because it abbreviates the purpose of the macro, "**f**ormat." Move left to cell C17, and enter the label: '\F. Finally, move right to cell F17, and enter a short description of the macro's function: Column numeric format = 3.

Now, move the cursor to cell C4, and strike the key combination [Alt-F]. Column C should be formatted to three decimal places. The final spreadsheet should look like Fig. 38.3.

There are a variety of other abbreviations in addition to {END} and {D}. These are summarized in Table 38.1. Note that certain commands such as {UP} or {DOWN} may need to be implemented several times in succession. For example, to move down four cells, you could use the commands

```
{DOWN}{DOWN}{DOWN}{DOWN}
```

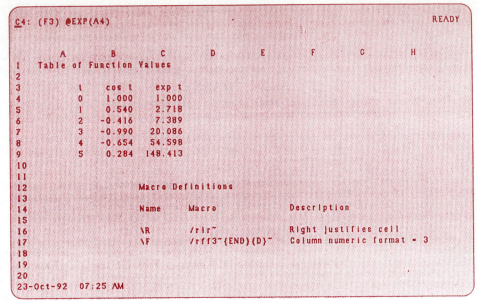

FIGURE 38.3
Spreadsheet screen showing
a short table that has been
modified using the two
command macros described
in Exercise 38.2. The macros
were used to change the
justification and number
format.

TABLE 38.1   Keyword Equivalents Used in Macros

Key	Macro Keyword
F1	{HELP}
F2	{EDIT}
F5	{GOTO}
F9	{CALC}
F10	{GRAPH}
Backspace	{BACKSPACE} or {BS}
Caps Lock	{CAPON} or {CAPOFF}
Ctrl–Break	{BREAK}
Ctrl ←	{BIGLEFT}
Ctrl →	{BIGRIGHT}
Del	{DELETE} or {DEL}
↓	{DOWN} or {D}
↑	{UP} or {U}
End	{END}
Enter	~
Esc	{ESCAPE} or {ESC}
Home	{HOME}
Ins	{INSERT} or {INS}
←	{LEFT} or {L}
Num Lock	{NUMON} or {NUMOFF}
PgDn	{PGDN}
PgUp	{PGUP}
→	{RIGHT} or {R}
Scroll Lock	{SCROLLON} or {SCROLLOFF}

However, for your convenience, the operation can also be represented in abbreviated form as

{D 4}

Other features of this type as well as other key equivalences are summarized in your spreadsheet's users manual or its help facility [F1].

38.2   MACRO PROGRAMS

Macros can become quite intricate. In such cases, it is advisable to break them up into more than one cell. This is done by continuing the macro into the cells directly below. The spreadsheet will proceed in this way until it reaches a blank cell, whereupon the macro terminates. This is very similar to the sequential execution of common high-level programming languages such as FORTRAN or QuickBASIC. Consequently, a *macro program* can be developed.

Just as with standard programming languages, such programs allow other structures beyond simple sequencing. For example, loops and decisions are also allowed.

Some of the more commonly used commands are summarized in Table 38.2. As described in the following exercise, such commands allow structured algorithms to be implemented within the spreadsheet environment.

HANDS-ON EXERCISE 38.3   A Macro Program to Determine the Square Root Iteratively

GOAL: To illustrate how an iterative numerical method for determining square roots can be implemented with a macro program.

BACKGROUND: The following iterative formula can be derived to determine the square root of a number, a,

$$x_{i+1} = \frac{1}{2}\left(x_i + \frac{a}{x_i}\right)$$

Starting with an initial guess of x_0, this formula can be repeated until it converges on the square root. After each iteration, the percent relative error can be estimated by the formula,

$$\epsilon_a = \left|\frac{x_{i+1} - x_i}{x_{i+1}}\right| \times 100\%$$

When ϵ_a falls below a certain tolerance, the computation can be terminated.

For example, if $a = 49$, $x_0 = 20$, and the tolerance is 0.1 percent, the iterations shown at the top of p. 789 result.

TABLE 38.2　Commonly Used Macro Programming Commands

{LET *location, value*}	The macro assignment statement. Places a number or label in *location. Value* can be a number, string, formula, or cell reference.
{IF *condition*}	Checks whether the *condition* is true or false. If true, continues executing next instruction in same cell. If false, proceeds to cell below the {IF} command. The *condition* is typically a logical expression using the operators $<$, $>$, $<=$, $>=$, $=$, $<>$, #NOT#, #AND#, #OR#, or a cell reference containing a logical expression.
{BRANCH *location*}	Passes execution control to the macro instructions of *location*.
{FOR *count, start, stop, step, macro*}	Creates a counter loop that repeats the designated *macro*. The *start, stop,* and *step* control the number of iterations, while the *count* keeps track of the repetitions.
{FORBREAK}	Immediately terminates a loop created by a {FOR} command. Control is transferred to the statement immediately following the {FOR} statement.
{GET *location*}	Pauses the macro, accepts one keystroke, and stores it in *location*.
{GETLABEL *message, location*}	Pauses the macro, accepts a *message* entered by the user, and stores it in *location*.
{GETNUMBER *message, location*}	Pauses the macro, accepts a *message* (which must be a number or numeric formula) entered by the user, and stores it in *location*.
{QUIT}	Terminates macro execution, and returns control to the keyboard.

i	x_i	x_{i+1}	ϵ_a
0	20.0000	11.2250	78.1737%
1	11.2250	7.7951	44.0002%
2	7.7951	7.0406	10.7176%
3	7.0406	7.0001	0.5776%
4	7.0001	7.0000	0.0017%
5	7.0000		

Therefore, after five iterations, the correct result of 7 is computed with an estimated error of 0.0017 percent. Employ a macro program to perform the same computation on a spreadsheet.

EXERCISE:

1. Set up some cells to hold the input data:

   [A1]: 'Input data: [A3]: "a=   [A4]: "xinit=
   [A5]: "tol=   [B3]: 49   [B4]: 20
   [B5]: 0.1   [C5]: '%

2. Set up some cells to hold the calculation results:

   [E1]: 'Calculation results:
   [E3]: "old x=   [E4]: "new x=   [E5]: "error=
   [G5]: '%   [E6]: 'iter=

3. Set up the macro definition area with the cell entries:

   [A8]: 'Macro Definitions
   [A10]: 'Name   [B10]: 'Macro   [F10]: 'Description
   [F12]: 'Iterative square root   [F13]: 'program

4. Move the cursor to cell B3, implement the commands:

   123: / r n c   QP: / e n c

When you are prompted for a name, enter a. Using the same procedure, give names to the following cells:

Cell	Name
B3	a
B4	xinit
B5	tol
F3	x
F4	xn
F5	er
F6	i

FIGURE 38.4
Spreadsheet setup for the
iterative square root problem
described in Exercise 38.3.

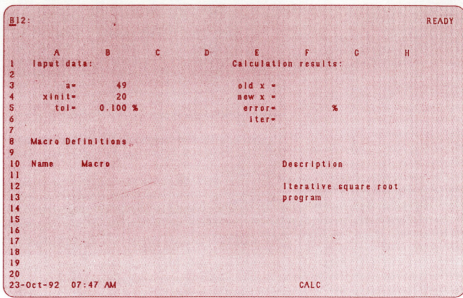

FIGURE 38.5
Flowchart of an algorithm to
compute the square root of a
number iteratively.

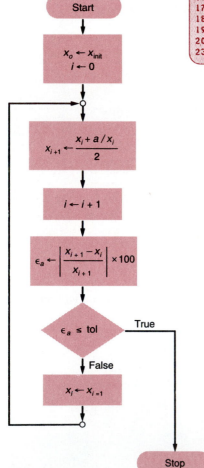

At this point, the spreadsheet should look like Fig. 38.4.

5. Now we are ready to enter the macro program. A flowchart for the algorithm is depicted in Fig. 38.5. The following commands can be entered to implement this algorithm:

```
[B12]: {let x,xinit}
[B13]: {let i,0}
[B14]: {let xn,.5*(x+a/x)}
[B15]: {let i,i+1}
[B16]: {let er,@abs((xn-x)/xn)*100}
[B17]: {if er<=tol}~{quit}
[B18]: {let x,xn}
[B19]: {branch cycle}
```

Now give the macro a name by moving to cell B12 and naming it \S. Move to cell B14, and name it cycle. Finally, make the following entries to document the macro:

```
[A12]: '\S                          [A14]: 'cycle
```

6. Strike [Alt-s] and the result should appear as in Fig. 38.6.

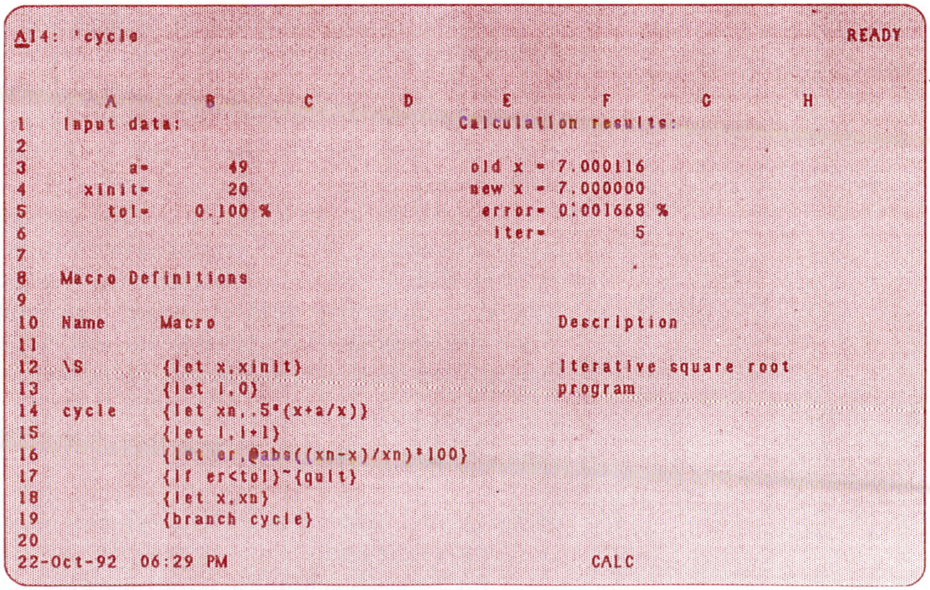

FIGURE 38.6
Spreadsheet to compute the
square root of a number
iteratively with a program
macro.

```
A14: 'cycle                                                              READY

          A        B        C        D        E          F        G      H
1    Input data:                          Calculation results:
2
3         a=       49                     old x = 7.000116
4     xinit=       20                     new x = 7.000000
5       tol=    0.100 %                   error= 0.001668 %
6                                         iter=         5
7
8    Macro Definitions
9
10   Name     Macro                                       Description
11
12   \S       {let x,xinit}                               Iterative square root
13            {let i,0}                                   program
14   cycle    {let xn,.5*(x+a/x)}
15            {let i,i+1}
16            {let er,@abs((xn-x)/xn)*100}
17            {if er<tol}~{quit}
18            {let x,xn}
19            {branch cycle}
20
22-Oct-92   06:29 PM                                      CALC
```

38.3 TO MACRO OR NOT TO MACRO?

At our campuses, we have rooms filled with personal computers for student use. Most of the time, all the machines are being used by students preparing reports, homework, and projects for their classes. However, every once in a while, you will see a nonoccupied computer with a note on it. At first glance you might conclude that the unit is out of order. Upon closer inspection, you will see that the monitor screen displays a spreadsheet and the sign says "Please do not disturb. Job in progress." At this point, we usually have the strong suspicion that the user has applied a spreadsheet to a problem for which it is not well-suited. Usually, this is an individual who has become so dazzled by the truly marvelous features of the spreadsheet that he or she believes it is the one and only tool for every computer-oriented task. It also usually means that there's a macro involved.

Macros are clearly a double-edged sword. By allowing the user to introduce logic and repetition into spreadsheets, they provide a way for more conventional programming operations to be integrated into the spreadsheet environment. Thus, the rich array of spreadsheet capabilities (particularly input-output features and built-in utilities) can be extended to a much wider range of problem contexts. However, when applied indiscriminately, they can result in the application of spreadsheets to areas where they become ineffective and counterproductive.

In particular, computationally intensive problems are almost always better approached with high-level algorithmic programming languages. There are two reasons for this argument. The principal consideration is that spreadsheets perform operations in an interpreted fashion. That is, the operations are translated into machine language as they are implemented. This is in contrast to most high-level languages, which are compiled. The net effect is that large computationally intensive algorithms take much longer to run on a spreadsheet.

The second reason relates to the modularity and structure of high-level languages. Although modular and structured programming techniques can be applied to macro-language programming, it is not as natural and convenient as for the algorithmic languages. Thus, the advantages of these approaches (large programming libraries, clear modifiable code, and the like) cannot be as readily exploited in the spreadsheet environment.

In summary, as we have tried to stress throughout this book, different programming tasks demand different programming tools. Just as mechanics would not be very effective if they were limited to employing a single tool, so computer users must be capable of marshaling different computer capabilities for particular tasks. If used wisely, the spreadsheets described in this part of the book can be an extremely useful and enjoyable addition to your programming repertoire.

PROBLEMS

38.1 Develop keyboard macros to implement the following:
(*a*) Center-justify a row of labels.
(*b*) Format a cell so that it uses scientific notation with three decimal places.

(c) Format a block of cells so that it uses a fixed numeric format with four decimal places.

(d) Print one copy of a block of cells.

(e) Save the active spreadsheet under its current file name.

(f) Place the phrase United States in the current cell, and move the cursor down three cells.

(g) Format a column so that its width is expanded to 12.

38.2 Develop a keyboard macro for your name, address, and telephone number. Have the macro type your name in the current cell, and print the address and phone number in the cells directly below.

38.3 Develop a keyboard macro for the course you are taking in conjunction with this book. Have the macro type the course title in the current cell. Then have it enter the course number, section number, and instructor's name in the three cells directly below. An example is:

	C	D	E	F
4	Introduction to Engineering Computing			
5	GEEN 1300			
6	Sec. 010			
7	Dr. Ima Prof			

38.4 Develop a macro program to determine the roots of a quadratic equation. Make sure that the program considers all possible outcomes, including complex results.

38.5 Develop a macro program that allows users to set the column width at a value of their choosing.

38.6 Your job is to improve the design of a parachute. The velocity of a falling parachutist is given by

$$v = \frac{gm}{c}[1 - e^{-(c/m)t}]$$

where velocity v is the dependent variable in m s^{-1}; time t is the independent variable; the gravitational constant $g = 9.81$ m s^{-2}; and the drag coefficient c in kg s^{-1} and mass m in kg are parameters. Suppose that we had to determine the drag coefficient for a parachutist of a given mass to attain a prescribed velocity in a set time period. Develop a macro program to solve this problem.

38.7 The upward velocity of a rocket can be computed by the following formula:

$$v = u \ln \frac{m_0}{m_0 - qt} - gt$$

where v is upward velocity, u is the velocity at which fuel is expelled relative to the rocket, m_0 is the initial mass of the rocket at time $t = 0$, q is the fuel consumption rate, and g is the downward acceleration of gravity (assumed constant $= 9.8$ m/s^2). Develop a macro that computes v as a function of parameter values and a specific time entered by the user. Employ the following values to test your macro: $u = 2200$ m/s, $m_0 = 160{,}000$ kg, $q = 2680$ kg/s, and $t = 15$ s.

38.8 The annual worth of an engineering project is characterized by the equation

$$A_t = \frac{-1900(1.18)^n}{1.18^n - 1} + \frac{45n}{1.18^n - 1} + 3000$$

where n is the number of years after the project begins operation. Develop a macro to compute a table of annual worths for $n = 1$ to 10 using increments of 1.

38.9 Set up a macro to solve the following differential equation for $y(50)$:

$$\frac{dy}{dx} = -\frac{y}{25}$$

Use Euler's method to obtain your solution,

$$y(x + \Delta x) = y(x) + \frac{dy(x)}{dx} \Delta x$$

Test the program with a step size of $\Delta x = 0.0001$ and with an initial condition of $y(0) = 10$.

38.10 Perform Prob. 38.9, and time a run. Then reprogram the algorithm in a high-level programming language and again time the run. Compare and discuss.

38.11 Given two columns of xy data, set up a macro program to perform linear regression and generate a plot of the data and best-fit line.

38.12 Given a column of numbers, develop a macro to determine the minimum and maximum values with @ functions. Start the macro with the cell positioned at the top of the column. Print the minimum and the maximum in the two cells just below the column.

38.13 Repeat Prob. 38.12, but do not employ the @MIN and @MAX functions.

38.14 Develop a macro to solve a system of linear algebraic equations in a fashion similar to Exercise 37.5.

APPENDIXES

Function	Equivalent
$\sec x$	$\dfrac{1}{\cos x}$
$\csc x$	$\dfrac{1}{\sin x}$
$\cot x$	$\dfrac{1}{\tan x}$
$\sin^{-1} x$	$\tan^{-1}\left(\dfrac{x}{\sqrt{-x^2+1}}\right)$
$\cos^{-1} x$	$\tan^{-1}\left(\dfrac{x}{\sqrt{-x^2+1}}\right)+2\pi$
$\sec^{-1} x$	$\tan^{-1}\left(\sqrt{x^2-1}\right)+(\operatorname{sgn} x-1)2\pi$
$\csc^{-1} x$	$\tan^{-1}\left(\dfrac{1}{\sqrt{x^2-1}}\right)+(\operatorname{sgn} x-1)2\pi$
$\cot^{-1} x$	$-\tan^{-1} x+2\pi$
$\sinh x$	$\dfrac{e^x-e^{-x}}{2}$
$\cosh x$	$\dfrac{e^x+e^{-x}}{2}$
$\tanh x$	$\dfrac{e^x-e^{-x}}{e^x+e^{-x}}$
$\operatorname{sech} x$	$\dfrac{2}{e^x+e^{-x}}$
$\operatorname{csch} x$	$\dfrac{2}{e^x-e^{-x}}$
$\coth x$	$\dfrac{e^x+e^{-x}}{e^x-e^{-x}}$
$\sinh^{-1} x$	$\log\left(x+\sqrt{x^2+1}\right)$
$\cosh^{-1} x$	$\log\left(x+\sqrt{x^2-1}\right)$
$\tanh^{-1} x$	$0.5\log\left(\dfrac{1+x}{1-x}\right)$
$\operatorname{sech}^{-1} x$	$\log\left(\dfrac{1+\sqrt{1-x^2}}{x}\right)$
$\operatorname{csch}^{-1} x$	$\log\left(\dfrac{1}{x}+\sqrt{1+\dfrac{1}{x^2}}\right)$
$\coth^{-1} x$	$0.5\log\left(\dfrac{x+1}{x-1}\right)$

[1]The function, sgn x, has the value 1, 0 or −1 if x is positive, zero, or negative, respectively.

Name	Description
ABS(*num-expr*)	Returns the absolute value of a numeric expression.
ASC(*str-expr*)	Returns a numeric value that is the ASCII code for the first character in a string expression.
ATN(*num-expr*)	Returns the arctangent (in radians) of a numeric expression.
CDBL(*num-expr*)	Converts a numeric expression to a double-precision number.
CHR$(*code*)	Returns a one-character string whose ASCII code is the argument.
CINT(*num-expr*)	Returns a numeric expression to an integer by rounding the fractional part of the expression.
CLNG(*num-expr*)	Converts a numeric expression to a long (4-byte) integer by rounding the fractional part of the expression.
COMMAND$	Returns the command line used to invoke the program.
COS(*num-expr*)	Returns the cosine of an angle given in radians.
CSNG(*num-expr*)	Converts a numeric expression to a single-precision value.
CSRLIN	Returns the current line (row) position of the cursor.
CVI(*2-byte-string*)	Converts a 2-byte string created with MKI$ back to an integer.
CVS(*4-byte-string*)	Converts a 4-byte string created with MKS$ back to a single-precision number.
CVL(*4-byte-string*)	Converts a 4-byte string created with MKL$ back to a long integer.
CVD(*8-byte-string*)	Converts an 8-byte string created with MKD$ back to a double-precision number.
CVSMBF(*4-byte-string*)	Converts strings containing Microsoft binary format numbers to a single-precision IEEE-format number.
CVDMBF(*8-byte-string*)	Converts strings containing Microsoft binary format numbers to a double-precision IEEE-format number.
DATE$	Returns a string containing the present date
ENVIRON$(*environmentstring*) ENVIRON(*n*)	Returns an environment string from the DOS environment-string table
EOF(*filenumber*)	Tests for the end-of-file condition.
ERDEV	Integer function that returns an error code from the last device to declare an error.
ERDEV$	String function that returns the name of the device generating the error.
ERR	Returns error status. After an error, returns the error code.
ERL	Returns error status. After an error, returns the line number where the error occurred.
EXP(*num-expr*)	Calculates the exponential function of the numeric expression.
FILEATTR(*filenumber, attribute*)	Returns information about an open file.
FIX	Returns the truncated integer part of the expression.
FRE(*num-expr*) FRE(*str-expr*)	Returns the amount of available memory.
FREEFILE	Returns the next free BASIC file number.
HEX$(*expr*)	Returns a string that represents the hexadecimal value of the decimal argument expression.
INKEY$	Reads a character from the keyboard.
INP(*port*)	Returns the byte read from an I/O port.
INPUT$(*n*[,[#]*filenumber*])	Returns a string of characters from the specified file.
INSTR([*start*,]*str-expr*1,*str-expr*2)	Returns the character position of the first occurrence of a string in another string.

INT(*num-expr*)	Returns the largest integer less than or equal to the numeric expression.
IOCTL$([#]*filenumber*)	Receives a control data string from a device operator.
LBOUND(*array*[,*dimension*])	Returns the lowest bound (smallest available subscript) for the indicated dimension of an array.
LCASE$(*str-expr*)	Returns a string expression with all letters in lower-case.
LEFT$(*str-expr*,*n*)	Returns a string consisting of the leftmost *n* characters of a string.
LEN(*str-expr*) LEN(*variable*)	Returns the number of characters in a string or the number of bytes required by a variable.
LOC(*filenumber*)	Returns the current position within the file.
LOF(*filenumber*)	Returns the length of the named file in bytes.
LOG(*num-expr*)	Returns the natural logarithm of the numeric expression.
LPOS(*n*)	Returns the current position of the line printer's print head within the printer buffer.
LTRIM$(*str-expr*)	Returns a copy of a string with leading spaces removed.
MID$(*str-expr*,*start*,[,*length*])	Returns a substring of a string.
MKI$(*integer-expr*)	Converts numeric values to string values.
MKS$(*single-precision-expr*)	Converts numeric values to string values.
MKL$(*long-integer-expr*)	Converts numeric values to string values.
MKD$(*double-precision-expr*)	Converts numeric values to string values.
MKSMBF$(*single-precision-expr*) MKDMBF$(*double-precision-expr*)	Converts an IEEE-format number to a string containing a Microsoft Binary format number.
OCT$(*num-expr*)	Returns a string representing the octal value of the numeric argument.
PEEK(*address*)	Returns the byte stored at a specified memory location.
PEN(*n*)	Reads the lightpen coordinates.
PLAY *commandstring*	Returns the number of notes currently in the background-music queue.
PMAP(*expression*,*function*)	Maps view-coordinate expressions to physical locations, or maps physical expressions to a view-coordinate location.
POINT(*x*,*y*) POINT(*number*)	Reads the color number of a pixel from the screen or returns the pixel's coordinates.
POS(0)	Reads the current horizontal position of the cursor.
RND[(*n*)]	Returns a single random number between 0 and 1.
RTRIM$(*str-exp*)	Returns a string with trailing (right-hand) spaces removed.
SADD(*str-var*)	Returns the address of a specified string expression.
SCREEN(*row*,*column*[,*colorflag*])	Reads a character's ASCII value or its color from a specified screen location.
SEEK[#]*filenumber*,*position*	Returns the current file position.
SETMEM(*num-expr*)	Changes the amount of memory used by the far heap -- the area where far objects and internal tables are stored.
SGN(*num-expr*)	Indicates the sign of a numeric expression (positive = 1, zero = 0, negative = -1).
SIN(*num-expr*)	Returns the sine of an angle given in radians.
SPACE$(*n*)	Returns a string of spaces of length *n*.
SPC(*n*)	Skips *n* spaces in a PRINT statement.
SQR(*n*)	Returns the square root of *n*.
STICK(*n*)	Returns the *x* and *y* coordinates of the two joysticks.
STR$(*num-expr*)	Returns a string representation of the value of a numeric expression.
STRIG(*n*)	Returns the status of a specified joystick trigger.

STRING$(*m, n*) STRING$(*m, str-exp*)	Returns a string whose characters all have a given ASCII code or whose characters are all the first characters of a string expression.
TAB(*column*)	Moves the print position.
TAN(*num-expr*)	Returns the tangent of an angle, where the angle is in radians.
TIME$	Returns the current time from the operating system.
TIMER	Returns the number of seconds past midnight.
UBOUND(*array*[,*dimension*])	Returns the upper bound (largest available subscript) for the indicated dimension of an array.
UCASE$(*str-expr*)	Returns a string expression with all letters in uppercase.
VAL(*str-expr*)	Returns the numeric value of a string of digits.
VARPTR(*variable-name*) VARSEG(*variable-name*)	Returns the address of a variable.
VARPTR$(*variable-name*)	Returns a string representation of a variable's address for use in DRAW and PLAY statements.

Mathematical Functions

Generic Name	Specific Name	Argument Number	Argument Type	Function Type	Description
ABS	IABS	1	Integer	Integer	Absolute value of argument
	ABS		Real	Real	
	DABS		Double	Double	
	CABS		Complex	Real	
ACOS	ACOS	1	Real	Real	Arccosine (in radians) of argument
	DACOS		Double	Double	
AINT	AINT	1	Real	Real	Truncation
	DINT		Double	Double	
ANINT	ANINT	1	Real	Real	Rounding to nearest integer
	DNINT		Double	Double	
ASIN	ASIN	1	Real	Real	Arcsine (in radians) of argument
	DASIN		Double	Double	
ATAN	ATAN	1	Real	Real	Arctangent (in radians) of argument.
	DATAN		Double	Double	
ATAN2	ATAN2	2	Real	Real	Arctangent (in radians) of first argument/second argument
	DATAN2		Double	Double	
CMPLX	--	1 or 2	Integer	Complex	Conversion of numeric to complex
	--		Real	Complex	
	--		Double	Complex	
	--		Complex	Complex	
--	CONJG	1	Complex	Complex	Conjugate of a complex number
COS	COS	1	Real	Real	Cosine in radians
	DCOS		Double	Double	
	CCOS		Complex	Complex	
COSH	COSH	1	Real	Real	Hyperbolic cosine of argument
	DCOSH		Double	Double	
DBLE	--	1	Integer	Double	Conversion of argument to double precision.
	--		Real	Double	
	--		Double	Double	
	--		Complex	Double	
DIM	IDIM	2	Integer	Integer	Positive difference. First argument minus second, if first argument greater than or equal to second. Otherwise zero.
	DIM		Real	Real	
	DDIM		Double	Double	
--	DPROD	2	Real	Double	Double precision product of arguments.
EXP	EXP	1	Real	Real	Exponential function
	DEXP		Double	Double	
	CEXP		Complex	Complex	

INT	--	1	Integer	Integer	Conversion of the argument to the integer type. Sign of argument (or the real part of the argument) times the greatest integer less than or equal to the absolute value of the argument.
	INT		Real	Integer	
	IFIX		Real	Integer	
	IDINT		Double	Integer	
	--		Complex	Integer	
LOG	ALOG	1	Real	Real	Natural logarithm
	DLOG		Double	Double	
	CLOG		Complex	Complex	
LOG10	ALOG10	1	Real	Real	Common logarithm
	DLOG10		Double	Double	
MAX	MAX0	≥2	Integer	Integer	Maximum of arguments
	AMAX1		Real	Real	
	DMAX1		Double	Double	
--	AMAX0		Integer	Real	
--	MAX1		Real	Integer	
MIN	MIN0	≥2	Integer	Integer	Minimum of arguments
	AMIN1		Real	Real	
	DMIN1		Double	Double	
--	AMIN0		Integer	Real	
--	MIN1		Real	Integer	
MOD	MOD	2	Integer	Integer	Remainder of division of first argument by second.
	AMOD		Real	Real	
	DMOD		Double	Double	
NINT	NINT	1	Real	Integer	Argument rounded to the nearest integer.
	IDNINT		Double	Integer	
REAL	REAL	1	Integer	Real	Conversion to REAL type.
	FLOAT		Integer	Real	
	--		Real	Real	
	SNGL		Double	Real	
	--		Complex	Real	
SIGN	ISIGN	2	Integer	Integer	Transfer of sign
	SIGN		Real	Real	
	DSIGN		Double	Double	
SIN	SIN	1	Real	Real	Sine in radians
	DSIN		Double	Double	
	CSIN		Complex	Complex	
SINH	SINH	1	Real	Real	Hyperbolic sine of argument
	DSINH		Double	Double	
SQRT	SQRT	1	Real	Real	Square root of argument.
	DSQRT		Double	Double	
	CSQRT		Complex	Complex	
TAN	TAN	1	Real	Real	Sine in radians
	DTAN		Double	Double	
	CTAN		Complex	Complex	
TANH	TANH	1	Real	Real	Hyperbolic sine of argument
	DTANH		Double	Double	

Character Functions

Generic Name	Specific Name	Argument Number	Argument Type	Function Type	Description
--	CHAR	1	Integer	Character	Conversion of integer to character
--	ICHAR	1	Character	Integer	Conversion of character to integer
	INDEX	2	Character	Integer	Location of substring (second argument) in string (first argument).
--	LEN	1	Character	Integer	Length of character string
--	LGE	2	Character	Logical	First argument is lexically greater than or equal to the second argument.
--	LGT	2	Character	Logical	First argument is lexically greater than the second argument.
--	LLE	2	Character	Logical	First argument is lexically less than or equal to the second argument.
--	LLT	2	Character	Logical	First argument is lexically less than the second argument.

Introduction. The Electronic TOOLKIT is a software package that contains programs that can be used to analyze data and functions and perform mathematical and statistical computations. It can be used by engineers and scientists to solve a variety of applied problems.

Hardware requirements. The TOOLKIT can be used on an IBM compatible computer with 384K or more memory. It automatically recognizes monochrome, CGA, EGA, or Hercules video modes.

Operation and capabilities. Insert the disk provided with this text into the drive unit of your computer. The program is run by typing the word "TOOLKIT" following the screen prompt. You advance through the program by following the directions that are usually located in a boxed area at the bottom of the screen. Figure D.1 shows the Main Menu of the TOOLKIT. This menu is divided into three main sections and an Exit line.

The first section contains spreadsheet, plotting, and statistical programs. These programs have general problem-solving utility and are often used for preliminary analysis of data or functions prior to implementation of numerical methods applications.

The second section contains six numerical methods programs which can be used to

1. find the root of a function
2. solve systems of linear algebraic equations
3. fit the trend of data using linear regression
4. interpolate among data points with a polynomial
5. integrate a mathematical function or tabular data
6. solve differential equations

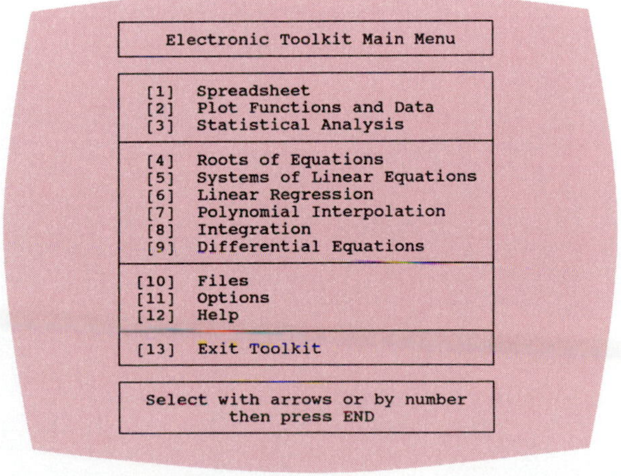

FIGURE D.1
The Main Menu of the
TOOLKIT.

```
        Electronic Toolkit Main Menu

   [1]   Spreadsheet
   [2]   Plot Functions and Data
   [3]   Statistical Analysis

   [4]   Roots of Equations
   [5]   Systems of Linear Equations
   [6]   Linear Regression
   [7]   Polynomial Interpolation
   [8]   Integration
   [9]   Differential Equations

   [10]  Files
   [11]  Options
   [12]  Help

   [13]  Exit Toolkit

       Select with arrows or by number
              then press END
```

Notice that all nine methods are similar in that they each analyze a mathematical function or tabular data. Often it is necessary to use more than one of the above methods for a single problem application. Therefore data and functions input to any one part of the TOOLKIT are shared among all the other parts. Furthermore, the results of spreadsheet calculations can be transferred to various other numerical methods programs. Therefore we say that the software has an integrated design.

The first and second sections of the Main Menu are described and applied for example problems in various sections of the text. Here we will describe general features of the third section on files and options.

Files. Selection [10] from the Main Menu activates a files submenu as shown in Figure D.2. This menu allows you to start a new file, open an existing file that was previously saved, or save the current application.

You must use care when starting a new file because the current application will be lost unless it is saved prior to opening a new file.

The TOOLKIT can only open existing files that end with the extension ".etk". These files are usually created by the TOOLKIT itself. However, ASCII spreadsheet data files created by another programs can also be opened by the TOOLKIT. Figure D.3 shows the Open File menu.

Files to be saved can be stored in either a special binary format or an ASCII format. The binary format is fast and efficient; however, it can only be read by the TOOLKIT. The ASCII format is less efficient but can be read by other commercial spreadsheet programs. Figure D.4 shows the menu used to save files. Note that you must supply both the name and format of files. The names of previously saved files are listed for convenience. If you save a current application under an old file name, the contents of the old file will be lost.

Options and set-up. Selection [11] from the Main Menu concerns options related to screen layout and printer set-up.

Figure D.5 shows the screen layout options submenu. This menu allows you to alter the decimal precision, size, and color of numbers on the screen.

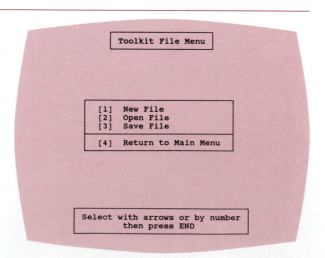

FIGURE D.2
The file sub-menu activated from selection [10] from the Main Menu.

```
              ┌─────────────────────┐
              │  Toolkit File Menu  │
              └─────────────────────┘



              ┌─────────────────────────┐
              │  [1]   New File         │
              │  [2]   Open File        │
              │  [3]   Save File        │
              ├─────────────────────────┤
              │  [4]   Return to Main Menu │
              └─────────────────────────┘




              ┌─────────────────────────┐
              │ Select with arrows or by number │
              │        then press END    │
              └─────────────────────────┘
```

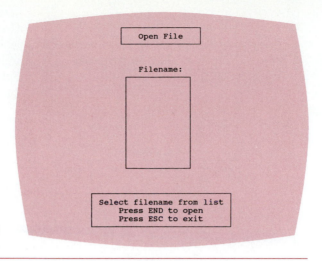

FIGURE D.3
Open File menu.

FIGURE D.4
Menu used to save files.

Figure D.6 shows the printer set-up submenu. You can choose among several printers. You must also specify the connection mode of your printer and computer. This is usually parallel for most dot-matrix and laser printers. The speed and quality of the graphs depend on the computer and printer hardware capabilities.

Hard copies of TOOLKIT graphics screens can also be produced on some printers using the [PrtScr] key and the GRAPHICS.COM program from DOS. Also some word processing and home publishing programs have a screen capture capability which can be used to transfer TOOLKIT graphics screens to a document. Consult appropriate user manuals for more details.

Help.   The TOOLKIT was designed so that it would be simple to operate. Instructions regarding various options and alternatives are always given directly on the screen. You only need general knowledge of the numerical methods as provided in this text to

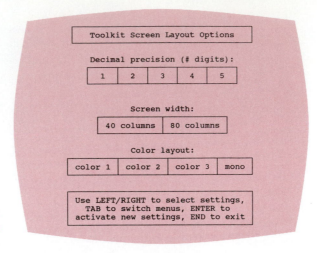

FIGURE D.5
Screen layout options
sub-menu.

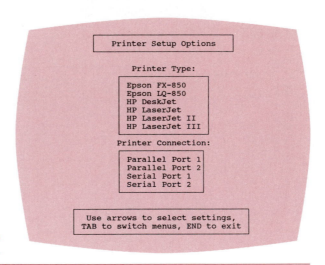

FIGURE D.6
Printer set-up sub-menu.

use the TOOLKIT for applied problems. Help screens are provided in a few places in the program where appropriate.

Errors and approximations.   Sometimes you may enter incorrect data into the TOOLKIT. For example, the program may ask you for the upper and lower limits for plot dimensions. If you reverse the input values, the lower limit will exceed the upper limit. In such cases the TOOLKIT will notify you about the inconsistency and wait for new data. The TOOLKIT has extensive testing capabilities to help you avoid data input, logic, and typing errors.

Errors or improper mathematical operations may occur during execution of the program. For example, suppose that you ask the TOOLKIT to plot the function $f(x) = 1/x$ over the range from $x = 0$ to $x = 1$. Obviously, this function is not defined at $x = 0$. However, the TOOLKIT does its best with your instructions by plotting the function where it is defined and alerting you with a message that a "floating point error was encountered." Although in many cases the TOOLKIT may perform well when presented with a singular or discontinuous problem, you should take special care to avoid incorrect interpretation of the TOOLKIT output.

Sometimes the TOOLKIT has problems with very small, very large, or noninteger numbers. Number representation by the computer is approximate: therefore rounding is often necessary. These problems can usually be avoided by proper scaling.

REFERENCES

Ang, A. H-S., and W. H. Tang, *Probability Concepts in Engineering Planning and Design, Vol. 1, Basic Principles,* Wiley, New York, 1975.

Barron, J. C., *Basic Programming with Structured Modules,* Holt, New York, 1983.

Beer, Ferdinand P., and E. Russell Johnston, Jr., *Vector Mechanics for Engineers: Statics and Dynamics,* 4th ed., McGraw-Hill, New York, 1984.

Box, G. E. P., W. G. Hunter, and J. S. Hunter, *Statistics for Experimenters: An Introduction to Design, Data Analysis and Model Building,* Wiley, New York, 1978.

Bradbeer, R., P. DeBono, and P. Laurie, *The Beginner's Guide to Computers,* Addison-Wesley, Reading, Mass., 1982.

Brainard, W. S., C. H. Goldberg, and J. C. Adams, *Programmer's Guide to FORTRAN 90,* McGraw-Hill, New York, 1990.

Brand, S., *Whole Earth Software Catalog,* Quantum/Doubleday, Garden City, N.Y., 1984.

Branscomb, L. M., "Electronics and Computers: An Overview," *Science,* **215**:755 (1982).

Burden, R. L. and J. D. Faires, *Numerical Analysis,* 4th ed., PWS Kent, Boston, 1989.

Chapra, S. C., and R. P. Canale, *Numerical Methods for Engineers with Personal Computer Applications,* 2d Ed. McGraw-Hill, New York, 1988.

Cougar, J. D., and F. R. McFadden, *First Course in Data Processing,* Wiley, New York, 1981.

Draper, N. R., and H. Smith, *Applied Regression Analyses,* 2d ed., Wiley, New York, 1981.

Gerald, C. F., and P. O. Wheatley, *Applied Numerical Analysis,* 3d ed., Addison-Wesley, Reading, Mass., 1984.

Kay, A., "A Personal Computer for Children of All Ages," *Proceedings of the ACM National Conference,* 1972.

Kelly-Bootle, S., *The Devil's DP Dictionary,* McGraw-Hill, New York, 1981.

Kemeny, J. G. *True BASIC,* Addison-Wesley, Reading, Mass., 1985.

Kemeny, J. G., and T. E. Kurtz, *Back to BASIC,* Addison-Wesley, Reading, Mass., 1985.

Knuth, D. E., *The Art of Computer Programming, Vol. 2, Seminumerical Algorithms,* 2d ed., Addison-Wesley, Reading, Mass., 1981.

——, *The Art of Computer Programming, Vol. 3, Sorting and Searching,* Addison-Wesley, Reading, Mass., 1973.

Milton, J. S., and J. C. Arnold, *Introduction to Probability and Statistics,* 2d ed., McGraw-Hill, New York, 1990.

Press, W. H., B. P. Flannery, S. A. Teukolsky, and W. T. Vetterling, *Numerical Recipes: The Art of Scientific Computing,* Cambridge University Press, New York, 1986.

Ralston, A., and P. Rabinowitz, *A First Course in Numerical Analysis,* 2d ed., McGraw-Hill, New York, 1978.

Snedecor, G. W., and W. G. Cochran, *Statistical Methods,* Iowa State University Press, Ames, Iowa, 1967.

Toong, H. D., and A. Gupta, "Personal Computers," *Scientific American,* **247**:86 (1982).

Wonnacott, T. H., and R. J. Wonnacott, *Introductory Statistics,* Wiley, New York, 1972.

INDEX

Abacus, 14
ABC (Atanasoff-Berry computer), 8, 17
ABS function:
    BASIC, 124–125
    FORTRAN, 317
Abscissa, 378, 559
Accumulator, 65–66
Accuracy, 523–524
ACOS function, 320
Address, 93, 230, 431
Aiken, Howard, 16
AINT function, 324
Algorithms, 59
    binary search, 513
    bisection, 597
    bubble sort, 507
    descriptive statistics, 548
    Euler's method, 650
    Gauss elimination, 611–612
    interpolation, 583
    linear regression, 567
    selection sort, 505
    sequential search, 512
    Shell sort, 508
    trapezoidal rule, 635
ALOG function, 319
ALOG10 function, 318–319
Alpha testing, 110, 301
Altair, 23
American simplified keyboard (ASK),
    26
Analog computer, 16
Analytical engine, 16
ANINT function, 324–325
ANSI (American National Standards
    Institute):
    flowchart symbols, 61
    FORTRAN, 274–275
APPEND, 262
Apple Computer Corporation, 23
    (See also Excel)
Apple Macintosh, 54
Applications programs, 33, 50, 57
Arithmetic IF, 336
Arithmetic-logic unit, 25
Arithmetic mean (see Mean)
Arithmetic operators:
    BASIC, 117

Arithmetic operators (cont.)
    FORTRAN, 308
    spreadsheets, 715
Arrays:
    BASIC, 223–235
    FORTRAN, 421–441
    storage, 229–232, 431–433
Artificial intelligence, 19
ASCII (American Standard Code for
    Information Interchange), 47–48
ASIN function, 320
ASK (American simplified keyboard), 26
Assembly language, 50
Assignment:
    BASIC, 93–96
    FORTRAN, 285–288
ATAN function, 320–322
Atanasoff-Berry computer (ABC), 8–17
ATN function, 127–129

Babbage, Charles, 16
Backward difference, 620
Bar diagrams, 494
    (See also Histogram)
BASIC (Beginner's All-purpose Symbolic
    Instruction Code), 5, 21, 83–270
    advantages, 85–86
    algebraic vs. computer equations, 95–96
    arithmetic expressions, 116–122
    arithmetic operators, 117–118
    arrays, 223–235
    assignment, 93–96
    CASE construct, 158, 160–162
    comments, 88–90
    constants, 90–93
    debugging, 108–109
    declaration, 98–99
    dialects, 86–88
    documentation, 111–113
    dummy variable, 137
    dynamic dimensioning, 226
    exponentiation, 122–123
    FOR/NEXT loops, 179–185, 188
    formatting output: commas, 104
        PRINT USING, 245–251
        semicolons, TAB, and SPC, 104–105
    FUNCTION procedures, 202–205
    GO TO (infinite) loop, 167–169

BASIC (cont.)
    graphics, 251–253, 268–269
    hyperbolic trigonometric functions,
        127–128
    intrinsic or library functions, 123–135
        ABS, 124–125
        ATN, 127–129
        CINT, 133
        COS, 126–127
        EXP, 126–127
        FIX, 129–132
        INKEY$, 178–179
        INSTR, 256
        INT, 129–132
        LCASE$, 253–254
        LEFT$, 256
        LEN, 256
        LOG, 125–126
        LTRIM$, 255
        MID$, 255
        RND, 133–135
        RIGHT$, 256
        RTRIM$, 255
        SGN, 131–132
        SIN, 126–127
        SQR, 124–125
        TAN, 126–127
        UCASE$, 253–254
    local and global variables, 205–209
    LOCATE statement, 252
    logical operators, 149–153
    loops, 167–173
        break (mid-test), 173–174
        DO UNTIL (post-test), 168, 172–173
        DO WHILE (pre-test), 168–171
        FOR/NEXT, 179–185, 188
        GO TO, 167–169
        WHILE/WEND, 171
    naming conventions, 96–97
    numeric constants, variables, and
        assignment, 90–97
    procedures, 194–218
    records, 236–240
    relational operators, 147–149, 151
    screen clearing, 100
    SELECT CASE construct, 158, 160–162
    selection, 144–165

BASIC (*cont.*)
  statements: APPEND, 262
    CALL, 195
    CASE, 158, 160–162
    CLOSE#, 264
    CLS, 100
    COMMON SHARED, 206
    CONST, 99–100
    DEF, 136–139
    DIM, 225–227
    END, 88
    EXIT DO, 173–174
    EXIT FOR, 188–189
    FOR/NEXT, 179–185, 188
    FUNCTION, 202–205
    GO SUB, 194–195
    GO TO, 144–146
    IF/THEN, 146–147
    IF/THEN/ELSE, 154
    IF/THEN/ELSEIF, 161–162
    INPUT, 100–103
    INPUT#, 263
    KILL, 266
    LINE INPUT, 103–104
    LOCATE, 252–253
    NAME, 266
    OPEN, 262
    PRINT, 103–106
    PRINT#, 263
    PRINT USING, 245–251
    PRINT# USING, 263
    RANDOMIZE, 134–135
    READ/DATA, 219–223
    REM, 90
    RESTORE, 221–222
    RETURN, 194–195
    SELECT CASE, 158, 160–162
    SHARED, 205–209, 233–234
    SPC, 105
    STATIC, 207–208
    SUB, 195–199
    TAB, 105
    TIMER, 135
    WHILE/WEND, 171
    WRITE #, 264
  static dimensioning, 226
  string constants, 97
  string manipulation, 253–258
  string variables, 97
  SUB procedures, 195–199
  testing, 108–110
  type declaration, 98–99
  user-defined functions and procedures:
    DEF (one-line) functions, 136–139

BASIC, user-defined functions and
    procedures (*cont.*)
      FUNCTION procedures, 202–205
      SUB procedures, 195–199, 204
    variable names, 96–97
    variable types, 98–99
Batch processing, 275–276
Beta testing, 301
Bias, 525–527
Biased exponent, 46
Bimodal distribution, 534
Binary chopping (*see* Bisection)
Binary digit (*see* Bit)
Binary minus operation, 118
Binary (base-2) number system, 34–38
Binary search, 244, 447, 512–513
Bisection, 519–599
Bit (binary digit), 17, 33
Bolzano's method (*see* Bisection)
Borland Corporation, 707
Bracketing methods, 599
Break-loop structure:
    BASIC, 173–174
    FORTRAN, 368–371, 386–387
Bricklin, Daniel, 707, 767
Bubble sort, 243, 447, 506–507
Bugs, 108
Bus, 38
Byte, 38

CAD/CAM, 9
Calibration:
    bias correction, 527
    plots, 482–483
CALL statement:
    BASIC, 195
    FORTRAN, 395–397
Capacitance, 699
Cartesian coordinate systems, 476
    polar, relation to, 80, 143, 165–166, 493
CASE construct:
    BASIC, 158, 160–162
    FORTRAN, 349–352
Cathode-ray tube (CRT), 28, 496
Cell (spreadsheet), 710
Central difference, 620
Central processing unit, 21, 24–25
Character representation, 47–48
CHARACTER statement, 290
Characteristic (*see* Exponent)
Chemical engineering, 664–675
Chips (*see* Microprocessor)
CINT, 133
Circuit analysis, 695–703

Civil engineering, 676–686
Clock, 26
CLOSE# statement, 264
CLS statement, 100
Coefficient of determination (r^2), 566
Coefficient of variation, 546
Commensurate number bases, 37
Comment statement, 282–283
COMMON statement:
    BASIC, 206
    FORTRAN, 406–409, 438–441
Compilers, 54–55, 85
Complex numbers, 217–218, 315–316,
    416–417
Computed GO TO, 336
Computer-aided design and manufacturing
    (CAD/CAM), 8–9, 501
Computer-aided drafting (CAD), 6, 8, 501
Computer algorithms (*see* Algorithms)
Computer calculus (*see* Differentiation;
    Integration; Rate equations)
Computer graphics, 9, 494–504
Computer literacy, 5
Concatenation, 253
Cone, 197, 398
Conservation laws, 655–663
CONST statement, 99–100
Constants:
    BASIC, 90–93
    FORTRAN, 284
CONTINUE statement, 336
Control unit, 24–25
Correlation coefficient (r), 566
COS function:
    BASIC, 126–127
    FORTRAN, 320
COSH function, 127–128, 320
Counter, 65–66
CPM (critical path method), 10
CPU (*see* Central processing unit), 24–26
Cray, 49
    (*See also* Supercomputers)
Critical path method, 10
CRT (*see* Cathode-ray tube)
Cubic splines, 584–586
Curve fitting (*see* Interpolation; Regres-
    sion)
Cylinder, 197, 398

Dartmouth College, 5, 21, 85
DATA statement, 219–223, 419–421, 434
Data structuring, 419
DATA/READ statements, 219–223

Database, 733
DATAN function, 321
Debugging, 108–109, 300–301
DEC (Digital Equipment Corporation), 21
Decisions (*see* Selection)
DEF statement, 136–139
Degrees of freedom:
  standard deviation, 546
  standard error of the estimate, 564–565
Dependent variable, 477, 559
Derivative, 617
Descriptive statistics, 445–446, 531–555,
    756–758
Diagnostics, 299
Difference, 620–621
Difference engine, 16
Differential equations (*see* Rate equations)
Differentiation, 617–626
  advanced methods, 623
  numerical differentiation, 620–623
  pitfalls, 622–623
  (*See also* Numerical differentiation)
Diffusion, 554–555, 625–626
Digital computer, 17
Digital Equipment Corporation (DEC), 21
DIM statement, 225–227
DIMENSION statement 423–425,
Direct-access file, 259–261, 463
Disk density, 28
Disk operating system (DOS), 53
  (*See also* Operating system)
Diskettes, 28–29
  maintenance, 113
Distributions:
  bimodal, 534
  normal, 533, 549–552
  skewed, 535
  uniform, 534
DO loop, 375–384
DO UNTIL structure:
  BASIC, 168, 172–173
  FORTRAN, 366–367
  pseudocode, 69
  structured flowcharts, 68
DO WHILE structure:
  BASIC, 168–171
  FORTRAN, 361–368
  pseudocode, 69
  structured flowcharts, 68
Documentation:
  external, 111–113, 303–304
  internal, 111, 302–303
DOS (*see* Disk operating system)
Dot-matrix printers, 28

Double precision, 46
  BASIC, 92
  FORTRAN, 290–291
DOUBLE PRECISION statement, 290–291
Double roots (*see* Roots of an equation)
Drunkard's walk, 554–555
Dummy variable, 137, 326
Dvorak simplified keyboard (DSK), 26

EBCDIC (extended-binary-coded decimal
    interchange code), 47–48
Eckert, J. Presper, Jr., 17
EDSAC (electronic delay storage automatic
    calculator), 19
EDVAC (electronic discrete variable
    automatic calculator), 19
Elastic range, 680
Electrical engineering, 695–703
Electromechanical computer, 16
Electronic delay storage automatic calculator
    (EDSAC), 19
Electronic discrete variable automatic
    calculator (EDVAC), 19
Electronic numerical integrator and calculator
    (ENIAC), 8, 17–18, 23
Electronic TOOLKIT, 13
  bisection, 596–597
  Euler's method, 650
  Gauss elimination, 611–612, 772–773
  histogram, 536
  graphics, 497–500, 768–771
  interpolation, 583–584, 773–775
  linear regression, 567–568, 768–771
  spreadsheet, 724–727
  statistics, 552–553
  trapezoidal rule, 635
END statement:
  BASIC, 88
  FORTRAN, 280, 396
Engineering problem-solving process,
    661–662
Engineering workstation (*see* Workstation)
ENIAC (electronic numerical integrator and
    calculator), 8, 17–18, 23
Equation solving (*see* Linear algebraic
    equations; Roots of an equation)
Error, 524–525
Errors, 108–109
  logic, 108, 300
  random, 525
  run-time, 108, 299
  syntax, 108, 299
  systematic, 525

Euler's method:
  computer algorithm, 649–650
  equation, 649
  graphical interpretation, 650
Excel, 707
Executable module, 54
EXIT DO, 173–174
EXIT FOR, 188–189
EXP function:
  BASIC, 126–127
  FORTRAN, 319–320
Explicit declaration, 291–292
Exponent, 43, 46, 92, 284
Exponentiation:
  BASIC, 122–123
  FORTRAN, 313–314
Extended-binary-coded decimal interchange
    code (EBCDIC), 47–48

Factorial, 202–203, 214, 404, 413–415
False position, 599
Faraday's law, 696–699
Fibonacci sequence, 192, 216–217, 392,
    417–418
Fick's first law of diffusion, 625–626
Fifth generation, 19
Files:
  BASIC, 259–267
  FORTRAN, 463–468
  naming convention, 261–262
Firmware, 25
First generation, 18–20
FIX function, 129–132
Fixed-length, binary-based code (*see* ASCII;
    EBCDIC)
Floating point, 43–47
Floppy disks (*see* Diskettes)
Flowchart, 60–63
  structured, 66–68
  symbols, 61
  template, 63
Flystra, Daniel, 707
FOR/NEXT loop, 179–185, 188
Force balance, 660
  civil engineering, 676–680
  mechanical engineering, 687–689
FORMAT statement, 453–455
FORTRAN (formula translation), 271–471
  advantages in using, 273–274
  algebraic vs. computer equations, 287–288
  arithmetic expressions, 307–316
    complex variables, 315–316

FORTRAN, arithmetic expressions (*cont.*)
    integer and mixed-mode arithmetic,
      314–315
  arithmetic IF, 336
  arithmetic operators, 308
  arrays, 421–441
  assignment, 285–288
  CASE construct, 349–352
  comments, 282–283
  compiling, 281
  constants, 284
  debugging, 300–301
  declaration, 425–426
  dialects, 274–275
  dummy variables, 326
  editing, 277, 280–281
  explicit type declaration, 291–292
  files, 463–468
  format codes, 452–463
    A, 457–459
    carriage control, 455–456
    E, 459, 461–463
    F, 459, 461–463
    I, 454, 460–461
    T, 456–457
    X, 456–457
  function subprograms, 403–406
  GO TO (infinite) loop, 360–361
  hyperbolic trigonometric functions, 320
  implicit declaration, 290–291
  implied DO loops, 426–428
  intrinsic or library functions, 316–325
    ABS, 317
    ACOS, 320
    ALOG, 319
    ALOG10, 318–319
    AINT, 324
    ANINT, 324–325
    ASIN, 320
    ATAN, 320–322
    COS, 320
    COSH, 320
    DATAN, 321
    EXP, 319–320
    INT, 322–324
    NINT, 324–325
    SIN, 320
    SINH, 320
    SQRT, 316–317
    TAN, 320
    TANH, 320
  job, 276
  linking, 281
  local and global variables, 406–409
  logical operators, 341–344

FORTRAN (*cont.*)
  loops, 360–394
    break (mid-test), 368–371
    DO, 375–384
    DO UNTIL (post-test), 366–367
    DO WHILE (pre-test), 361–368
    GO TO, 360–361
  naming conventions, 289–290
  numeric constants, variables, and
    assignment, 283–293
  repetition, 360–394
  relational operators, 338–341
  selection, 334–359
  specification statements, 425–426
  statements: CALL, 395–397
    CHARACTER, 290
    comment, 282–283
    COMMON, 406–409, 438–441
    COMPLEX, 315–316
    computed GO TO, 336
    CONTINUE, 336
    DATA, 419–421, 434
    DIMENSION, 423–425
    DO/END DO, 379
    DO WHILE, 365–368
    DOUBLE PRECISION, 290–291
    END, 280, 396
    FORMAT, 453–455
    FUNCTION, 403–406
    GO TO, 335–336, 396–397
    IF/THEN, 347
    IF/THEN/ELSE, 345–349
    IF/THEN/ELSEIF, 351–352
    IMPLICIT, 290–291
    INTEGER, 290–291
    LOGICAL, 338, 373–374
    logical IF, 337
    PARAMETER, 292–293
    PROGRAM, 292–293
    PRINT, 295–296
    READ, 293–295
    REAL, 290–291
    RETURN, 396
    SUBROUTINE, 396–397
    WRITE, 465
  string constants, 289
  string variables, 289
  subroutines, 395–402
  testing, 301
  user-defined functions and procedures:
    FUNCTION subprograms, 403–406
    statement functions, 325–328
    subroutines, 396–397
  variable names, 288
    variable types, 289–292

FORTRAN I, 274
FORTRAN IV, 274
FORTRAN 66, 274
FORTRAN 77, 274–275
FORTRAN 90, 275
  CASE, 352–354
  DO/END DO, 379
  DO WHILE, 365–368
  mid-test (break loop), 370–371
  records, 441–444
  recursion, 413–415
  repetition, 385–388
Forward difference, 620
Fourth generation, 21–23
Frankston, Robert, 707, 767
Free format, 449–452
FUNCTION procedure, 202–205
Function subprogram, 403–406

Gates, Bill, 85
Gauss elimination, 604–616, 667
  algorithm, 611–612
  computer algorithm, 611–612
Gauss-Seidel, 613
General Electric, 21
Generations, 19–23
  fifth, 19
  first, 18–20
  fourth, 21–23
  second, 18–21
  third, 18–21
Geographical information system (GIS), 6
Geometric mean, 541–542
GIS (geographical information system), 6
Global variables:
  BASIC, 205–209
  FORTRAN, 406–409
Golden Ratio, 192, 216–217, 392, 417–418
GOSUB statement, 194–195
GO TO, 144–146, 335–336, 396–397
Graduations, 482–483
Graphical user interface (GUI), 54
Graphics, 9, 494–504
GUI (graphical user interface), 54

Hard disk, 29–30
Hardware, 14
Heap sort, 512
Henry's law, 696, 698
Hexadecimal (base-16) number system,
  36–37

High-level language, 50
Histogram, 532–536
    Electronic TOOLKIT, 536
Hollerith, Herman, 16
Hooke's law, 680, 687
Hopper, Capt. Grace, 108
Hyperbolic trigonometric functions:
    BASIC, 127–128
    FORTRAN, 320

IBM, 16, 23
IC (integrated circuit), 21
IEEE convention, 43–46
IF/THEN statement, 146–147, 347
IF/THEN/ELSE statement, 154, 345–349
IF/THEN/ELSEIF statement, 161–162,
    351–352
IMPLICIT declaration, 290–291
IMSL, 413
Independent variable, 477, 559
Inductance, 698
Inference engine, 16
Infinite loop, 65
INKEY$ function, 178–179
INPUT statement, 100–103
INPUT# statement, 263
INSTR function, 256
Instruction set, 50
INT function:
    BASIC, 129–132
    FORTRAN, 322–324
INTEGER statement, 290–291
Integer number representation, 39–43
Integrated circuit (IC), 21
Integration, 627–641
    advanced methods, 636
    algorithm, 635
    Electronic TOOLKIT, 635
    Simpson's 1/3 rule, 636
    trapezoidal rule, 630–635
Intel (see Microprocessor)
Internal representation, 33–50
International Business Machine Corporation
    (IBM), 16–23
Interpolation, 482, 560, 575–588
    advanced methods, 584–586
    algorithm, 583
    Electronic TOOLKIT, 583–584
    Lagrange polynomial, 581
    linear algebraic equations, 616
    pitfalls, 584–585
    (See also Lagrange interpolation)
Interpreters, 54–55, 85
Interval halving (see Bisection)

Intrinsic functions:
    BASIC, 123–135
    FORTRAN, 316–325
Iowa State University, 17

Jacquard, Joseph, 15
Job-control language (JCL), 51
Job-control program, 51
Jobs, Steve, 23

Kb (kilobyte), 25
Kay, Alan, 22
Kemeny, J., 5, 21, 85
Keyboards, 26
Keypunch, 275
Kilobyte (Kb), 25
KILL, 266
Kirchhoff's laws:
    current rule, 695
    voltage rule, 696
Kurtz, T., 21, 85

Lagrange interpolation, 577–588
    computer algorithm, 583
    differentiation, 623
    general form, 581
    linear, 577
    parabolic, 579
Language translators, 54–55
Large-scale integration (LSI), 21
LCASE$ function, 253–254
LCD (Liquid crystal display), 28
Least significant digit, 37
Least-squares (see Regression)
LEFT$ function, 256
Leibniz, Gottfried, 15
LEN function, 256
Library functions (see Intrinsic functions)
Line of action, 682
Line graphs, 475–477
LINE INPUT statement, 103–104
Linear algebraic equations, 604–616
    advanced methods, 613
    computer algorithm, 611–612
    Electronic TOOLKIT, 611–612
    Gauss elimination, 607–611
    pitfalls, 611
    (See also Gauss elimination; Spread-
    sheets)
Linear interpolation, 575, 577–581
Linear regression, 561–569
    computer algorithm, 567

Linking, 54
Liquid crystal display (LCD), 28
List directed, 449–452
Literals (see Constants)
Load module, 280–281
Local variables:
    BASIC, 205–209
    FORTRAN, 406–409
LOCATE statement, 252
LOG function, 125–126
    (See also ALOG function; ALOG10
    function)
Log-log plots, 490–491
Logarithms, 15, 125–126
Logic element, 17–19
Logic errors, 108, 300
Logical IF statements, 337
Logical operators, 149–153,
    341–344
LOGICAL statement, 338, 373–374
Loops, 65
    BASIC, 167–173
    control structures, 68–69
    FORTRAN, 360–394
Lotus 1-2-3, 707
    (See also Spreadsheets)
LSI (Large-scale integration), 21
LTRIM$ function, 255

Mb (megabyte), 25
Machine language, 50
Macintosh, 54
Macros:
    command, 783–787
    keyboard, 781–794
    keywords, 786
    program, 787–792
Mainframe computer, 21
    operating system, 51–52
Maintenance, 113, 304
Mantissa, 43, 92, 284
Mark I, 16
Mass balance, 664–672
Math coprocessor, 46
Matrix, 227
    algebra, 761
    equation solving with, 763–767
    identity, 762
    inverse, 763
    product, 762–763
Mauchly, John, W., 17
Mean, 537–538
Mechanical engineering, 687–694
Median, 540–541

Megabyte (Mb), 25
Megahertz (MHz), 26
Megatrends, 75
Memory (*see* RAM; Secondary memory)
Menus, 209–213, 410
Microcomputer, 21
   motherboard, 27
   operating system, 53–54
   (*See also* Personal computer)
Microminiaturization, 21
Microprocessor, 9, 21
   Intel 4004, 21
   Intel 8008, 22
   Intel 8080, 23
   Intel 8086, 23
   Intel 8088, 23
   Intel iAPX-87, 46
Microsoft, 53
   BASIC, 86–87
   Excel, 707
   QuickBASIC, 88
   Windows, 54
Mid-test loop (*see* Break-loop structure)
MID$ function, 255
Minicomputer, 21
   operating system, 53
   (*See also* Workstation)
MITS Altair, 23
Mode, 541
Modular design, 72–73
Module, 63
Modulus of elasticity, 680
Monitor, 28
Motherboard, 27
Mouse, 27
MS-DOS, 53–54
Multiple roots, 499, 598–599
Multiprogramming, 52

NAG, 413
Naisbett, Robert, 75
NAME, 266
Napier, John, 14
   (*See also* Logarithms)
Napier's bones, 15
Nesting, 380–385
Network analysis, 695–703
Newton-Raphson, 599
Newton's interpolating polynomial, 584
Newton's law of cooling, 624
Newton's second law, 644–645, 688

Next, 54
NINT function, 324–325
Nomographs, 493
Normal distribution, 533, 549–552
Number circle, 39–40
Number crunching, 167, 360
Number systems, 33–38
Numerical differentiation:
   difference approximations, 620
   integration (contrast with), 630
   Lagrange polynomial, 623
Numerical integration (*see* Integration)
Numerical methods, 12, 557–654
   differentiation, 617–626
   integration, 627–641
   interpolation, 482, 560, 575–588
   linear algebraic equations, 604–616
   rate equations, 642–653
   regression, 560–569
   roots of an equation, 589–603
   (*See also* Electronic TOOLKIT;
      Spreadsheets)

Object code, 54
Octal (base 8) number system, 35
Ohm's law, 696, 698
Open methods, 599
OPEN statement, 262
Operating system, 50–54
   mainframe, 51–53
   personal (micro-) computer, 53–54
   workstation, 53
Optical storage, 30
Order of magnitude, 125
Ordinary differential equation (*see* Rate
   equations)
Ordinate, 378, 559
Organizing principles, 658–661
   conservation laws, 660
   force balance, 660, 676–680, 687–689
   Kirchhoff's laws, 695–696
   mass balance, 665
   Newton's second law, 644–645, 688

PARAMETER statement, 292–293
Parameters, procedures, and subroutines,
   194–202, 397–402
Pascal, Blaise, 15
Pascal (language), 95
Pascaline, 15

Pen, 27
Personal computer, 4, 21
   motherboard, 27
   (*See also* Microcomputer)
PERT (program evaluation and review
   technique), 10
Pharmacokinetics, 674–675
Picture elements (pixels), 497
Pie charts, 494
Pivoting, 610–611
Pixels, 497
Point-slope method (*see* Euler's method)
Pointing devices, 26–28
Polar coordinate systems, relation to
   cartesian, 80, 143, 165–166, 493
Polar graphs, 476–477, 491–493
Polynomial interpolation (*see* Interpolation)
Polynomial regression, 568–569
Population:
   prediction, 652
   statistical, 532
Positional notation, 34
Post-test loop (*see* DO UNTIL structure)
Precision:
   computer, 46
   measurement, 524
Pre-test loop (*see* DO WHILE structure)
Primary memory (*see* RAM), 25
PRINT statement:
   BASIC, 103–106
   FORTRAN, 295–296
PRINT# statement, 263
PRINT USING statement, 245–251
PRINT# USING statement, 263
Printers, 28
Prism, 197, 398
Procedures:
   BASIC, 194–218
   FORTRAN, 395–406
Program, 33, 57–58
   (*See also* Algorithms)
Program evaluation and review techniques
   (PERT), 10
PROGRAM statement, 292–293
Pseudocode, 67, 69–71
   (*See also* Algorithms)
Punched cards, 15–16, 20, 275

Quadratic formula, 589
Quattro Pro, 707
   (*See also* Spreadsheets)
QBASIC (QuickBASIC), 88

Quantizing errors, 49
QuickBASIC, 88
   (*See also* BASIC)
Quick sort, 512
QWERTY keyboard, 26

Radio Electronics, 22
RAM (Random-access memory), 25, 53
Random access file, 259–261, 463
Random walk, 554–555
RANDOMIZE statement, 134–135
Range, 543
Raster display, 496–497
Rate equations, 642–653
   advanced methods, 650
   computer algorithm, 649–650
   Electronic TOOLKIT, 650
   Euler's method, 647–649
READ statement, 293–295
READ/DATA statements, 219–223
Read-only memory, 25
Real numbers, 43–47
REAL statement, 290–291
Records:
   BASIC, 236–240
   FORTRAN 90, 441–444
Rectilinear graphs, 477–486
Recursion:
   BASIC, 213–214
   FORTRAN 90, 413–415
Redlich and Kwong equation, 236, 443
Regression, 560
   advanced methods, 568–569
   computer algorithm, 567
   Electronic TOOLKIT, 567–568
   goodness of fit, 564–566
   (*See also* Linear regression)
Relational database model, 733
Relational operators:
   BASIC, 147–149, 151
   FORTRAN, 338–341
Reliability, 517
REM statement, 90
Remington-Rand Corporation, 20
Repetition, 65, 167–193, 360–394
   (*See also* Loops)
Residual, 562
Resistance, 698
Resolution, 497
RESTORE statement, 221–222
RETURN statement, 194–195
Richardson's extrapolation, 640
RIGHT$ function, 256

RND function, 133–135
ROM (Read-only memory), 25
Roots of an equation, 589–603
   advanced methods, 599
   bisection, 591–599
   computer algorithm, 596–597
   Electronic TOOLKIT, 596–597
   graphical method, 590–591
   multiple roots, 599
   pitfalls, 598–599
   (*See also* Bisection)
Rounding, 143, 517–518
   with intrinsic functions, 322–325,
      332–333
   (*See also* FORMAT statement, PRINT
      USING statement)
Round-off error, 49
RTRIM$, 255
Runge-Kutta methods, 650
Run-time errors, 108, 299

Sample, 532
Scientific notation, 284
Screen clearing, 100
Searching, 244, 447, 512–513
Secant method, 599
Second generation, 18–21
Secondary memory, 28–30
SELECT CASE, 158, 160–162
Selection, 65
   BASIC, 144–165
   case construct, 158, 160–162
   control structures, 68–69
   FORTRAN, 334–359
Selection sort, 243, 446–447, 505–506
Semilog plots, 486–490
Sentinel value, 170
Sequential files, 259–267, 463–468
Sequential search, 512
SGN function, 131–132
SHARED, 205–209, 233–234
Shell sort, 507
Signed magnitude method, 39
Significand (*see* Mantissa)
Significant figures, 518–523
Simpson's 1/3 rule, 636
Simulation, 11
SIN function:
   BASIC, 126–127
   FORTRAN, 320
Single precision, 46, 92
Skewed distribution, 535
Slide rule, 15

Software, 4, 33
Sorting, 505–512
   (*See also* Bubble sort; Selection sort;
      Shell sort)
Source code, 54, 88
Spaghetti code, 334, 345
SPC statement, 105
Sphere, 197, 398
Splines, 584–586
Spooling, 52
Spreadsheets, 6, 705–794
   cell editing, 717–719
   cell move and copy, 741–743
   cell references, 743
   database, 733
   data fill, 752
   dynamic tables, 743–746
   Electronic TOOLKIT, 724–727, 767–775
   environment, 709–711
   graphics, 751–756
   linear algebraic equations, 760–767
   macros:
      command, 783–787
      keyboard, 781–794
      keywords, 786
      program, 787–792
   matrices, 762–767
   menu operation, 710–714
   numbers, labels, and formulas, 714–717
   pointing, anchoring, and painting, 717
   regression, 758–760
   relative and absolute cell references, 743
   sensitivity analysis, 736–741, 743–746
   sorting, 733
   static tables, 736–741
   statistics, 756–758
SQR function, 124–125
SQRT function, 316–317
Staircasing, 497
Standard deviation, 543–546
Standard error of the estimate, 564–565
Statement functions, 325–328
STATIC statement and attribute, 207–208
Statistics, 445–446, 531–555
Steady state, 659, 664
Storage, 113, 304
Stored-program computers, 19
String constants and variables:
   BASIC, 97–98
   FORTRAN, 289
Strings, 47
Structured design and programming,
      72–75
Structured programming, 73–75

SUB procedure, 195–199
Subroutine:
 BASIC, 194–218
 FORTRAN, 395–402
SUBROUTINE statement, 396–397
Successive approximation, 35
Sum of squares, 562
Supercomputers, 30–31
Syntax errors, 108, 299
System programs, 50–54
Systems of equations (*see* Linear algebraic
  equations)

TAB statement, 105
Tabulating Machine Company, 16
TAN function:
 BASIC, 126–127
 FORTRAN, 320
Tape storage, 30
Testing, 108–110, 301
Third generation, 18–21
Time variable (*see* Transient)
Timesharing, 21, 52, 276
Top-down design, 64, 73
Touch screen, 28

Transient, 659
Transistors, 20
Trapezoidal rule, 630–635
Trimmed mean, 538–540
True BASIC, 88
Turbo BASIC, 88
Tutorial software, 11
Two's complement method, 39–41

UCASE$, 253–254
Unary minus operation, 118
Uncertainty, 527–528
Uniform distribution, 534
UNIVAC I, 20
University of Pennsylvania, 17
UNIX, 53
User-defined function, (*see* FUNCTION
  procedure; Function subprogram;
  Statement functions)
Utility programs, 54

Vacuum tubes, 18–20
Variables, 90, 96–99, 288–292
Variance, 543–546

Vector, 227, 428
Vector display, 496
Very large-scale integration (VLSI), 19–21
Virtual storage, 52
VLSI (very large-scale integration), 19–21
von Neumann, John, 19

What if?, 175–178, 371–375, 708
WHILE/WEND statements, 171
Winchester disk (*see* Hard disk)
Windows, 54
Word processing, 6
Word size, 38
Worksheets (*see* Spreadsheets)
Workstation, 21, 54
Wozniak, Steve, 23
Wraparound, 42
WRITE statement, 465
WRITE# statement, 264

X-windows, 54

Young's modulus (Modulus of elasticity), 680